MW00814617

# Modelling and Analysis of Active Biopotential Signals in Healthcare, Volume 2

IPEM–IOP Series in Physics and Engineering in Medicine and Biology



## About the Series

The Series in Physics and Engineering in Medicine and Biology will allow IPEM to enhance its mission to 'advance physics and engineering applied to medicine and biology for the public good.'
The series will focus on key areas including, but not limited to:

- clinical engineering
- diagnostic radiology
- informatics and computing
- magnetic resonance imaging
- nuclear medicine
- physiological measurement
- radiation protection
- radiotherapy
- rehabilitation engineering
- ultrasound and non-ionizing radiation.

A number of IPEM–IOP titles are published as part of the EUTEMPE Network Series for Medical Physics Experts.

# Modelling and Analysis of Active Biopotential Signals in Healthcare, Volume 2

**Edited by**
**Varun Bajaj**
*PDPM-Indian Institute of Information Technology, Design and Manufacturing, Jabalpur, India*

**G R Sinha**
*Myanmar Institute of Information Technology, Mandalay, Myanmar*

**IOP** Publishing, Bristol, UK

ISBN 978-0-7503-3411-2 (ebook)
ISBN 978-0-7503-3409-9 (print)
ISBN 978-0-7503-3412-9 (myPrint)
ISBN 978-0-7503-3410-5 (mobi)

DOI 10.1088/978-0-7503-3411-2

Version: 20201201

IOP ebooks

British Library Cataloguing-in-Publication Data: A catalogue record for this book is available from the British Library.

Published by IOP Publishing, wholly owned by The Institute of Physics, London

IOP Publishing, Temple Circus, Temple Way, Bristol, BS1 6HG, UK

US Office: IOP Publishing, Inc., 190 North Independence Mall West, Suite 601, Philadelphia, PA 19106, USA

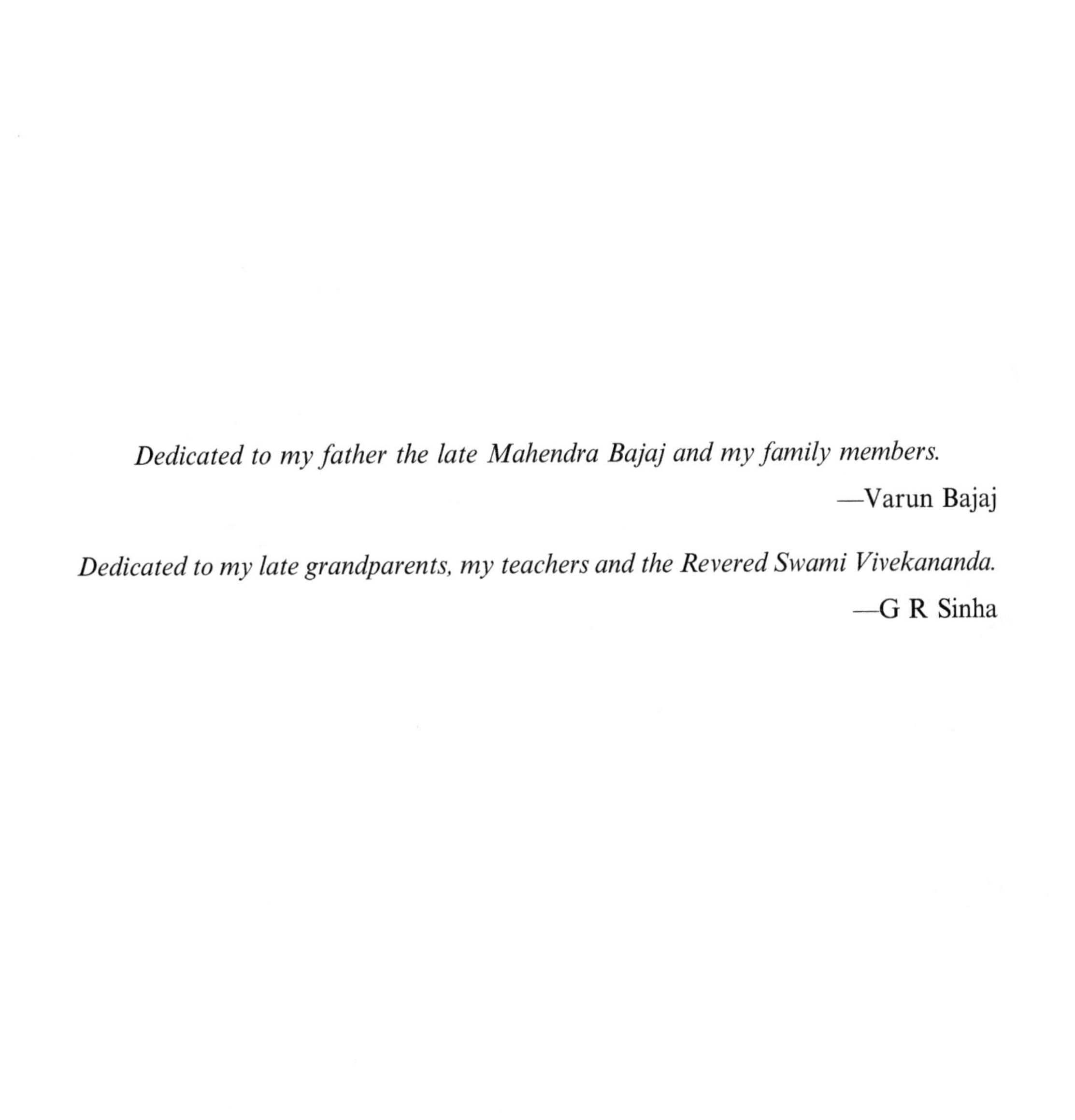

*Dedicated to my father the late Mahendra Bajaj and my family members.*

—Varun Bajaj

*Dedicated to my late grandparents, my teachers and the Revered Swami Vivekananda.*

—G R Sinha

# Contents

# Preface

This volume provides the framework for the modeling, concepts and applications of biopotential signal processing. Biopotential signals are measures of the different activities of neurons, organs, genes, cardiac rhythms, tissues, protein sequences, etc. We provide information on the sources of biopotential signals and also on how the signals are processed, manipulated and transformed in different domains and how the physiological and pathological conditions of human organs are extracted. This volume emphasizes the real-time challenges in biopotential signal processing due to the complex and non-stationary nature of signals for a variety of applications for the analysis, classification and identification of different states for the improvement of healthcare systems.

Each chapter begins with a description of a biomedical example and the significance of the biopotential signal processing methods with discussion to connect the technology with our understanding of human organs. We also provide information on the identification of conditions, including schizophrenia, epileptic seizures, physical actions, cardiac health monitoring, etc. Moreover, the chapters can be read independently by researchers, practicing physicians, R&D engineers and graduate students who wish to explore research in the field of biomedical engineering.

Chapter 1 proposes rational-dilation wavelet transform (RDWT) based features for the classification of motor-imagery tasks using EEG signals. Chapter 2 presents deep learning models for emotion recognition based on improved time–frequency image analysis of electroencephalography signals. In chapter 3 multivariate phase synchrony based on fuzzy statistics for post-traumatic stress disorder using EEG signals is described. Chapter 4 presents a review of the effects of meditation and music therapy on the vital parameters of the human body using EEG signals. Chapter 5 employs deep feature extraction based on cross wavelet transform for focal and non-focal electroencephalography signal classification. Chapter 6 introduces a local binary pattern based feature extraction with a machine learning algorithm for epileptic seizure prediction and detection. Chapter 7 demonstrates the usability of the Devanagari script input based P300 speller. Chapter 8 presents a comprehensive review of the fabrication and performance evaluation of dry electrodes for long term ECG monitoring. Chapter 9 describe event-driven ECG processing with sub-band feature extraction with machine learning techniques for cardiac health diagnosis. Tree based ensemble models are explored for the analysis of heart patients in chapter 10. Chapter 11 presents frequency and time–frequency based feature extraction methods for ECG beat classification. Chapter 12 describes a segmentation approach for the reliable and robust detection of critical ECG waves using long short-term memory (LSTM) networks. Chapter 13 presents a deep convolutional neural network (CNN) approach based diagnosis of COVID-19 using x-ray images. In chapter 14 an original dataset is created by collecting tympanic membrane images from volunteer patients using an otoscope device. The AlexNet, VGG16, GoogLeNet and ResNet50 deep transfer learning models, are proven successful in experimental studies and are used to classify normal tympanic membrane and abnormal tympanic membrane images in this dataset. Chapter 15 deals with numerical analysis for the active infrared thermographic technique to spot the presence of hidden tumors inside a simulated breast sample and frequency domain based phase and time domain based phase and amplitude thermal images are produced using post-processing analysis of the obtained data from the numerical simulation of the proposed models. In chapter 16 the evolution and significant advances of photoacoustic microscopy (PAM) imaging in the past two decades are described. The studies demonstrate that the achievable spatial resolution reaches a sub-microscopic scale (~100 nm) and photoacoustic imaging is considered, specifically photoacoustic nanoscopy (PAN). In chapter 17 the reliability of a diagnosis system for medical doctors, its proper comparison with the state-of-the-art, avoiding incorrect interpretation and the reporting of proper performance indices are linked with pattern recognition, and the typical biostatistics are also characterized.

# Acknowledgements

Dr Bajaj expresses his heartfelt appreciation to his mother Prabha, his wife Anuja and his daughter Avadhi for their wonderful support and encouragement throughout the completion of this important book on the *Modeling and Analysis of Active Biopotential Signals in Healthcare* (volume 2). His deepest gratitude goes to his mother-in-law and father-in-law for their constant motivation. This book is the outcome of committed effort that could only be achieved due to the great support of Dr Bajaj's family. He is also grateful to Professor Sanjeev Jain, Director of PDPM IIITDM Jabalpur, for his support and encouragement.

Dr Sinha also expresses his gratitude and sincere thanks to his family members, his wife Shubhra, his daughter Samprati, his parents and his teachers.

We would like to thank all our friends, well-wishers and all those who keep us motivated to do more and more, better and better. We sincerely thank all the contributors for their writing on the relevant theoretical background and real-time applications of biopotential signals in healthcare. We are also deeply grateful to many whose names are not mentioned here but whose help during this work we appreciate and wish to acknowledge.

We express our humble thanks to Michael Slaughter, Senior Commissioning Editor, Sarah Armstrong and all the editorial staff (IPEM–IOP) of IOP Publishing for their great support, necessary help, appreciation and quick responses. We also wish to thank IOP Publishing for giving us this opportunity to contribute on a relevant topic with a reputed publisher. Finally, we want to thank everyone who, in one way or another, helped us in editing this book.

Dr Bajaj particularly thanks his family from their encouragement throughout the editing of this book. This book is dedicated from the heart to his father who took the lead to heaven before the completion of this book.

Last but not least we would also like to thank God for showering us with his blessings and strength to perform this type of novel and quality work.

Varun Bajaj

G R Sinha

# Editor biographies

## Varun Bajaj

Dr Varun Bajaj has been working as a faculty member in the Department of Electronics and Communication Engineering at the Indian Institute of Information Technology, Design and Manufacturing (IIITDM) Jabalpur, India, since 2014. He worked as visiting faculty at IIITDM from September 2013 to March 2014. He served as an Assistant Professor in the Department of Electronics and Instrumentation, Shri Vaishnav Institute of Technology and Science, Indore, India, during 2009–10. He received his BE degree in electronics and communication engineering from Rajiv Gandhi Technological University, Bhopal, India, in 2006, his MTech degree with Honors in microelectronics and VLSI design from Shri Govindram Seksaria Institute of Technology and Science, Indore, India, in 2009. He received his PhD degree in the discipline of electrical engineering from the Indian Institute of Technology Indore, India, in 2014.

He serves as a Subject Editor of *IET Electronics Letters*. He is an IEEE member and also contributes as an active technical reviewer to leading international journals of IEEE, IET, Elsevier, etc. He has authored more than 90 research papers in various reputed international journals and conference proceedings of publishers such as IEEE Transactions, Elsevier, Springer, IOP, etc. The citation impact of his publications is around 1600 citations, with an h-index of 18 and an i10 index of 33 (Google Scholar, January 2020). He has supervised six PhD scholars and five MTech scholars. He is the recipient of various reputed national and international awards. His research interests include biomedical signal processing, image processing, time–frequency analysis and computer-aided medical diagnosis.

## G R Sinha

G R Sinha is an Adjunct Professor at the International Institute of Information Technology Bangalore (IIITB) and is currently deputed as a Professor at the Myanmar Institute of Information Technology (MIIT), Mandalay, Myanmar. He obtained his BE in electronics engineering and his MTech in computer technology with a Gold Medal from the National Institute of Technology Raipur, India. He received his PhD in electronics and telecommunication engineering from Chhattisgarh Swami Vivekanand Technical University (CSVTU) Bhilai, India. He was a Visiting Professor (Honorary) at the Sri Lanka Technological Campus, Colombo, for one year (2019–20). He has published 254 research papers, book chapters and books at the international and national level, that include *Biometrics* (published by Wiley India, a subsidiary of John Wiley), *Medical Image Processing* (published by Prentice Hall of India) and five edited books with IOP, Elsevier and Springer. He is an active reviewer and editorial member of more than twelve reputed international journals of IEEE, IOP, Springer,

Elsevier, etc. He has 21 years of teaching and research experience. He has been the Dean of Faculty and an Executive Council Member of CSVTU and is currently a member of the Senate of MIIT. Dr Sinha has been delivering ACM lectures as an ACM Distinguished Speaker in the field of DSP since 2017 across the world. A few of his more important roles include being an Expert Member of the vocational training program of the Tata Institute of Social Sciences (TISS) for two years (2017–19), the Chhattisgarh Representative of the IEEE MP Sub-Section Executive Council (2016–19) and a Distinguished Speaker in the field of digital image processing of the Computer Society of India (2015). He is the recipient of many awards and recognitions, such as the TCS Award 2014 for Outstanding Contributions in the Campus Community of TCS, the Rajaram Bapu Patil ISTE National Award 2013 for Promising Teacher in Technical Education by ISTE New Delhi, the Emerging Chhattisgarh Award 2013, the Engineer of the Year Award 2011, the Young Engineer Award 2008, the Young Scientist Award 2005, the IEI Expert Engineer Award 2007, the ISCA Young Scientist Award 2006 Nomination and the Deshbandhu Merit Scholarship for five years. He has served as a Distinguished IEEE Lecturer in the IEEE India Council for the Bombay section. He is a Senior Member of IEEE, a Fellow of the Institute of Engineers India and a Fellow of IETE India.

He has delivered more than 50 keynote/invited talks and has chaired many technical sessions at international conferences across the world. His Special Session on 'Deep Learning in Biometrics' was included in the IEEE International Conference on Image Processing 2017. He is also a member of many national professional bodies such as ISTE, CSI, ISCA and IEI. He is member of various committees at his university and has been the Vice President of the Computer Society of India for the Bhilai Chapter for two consecutive years. He is a consultant for various skill development initiatives of the NSDC, Government of India. He is regular referee of project grants under the DST-EMR scheme and several other schemes of the Government of India. He has received a few important consultancy supports such as grants and travel support. Dr Sinha has Supervised eight PhD scholars and 15 MTech scholars, and is currently supervising another PhD scholar. His research interests include biometrics, cognitive science, medical image processing, computer vision, outcome based education (OBE) and ICT tools for developing employability skills.

# Contributors

**Mithilesh Atulkar**
National Institute of Technology Raipur, India

**Vanita Arora**
School of Electronics, Indian Institute of Information Technology Una, India

**Varun Bajaj**
PDPM-Indian Institute of Information Technology, Design and Manufacturing, Jabalpur, India

**Erdal Başaran**
Computer and Instructional Technologies Education Department, Agri İbrahim Cecen University, Turkey

**Harald Binder**
Freiburg Center for Data Analysis and Modeling, University of Freiburg, Germany

**Rohit Bose**
University of Pittsburgh, PA, USA

**Ümit Budak**
Electrical and Electronics Engineering Department, Bitlis Eren University, Turkey

**Soumya Chatterjee**
Techno India University, West Bengal, India

**Yüksel Çelik**
Engineering Faculty, Computer Engineering Department, Karabuk University, Turkey

**Zafer Cömert**
Software Engineering Department, Samsun University, Turkey

**Bijit Choudhuri**
Department of Electronics and Communication Engineering, NIT Silchar, Silchar, India

**Sengul Dogan**
Firat University, Department of Digital Forensics Engineering, Elazig, Turkey

**Geetika Dua**
Electronics and Communication Engineering, Thapar Institute of Engineering and Technology, India

**Zahra Ghanbari**
Amirkabir University of Technology, Tehran, Iran

**Smith K Khare**
PDPM-Indian Institute of Information Technology, Design and Manufacturing, Jabalpur, India

**Ghanahshyam Kshirsagar**
National Institute of Technology Raipur, India

**Aboli Londhe**
National Institute of Technology Raipur, India

**Narendra Londhe**
National Institute of Technology Raipur, India

**Yogita Maithani**
Department of Physics, Indian Institute of Technology Delhi, India

**Suheshkumar Singh Mayanglambam**
School of Physics, Indian Institute of Science Education and Research Thiruvananthapuram (IISER-TVM), India

**Marjan Mansourian**
Universitat Politècnica de Catalunya-Barcelona Tech, Barcelona, Spain

**Hamid Reza Marateb**
Biomedical Engineering Department, University of Isfahan, Iran

**Mahsa Mansourian**
Medical Physics Department, Isfahan University of Medical Sciences, Iran

**Bodh Raj Mehta**
Department of Physics, Indian Institute of Technology Delhi, India

**Mohammad Hassan Moradi**
Amirkabir University of Technology, Tehran, Iran

**Mohammad Reza Mohebbian**
Department of Electrical and Computer Engineering, University of Saskatchewan, Canada

**Sudip Modak**
Techno India University, West Bengal, India

**Ravibabu Mulaveesala**
Department of Electrical Engineering, Indian Institute of Technology Ropar, India

**Neelamshobha Nirala**
NIT, Raipur, India

**Saurabh Pal**
Dept of MCA, VBS Purvanchal University, Jaunpur, India

**Souradip Paul**
School of Physics, Indian Institute of Science Education and Research Thiruvananthapuram (IISER-TVM), India

**Monoj Kumar Pradhan**
Department of Agricultural Statistics and Social Sciences (L), Indira Gandhi Agricultural University, Raipur, India

**Saeed Mian Qaisar**
Effat University, College of Engineering, Jeddah, Saudi Arabia

**Sayanjit Singha Roy**
Techno India University, West Bengal, India

**Somshubhra Roy**
Institute of Engineering and Management, India

**Kaniska Samanta**
Techno India University, West Bengal, India

**Abdulkadir Şengür**
Electrical and Electronics Engineering Department, Firat University, Turkey

**Anurag Shrivastava**
NIT, Raipur, India

**Anshul Sharma**
Centre for Biomedical Engineering, Indian Institute of Technology Ropar, India

**Vimal Kumar Shrivastava**
School of Electronics Engineering, Kalinga Institute of Industrial Technology (KIIT), Bhubaneswar, India

**G R Sinha**
Myanmar Institute of Information Technology (MIIT) Mandalay, Myanmar

**Bikesh Kumar Singh**
NIT, Raipur, India

**Jitendra Pratap Singh**
Department of Physics, Indian Institute of Technology Delhi, India

**Abdulhamit Subasi**
Institute of Biomedicine, Faculty of Medicine, University of Turku, Turku, Finland
Effat University, College of Engineering, Department of Computer Science, Jeddah, Saudi Arabia

**Fawad Hussain Syed**
Ghulam Ishaq Khan Institute of Engineering Sciences and Technology, Topi, Pakistan

**Sachin Taran**
Delhi Technological University (DTU), New Delhi, India

**Turker Tuncer**
Firat University, Department of Digital Forensics Engineering, Elazig, Turkey

**Dahiru Tanko**
Firat University, Department of Digital Forensics Engineering, Elazig, Turkey

**Anjali Thomas**
School of Physics, Indian Institute of Science Education and Research Thiruvananthapuram (IISER-TVM), India

**Miguel Ángel Mañanas Villanueva**
Universitat Politècnica de Catalunya-Barcelona Tech, Barcelona, Spain

**Dhyan Chandra Yadav**
Department of MCA, VBS Purvanchal University, Jaunpur, India

**IOP** Publishing

Modelling and Analysis of Active Biopotential Signals in Healthcare, Volume 2

**Varun Bajaj and G R Sinha**

# Chapter 1

# Classification of motor-imagery tasks from EEG signals using the rational dilation wavelet transform

**Sachin Taran[1], Smith K Khare[2], Varun Bajaj[2] and G R Sinha[3]**

[1]*Delhi Technological University (DTU), Shahbad Daulatpur, New Delhi, 110042, India*

[2]*Indian Institute of Information Technology Design and Manufacturing, Jabalpur, MP, India*

[3]*Myanmar Institute of Information Technology Mandalay, Myanmar*

Motor-imagery (MI) tasks are important activities of the human brain which are extracted from EEG signals. These tasks play an important role in the brain–computer interface (BCI) technology that helps in communication between a computer and the brain. Electroencephalogram (EEG) signals are commonly used for the assessment of MI tasks associated with the brain and, as a non-invasive method, are the most popular choice. The performance and reliability of BCI functions using EEG signals are controlled by the identification of different types of motor-imagery tasks. This chapter proposes rational dilation wavelet transform (RDWT) based features for classifying the tasks with the help of EEG signals. The oscillatory characteristics of EEG signals are utilized by RDWT as sub-bands (SBs). Further, these band-limited SBs are used for the extraction of features. These features are examined using different machine learning algorithms. The quadratic support vector machine (SVM) is employed for evaluating the best classification performance measures compared to other classifiers. The proposed work produces better accuracy, sensitivity and specificity compared to other methods tested on the same dataset.

## 1.1 Introduction

The brain–computer interface (BCI) plays a key role in interfacing the neuro-electrical signals produced in the human brain to communicate with computers. The

doi:10.1088/978-0-7503-3411-2ch1

signal taken from the scalp surfaces originates when a person moves (or imagines moving) different parts of their body, makes gestures, expresses emotion, etc, and this signal is also known as motor imagery (MI) [1–4]. Differently abled people can find a way of communicating with and controlling the real world with the help of an MI-based BCI. The computer captures the electrical activity of the MI task from the scalp and then converts it into commands for communication and control without requiring any physical movement from the user [5]. The different steps in a BCI to carry out the brain commands are signal acquisition, feature extraction and feature classification [6]. There are various methods in the literature to capture the activities of the brain for MI, such as electroencephalography (EEG), electrocorticography (ECoG), positron emission tomography (PET), functional magnetic resonance imaging (fMRI), magnetoencephalography (MEG) and electrocardiogram (ECG). EEG is the best method to classify the MI signals for different brain activities, due to its ability to provide fast feedback in real time and its accurate detection ability, low cost, simplicity, ease of use, excellent resolution and non-invasive nature [7–19]. EEG signals are recorded by wearing an electrode cap on the scalp. The EEG electrodes are placed over areas of the brain that are highly sensitive for discrimination [20–22].

To date, several studies have been proposed in the literature on EEG-based MI classification. Feature extraction based on time, frequency, chaotic and time–frequency characteristics have been suggested for the identification and classification of MI tasks. The extraction of frequency domain features using fast Fourier transform is used in [23]. The adaptive autoregressive model is used in [24] and the power spectral density has been employed [25] with analysis of time and frequency based features. In [26, 27] informative samples are selected using optimal allocation, followed by feature extraction and classification using a least squares support vector machine (LS-SVM) and logistic regression (LR). The *Z*-score and linear discriminant analysis are employed in [28] for MI classification. The selection of channel and feature extraction based on canonical cross-correlation is employed in [29–33]. These features are classified using LR and LS-SVM to detect MI. A multistate canonical method has been used to determine the frequency changes during MI tasks [34]. Kolmogorov complexity and Adaboost classifiers have also been used to detect MI tasks [35]. Features extracted from the fast Walsh–Hadamard transform are classified using artificial neural networks (ANNs) in [36].

Wavelet methods have also been employed to study the changes during MI tasks, with the signal being split into frequency sub-bands using the Haar wavelet, wavelet packets, Daubechies wavelet and wavelet coefficients. The features extracted and selected using statistical analysis and higher-order statistics are classified with quadratic discriminant analysis, support vector machines (SVMs), *k*-nearest neighbors (*k*NN), linear discriminant analysis (LDA) and ANNs [37–42]. EEG signals have been decomposed into amplitude and frequency modulated instantaneous mode functions using empirical mode decomposition (EMD) and its different variants such as noisy EMD and multivariate EMD. Classification of the moment, Hjorth and analytical features have been carried out using LDA, ANN and LS-SVM [1, 43–46]. Classification of MI tasks has been explored using common spatial patterns (CSPs). The CSPs and their different variants, namely regularized patterns, Tikhonov regularized patterns, sparse patterns, spatially regularized patterns, common spatial-spectral boosting patterns, filterbanks, local temporal correlation, strong-uncorrelated transform and sparse time–frequency segment have been used for the extraction of features. These features were classified using LDA, LR, *k*-NN, SVM, naïve Bayes and ANN classifiers [47–62]. In [63] common spatial subspace decomposition with Fisher discriminant analysis and in [64] regularized tensor discriminant analysis are explored for analysis of MI tasks. Channel selection for the classification of MI EEG has been explored by the combination of the genetic algorithm with the Rayleigh coefficient [65]. Recursive feature elimination and SVM [66], adaptive neuro-fuzzy and LDA [67], and a recurrent quantum neural network [68] have been used in MI tasks. In [69, 70] sparse Bayesian and sparse kernel machines are explored.

The utility of tunable Q wavelet transforms and long short-term memory have also been used for MI tasks [71, 72]. Recently, time–frequency analysis and deep convolutional neural networks have been used [73–78].

## 1.2 Methodology

The workflow of the method proposed in this chapter is shown in figure 1.1. The major components of this methodology are as follows: the dataset, the rational dilation wavelet transforms (RDWT), feature extraction, and machine learning or classification.

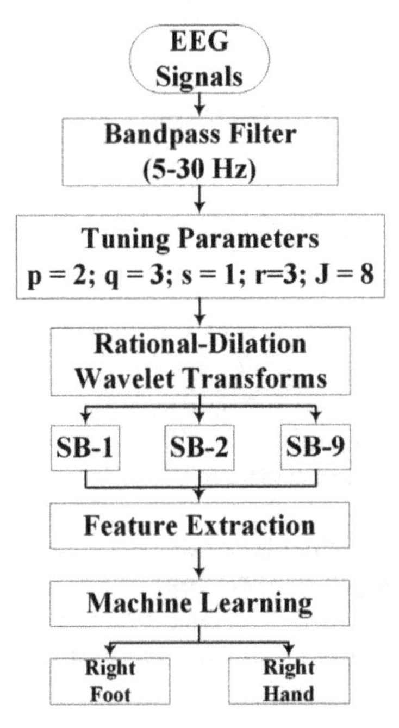


**Figure 1.1.** Classification of MI tasks from EEG signals using RDWT-based features.

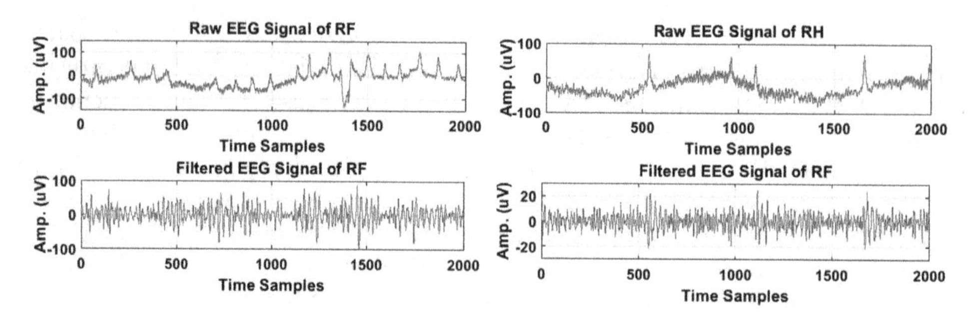


**Figure 1.2.** RF and RH MI tasks from raw and filtered EEG signals, respectively.

### 1.2.1 The dataset

We perform the MI task classification using the IV-A EEG dataset of BCI Competition-III [2, 3]. The dataset includes EEG signals of the right hand (RF) and right foot (RF) MI activities. It is a multichannel EEG dataset recorded at 1000 Hz. For our research purposes the EEG signals are down-sampled to 100 Hz. Out-band artifacts are removed by a bandpass filter. The proposed method is tested on a number of EEG signals taken from the first ten channels. The MI tasks form the raw and filtered EEG signals of the RF and RH, as shown in figure 1.2.

### 1.2.2 Rational dilation wavelet transforms

The rational dilation wavelet transforms (RDWTs) provide for oscillatory behavior of the signal, with the help of an adjusted Q-factor of wavelet bases. The RDWT can be expressed as [79]

$$\varphi(t) = \frac{q}{p}\sum_{n} h_0(n)\varphi\left(\frac{q}{p} - n\right) \tag{1.1}$$

$$\psi(t) = \frac{q}{p}\sum_{n} g_0(n)\psi\left(\frac{q}{p} - n\right), \tag{1.2}$$

where $h_0(n)$ and $g_0(n)$ denote the scaling and wavelet function filters, respectively. The Q-factor of the RADWT is represented in terms of $p$, $q$ and $s$. The dilation factor of the RADWT is $q/p$ where the terms $q$ and $p$ are coprime assuring $q > p$, $1 < q/p < 2$, $q/p$ must be almost 1 and $s > 1$ [79].

In this work, the tuning parameters for RDWT are chosen as $p = 2$, $q = 3$, $s = 1$, $r = 3$ and $j = 8$.

For these parameters, the original RF and RH signal decomposed SBs are shown in figures 1.3 and 1.4, respectively. The lower SBs exhibit higher frequency oscillation and the higher SBs show lower frequency oscillation. This is similar to normal wavelet-based decomposition. These SBs are further used for the extraction of features.

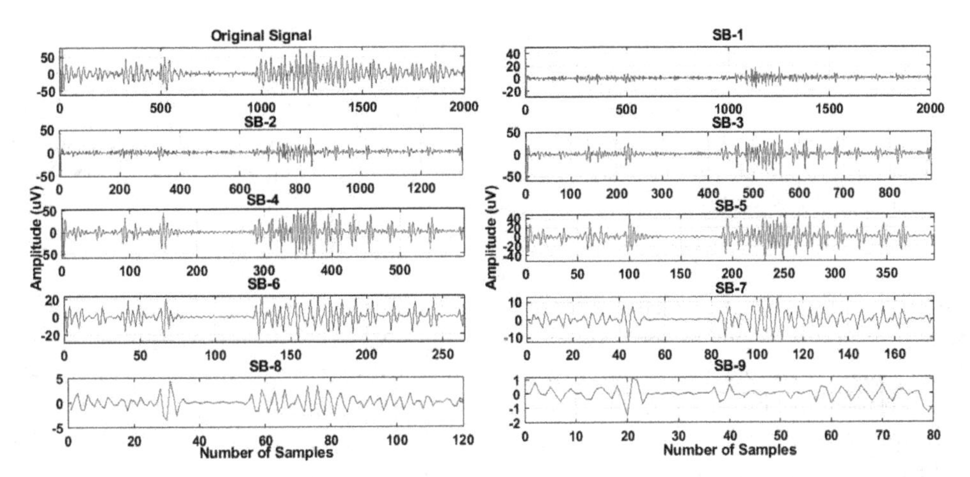


**Figure 1.3.** RF MI task based EEG signal sub-bands.

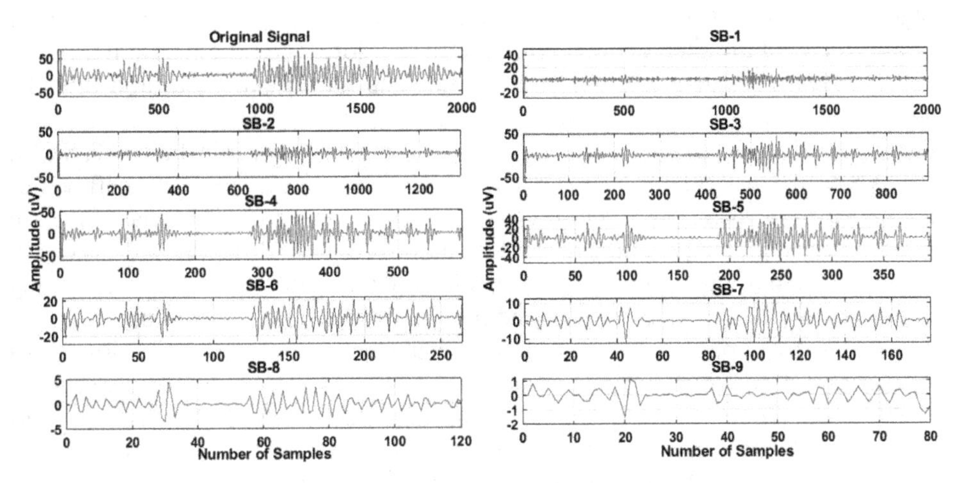


**Figure 1.4.** RH MI task based EEG signal sub-bands.

### 1.2.3 Feature extraction

The SBs produced by RDWT represent the oscillatory characteristics of different MI tasks from EEG signals. These SBs are used for extraction of the following features: modified mean absolute value type 2 (MMAV2), average amplitude change (AAC), root mean square (RMS), simple square integral (SSI) and waveform length (WL). These features are used for the classification of the RF and RH MI task EEG signals. In [80–82] a detailed description of these features is available.

### 1.2.4 Classification and performance measures

In this work, the decision tree (DT), discriminant analysis (DA), support vector machine (SVM), *k*-nearest neighbors (*k*NN) and ensemble classifiers variants are tested. The DT variants are fine tree (FT), medium tree (MT) and coarse tree (CT).

The DA variants are linear DA (LDA) and quadratic DA (QDA). The *k*NN variants are fine (F), medium (M), coarse (Co), cosine (Cos), cubic (Cu) and weighted (W). The SVM variants are linear (L), quadratic (Q), cubic (Cu), fine Gaussian (FG), medium Gaussian (MG) and coarse Gaussian (CoG). The ensemble variants are boosted trees (BoTs), bagged trees (BTs), subspace discriminant (SDE), subspace *k*NN (SS-*k*NN) and RUSBoosted trees (RUSBoT).

Many human activities involve the brain, related to cognition, language, reasoning, etc, and these activities can be recorded in the form of EEG signals. The different waves of EEG signals that represent different activities are recorded and sent to computers where the EEG data are interpreted. The interpretation of the EEG signal aims to generate suitable commands for motor tasks if the EEG signals are intended to be used to move objects or operate machines. An appropriate description of the EEG data is carried out and the subsequent classification of the data requires suitable machine learning methods and classifiers [83–85]. There are numerous classifiers available in the literature that are used in various applications of signal processing. These classifiers are mainly used to predict different classes of data and the data are then divided into several categories based on certain criteria. These criteria are given as attributes to the machine learning methods and thus the classification of data points is achieved.

The most commonly used classifiers for EEG signal processing include DT, SVM, *k*NN, ANN, naïve Bayes, etc. Generally, the classifiers are supervised methods of machine learning where the input data and target are given, and the classifiers are trained in such a manner that the target function of the output is achieved. Decision trees [86] are used to build the regression as well as classification models for separating the data into several clusters with the help of an appropriate training process. The training rules are applied sequentially in a such a way that the tree breaks down into its branches in a top-down approach until the convergence where no further clusters are obtained. The SVM [83, 85, 86] is also considered as a classification and regression model of supervised learning and is widely accepted as a robust classifier. It was conceptualized by the AT&T Lab and, since its development, has been preferred for the classification of image-related data. The SVM is now found in several variants, such as linear, quadratic, cubic, Gaussian, etc, which are used for specific applications of signal processing. A non-parametric method, also known as *k*NN [87–90], is used both for classification and regression depending on the type of data to be segregated.

As we have seen, there are various types of classifiers available in the current literature which can be employed for the classification of EEG data. Choosing an appropriate classifier for a specific application is based on few evaluation parameters, such as the region of convergence (ROC) curve, cross-evaluation, precision and recall. Sometimes a framework of classifiers is used rather than choosing a particular classifier. The accuracy, sensitivity and specificity performance measures are used to assess the classification performance. These measures are defined in [91–93].

**Table 1.1.** KW test *p*-values of the MMAV2 feature.

| MMAV2 | CH-1 | CH-2 | CH-3 | CH-4 | CH-5 | CH-6 | CH-7 | CH-8 | CH-9 | CH-10 |
|---|---|---|---|---|---|---|---|---|---|---|
| SB-1 | $3.10 \times 10^{-6}$ | $5.18 \times 10^{-7}$ | $1.14 \times 10^{-6}$ | $1.08 \times 10^{-6}$ | $3.37 \times 10^{-6}$ | $2.32 \times 10^{-5}$ | $9.41 \times 10^{-7}$ | $1.43 \times 10^{-6}$ | $1.92 \times 10^{-6}$ | $1.44 \times 10^{-6}$ |
| SB-2 | $3.09 \times 10^{-6}$ | $5.33 \times 10^{-7}$ | $1.08 \times 10^{-6}$ | $1.16 \times 10^{-6}$ | $3.60 \times 10^{-6}$ | $2.23 \times 10^{-5}$ | $9.20 \times 10^{-7}$ | $1.43 \times 10^{-6}$ | $1.84 \times 10^{-6}$ | $1.53 \times 10^{-6}$ |
| SB-3 | $2.77 \times 10^{-6}$ | $5.25 \times 10^{-7}$ | $9.65 \times 10^{-7}$ | $1.20 \times 10^{-6}$ | $3.60 \times 10^{-6}$ | $2.21 \times 10^{-5}$ | $8.80 \times 10^{-7}$ | $1.39 \times 10^{-6}$ | $1.69 \times 10^{-6}$ | $1.46 \times 10^{-6}$ |
| SB-4 | $2.67 \times 10^{-6}$ | $5.49 \times 10^{-7}$ | $9.25 \times 10^{-7}$ | $1.16 \times 10^{-6}$ | $3.65 \times 10^{-6}$ | $2.19 \times 10^{-5}$ | $8.85 \times 10^{-7}$ | $1.35 \times 10^{-6}$ | $1.57 \times 10^{-6}$ | $1.40 \times 10^{-6}$ |
| SB-5 | $2.59 \times 10^{-6}$ | $5.59 \times 10^{-7}$ | $9.25 \times 10^{-7}$ | $1.10 \times 10^{-6}$ | $3.68 \times 10^{-6}$ | $2.14 \times 10^{-5}$ | $9.20 \times 10^{-7}$ | $1.42 \times 10^{-6}$ | $1.46 \times 10^{-6}$ | $1.39 \times 10^{-6}$ |
| SB-6 | $1.27 \times 10^{-5}$ | $9.09 \times 10^{-7}$ | $1.18 \times 10^{-5}$ | $9.89 \times 10^{-6}$ | $1.70 \times 10^{-5}$ | $1.52 \times 10^{-5}$ | $5.12 \times 10^{-6}$ | $6.53 \times 10^{-6}$ | $6.59 \times 10^{-6}$ | $4.08 \times 10^{-6}$ |
| SB-7 | $3.39 \times 10^{-6}$ | $5.54 \times 10^{-7}$ | $1.43 \times 10^{-6}$ | $1.17 \times 10^{-6}$ | $4.41 \times 10^{-6}$ | $2.16 \times 10^{-5}$ | $1.01 \times 10^{-6}$ | $1.69 \times 10^{-6}$ | $1.23 \times 10^{-6}$ | $1.66 \times 10^{-6}$ |
| SB-8 | $3.31 \times 10^{-6}$ | $6.26 \times 10^{-7}$ | $1.42 \times 10^{-6}$ | $1.29 \times 10^{-6}$ | $4.67 \times 10^{-6}$ | $2.07 \times 10^{-5}$ | $1.11 \times 10^{-6}$ | $1.92 \times 10^{-6}$ | $1.23 \times 10^{-6}$ | $1.67 \times 10^{-6}$ |

**Table 1.2.** KW test *p*-values of the AAC feature.

| AAC | CH-1 | CH-2 | CH-3 | CH-4 | CH-5 | CH-6 | CH-7 | CH-8 | CH-9 | CH-10 |
|---|---|---|---|---|---|---|---|---|---|---|
| SB-1 | $2.49 \times 10^{-06}$ | $3.53 \times 10^{-08}$ | $2.22 \times 10^{-07}$ | $3.01 \times 10^{-07}$ | $8.26 \times 10^{-07}$ | $1.14 \times 10^{-05}$ | $7.43 \times 10^{-07}$ | $7.44 \times 10^{-07}$ | $3.61 \times 10^{-06}$ | $1.70 \times 10^{-06}$ |
| SB-2 | $2.42 \times 10^{-06}$ | $3.48 \times 10^{-08}$ | $2.83 \times 10^{-07}$ | $3.57 \times 10^{-07}$ | $9.77 \times 10^{-07}$ | $1.21 \times 10^{-05}$ | $8.36 \times 10^{-07}$ | $8.54 \times 10^{-07}$ | $3.63 \times 10^{-06}$ | $1.79 \times 10^{-06}$ |
| SB-3 | $2.46 \times 10^{-06}$ | $3.47 \times 10^{-08}$ | $3.65 \times 10^{-07}$ | $4.24 \times 10^{-07}$ | $1.19 \times 10^{-06}$ | $1.28 \times 10^{-05}$ | $8.63 \times 10^{-07}$ | $9.62 \times 10^{-07}$ | $3.53 \times 10^{-06}$ | $1.81 \times 10^{-06}$ |
| SB-4 | $2.49 \times 10^{-06}$ | $3.42 \times 10^{-08}$ | $4.56 \times 10^{-07}$ | $4.87 \times 10^{-07}$ | $1.40 \times 10^{-06}$ | $1.37 \times 10^{-05}$ | $9.10 \times 10^{-07}$ | $1.09 \times 10^{-06}$ | $3.50 \times 10^{-06}$ | $1.82 \times 10^{-06}$ |
| SB-5 | $2.55 \times 10^{-06}$ | $3.57 \times 10^{-08}$ | $5.83 \times 10^{-07}$ | $5.51 \times 10^{-07}$ | $1.57 \times 10^{-06}$ | $1.44 \times 10^{-05}$ | $9.63 \times 10^{-07}$ | $1.19 \times 10^{-06}$ | $3.44 \times 10^{-06}$ | $1.82 \times 10^{-06}$ |
| SB-6 | $7.53 \times 10^{-06}$ | $9.88 \times 10^{-07}$ | $7.22 \times 10^{-06}$ | $6.68 \times 10^{-06}$ | $1.04 \times 10^{-05}$ | $9.45 \times 10^{-06}$ | $5.54 \times 10^{-06}$ | $5.21 \times 10^{-06}$ | $4.02 \times 10^{-06}$ | $5.51 \times 10^{-06}$ |
| SB-7 | $2.40 \times 10^{-06}$ | $4.30 \times 10^{-07}$ | $8.36 \times 10^{-07}$ | $8.22 \times 10^{-07}$ | $3.70 \times 10^{-06}$ | $1.54 \times 10^{-05}$ | $6.72 \times 10^{-07}$ | $8.36 \times 10^{-07}$ | $6.70 \times 10^{-07}$ | $1.11 \times 10^{-06}$ |
| SB-8 | $2.59 \times 10^{-06}$ | $4.55 \times 10^{-07}$ | $9.90 \times 10^{-07}$ | $9.45 \times 10^{-07}$ | $4.37 \times 10^{-06}$ | $1.43 \times 10^{-05}$ | $6.76 \times 10^{-07}$ | $9.82 \times 10^{-07}$ | $6.39 \times 10^{-07}$ | $1.10 \times 10^{-06}$ |

**Table 1.3.** KW test *p*-values of the RMS feature.

| RMS | CH-1 | CH-2 | CH-3 | CH-4 | CH-5 | CH-6 | CH-7 | CH-8 | CH-9 | CH-10 |
|---|---|---|---|---|---|---|---|---|---|---|
| SB-1 | $3.02 \times 10^{-05}$ | 0.001 567 | 0.000 101 | $1.22 \times 10^{-05}$ | 0.015 45 | $7.04 \times 10^{-05}$ | 0.000 24 | $8.49 \times 10^{-06}$ | $1.06 \times 10^{-05}$ | 0.001 62 |
| SB-2 | $4.83 \times 10^{-05}$ | 0.003 077 | $9.27 \times 10^{-05}$ | $2.81 \times 10^{-05}$ | 0.011 91 | $5.49 \times 10^{-05}$ | 0.000 918 | $4.89 \times 10^{-05}$ | $3.98 \times 10^{-05}$ | 0.009 795 |
| SB-3 | 0.000 113 | 0.007 89 | 0.000 236 | $7.89 \times 10^{-05}$ | 0.017 85 | 0.000 13 | 0.002 385 | 0.000 216 | 0.000 203 | 0.022 35 |
| SB-4 | 0.000 197 | 0.0192 | 0.000 437 | 0.000 209 | 0.024 45 | 0.000 257 | 0.005 115 | 0.000 767 | 0.000 836 | 0.044 25 |
| SB-5 | 0.000 371 | 0.038 027 | 0.000 731 | 0.000 441 | 0.033 | 0.000 524 | 0.012 21 | 0.001 965 | 0.002 655 | 0.0708 |
| SB-6 | 0.001 14 | 0.035 41 | 0.000 209 | $9.21 \times 10^{-05}$ | 0.001 53 | 0.000 152 | 0.0207 | 0.000 42 | 0.005 61 | 0.125 55 |
| SB-7 | 0.0198 | 0.000 28 | 0.002 94 | 0.003 42 | 0.009 555 | 0.1029 | 0.006 96 | 0.007 755 | 0.0285 | 0.014 58 |
| SB-8 | 0.0201 | 0.000 276 | 0.003 675 | 0.003 93 | 0.011 295 | 0.110 25 | 0.007 335 | 0.008 805 | 0.0282 | 0.014 715 |

**Table 1.4.** KW test *p*-values of the SSI feature.

| SSI | CH-1 | CH-2 | CH-3 | CH-4 | CH-5 | CH-6 | CH-7 | CH-8 | CH-9 | CH-10 |
|---|---|---|---|---|---|---|---|---|---|---|
| SB-1 | $3.47 \times 10^{-05}$ | 0.003 771 | $5.85 \times 10^{-06}$ | $5.07 \times 10^{-06}$ | 0.002 273 | $8.32 \times 10^{-05}$ | $4.02 \times 10^{-06}$ | $2.08 \times 10^{-05}$ | $1.73 \times 10^{-05}$ | $8.88 \times 10^{-06}$ |
| SB-2 | $4.14 \times 10^{-05}$ | 0.003 031 | $6.19 \times 10^{-06}$ | $7.11 \times 10^{-06}$ | 0.002 625 | 0.000 109 | $7.69 \times 10^{-06}$ | $3.51 \times 10^{-05}$ | $2.11 \times 10^{-05}$ | $1.72 \times 10^{-05}$ |
| SB-3 | 0.000 146 | 0.001 336 | $2.68 \times 10^{-05}$ | $4.33 \times 10^{-05}$ | 0.004 758 | 0.000 388 | $4.16 \times 10^{-05}$ | 0.000 151 | $7.89 \times 10^{-05}$ | 0.000 113 |
| SB-4 | 0.000 456 | 0.000 56 | $9.27 \times 10^{-05}$ | 0.000 187 | 0.008 07 | 0.000 987 | 0.000 222 | 0.000 515 | 0.000 238 | 0.000 548 |
| SB-5 | 0.001 188 | 0.000 218 | 0.000 264 | 0.000 594 | 0.012 229 | 0.002 379 | 0.000 821 | 0.001 464 | 0.000 68 | 0.001 751 |
| SB-6 | $1.27 \times 10^{-05}$ | $9.76 \times 10^{-16}$ | $4.81 \times 10^{-05}$ | $1.31 \times 10^{-06}$ | $1.79 \times 10^{-06}$ | $2.34 \times 10^{-06}$ | $3.72 \times 10^{-08}$ | $6.49 \times 10^{-08}$ | $1.30 \times 10^{-06}$ | $1.79 \times 10^{-09}$ |
| SB-7 | 0.000 146 | 0.001 336 | $2.68 \times 10^{-05}$ | $4.33 \times 10^{-05}$ | 0.004 758 | 0.000 388 | $4.16 \times 10^{-05}$ | 0.000 151 | $7.89 \times 10^{-05}$ | 0.000 113 |
| SB-8 | 0.000 456 | 0.000 56 | $9.27 \times 10^{-05}$ | 0.000 187 | 0.008 07 | 0.000 987 | 0.000 222 | 0.000 515 | 0.000 238 | 0.000 548 |

**Table 1.5.** KW test *p*-values of the WL feature.

| WL | CH-1 | CH-2 | CH-3 | CH-4 | CH-5 | CH-6 | CH-7 | CH-8 | CH-9 | CH-10 |
|---|---|---|---|---|---|---|---|---|---|---|
| SB-1 | $9.38 \times 10^{-16}$ | $4.87 \times 10^{-16}$ | $2.93 \times 10^{-15}$ | $2.42 \times 10^{-18}$ | 7.13 × 10–20 | $1.87 \times 10^{-16}$ | $9.14 \times 10^{-21}$ | $3.22 \times 10^{-18}$ | $4.85 \times 10^{-17}$ | $1.70 \times 10^{-18}$ |
| SB-2 | 2.76 × 10–08 | $2.83 \times 10^{-08}$ | $7.75 \times 10^{-09}$ | $2.90 \times 10^{-07}$ | 9.84 × 10–06 | 4.80 × 10–07 | $3.65 \times 10^{-06}$ | $8.90 \times 10^{-07}$ | $1.37 \times 10^{-07}$ | $6.36 \times 10^{-06}$ |
| SB-3 | 0.008 602 | 0.012 714 | 0.004 838 | 0.001 82 | $8.20 \times 10^{-05}$ | 0.004 677 | 0.000 427 | 0.000 852 | 0.000 696 | 0.000 492 |
| SB-4 | $6.24 \times 10^{-07}$ | $3.86 \times 10^{-06}$ | $1.10 \times 10^{-07}$ | $4.49 \times 10^{-08}$ | $3.19 \times 10^{-10}$ | $1.72 \times 10^{-07}$ | $1.29 \times 10^{-08}$ | $1.35 \times 10^{-08}$ | $6.19 \times 10^{-09}$ | $1.13 \times 10^{-08}$ |
| SB-5 | $1.54 \times 10^{-07}$ | $2.40 \times 10^{-06}$ | $2.66 \times 10^{-08}$ | $2.78 \times 10^{-08}$ | $1.90 \times 10^{-10}$ | $1.72 \times 10^{-08}$ | $1.61 \times 10^{-08}$ | $7.87 \times 10^{-09}$ | $1.43 \times 10^{-09}$ | $1.67 \times 10^{-08}$ |
| SB-6 | $3.67 \times 10^{-12}$ | $8.16 \times 10^{-09}$ | $6.58 \times 10^{-14}$ | $9.26 \times 10^{-12}$ | $1.45 \times 10^{-12}$ | $4.99 \times 10^{-15}$ | $2.00 \times 10^{-14}$ | $7.70 \times 10^{-14}$ | $3.41 \times 10^{-16}$ | $3.00 \times 10^{-15}$ |
| SB-7 | $1.99 \times 10^{-55}$ | $1.00 \times 10^{-14}$ | $1.50 \times 10^{-52}$ | $2.95 \times 10^{-56}$ | $8.54 \times 10^{-45}$ | $9.24 \times 10^{-35}$ | $9.50 \times 10^{-58}$ | $4.42 \times 10^{-45}$ | $1.44 \times 10^{-44}$ | $1.92 \times 10^{-57}$ |
| SB-8 | $1.51 \times 10^{-50}$ | $9.36 \times 10^{-13}$ | $5.83 \times 10^{-48}$ | $7.63 \times 10^{-52}$ | $3.19 \times 10^{-41}$ | $1.50 \times 10^{-32}$ | $1.05 \times 10^{-52}$ | $6.14 \times 10^{-42}$ | $1.24 \times 10^{-41}$ | $3.07 \times 10^{-53}$ |

## 1.3 Results and discussion

RDWT decomposes the MI task EEG signal into sub-bands. The sub-bands show different oscillatory characteristics corresponding to the MI task. The sub-bands' oscillatory characteristics are extracted in terms of the following features:

**Table 1.6.** The SB-wise classification accuracy with the DT and DA variants.

| SB | FT | MT | CT | LDA | QDA |
|---|---|---|---|---|---|
| SB-1 | 91.3 | 91.3 | 83.1 | 97.3 | 93.3 |
| SB-2 | 80.7 | 80.2 | 73.1 | 89.8 | 75.8 |
| SB-3 | 80.7 | 79.6 | 68.7 | 86.9 | 70.2 |
| SB-4 | 77.6 | 77.1 | 71.3 | 80.4 | 63.6 |
| SB-5 | 82.9 | 81.6 | 74.4 | 90 | 73.6 |
| SB-6 | 82.2 | 81.1 | 82.9 | 88.7 | 71.6 |
| SB-7 | 78.2 | 75.3 | 70 | 94.2 | 68.9 |
| SB-8 | 57.3 | 57.6 | 58.7 | 80.9 | 52.4 |

**Table 1.7.** The SB-wise classification accuracy with the SVM variants.

| SB | L-SVM | Q-SVM | CU-SVM | FG-SVM | MG-SVM | CoG-SVM |
|---|---|---|---|---|---|---|
| SB-1 | 96.7 | 98.9 | 98 | 87.8 | 97.8 | 94.4 |
| SB-2 | 89.6 | 97.6 | 96.9 | 82.9 | 94.7 | 70.7 |
| SB-3 | 86.9 | 99.1 | 97.6 | 83.1 | 94 | 66.9 |
| SB-4 | 83.3 | 98.7 | 97.6 | 80 | 95.1 | 68.4 |
| SB-5 | 89.1 | 98.7 | 99.1 | 82 | 96.7 | 72.2 |
| SB-6 | 91.8 | 98 | 96.9 | 86 | 96.9 | 61.6 |
| SB-7 | 94.4 | 97.3 | 96.7 | 87.8 | 95.3 | 55.3 |
| SB-8 | 83.3 | 97.1 | 95.3 | 85.3 | 93.8 | 55.6 |

**Table 1.8.** The SB-wise classification accuracy with the *k*NN variants.

| SB | F-*k*NN | M-*k*NN | Co-*k*NN | Cos-*k*NN | Cu-*k*NN | W-*k*NN |
|---|---|---|---|---|---|---|
| SB-1 | 95.1 | 92.2 | 79.8 | 91.3 | 91.8 | 94.9 |
| SB-2 | 87.6 | 82.9 | 70.7 | 74.2 | 79.1 | 87.6 |
| SB-3 | 83.3 | 77.8 | 65.8 | 71.6 | 78.2 | 82.7 |
| SB-4 | 81.1 | 76.4 | 65.3 | 73.1 | 76.2 | 83.6 |
| SB-5 | 82.7 | 78.4 | 70 | 76.2 | 78.9 | 85.6 |
| SB-6 | 84.7 | 78.2 | 64 | 76.7 | 75.6 | 86 |
| SB-7 | 81.6 | 78.7 | 62.2 | 78 | 77.3 | 82 |
| SB-8 | 74.7 | 68.9 | 59 | 67.7 | 68 | 73.3 |

**Table 1.9.** The SB-wise classification accuracy with ensemble classifier variants.

| SB | BoT | BaT | SDE | SS-*k*NN | RUSBoT |
|---|---|---|---|---|---|
| SB-1 | 63.8 | 91.6 | 95.3 | 96.4 | 88.4 |
| SB-2 | 84 | 86 | 83.1 | 92.2 | 86.2 |
| SB-3 | 85.3 | 85.6 | 81.3 | 92.4 | 84.4 |
| SB-4 | 85.8 | 86.7 | 77.1 | 91.3 | 81.6 |
| SB-5 | 88.2 | 88.7 | 88 | 93.3 | 84 |
| SB-6 | 82 | 87.8 | 86.4 | 93.1 | 82 |
| SB-7 | 82 | 85.6 | 92.2 | 92.9 | 78.7 |
| SB-8 | 65.1 | 67.3 | 74.7 | 77.1 | 62.2 |

**Table 1.10.** SB-wise confusion matrix of the Q-SVM classifier.

| SB | SB-1 | | SB-2 | | SB-3 | | SB-4 | |
|---|---|---|---|---|---|---|---|---|
| Class | RH | RF | RH | RF | RH | RF | RH | RF |
| RH | 100 | 0 | 99 | 1 | 100 | 0 | 100 | 0 |
| RF | 2 | 98 | 3.6 | 96.4 | 1.6 | 98.4 | 2.4 | 97.6 |
| SB | SB-1 | | SB-2 | | SB-3 | | SB-4 | |
| Class | RH | RF | RH | RF | RH | RF | RH | RF |
| RH | 100 | 0 | 99 | 1 | 98.5 | 1.5 | 97.5 | 2.5 |
| RF | 1.6 | 98.4 | 2.8 | 97.2 | 3.6 | 96.4 | 3.2 | 96.8 |

MMAV2, AAC, RMS, SSI and EWL. These features' statistical significance is first examined using the Kruskal–Wallis (KW) test. The KW-test probabilistic ($p$)-values for different features are shown in tables 1.1–1.5. All the features for all the channels of EEG sub-bands show significantly smaller $p$-values. This demonstrates the statistical significance of the RDWT based features.

The MI task EEG signal classification performance of RDWT based features is examined using several classification algorithms. The feature dimensions of the RF and RH MI tasks are $50 \times 250$ and $50 \times 200$, respectively. Here, the rows represent the features extracted from the ten channels of recorded EEG signals and the columns represent the instances of each class. Tables 1.6–1.9 list the classification accuracy of the extracted feature-set with different machine learning algorithms. Among the machine learning algorithms used, the SVM and *k*NN variants exhibit better performance compared to the other algorithms used.

Among all the classification algorithms, the SVM classifier variant Q-SVMs provides the highest classification accuracy for most of the sub-bands. Table 1.10 further presents the classification performance of Q-SVM using confusion matrices.

**Table 1.11.** The proposed method's performance measures with the Q-SVM classifier.

| SB | Accuracy (%) | Sensitivity (%) | Specificity (%) |
|---|---|---|---|
| SB-1 | 98.89 | 97.56 | 100 |
| SB-2 | 97.56 | 95.65 | 99.18 |
| SB-3 | 99.11 | 98.04 | 100 |
| SB-4 | 98.67 | 97.09 | 100 |
| SB-5 | 99.11 | 98.04 | 100 |
| SB-6 | 98.00 | 96.59 | 99.18 |
| SB-7 | 97.33 | 95.63 | 98.77 |
| SB-8 | 97.11 | 96.06 | 97.98 |

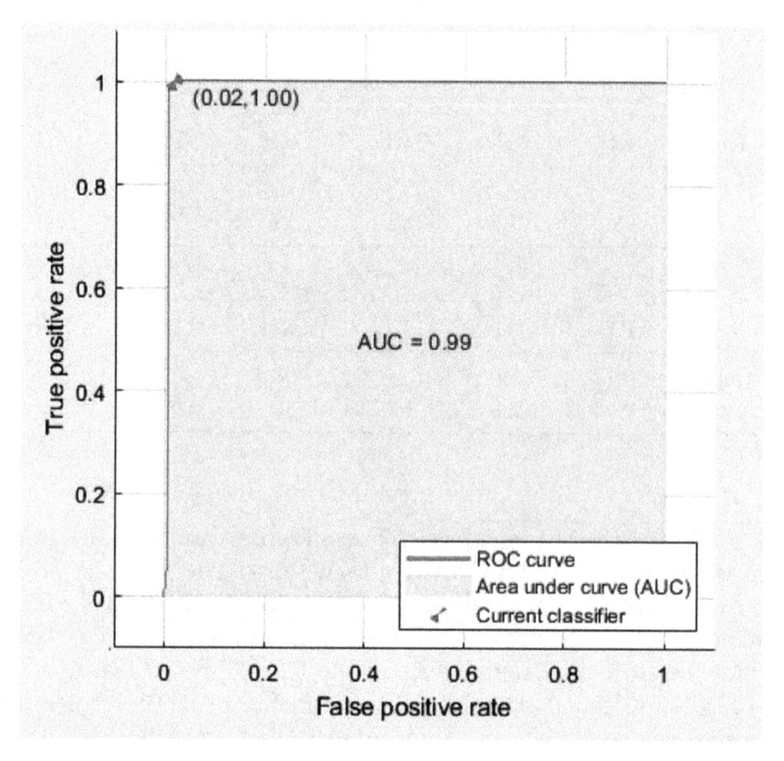


**Figure 1.5.** The ROC curve of the Q-SVM classifier for SB-5-based features.

The diagonal values of the confusion matrices show the detection rates of the RH and RF MI tasks, respectively. The SB-3 based features achieve the best detection rates, 100% and 98.4% for the RH and RF MI tasks, respectively. The other classification performance parameters of Q-SVM with different SBs are presented in table 1.11, in which the SB-3 based features achieved the best values of accuracy (99.11%), sensitivity (98.04%) and specificity (99.18%).

**Table 1.12.** Performance of different methods using the same dataset.

| Authors | Method | Classifier | Accuracy (%) |
|---|---|---|---|
| Zhang *et al* [28] | Z-score | LDA | 81.1 |
| Lu *et al* [47] | R-CSP with aggregation | R-CSP | 83.90 |
| Kevric *et al* [45] | MSPCA, WPD and HOS | *k*NN | 92.8 |
| Siuly and Li [29] | CC | LS-SVM | 95.72 |
| Taran *et al* [71] | TQWT | LS-SVM | 96.89 |
| Taran *et al* [2] | AFE and FN | LS-SVM | 97.56 |
| Proposed approach | RDWT | Q-SVM | 99.11 |

In figure 1.5, the Q-SVM performance is also assessed using the receiver operating characteristic (ROC) curve. In this figure, a drastic increase in true positive rate (TPR) is observed for small variations in the false positive rate (FPR). Also, it can be observed that the area under the curve (AUC = 0.99) is also close to unity. Both ROC characteristics of the Q-SVM are the desired characteristics, which confirm the stable classification of the RF and RH MI tasks.

Table 1.12 presents the performance report for the same dataset methods. The table compares the classification accuracies of the different suggested methods and classifiers. The accuracy range of the previous methods is 81.1%–97.56%. Of all the previous methods, the analytic feature extraction (AFE) and feature normalization (FN) with LS-SVM classifier suggested by Taran *et al* [2] achieved the highest classification accuracy (97.56%). The proposed method also outperforms Taran *et al* [2] and obtained the highest classification accuracy of 99.11%. This can be advantageous for accurate classification of RF and RF MI tasks in practical BCI systems such as robotic arms, mice and keyboard control, etc.

## 1.4 Conclusion

A reliable BCI system requires the accurate classification of different MI tasks. In this chapter MI tasks are classified using EEG signals. RDWT-based features are proposed for the classification of RF and RH MI task EEG signals. RDWT decomposes the EEG signals into SBs. The different time-domain features MMAV2, AAC, RMS, SSI and WL are extracted from the RDWT-provided SBs. These features are examined using different machine learning algorithms. The Q-SVM provides the best classification performance compared to the other classifiers. The proposed method's classification accuracy with Q-SVM is 99.11% and this is also the highest value compared to other existing methods tested on the same dataset. This demonstrates that the proposed method can provide an accurate solution for the classification of MI tasks in a BCI system.

## References

[1] Demir F, Bajaj V, Ince M C, Taran S and Şengür A 2019 Surface EMG signals and deep transfer learning-based physical action classification *Neural. Comput. Appl.* **31** 8455–62

[2] Taran S, Bajaj V, Sharma D, Siuly S and Sengur A 2018 Features based on analytic IMF for classifying motor imagery EEG signals in BCI applications *Measurement* **116** 68–76

[3] Chaudhary S, Taran S, Bajaj V and Sengur A 2019 Convolutional neural network based approach towards motor imagery tasks EEG signals classification *IEEE Sensors J.* **19** 4494–500

[4] Khare S K, Bajaj V and Sinha G R 2020 Automatic drowsiness detection based on variational nonlinear chirp mode decomposition using electroencephalogram signals *Modelling and Analysis of Active Biopotential Signals in Healthcare vol 1* (Bristol: IOP Publishing) pp 5.1–5.25

[5] Chaudhary S, Taran S, Bajaj V and Siuly S 2020 A flexible analytic wavelet transform based approach for motor-imagery tasks classification in BCI applications *Comput. Methods Programs Biomed.* **187** 105325

[6] Wang H and Zhang Y 2016 Detection of motor imagery EEG signals employing naïve Bayes based learning process *Measurement* **86** 148–58

[7] Khan M J and Hong K S 2017 Hybrid EEG–fNIRS-based eight-command decoding for BCI: application to quadcopter control *Front. Neurorobotics* **11** 6

[8] Bajaj V, Taran S, Khare S K and Sengur A 2020 Feature extraction method for classification of alertness and drowsiness states EEG signals *Appl. Acoust.* **163** 107224

[9] Taran S, Bajaj V and Sharma D 2017 TEO separated AM–FM components for identification of apnea EEG signals *2nd Int. Conf. on Signal and Image Processing (ICSIP)* (Piscataway, NJ: IEEE) pp 391–5

[10] Bajaj V, Taran S, Tanyildizi E and Sengur A 2019 Robust approach based on convolutional neural networks for identification of focal EEG signals *IEEE Sens. Lett.* **3** 1–4

[11] Taran S and Bajaj V 2019 Emotion recognition from single-channel EEG signals using a two-stage correlation and instantaneous frequency-based filtering method *Comput. Methods Programs Biomed.* **173** 157–65

[12] Khare S K, Bajaj V and Sinha G R 2020 Adaptive tunable Q wavelet transform based emotion identification *IEEE Trans. Instrum. Meas.* **69** 9609–17

[13] Taran S and Bajaj V 2019 Sleep apnea detection using artificial bee colony optimize Hermite basis functions for EEG signals *IEEE Trans. Instrum. Meas.* **69** 608–16

[14] Khare S K and Bajaj V 2020 Constrained based tunable Q wavelet transform for efficient decomposition of EEG signals *Appl. Acoust.* **163** 107234

[15] Taran S and Bajaj V 2018 Drowsiness detection using adaptive Hermite decomposition and extreme learning machine for electroencephalogram signals *IEEE Sensors J.* **18** 8855–62

[16] Khare S K and Bajaj V 2020 Time–frequency representation and convolutional neural network based emotion recognition *IEEE Trans Neural Netw. Learn. Syst.* 1

[17] Taran S and Bajaj V 2018 Clustering variational mode decomposition for identification of focal EEG signals *IEEE Sens. Lett.* **2** 1–4

[18] Yadav P, Priya A, Taran S, Bajaj V and Sharma D 2017 Discrimination of alcohol and normal EEG signal using EMD *4th Int. Conf. on Signal Processing and Integrated Networks (SPIN)* (Piscataway, NJ: IEEE) pp 410–3

[19] Taran S and Bajaj V 2017 Rhythm-based identification of alcohol EEG signals *IET Sci. Meas. Technol.* **12** 343–9

[20] Khare S K and Bajaj V 2020 Optimized tunable Q wavelet transform based drowsiness detection from electroencephalogram signals *Innov. Res. Biomed. Eng.* in press https://doi.org/10.1016/j.irbm.2020.07.005

[21] Taran S, Bajaj V and Sharma D 2017 Robust Hermite decomposition algorithm for classification of sleep apnea EEG signals *Electron. Lett.* **53** 1182–4

[22] Khare S K, Bajaj V, Siuly S and Sinha G R 2020 Classification of schizophrenia patients through empirical wavelet transformation using electroencephalogram signals *Modelling and Analysis of Active Biopotential Signals in Healthcare* vol 1 (Bristol: IOP Publishing) 1.1–1.26

[23] Polat K and Gunes S 2007 Classification of epileptiform EEG using a hybrid system based on decision tree classifier and fast Fourier transform *Appl. Math. Comput.* **187** 1017–26

[24] Pfurtscheller G, Neuper C, Guger C, Harkam W A, Ramoser H, Schlogl A, Obermaier B A and Pregenzer M A 2000 Current trends in Graz brain–computer interface (BCI) research *IEEE Trans. Rehabil. Eng.* **8** 216–9

[25] Herman P, Prasad G, McGinnity T M and Coyle D 2008 Comparative analysis of spectral approaches to feature extraction for EEG-based motor imagery classification *IEEE Trans. Neural Syst. Rehabil. Eng.* **16** 317–26

[26] Siuly S, Wang H and Zhang Y 2016 Detection of motor imagery EEG signals employing naive Bayes based learning process *Measurement* **86** 148–58

[27] Siuly S and Li Y 2015 Discriminating the brain activities for brain–computer interface applications through the optimal allocation-based approach *Neural Comp. Appl.* **26** 799–811

[28] Zhang R, Xu P, Guo L, Zhang Y, Li P and Yao D 2013 *Z*-score linear discriminant analysis for EEG based brain–computer interfaces *PLoS One* **8** e74433

[29] Siuly S and Li Y 2012 Improving the separability of motor imagery EEG signals using a cross correlation-based least square support vector machine for brain–computer interface *IEEE Trans. Neural Syst. Rehabil. Eng.* **20** 526–38

[30] Hsu W Y 2011 Continuous EEG signal analysis for asynchronous BCI application *Int. J. Neural Syst.* **21** 335–50

[31] Siuly S, Li Y and Wen P 2013 Identification of motor imagery tasks through CCLR algorithm in brain computer interface *Int. J. Bioinformat. Res. Appl.* **9** 156–72

[32] Siuly, Li Y and Wen P P 2014 Modified CC-LR algorithm with three diverse feature sets for motor imagery tasks classification in EEG based brain–computer interface *Comput. Methods Programs Biomed.* **113** 767–80

[33] Spüler M, Walter A, Rosenstiel W and Bogdan M 2013 Spatial filtering based on canonical correlation analysis for classification of evoked or event-related potentials in EEG data *IEEE Trans. Neural Syst. Rehabil. Eng.* **22** 1097–103

[34] Zhang Y U, Zhou G, Jin J, Wang X and Cichocki A 2014 Frequency recognition in SSVEP-based BCI using multiset canonical correlation analysis *Int. J. Neural Syst.* **24** 1450013

[35] Gao L, Cheng W, Zhang J and Wang J 2016 EEG classification for motor imagery and resting state in BCI applications using multi-class Adaboost extreme learning machine *Rev. Sci. Instrum.* **87** 085110

[36] Saka K, Aydemir Ö and Öztürk M 2016 Classification of EEG signals recorded during right/left hand movement imagery using fast Walsh Hadamard transform based features *39th Int. Conf. on Telecommunications and Signal Processing (TSP)* (Piscataway, NJ: IEEE) pp 413–6

[37] Iacoviello D, Petracca A, Spezialetti M and Placidi G 2015 A real-time classification algorithm for EEG-based BCI driven by self-induced emotions *Comput. Methods Programs Biomed.* **122** 293–303

[38] Kołodziej M, Majkowski A and Rak R J 2011 A new method of EEG classification for BCI with feature extraction based on higher order statistics of wavelet components and selection

with genetic algorithms *Int. Conf. on Adaptive and Natural Computing Algorithms* (Berlin: Springer) pp 280–9

[39] Hu D, Li W and Chen X 2011 Feature extraction of motor imagery EEG signals based on wavelet packet decomposition *The 2011 IEEE/ICME Int. Conf. on Complex Medical Engineering* vol 22 (Piscataway, NJ: IEEE) pp 694–7

[40] Bhattacharyya S, Khasnobish A, Chatterjee S, Konar A and Tibarewala D N 2010 Performance analysis of LDA, QDA and KNN algorithms in left–right limb movement classification from EEG data *2010 Int. Conf. on Systems in Medicine and Biology* (Piscataway, NJ: IEEE) pp 126–31

[41] Sai C Y, Mokhtar N, Arof H, Cumming P and Iwahashi M 2017 Automated classification and removal of EEG artifacts with SVM and wavelet-ICA *IEEE J. Biomed. Health Informat.* **22** 664–70

[42] Lekshmi S S, Selvam V and Rajasekaran M P 2014 EEG signal classification using principal component analysis and wavelet transform with neural network *2014 Int. Conf. on Communication and Signal Processing* vol 3 (Piscataway, NJ: IEEE) pp 687–90

[43] She Q S, Ma Y L, Meng M, Xi X G and Luo Z Z 2017 Noise-assisted MEMD based relevant IMFs identification and EEG classification *J. Central South Univ.* **24** 599–608

[44] Gaur P, Pachori R B, Wang H and Prasad G 2015 An empirical mode decomposition based filtering method for classification of motor-imagery EEG signals for enhancing brain–computer interface *2015 Int. Joint Conf. on Neural Networks (IJCNN)* (Piscataway, NJ: IEEE) pp 1–7

[45] Kevric J and Subasi A 2017 Comparison of signal decomposition methods in classification of EEG signals for motor-imagery BCI system *Biomed. Signal Process. Control* **31** 398–406

[46] Park C, Looney D, Naveed ur Rehman N, Ahrabian A and Mandic D P 2012 Classification of motor imagery BCI using multivariate empirical mode decomposition *IEEE Trans. Neural Syst. Rehabil. Eng.* **21** 10–22

[47] Lu H, Eng H L, Guan C, Plataniotis K N and Venetsanopoulos A N 2010 Regularized common spatial pattern with aggregation for EEG classification in small-sample setting *IEEE Trans. Biomed. Eng.* **57** 2936–46

[48] Lotte F and Guan C 2010 Regularizing common spatial patterns to improve BCI designs: unified theory and new algorithms *IEEE Trans. Biomed. Eng.* **58** 355–62

[49] Arvaneh M, Guan C, Ang K K and Quek C 2011 Optimizing the channel selection and classification accuracy in EEG-based BCI *IEEE Trans. Biomed. Eng.* **58** 1865–73

[50] Guger C, Ramoser H and Pfurtscheller G 2000 Real-time EEG analysis with subject-specific spatial patterns for a brain–computer interface (BCI) *IEEE Trans. Rehabil. Eng.* **8** 447–56

[51] Lotte F and Guan C 2010 Spatially regularized common spatial patterns for EEG classification *2010 20th Int. Conf. on Pattern Recognition* (Piscataway, NJ: IEEE) pp 3712–5

[52] Sturm I, Lapuschkin S, Samek W and Müller K R 2016 Interpretable deep neural networks for single-trial EEG classification *J. Neurosci. Methods* **274** 141–5

[53] Liu Y, Zhang H, Chen M and Zhang L 2015 A boosting-based spatial–spectral model for stroke patients' EEG analysis in rehabilitation training *IEEE Trans. Neural Syst. Rehabil. Eng.* **24** 169–79

[54] Song X, Yoon S C and Perera V 2013 Adaptive common spatial pattern for single-trial EEG classification in multisubject BCI *2013 6th Int. IEEE/EMBS Conf. on Neural Engineering (NER)* (Piscataway, NJ: IEEE) pp 411–4

[55] Leamy D J, Kocijan J, Domijan K, Duffin J, Roche R A, Commins S, Collins R and Ward T E 2014 An exploration of EEG features during recovery following stroke–implications for BCI-mediated neurorehabilitation therapy *J. Neuroeng. Rehab* **11** 9

[56] Caramia N, Lotte F and Ramat S 2014 Optimizing spatial filter pairs for EEG classification based on phase-synchronization *2014 IEEE Int. Conf. on Acoustics, Speech and Signal Processing (ICASSP)* (Piscataway, NJ: IEEE) pp 2049–53

[57] Xu P, Liu T, Zhang R, Zhang Y and Yao D 2014 Using particle swarm to select frequency band and time interval for feature extraction of EEG based BCI *Biomed. Signal Process. Control* **10** 289–95

[58] Bentlemsan M, Zemouri E T, Bouchaffra D, Yahya-Zoubir B and Ferroudji K 2014 Random forest and filter bank common spatial patterns for EEG-based motor imagery classification *2014 5th Int. Conf. on Intelligent Systems, Modelling and Simulation* (Piscataway, NJ: IEEE) pp 235–8

[59] Robinson N, Vinod A P, Guan C, Ang K K and Peng T K 2011 A wavelet-CSP method to classify hand movement directions in EEG based BCI system *2011 8th Int. Conf. on Information, Communications and Signal Processing* (Piscataway, NJ: IEEE) pp 1–5

[60] Zhang R, Xu P, Liu T, Zhang Y, Guo L, Li P and Yao D 2013 Local temporal correlation common spatial patterns for single trial EEG classification during motor imagery *Comput Math. Methods Med.* **2013** 1–7

[61] Park C, Took C C and Mandic D P 2013 Augmented complex common spatial patterns for classification of noncircular EEG from motor imagery tasks *IEEE Trans. Neural Syst. Rehabil. Eng.* **22** 1

[62] Miao M, Zeng H, Wang A, Zhao C and Liu F 2017 Discriminative spatial–frequency–temporal feature extraction and classification of motor imagery EEG: an sparse regression and weighted naïve Bayesian classifier-based approach *J. Neurosci. Methods* **278** 13–24

[63] Wang Y, Zhang Z, Li Y, Gao X, Gao S and Yang F 2004 BCI competition 2003—data set IV: an algorithm based on CSSD and FDA for classifying single-trial EEG *IEEE Trans. Biomed. Eng.* **51** 1081–6

[64] Li J and Zhang L 2010 Regularized tensor discriminant analysis for single trial EEG classification in BCI *Pattern Recognit. Lett.* **31** 619–28

[65] Herman P A, Prasad G and McGinnity T M 2016 Designing an interval type-2 fuzzy logic system for handling uncertainty effects in brain–computer interface classification of motor imagery induced EEG patterns *IEEE Trans. Fuzzy Syst.* **25** 29–42

[66] Lal T N, Schroder M, Hinterberger T, Weston J, Bogdan M, Birbaumer N and Scholkopf B 2004 Support vector channel selection in BCI *IEEE Trans. Biomed. Eng.* **51** 1003–10

[67] Hsu W Y 2010 EEG-based motor imagery classification using neuro-fuzzy prediction and wavelet fractal features *J. Neurosci. Methods* **189** 295–302

[68] Gandhi V, Prasad G, Coyle D, Behera L and McGinnity T M 2013 Quantum neural network-based EEG filtering for a brain–computer interface *IEEE Trans Neural Netw. Learn. Syst.* **25** 278–88

[69] Oikonomou V P, Nikolopoulos S, Petrantonakis P and Kompatsiaris I 2018 Jul 18 Sparse kernel machines for motor imagery EEG classification *2018 40th Annual Int. Conf. of the IEEE Engineering in Medicine and Biology Society (EMBC)* (Piscataway, NJ: IEEE) pp 207–10

[70] Zhang Y, Zhou G, Jin J, Zhao Q, Wang X and Cichocki A 2015 Sparse Bayesian classification of EEG for brain–computer interface *IEEE Trans Neural Netw. Learn. Syst.* **27** 2256–67

[71] Taran S and Bajaj V 2019 Motor imagery tasks-based EEG signals classification using tunable-Q wavelet transform *Neural Comput. Appl.* **31** 6925–32

[72] Wang P, Jiang A, Liu X, Shang J and Zhang L 2018 LSTM-based EEG classification in motor imagery tasks *IEEE Trans. Neural Syst. Rehabil. Eng.* **26** 2086–95

[73] Lee H, Kim Y D, Cichocki A and Choi S 2007 Nonnegative tensor factorization for continuous EEG classification *Int. J. Neural Syst.* **17** 305–17

[74] Cecotti H and Graeser A 2008 Convolutional neural network with embedded Fourier transform for EEG classification *2008 19th Int. Conf. on Pattern Recognition* (Piscataway, NJ: IEEE) pp 1–4

[75] Tang Z, Li C and Sun S 2017 Single-trial EEG classification of motor imagery using deep convolutional neural networks *Optik* **130** 11–8

[76] Manor R and Geva A B 2015 Convolutional neural network for multi-category rapid serial visual presentation BCI *Front. Comput. Neurosci.* **9** 146

[77] Tabar Y R and Halici U 2016 A novel deep learning approach for classification of EEG motor imagery signals *J. Neural Eng.* **14** 016003

[78] Tan C, Sun F, Zhang W, Chen J and Liu C 2017 Multimodal classification with deep convolutional-recurrent neural networks for electroencephalography *Int. Conf. on Neural Information Processing* (Cham: Springer) pp 767–76

[79] Bayram I and Selesnick I W 2009 Frequency-domain design of overcomplete rational-dilation wavelet transforms *IEEE Trans. Signal Process.* **57** 2957–72

[80] Krishna A H, Sri A B, Priyanka K Y V S, Taran S and Bajaj V 2018 Emotion classification using EEG signals based on tunable-Q wavelet transform *IET Sci. Meas. Technol.* **13** 375–80

[81] Chada S, Taran S and Bajaj V 2020 An efficient approach for physical actions classification using surface EMG signals *Health Inf. Sci. Syst.* **8** 3

[82] Sravani C, Bajaj V, Taran S and Sengur A 2020 Flexible analytic wavelet transform based features for physical action identification using sEMG signals *IRBM* **41** 18–22

[83] SVMS.org 2010 Introduction to support vector machines http://svms.org/introduction.html

[84] Guenther N and Schonlau M 2016 Support vector machines *Stata J.* **16** 917–37

[85] Bennett K P and Campbell C 2000 Support vector machines: hype or hallelujah? *SIGKDD Explor. Newsl.* **2** 1–13

[86] Akhil jabbar M, Deekshatulu B L and Chandra P 2013 Classification of heart disease using *k*-nearest neighbor and genetic algorithm *Procedia Technol.* **10** 85–94

[87] Taran S, Sharma P C and Bajaj V 2020 Automatic sleep stages classification using optimize flexible analytic wavelet transform *Knowl.-Based Syst.* **192** 105367

[88] Taran S and Bajaj V 2018 Drowsiness detection using instantaneous frequency based rhythms separation for EEG signals *2018 Conf. on Information and Communication Technology (CICT)* (Piscataway, NJ: IEEE) pp 1–6

[89] Bajaj V, Taran S and Sengur A 2018 Emotion classification using flexible analytic wavelet transform for electroencephalogram signals *Health Inform. Sci. Syst.* **6** 12

[90] Kiran P U, Abhiram N, Taran S and Bajaj V 2018 TQWT based features for classification of ALS and healthy EMG signals *Am. J. Comput. Sci. Inf. Technol* **6** 19

[91] Pareta A, Taran S, Bajaj V and Sengur A 2019 Automatic environment sounds classification using optimum allocation sampling *2019 4th Int. Conf. on Robotics and Automation Engineering (ICRAE)* (Piscataway, NJ: IEEE) pp 69–73

[92] Taran S, Bajaj V and Siuly S 2017 An optimum allocation sampling based feature extraction scheme for distinguishing seizure and seizure-free EEG signals *Health Inform. Sci. Syst.* **5** 7

[93] Nagineni S, Taran S and Bajaj V 2018 Features based on variational mode decomposition for identification of neuromuscular disorder using EMG signals *Health Inform. Sci. Syst.* **6** 13

**IOP** Publishing

Modelling and Analysis of Active Biopotential Signals in Healthcare, Volume 2

**Varun Bajaj and G R Sinha**

# Chapter 2

# A deep learning framework for emotion recognition using improved time–frequency image analysis of electroencephalography signals

**Kaniska Samanta[1], Soumya Chatterjee[1] and Rohit Bose[2]**

[1]*Department of Electrical Engineering Techno India University Kolkata, India*

[2]*Department of Bio-Engineering University of Pittsburgh Pittsburgh, PA, USA*

The recognition of human emotions based on the analysis of non-stationary electroencephalogram (EEG) signals is important to design a robust human–computer interaction (HCI) model. This contribution investigates the feasibility of using Stockwell transform (ST) based time–frequency (T–F) analysis of non-stationary EEG signals for emotion recognition. To improve the resolution of EEG signals in the T–F plane, the traditional Gaussian window present in the ST was replaced by an adaptive signal dependent modified window whose parameters are optimally selected using maximum energy concentration measure based constrained optimization. The T–F images of the processed EEG signals obtained by employing the proposed modified ST were subsequently fed to a pretrained highly dense convolution neural network (CNN) in order to classify different emotional states. In this work, both subject-specific as well as cross-subject based classification was performed to test the robustness of the proposed technique. The proposed methodology was verified on two publicly available databases, named the SJTU Emotion EEG Dataset (SEED) and SEED-IV. It was observed that the highest mean accuracies of 98.6% and 89.9% for the SEED and the SEED-IV databases, respectively, were obtained considering subject-specific classification. In the case of cross-subject classification, the proposed method delivered an average accuracy of 93.26% for the SEED and 78.6% for the SEED-IV database, respectively. We observed that the performance of the proposed method was found to be significantly

doi:10.1088/978-0-7503-3411-2ch2

better than the existing literature studies on the same datasets which validates the efficiency of the proposed method to develop an efficient human–computer machine interface system.

## 2.1 Introduction

Emotion is considered as one of the most fundamental characteristics of human nature as it plays a crucial role in every human life with significant impacts on interactive behavior, decision making, work performance, etc. In recent times, the recognition of the emotional state of the human brain has become a growing field of research as it has many applications in neuroscience and psychology. The recognition of emotional states can be used as a feedback in order to develop state-of-the-art human–computer interaction (HCI) models [1–3]. Many psychologists have confirmed that human emotions play a critical role in the mental and physical health of a human being. Various states of the human mind, such as stress, depression, anxiety, etc, are regulated by emotional drives and can cause adverse effects for human health. Hence, it can be said that emotion recognition also plays a key role in the development of mental health monitoring systems [4].

In medical technology, several techniques are available for monitoring brain activity, including EEG, diffusion magnetic resonance imaging (dMRI), functional magnetic resonance imaging (fMRI), etc [5, 6]. However, the image based brain activity recognition techniques such as fMRI and dMRI are not only expensive but also suffer from the drawback of poor temporal resolution. In contrast, human brain activity recognition employing EEG signal screening is cheap and has been accepted worldwide by medical experts as it returns better time resolution. EEG signals are a type of complex, non-stationary and aperiodic biopotential signal generated from the neurons, i.e. nerve cells located inside the brain which is a part of the central nervous system. In clinics and pathology labs EEG recordings are used for the diagnosis of various brain-related disorders such as epilepsy [7], schizophrenia [8], Alzheimer's disease [9], autism [10], etc. Apart from the diagnosis of brain disorders, EEG signals are also widely used to develop efficient brain–computer interface systems. The application of EEG for motor imagery (hand and foot) movement recognition [11–13], eye movement detection [14], alcoholism detection [15, 16] and also emotion recognition [2, 4, 17–20] has been reported in existing works. In the context of human emotion recognition, recording EEG data from the scalp is considered to be more scientific compared to other methods such as facial expression or speech based approaches as the neural oscillations caused by the emotional drive are much more spontaneous and immune to any manipulation by the test subject [4]. Because of these aforesaid advantages, EEG signals have proven to be useful to trace the emotional state of a person [18] as they directly represents the activation of the central nervous system which varies from person to person and also with changes in the emotional drives [4].

In the published literature, the application of various techniques to classify human emotions from EEG signals has been reported [4, 18–29]. In [4] the authors explored different types of features from time domain, frequency domain and non-linear dynamical systems and used an automatic feature selection process to detect the most robust features across the subjects. Gupta [25] and Bajaj [26] used a flexible analytic wavelet transform based approach to perform channel specific cross-subject classification. Krishna [27] and Khare [28] successfully implemented a tunable Q-parameter wavelet transform (TQWT) and adaptive TQWT for emotion classification. Two-stage correlation and instantaneous frequency based filtering has proved to be a useful tool to develop automated emotion recognition systems [29]. In [22] and [30] the researchers explored different domain adaptation techniques to classify emotional states. A deep adaptation network based cross-subject emotion recognition system and a bi-hemisphere domain adversarial neural network model for the recognition of emotions is discussed in [20, 21]. In [24] Zhang proposed a spatial–temporal recurrent neural network for emotion recognition and a dynamical graph convolution network is proposed in [23]. The use of a smoothed pseudo-Wigner–Ville distribution aided joint time–frequency analysis and convolutional neural network (CNN) for the classification of different human emotions is reported in [31]. It is thus clear from the above literature survey that many non-stationary signal processing techniques have been successfully implemented by researchers for the recognition of human emotions from EEG signals. In this vein, in the current contribution a novel technique for the analysis of non-stationary EEG signals in the joint time–frequency (T–F) domain employing the Stockwell transform (ST) is proposed for human emotion recognition using EEG signals.

The Stockwell transform (ST) was proposed by Robert G Stockwell [32] as a phase corrected version of the continuous wavelet transform (CWT). The main advantage of the ST over the CWT is its ability to retain the phase information at each frequency, whereas the CWT provides only a locally referenced phase. Thus, the ST provides better information during the analysis of EEG signals in the time–frequency frame than the CWT. In addition, compared to other T–F domain methods, the ST offers certain advantages. First, the ST presents noise immunity and is also independent of the type of mother wavelet used. Previously, the ST has been successfully applied for non-stationary signal analysis in different fields of science such as oceanography [33], geophysics [34, 35], atmospheric physics [32, 35–37], biomedical engineering [38–40], power systems [41–43], mechanical engineering [44], hydrogeology [45], etc. Although the application of the ST has been previously used for analysis of non-stationary EEG signals for epilepsy detection [46], in the existing literature the application of the ST for emotion recognition from EEG signals has not yet been reported. This chapter therefore investigates the feasibility of using ST based T–F analysis for the recognition of human emotion EEG signals.

Although the generalized version of the ST offers certain advantages over existing methods for non-stationary signal analysis in the T–F plane, from the point of view of non-stationary signal analysis the conventional ST has a major limitation since it uses a Gaussian window (GW) whose standard deviation varies inversely with the frequency. Therefore, for high frequency components of a signal, the width of the GW decreases which imposes a serious problem in dealing with non-stationary signals. In addition, for both the high and low frequency components present in a signal, the traditional ST results in higher time and frequency resolutions, respectively. For better analysis of non-stationary signals in the T–F plane it is necessary to have a signal dependent GW rather than a frequency dependent one, which can adjust itself with the nature of the signal to be analysed and, at the same time, can improve the resolution in the T–F plane. To address this issue, a novel signal dependent adaptive GW is proposed in this chapter for better temporo-spectral representation of EEG signals in the T–F plane. The components of the proposed GW are determined so as to achieve the highest energy concentration measure (ECM) in the T–F plane. For this purpose, a differential evolution (DE) algorithm has been used to determine the parameters of the GW.

Another major challenge in emotion recognition is meaningful feature extraction from the EEG signals for the classification of human emotions. This is because EEG patterns vary greatly from subject to subject [25] which makes it difficult to generalize the EEG features containing emotion related information across subjects. Thus, the extraction of handcrafted features from the time–frequency matrix of the ST of the EEG signals may not provide accurate information regarding a particular emotional state for all subjects. In order to address the problem of feature extraction, a convolution neural network (CNN) aided deep learning framework has been implemented which eliminates the need for a separate feature extraction stage. Thus, the problem of manual feature extraction which is tedious, unsophisticated and often leads to erroneous information regarding subject-specific emotion recognition has been eliminated. In addition, since the nature of the variation of EEG signals for a specific emotion are different from subject to subject, along with subject-specific classification, cross-subject classification across different channels is also performed in this study which makes the proposed system effective and robust.

In this study, EEG signals representing different emotions were procured from two available online databases. After that, the ST with a modified Gaussian window was applied on the acquired EEG signals to obtain their respective time–frequency representations. The T–F images of the EEG signals with improved resolution obtained by using the proposed modified GW were finally served as inputs to an appropriate CNN model for the classification of human emotions. In the present framework, a pretrained network known as 'DenseNet201' CNN architecture with a transfer learning process has been employed for training the network architecture. We achieved very high recognition accuracy for both subject-specific as well as cross-subject classification in comparison to the existing literature. A comparative study with other T–F based techniques revealed that the proposed $ST_M$ employing a novel GW provides better results, indicating the superiority of this proposed method over existing ones. The rest of the chapter is organized as follows.

In section 2.2 a detailed description of the dataset used in this study is presented. The theoretical background of the ST, the proposed modification of the GW, the theory of the differential evolution (DE) optimization algorithm and the convolutional neural network are discussed in section 2.3. EEG signal analysis in the time–frequency frame using the proposed optimally parameterized GW of the ST and the classification performance (both subject-specific and cross-subject) using the convolutional neural network are reported in section 2.4. Finally, section 2.5 concludes the paper. A flowchart showing the proposed method of emotion recognition from EEG signals is shown in figure 2.1.

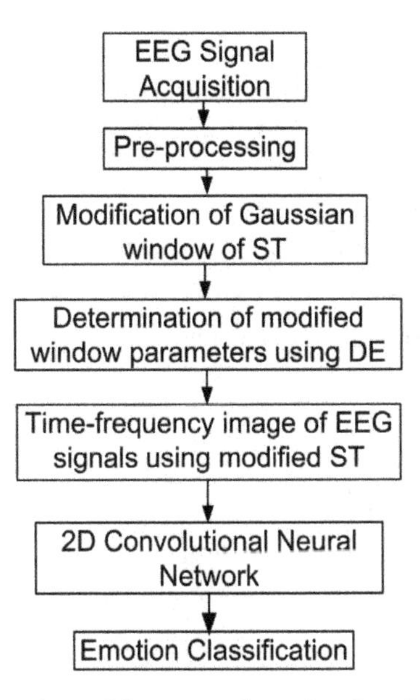


**Figure 2.1.** Flowchart of the proposed emotion detection framework.

## 2.2 Description of the EEG dataset

In this chapter, two open-access benchmark EEG datasets, namely the SJTU Emotion EEG Dataset (SEED) [25] and SEED-IV [19], are used for emotion recognition. Brief descriptions of the two datasets used in this study are given below.

In the SEED dataset, three classes of emotion EEG signals, namely neutral, positive and negative, were collected from fifteen different subjects, eight female and seven male, with a mean age of 23.27 years. These subjects were labeled with a number as 1–15. According to the 10–20 international electrode system, EEG signals were recorded from 62 channels. After recording the original signal, a band-pass filter with a frequency band of 0–75 Hz was applied to eliminate the noise present in the EEG signals. The sampling rate of the EEG signal was kept at 200 Hz. Fifteen Chinese movie clips were chosen to present to the subjects, which were carefully selected to generate distinct classes of emotion. The film clips were selected based on the following criteria. The length of the experiment was kept short enough to prevent the subjects from becoming fatigued, the video clips were easily understood without any clarification and finally the video clips should evoke an individual desired target emotion. Data were collected from each subject after performing three sessions with a gap of about one week. Each session consisted of 15 trials. Every trial began with a 5 s hint before playing each clip. The video clips played were around 4 min long. After watching the clips there was 45 s for self-assessment and a 15 s rest, followed by a feedback session, where the participants were urged to address their emotional feelings. A schematic diagram of the trials is shown in figure 2.2.

The SEED-IV database is an extension of the above SEED dataset. This dataset consists of four emotional brain states based on the emotions happy, fearful, sad and neutral. Here fifteen subjects also took part in the experiment. The participants wore

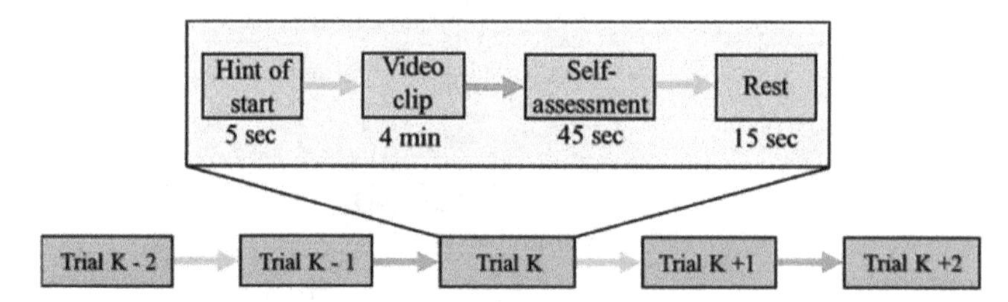


**Figure 2.2.** Schematic diagram of the trials for the SEED database.

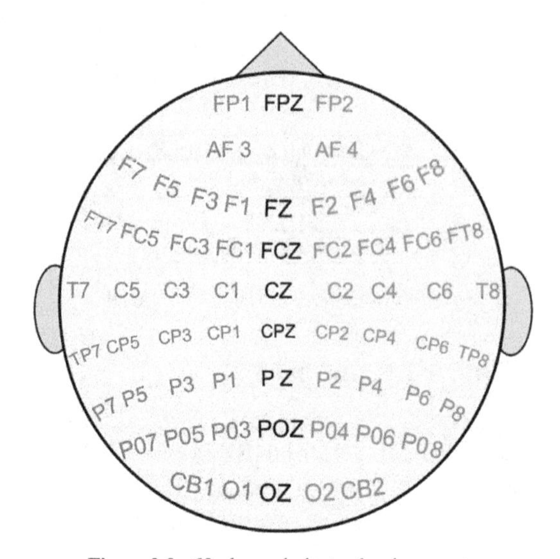


**Figure 2.3.** 62 channel electrode placement.

an EEG cap and eye-tracking glasses during the experiment. The EEG signals were recorded using 62 channels as per the 10–20 international electrode system. The placement of the recording electrodes is shown in figure 2.3.

The recorded signals were down-sampled at a frequency of 200 Hz. After data acquisition, band-pass filtering with cut off frequencies between 0–75 Hz was carried out in order to remove the noise and artefacts that were present in the signals. Seventy-two film clips were chosen for recording purposes. These video clips were carefully chosen based on various preliminary studies. Each participant performed a total of three sessions on different days and each session included 24 trials. There was a 5 s hint before each clip and the video clips were played for approximately 2 min followed by 45 s for self-assessment. The participants were requested to give some feedback on their emotional state after watching each clip. A schematic diagram of the trials is shown in figure 2.4. In this contribution only 12 of 62 channels are selected as it has been reported in the existing literature that these aforesaid channels have been found to be more effective for emotion classification. These channels were designated as C5, C6, CP5, CP6, FT7, FT8, P7, P8, TP7, TP8, T7 and T8.

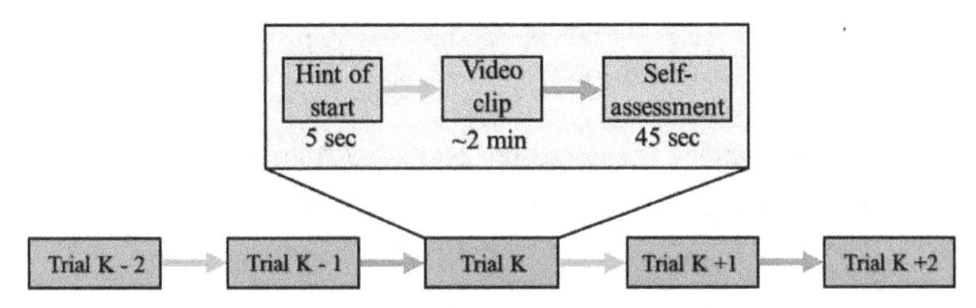


**Figure 2.4.** Schematic diagram of the SEED-IV trials.

## 2.3 Methodology

### 2.3.1 The Stockwell transform

The Stockwell transform (ST) is a popular method to represent a non-stationary signal in the T–F frame. The mathematical expression of the ST of a signal $g(t)$ can be expressed as

$$S(\tau, f) = \int_{-\infty}^{+\infty} g(t)W(t - \tau, f)\mathrm{e}^{-2\pi jft}\mathrm{d}t. \tag{2.1}$$

In (2.1) $W(t - \tau, f)$ is the Gaussian window (GW), defined as

$$W(t, f) = \frac{1}{\sigma(f)\sqrt{2\pi}}\mathrm{e}^{\frac{-t^2}{2\sigma(f)^2}}. \tag{2.2}$$

In (2.2) $\sigma(f)$ is the standard deviation of the GW, which is inversely proportional to the signal frequency and can be expressed as

$$\sigma(f) = \frac{1}{|f|}. \tag{2.3}$$

Now, from (2.3) it can be observed that in the conventional ST the standard deviation, i.e. the shape of the GW, is inversely proportional to the frequency, whatever type signal is analysed, and as a result the width of the GW is decreased with an increase in frequency. This creates a problem in dealing with non-stationary signal analysis as there may be a possibility that the conventional GW of the generalized version of the ST fails to return the optimum localized T–F representation of the input signal. It is therefore desirable to have an adaptive GW that can adjust itself to the type of analyzed signal for improved non-stationary signal analysis in the T–F frame. In order to achieve better control of the GW, the standard deviation of the GW needs to be modified by introducing several new parameters. It is to be noted here that several modifications of the standard deviation of the GW for analysis of heart sounds [47], power quality detection [48], EMG signal analysis [49], etc, have already been reported in the existing literature. However, such modifications of the existing GW are signal specific and hence in general cannot be applied to EEG signal analysis. Considering this issue, a new standard deviation of the GW specific to the EEG signal being analysed is proposed in this article, the details of which are discussed below.

### 2.3.2 The proposed modified version of the Stockwell transform

In this chapter, the proposed $\sigma(f)$ of the GW is expressed as

$$\sigma(f) = \sqrt{\frac{(xf^{y} + z)}{|f|^{2w}}}. \tag{2.4}$$

Here $x$, $y$, $z$ and $w$ are four different window parameters which control the shape of the GW. Interestingly, using the values $x = 1$, $y = 0$, $z = 0$ and $w = 1$, the standard deviation of the conventional ST given by (2.3) can be obtained. Therefore, the proposed window is a generalized version of the GW, which can be used for better analysis of the EEG signals in the T–F plane. The introduction of several parameters makes the GW more resilient and at the same time provides more degrees of freedom to tune the size and shape of the GW, thereby making it more suitable to adapt to the analyzed EEG signal, which is a major limitation of the conventional ST. Now, using equations (2.2) and (2.4) the modified GW, i.e. $W_{\mathrm{M}}(\tau - t, f)$, can be expressed as follows:

$$W_{\mathrm{M}}(\tau - t, f) = \frac{|f|^{w}}{\sqrt{2\pi(xf^{y} + z)}} \mathrm{e}^{\frac{-(\tau-t)^2 f^{2w}}{2(xf^{y}+z)}}. \tag{2.5}$$

Substituting (2.5) in (2.1), the final mathematical expression for modified ST, $\mathrm{ST_M}(\tau - t, f)$ of any input signal $x(t)$ can be written as

$$\mathrm{ST_M}(\tau - t, f) = \int_{-\infty}^{+\infty} g(t) \frac{|f|^{w}}{\sqrt{2\pi(xf^{y} + z)}} \mathrm{e}^{\frac{-(\tau-t)^2 f^{2w}}{2(xf^{y}+z)}} \mathrm{e}^{-2\pi jft} \mathrm{d}t. \tag{2.6}$$

Now the main aim is to optimally select the parameters of the modified Gaussain window in such a way that an improved resolution in the T–F plane is achieved. In order to achieve better resolution of the T–F images of EEG signals, maximization of the energy concentration measure (ECM) is necessary, as reported in [47]. Therefore, the proposed GW parameters were selected using the maximum energy constraint based optimization. The objective function can be expressed as

$$\mathrm{ECM}(x, y, z, w) = \frac{1}{\int_{-\infty}^{+\infty} \int_{-\infty}^{+\infty} |\overline{\mathrm{ST_M}^{x,y,z,w}(t, f)}| \, \mathrm{d}t\mathrm{d}f}. \tag{2.7}$$

Here, $\overline{\mathrm{ST_M}^{x,y,z,w}(t, f)}$ is the normalized modified ST which can be expressed as (2.8)

$$\overline{\mathrm{ST_M}^{x,y,z,w}(t, f)} = \frac{\mathrm{ST_M}^{x,y,z,w}(t, f)}{\sqrt{\int_{-\infty}^{+\infty} \int_{-\infty}^{+\infty} |\overline{\mathrm{ST_M}^{x,y,z,w}(t, f)}|^2 \, \mathrm{d}t\mathrm{d}f}}. \tag{2.8}$$

Finally, the overall optimization problem is expressed as (2.9)

$$\max \arg_{x,y,z,w} = \frac{1}{\int_{-\infty}^{+\infty}\int_{-\infty}^{+\infty} |\overline{ST_M{}^{x,y,z,w}(t,f)}|\, dt df} \tag{2.9}$$

s.t.

$$\begin{aligned} &\left(nT_s f_{\max}^{w}\right)^2 - x f_{\min}^{y} - z \leqslant 0 \\ &x f_{\min}^{y} + z \leqslant (lT_s)^2 \\ &0 \leqslant x,\, y,\, z,\, w \leqslant 2. \end{aligned}$$

Here $T_s$ is the sampling rate ($T_s = 0.005$ s) and $f \in [f_{\min}, f_{\max}]$. Now, $f_{\min}$ and $f_{\max}$ are the minimum and maximum frequency ranges of the input EEG signal, respectively. In this study, $f_{\min}$ and $f_{\max}$ are chosen as 1 Hz and 70 Hz, respectively, since frequencies above 70 Hz are not relevant for EEG analysis and can be discarded as noise. The time resolution of the modified GW is controlled by $nT_s$ and $lT_s$, respectively. Care should be taken that the proposed GW is not so narrow as to alter the time resolution and at the same time must not be so broad as to hamper the frequency resolution. Therefore, the value $n$ in the lower bound of the GW (i.e. $nT_s$) should ideally be kept greater than 1 for minimum time resolution. In this chapter, the value of $n$ is empirically selected as 5. Similarly, the value of $l$ in the upper bound of GW is kept fixed at 200 (i.e. $lT_s = 1$ s). It is to be mentioned here that the objective function used in this paper for computing the ECM to determine the parameters of the modified GW was similar to that used in [49]. However, the GW function has been modified in the present study for the purpose of analysing EEG signals of emotion.

### 2.3.3 Differential evolution optimization algorithm

The the parameters of the novel $ST_M$ proposed in (2.4) can be selected either empirically or by using some evolutionary numerical optimization techniques. In this context, several methods to optimize the parameters of the GW have been reported in the existing literature. In [48] the authors use an active set algorithm to select the parameters of the GW for the classification of PQ signals. A heuristic search process employing a genetic algorithm (GA) for the selection of GW parameters for the classification of heart sounds is reported in [50]. In [49] particle swarm based optimization (PSO) is used to determine the optimum window parameters for EMG signal classification. In this article, a differential evolution (DE) based optimization technique [51] is adopted to optimally select the GW parameters. A brief theory of the DE algorithm is given below.

DE is a very popular evolutionary technique which is used to handle different global optimization problems. Inspired by the natural process of evolution, this stochastic search algorithm is designed to deduce a number of trial vectors out of which a few are suitable for solving the particular optimization problem [52]. The DE algorithm was proposed by Storn and Price in [51] and has successfully been applied to handle different optimization problems in various domains such as mechanical engineering [53, 54], communication [55] and pattern recognition [56]. In the DE algorithm, a population ($n_p$) of multi-dimensional parameter vectors is evolved, which encodes the desired solutions for the global optimization. The population should be initialized in such a way as to cover the entire search space by uniformly randomizing the individuals, bounded by the maximum and minimum values of the parameter constraints, depending upon the problem. After initialization, the algorithm goes through a loop of evolutionary operation that involves three steps: mutation, crossover and selection.

*Mutation.* In the mutation operation, the initialized population is updated and produces a new population by recombination of the initial random solutions. For a trial vector $\vec{A}_{i,G}$ at generation $G$, the mutant vector associated with it can be written as, $\vec{B}_{i,G} = \{b_{1,i,G}, b_{2,i,G} \ldots\ldots b_{D,i,G}\}$ where $D$ indicates the dimension of the target parameter vector. The mutant vector can be obtained by employing different mutation strategies. The strategy used in the present contribution is

$$\text{"DE/rand-to-best/1"}: \vec{B}_{i,G} = \vec{A}_{i,G} + F\left(\vec{A}_{\text{best},G} - \vec{A}_{i,G}\right) + F\left(\vec{A}_{r_1,G} - \vec{A}_{r_2,G}\right). \quad (2.10)$$

The indices $r_1$ and $r_2$ are two randomly chosen integers, mutually exclusive to each other and lie within the range from 1 to $n_p$ and both are different from the base index $i$. These indices are chosen only once for each mutant vector. To scale the difference vector, a positive control parameter $F$ (known as scale factor) is used in (2.10). $\vec{A}_{\text{best},G}$ denotes the individual vector having the best fitness value at generation $G$ within the population.

*Crossover.* The next step after mutation is the crossover, where the components of $\vec{B}_{i,G}$ are exchanged with $\vec{A}_{i,G}$ to generate a trial vector $\vec{C}_{i,G} = \{c_{1,i,G}, c_{2,i,G}, \ldots\ldots c_{D,i,G}\}$ by using a crossover strategy. In this work the binomial crossover method has been employed to obtain the trail vector. The crossover operation increases the potential diversity in the newly defined population.

*Selection.* To maintain the size of the population as constant over subsequent generations, the algorithm performs selection operation to determine if the trial or the target vector live on to the next generation $G + 1$. Mathematically, the selection operation is given by

$$\vec{A}_{i,G+1} = \vec{C}_{i,G} \text{ if } f(\vec{C}_{iG}) \leqslant f(\vec{A}_{i,G}) \quad (2.11)$$

else, $\vec{A}_{i,G}$ otherwise.

In equation (2.11) $f(\cdot)$ is the objective function which is to be minimized. The target vector is replaced by the new trial vector in the next generation if the objective function attains either an equal or lower value. Otherwise, the target does not become updated in the population. Hence, the population either obtains better fitness or maintains its previous fitness condition without any deterioration. The above three evolutionary operations are repeated for a few iterations until the

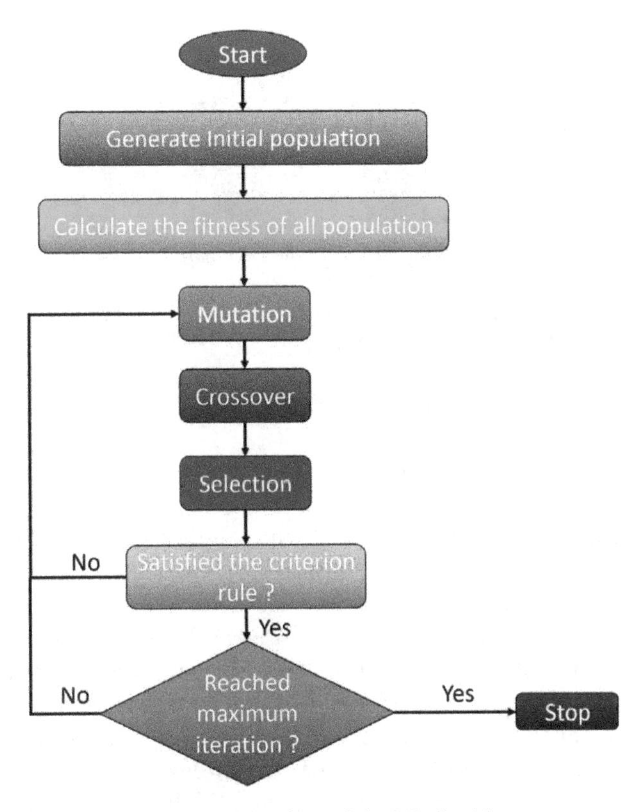


**Figure 2.5.** Flowchart of the DE algorithm.

maximum iteration has been reached. The outline of the DE algorithm has been provided in figure 2.5.

### 2.3.4 Convolutional neural network (CNN)

The selection of features plays an important role in any classification task. The manual selection and extraction of features often leads to misclassification because of redundant feature values. A deep learning algorithm is capable of extracting relevant features automatically from the input image and performs the classification task accordingly. Therefore, it offers a distinct advantage over the conventional handcrafted feature extraction method. In this work a convolutional neural network (CNN) based deep learning algorithm has been utilized for the classification of different emotional conditions of the human brain. Inspired by the learning process of the natural brain, the CNN requires rigorous training of many hidden layers to learn and extract significant features through multilayer perception that can clearly discriminate different categories of input classes. The basic CNN architecture consists of an input layer, a few hidden blocks that include convolution, pooling and fully connected layers, and one softmax layer as an output. In figure 2.6 the basic architecture of a CNN model is shown.

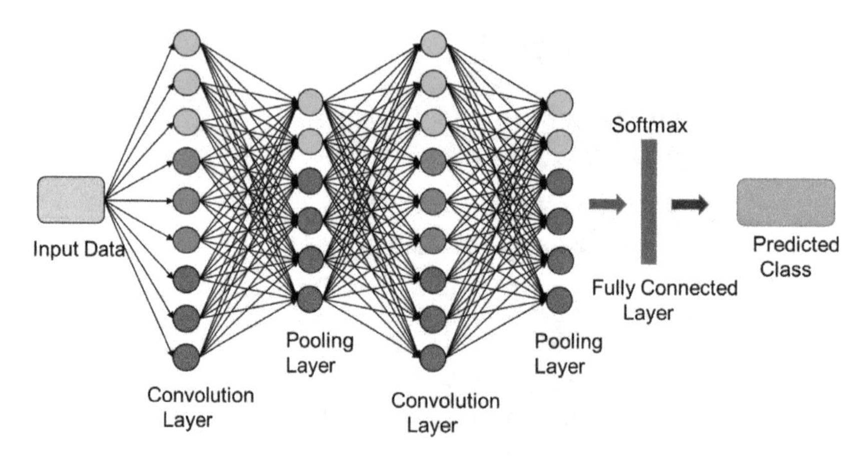


**Figure 2.6.** Basic architecture of the CNN.

In the hidden layer of the CNN model, convolution layers extract high dimensional meaningful features from the input data by utilizing a fixed number of independent filters, commonly known as edge detectors or kernels. The rectified linear unit (ReLU) blocks are the commonly used activation functions which remain integrated in the convolution layer itself and provide non-linearity to the CNN architecture. The pooling layer reduces the reduce dimension and computation complexity of the network and also avoids the model over-fitting and loss of useful information. The dimension reduction is performed with the help of applying non-linear down-sampling on the previously generated high dimensional feature output and hence the computation in the CNN model is reduced. This layer also regulates the over-fitting during the training process. The spatial volume of the output is reduced and also by reducing the network parameters, the training time of the CNN architecture is shortened. Finally, the fully connected layer feeds forward the neural network by the same forward function as the convolution layer. This layer converts the two-dimensional feature maps into one-dimensional feature vectors based on which the Softmax layer classifies the data into different categories. Lastly, the output layer returns the predicted class for random validation data [57].

*2.3.4.1 DenseNet*

DenseNet is a specific type of convolutional neural network (CNN) where each layer is connected to its all preceding and succeeding layers in a feed-forward way. While the normal CNNs with $L$ layers have $L$ number of connections, the DenseNet forms $\frac{L(L+1)}{2}$ connections. This particular type of CNN structure enables each layer of the model to use the feature maps extracted by all preceding layers as the input. DenseNet has several compelling advantages such as alleviation of the vanishing gradient problem, strong feature propagation, extensive reuse of features and subsequently a reduction of the number of parameters. In the DenseNet architecture, all the layers with the same feature map sizes are directly connected to each other and, to maintain the feed-forward characteristics, each layer obtains

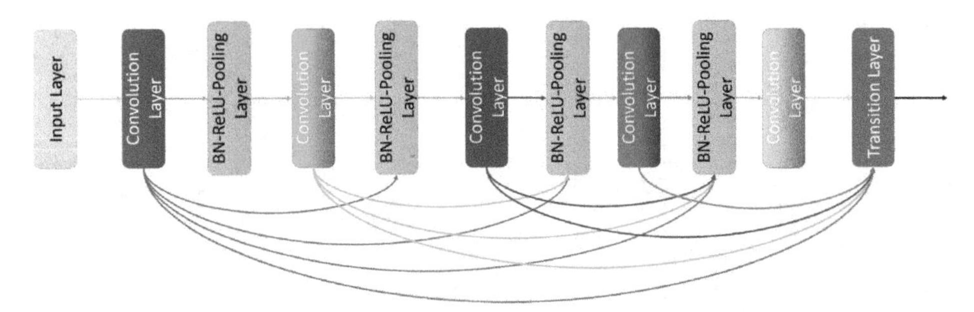


**Figure 2.7.** Architecture of a single dense block with five convolution layers.

additional inputs from all the previous layers and passes the same to all subsequent layers [58]. Figure 2.7 shows an illustration of a dense block comprising five convolution layers of the DenseNet architecture.

The DenseNet model involves fewer parameters than the traditional CNN models since it does not require relearning of the redundant feature maps. In the conventional feed-forward architectures, a state is passed from one layer to its subsequent layer with the latter writing the state read from its previous layers, thereby changing the state and preserving the information. The residual network proposed in [59] used additive identity mapping for the preservation of information. Recent studies have showed that many layers do not contribute equally in training and can be discarded, which makes the ResNet identical to the recurrent neural network (RNN). However, since each layer has its own weights, the number of parameters in the ResNet model inadvertently increases. The DenseNet can differentiate between the added and the preserved information within the network. The layers of the DenseNet are narrow which adds only a small feature map set to the 'collective knowledge' of the layer while the rest are kept unaltered. The final decision is made by the classification layer depending on the feature maps obtained from the entire feature space in the network.

Apart from better parameter efficiency, another major advantage of the DenseNet model is its efficient information flow procedure throughout the network. This enables the network to be trained easily. In DenseNet, the original data and the loss function gradients can be accessed directly from each layer of the network, leading to an implicit deep supervision [58]. It has also been observed that the dense connection has a regularizing effect which reduces the over-fitting on tasks with smaller training datasets. All the aforesaid advantages make the DenseNet model a perfect fit for the present emotion recognition framework. In this contribution, a pretrained 201 layer network DenseNet201 model has been utilized for classification. The architecture of DenseNet201 is shown in figure 2.8.

*2.3.4.2 Transfer learning using DenseNet*

Although CNN provides an end-to-end automated deep learning framework, yet one of the major problems is the time required to train any CNN model. Training any CNN architecture from scratch can take a huge amount of time. To handle this

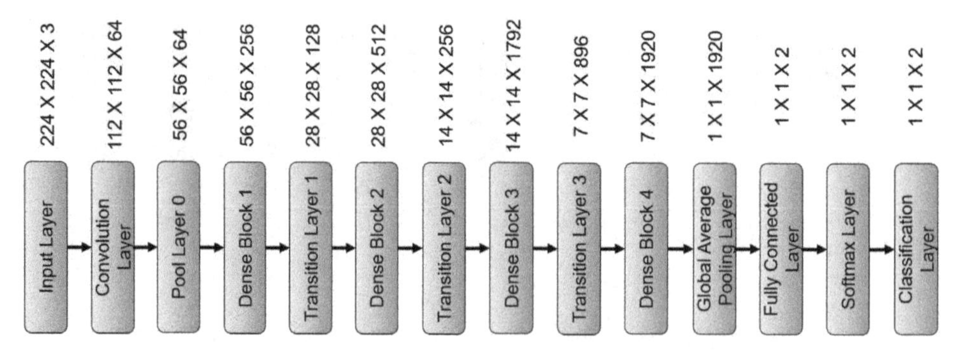


**Figure 2.8.** Architecture of the DenseNet201 model.

issue, a transfer learning (TL) technique has been implemented in this study. TL is widely used, through which a predefined network, trained for one task, can be trained for another task by modifying the hyper-parameters of the deeper layers. Through this process a predefined CNN model can be trained in much less time (from minutes to hours) depending on the input size. Using TL, the last few operating layers of the CNN, i.e. the fully connected layers, softmax classification layer and the output layers, were discharged and the rest of the CNN model can be trained on different image datasets for detecting and extracting meaningful features as well as to perform various classification tasks. In the other words, TL allows fine-tuning of a pretrained CNN architecture to operate on a fresh classification task. In this chapter, TL is implemented to train the DenseNet201 CNN for the automated feature extraction and classification of emotion EEG signals.

## 2.4 Results and discussion

### 2.4.1 Time–frequency analysis of EEG signals using a modified ST

The analysis of EEG signals in the time–frequency frame obtained using the proposed $ST_M$ is discussed in this section. The EEG signals corresponding to different emotional states were acquired from the SEED and the SEED-IV datasets. The last one second of EEG was chosen in this study, since it has been reported in the existing literature [25] that the last once second EEG data are more sensitive to the change in emotional drives of a human being. The acquired EEG signals from two datasets were transformed into their respective time–frequency domains by applying the modified ST ($ST_M$). The time variation of the EEG signals as well as their respective T–F image plots obtained using traditional ST and proposed $ST_M$ are presented in figures 2.9 and 2.10, respectively.

From the T–F images of the EEG signals, it is evident that significant differences between different classes exist for both the SEED and the SEED-IV datasets. Also, the resolution of the T–F images obtained using $ST_M$ was found to be better than that obtained using conventional ST. This improvement in the resolution of the T–F images was caused due to modification of the traditional GW whose parameters were optimized using the DE algorithm, as mentioned earlier. The parameters of the DE algorithm used to optimize the Gaussian window were selected empirically. The

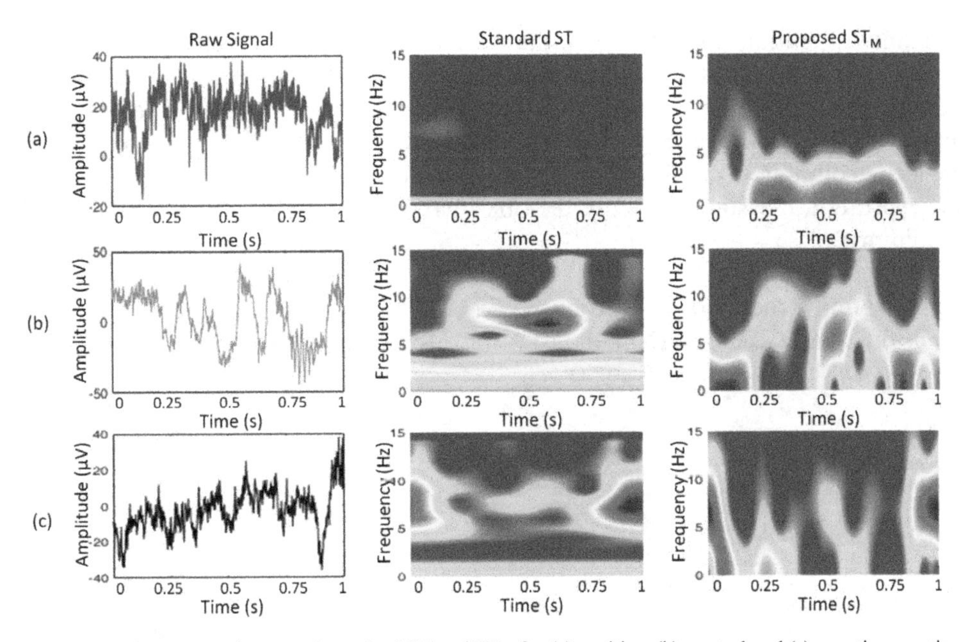


**Figure 2.9.** T–F spectrum images of standard ST and $ST_M$ for (a) positive, (b) neutral and (c) negative emotion EEG signals obtained from the SEED database.

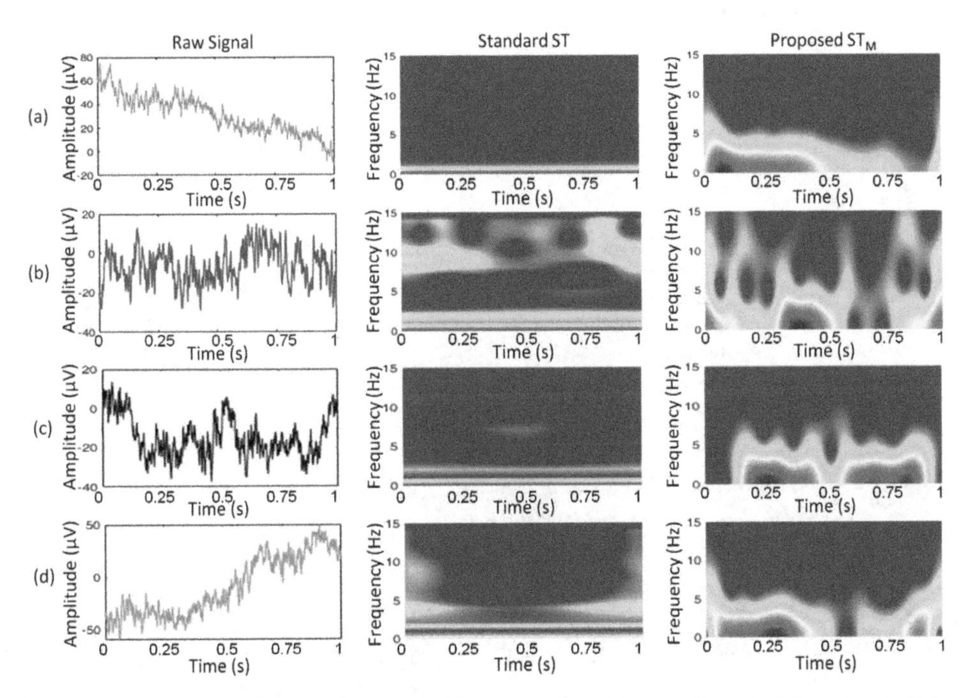


**Figure 2.10.** T–F spectrum images of standard ST and $ST_M$ for (a) happy, (b) neutral, (c) sad and (d) fear emotion EEG signals from the SEED-IV database.

maximum iteration was set to 500, the number of population members was chosen to be 30, and the DE-step size and the crossover probability were set at 0.8 and 0.7, respectively. The scale factor for DE optimization was set to 0.8. The lower and upper bounds of the decision variables were set to 0 and 2. In order to achieve a better resolution of the T–F images, it is necessary to maximize the energy concentration measure, therefore the objective function described in equation (2.7) is minimized.

In figures 2.11 and 2.12, the minimization of the cost function during the optimization process of the proposed $ST_M$ of a randomly selected EEG signal from the SEED and SEED-IV datasets for different emotional drives is shown.

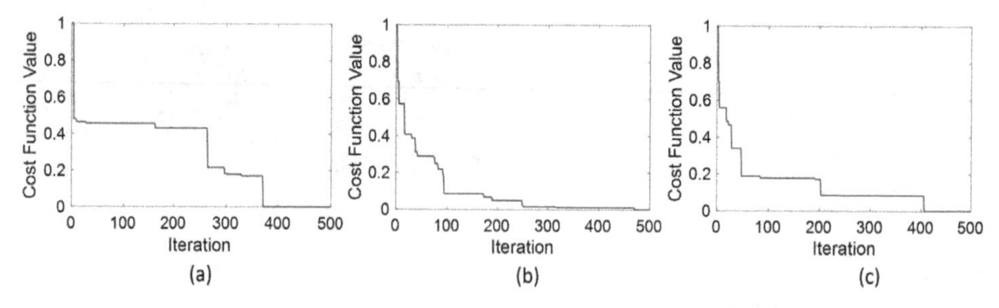


**Figure 2.11.** Minimization of loss in the DE based optimization process for (a) positive, (b) neutral and (c) negative emotions from database I.

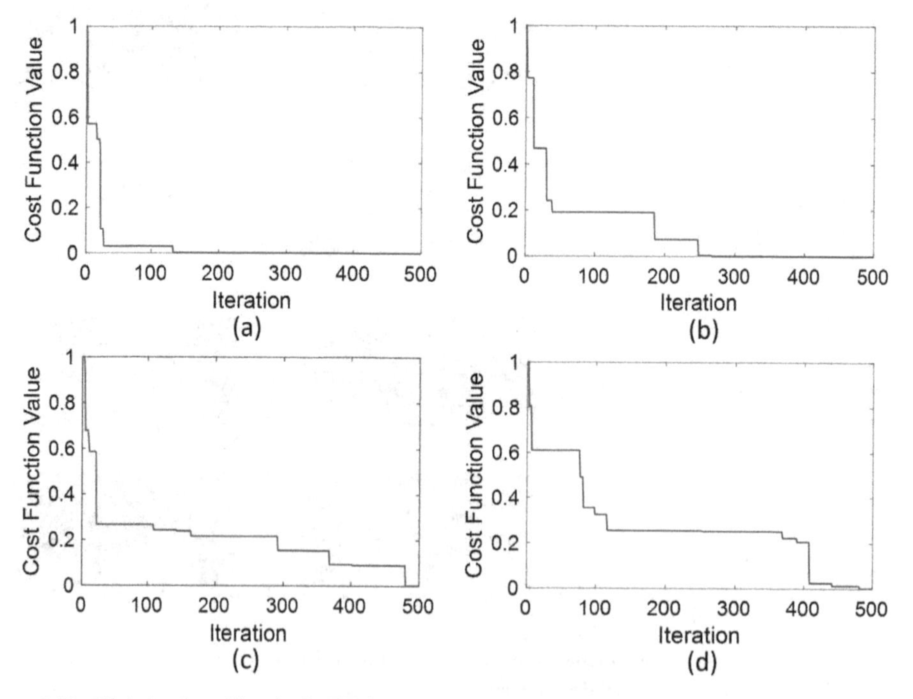


**Figure 2.12.** Minimization of loss in the DE based optimization process for (a) happy, (b) neutral, (c) sad and (d) fear emotions from database II.

From figures 2.11 and 2.12, it can be observed that the loss of the cost function gradually decreases and becomes nearly equal to zero at the final iteration. This in turn produces the maximum resolution of the EEG spectrograms in the time–frequency plane. According to the present hypothesis, the resolution of the spectrum images will play a crucial role in the classification of different emotions as the present framework does not deal with any manual feature extraction stage. Table 2.1 presents the comparison of mean ECMs, obtained for standard ST and the proposed $ST_M$ for different emotion EEG signals obtained from both the SEED and the SEED-IV datasets. From table 2.1 it can clearly be seen that ECM values for each category of emotion significantly improved for the DE optimized $ST_M$ compared to the standard ST.

### 2.4.2 Performance analysis of DenseNet CNN

The T–F images of the EEG signals obtained using $ST_M$ had dimensions of $300 \times 300 \times 3$. These images were further augmented to $224 \times 224 \times 3$ in order to feed them into the DenseNet201 CNN model. During the training of the network through TL, a learning rate of 0.000 03 having a drop factor of 0.1 was used. Both the bias learning rate factor and the weight learning rate factors were set to 10 and an adaptive moment estimation technique was used as the solver function of the CNN model. The gradient decay factor and squared gradient decay factor were set to 0.9 and 0.999, respectively. The learn rate drop period was set to 10. L2 normalization was considered as the gradient threshold method and the L2 regularization parameter was set to 0.000 01. During training of this model the mini-batch size was kept as 20 and the training process has been performed with 1000 iterations with random shuffling of data at every iteration. The entire work was performed using MATLAB R2019a in a computer with a processor of Intel core i5 (sixth generation) with a clock speed of 2.4 GHz, 8 GB memory, 500 GB SSD storage and a Nvidia 930 GPU.

In this contribution, both subject-specific as well as cross-subject emotion recognition tasks are reported. From the SEED database, 180 T–F spectrum images corresponding to 12 EEG channels and 15 epochs were obtained for each category of emotional state from each subject. Considering three emotional states, a total of

**Table 2.1.** Comparison of ECMs obtained using the ST and $ST_M$.

| Database | Emotion | Standard ST | Proposed $ST_M$ | Percentage increment |
|---|---|---|---|---|
| SEED | Positive | 0.0125 | 0.0189 | 51.2 |
| | Neutral | 0.0029 | 0.0043 | 48.27 |
| | Negative | 0.06 | 0.097 | 61.67 |
| SEED-IV | Happy | 0.013 | 0.021 | 61.9 |
| | Neutral | 0.0051 | 0.0085 | 66.67 |
| | Sad | 0.0046 | 0.0083 | 66.67 |
| | Fear | 0.037 | 0.061 | 64.86 |

540 spectrum images from one subject were obtained. Similarly, for the SEED-IV dataset which consists of four emotion classes, 864 spectrum images corresponding to 12 channels and 18 epochs have been obtained (216 images from each category) from a single subject. In subject-specific classification, the spectrum images obtained from all the selected EEG channels corresponding to each class of emotional state were divided into two parts, namely the training data and the validation data. In this type of classification task, the CNN was trained by the training data of one subject and was tested using the validation data, acquired from the same subject. In this contribution, the input data from each subject has been segmented in a ratio of 80% and 20% for training and testing purposes, respectively. In table 2.2 and table 2.3 the classification performance of the subject-specific emotion recognition scheme is reported in terms of four different performance indices, namely accuracy, recall, precision and F-measure. All these aforesaid parameters (expressed in percentages) were computed for the confusion matrix obtained for emotion EEG signal classification for both datasets.

From both tables, it can be observed clearly that the classification performance of the proposed scheme presents satisfactory results with a significantly low standard deviation for all subjects. Further, it can be observed from tables 2.2–2.3 that a better performance is achieved for the SEED database compared to the SEED-IV database. Nevertheless, the performance of the proposed subject-specific emotion recognition framework is reasonably satisfactory for all subjects.

Cross-subject emotion recognition is extremely important to develop a robust HCI system due to the dearth of generalized features across different subjects. Also the EEG signals vary among the subjects considerably with the change in emotional drive. While performing cross-subject emotion recognition the CNN model was

**Table 2.2.** Subject-specific emotion classification accuracy for the SEED dataset.

| Subject | Accuracy (%) | Recall (%) | Precision (%) | F-measure (%) |
|---|---|---|---|---|
| 1 | 95.24 ± 0.1 | 96.15 ± 0.2 | 95.75 ± 0.1 | 93.25 ± 0.2 |
| 2 | 94.41 ± 0.1 | 98.35 ± 0.2 | 93.38 ± 0.1 | 92.48 ± 0.1 |
| 3 | 94.60 ± 0.2 | 95.47 ± 0.2 | 94.46 ± 0.2 | 92.10 ± 0.3 |
| 4 | 97.63 ± 0.1 | 98.60 ± 0.1 | 98.72 ± 0.1 | 97.35 ± 0.2 |
| 5 | 95.65 ± 0.1 | 97.33 ± 0.1 | 94.00 ± 0.2 | 93.50 ± 0.2 |
| 6 | 93.91 ± 0.1 | 93.75 ± 0.3 | 94.36 ± 0.1 | 91.38 ± 0.2 |
| 7 | 95.26 ± 0.1 | 97.19 ± 0.2 | 95.54 ± 0.1 | 93.24 ± 0.1 |
| 8 | 96.80 ± 0.1 | 98.58 ± 0.2 | 96.80 ± 0.1 | 95.72 ± 0.1 |
| 9 | 95.15 ± 0.1 | 94.00 ± 0.2 | 96.52 ± 0.1 | 93.15 ± 0.1 |
| 10 | 95.98 ± 0.1 | 98.48 ± 0.1 | 95.46 ± 0.2 | 94.24 ± 0.1 |
| 11 | 97.43 ± 0.1 | 98.63 ± 0.1 | 97.25 ± 0.1 | 96.00 ± 0.1 |
| 12 | 96.35 ± 0.1 | 94.57 ± 0.2 | 97.68 ± 0.1 | 95.15 ± 0.1 |
| 13 | 94.57 ± 0.1 | 93.45 ± 0.3 | 95.30 ± 0.2 | 92.47 ± 0.2 |
| 14 | 98.16 ± 0.1 | 99.14 ± 0.1 | 98.24 ± 0.1 | 97.32 ± 0.1 |
| 15 | 98.64 ± 0.1 | 100.00 ± 0.0 | 98.56 ± 0.2 | 98.78 ± 0.3 |

**Table 2.3.** Subject-specific emotion classification accuracy for the SEED-IV dataset.

| Subject | Accuracy (%) | Recall (%) | Precision (%) | F-measure (%) |
|---|---|---|---|---|
| 1 | 85.21 ± 0.3 | 85.25 ± 0.2 | 85.58 ± 0.1 | 83.45 ± 0.2 |
| 2 | 84.45 ± 0.1 | 88.46 ± 0.2 | 83.35 ± 0.1 | 82.56 ± 0.1 |
| 3 | 84.65 ± 0.2 | 85.35 ± 0.2 | 84.56 ± 0.2 | 82.00 ± 0.3 |
| 4 | 77.61 ± 0.1 | 88.58 ± 0.1 | 88.42 ± 0.1 | 77.45 ± 0.2 |
| 5 | 85.27 ± 0.1 | 87.12 ± 0.1 | 84.68 ± 0.2 | 83.56 ± 0.2 |
| 6 | 89.92 ± 0.1 | 93.24 ± 0.3 | 94.50 ± 0.1 | 81.15 ± 0.2 |
| 7 | 85.20 ± 0.1 | 87.50 ± 0.2 | 85.14 ± 0.1 | 83.40 ± 0.1 |
| 8 | 86.85 ± 0.1 | 88.46 ± 0.2 | 86.72 ± 0.1 | 85.35 ± 0.1 |
| 9 | 75.14 ± 0.3 | 79.52 ± 0.2 | 76.54 ± 0.1 | 73.50 ± 0.1 |
| 10 | 85.91 ± 0.4 | 88.20 ± 0.1 | 85.25 ± 0.2 | 84.36 ± 0.1 |
| 11 | 87.45 ± 0.4 | 88.32 ± 0.1 | 87.05 ± 0.1 | 86.58 ± 0.1 |
| 12 | 76.32 ± 0.3 | 85.26 ± 0.2 | 87.15 ± 0.1 | 75.12 ± 0.1 |
| 13 | 84.55 ± 0.5 | 89.10 ± 0.3 | 85.48 ± 0.2 | 82.45 ± 0.2 |
| 14 | 78.12 ± 0.9 | 79.05 ± 0.1 | 78.00 ± 0.1 | 77.57 ± 0.1 |
| 15 | 78.64 ± 0.4 | 80.15 ± 0.4 | 78.52 ± 0.2 | 88.68 ± 0.3 |

**Table 2.4.** Cross-subject emotion classification accuracy using the proposed $ST_M$.

| Database | Accuracy (%) | Recall (%) | Precision (%) | F-measure (%) |
|---|---|---|---|---|
| SEED | 93.26 ± 0.25 | 90.24 ± 0.10 | 88.93 ± 0.10 | 87.64 ± 0.10 |
| SEED-IV | 78.60 ± 0.48 | 81.10 ± 0.52 | 79.43 ± 0.28 | 88.04 ± 0.35 |

trained with all the classes of data acquired from all the selected channels of some specific number of subjects and then the trained model was validated by data obtained from other subjects that were not used for training purposes. In this work, all the subjects are divided into two groups, namely the training set and validation set with a ratio of 80%–20% for training and testing purposes, respectively. In this contribution the DenseNet201 CNN model was trained using data recorded from 12 randomly selected subjects out of 15 and the trained model was tested using the remaining three subjects' data for both the SEED and SEED-IV datasets. In table 2.4 the classification performances of the DenseNet201 model has been reported for the proposed channel ensembled cross-subject emotion recognition framework in terms of the previously mentioned four statistical performance parameters.

It is evident from table 2.4 that our proposed methodology returned acceptable classification performance for cross-subject emotion recognition. Here also, the performance of the SEED database was found to be better than the SEED-IV database. Compared to subject-specific classification, the classification performance of cross-subject emotion was found to be inferior. This is due to the fact that, in the subject-specific classification we used the input image data from the same subject to train the DenseNet201 model and for validation we used the same subject. So, the accuracy was higher for the subject-specific classification. In contrast, in the cross-subject classification, the CNN model was trained with the data from the different subjects and validated on those subjects which were not used for training. Thus, the cross-subject recognition is more robust compared to the subject-specific classification.

### 2.4.3 Comparison to the conventional ST

The performance of the proposed subject-specific as well as cross-subject emotion recognition is also compared to the conventional ST. For this purpose, the T–F images of different emotion EEG signals were obtained by applying the conventional ST as described in (2.1), using the standard GW. The T–F images obtained using conventional ST were subsequently used as inputs to the DenseNet201 model for classification of emotion EEG signals. The variation of the mean classification accuracies for subject-specific as well as for cross-subject classification using ST and $ST_M$ is shown in figures 2.13(a)–(b).

It is evident from the variation in classification accuracies shown in figures 2.13(a)–(b) that for both the subject-specific as well as cross-subject classification, the performance of $ST_M$ is better than for the conventional ST. This

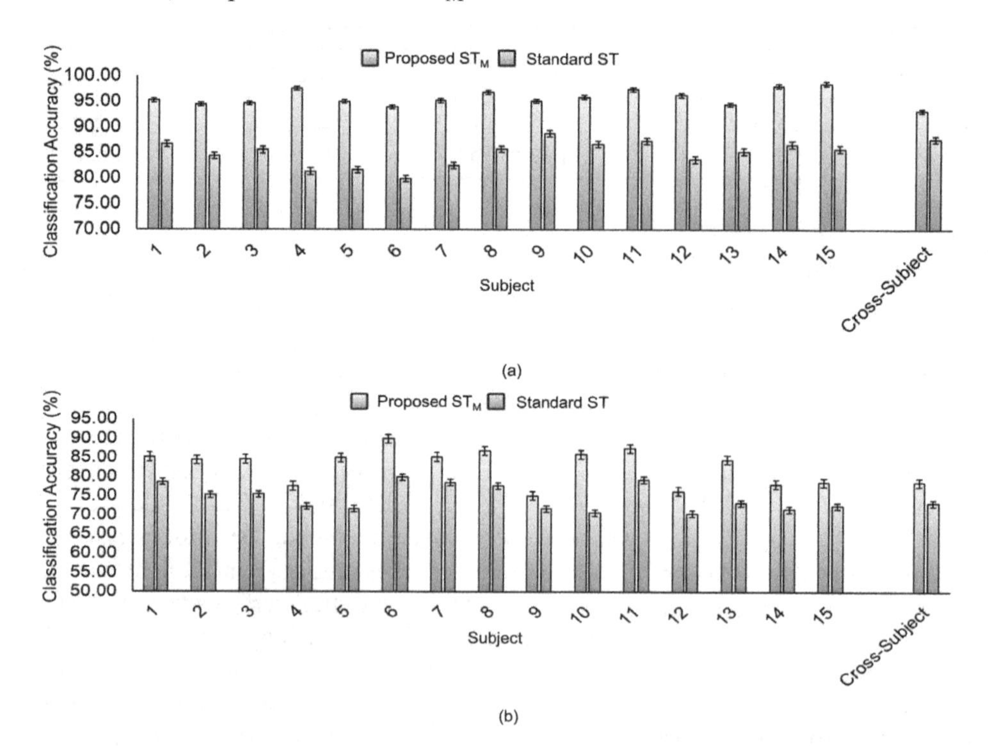


**Figure 2.13.** Comparison of the classification performance between ST and $ST_M$ for (a) the SEED and (b) SEED-IV database.

is due to the fact that the modification of the GW has resulted in better resolution of the T–F images compared to conventional ST, which enables the superior performance of the DenseNet201 model for feature extraction and classification of emotion EEG signals.

### 2.4.4 Performance analysis using other time–frequency images

The classification task for cross-subject emotion recognition was also repeated using different T–F analysis techniques such as the STFT, standard ST and CWT. The results in terms of classification accuracy are reported in this section. For analysis of the EEG signals using the STFT, the Hamming window with 75% overlap was used. In the case of the CWT, different mother wavelets were initially used to obtain the T–F representations of the EEG signals. However, it was observed that the Morlet mother wavelet delivered the best results among different mother wavelets. Thus, the classification accuracy using the Morlet wavelet and DenseNet201 CNN model is reported here. The variation of the mean classification accuracies obtained using the STFT, CWT, ST and proposed $ST_M$ for cross-subject emotion EEG recognition is shown in table 2.5 for both the SEED and the SEED-IV databases. From the classification accuracy reported in table 2.5, it can be observed that among the different T–F analysis methods the modified version of the Stockwell transform, $ST_M$, proposed in this chapter returned a better performance for cross-subject emotion recognition from EEG signals.

### 2.4.5 Performance analysis using different train–test ratios

In this section, further investigation was carried out to observe the variation in classification accuracy obtained for the cross-subject classification task when the train–test ratio of the acquired input spectrum images has been changed. Figure 2.14(a)–(b) indicates the variation in classification accuracy due to variation in the train–test images for the SEED and SEED-IV databases, respectively, for cross-subject classification of different emotions. It can be clearly observed that the classification performance tends to improve with the increase in train–test ratio for both datasets.

### 2.4.6 Performance analysis using other CNN architectures

In this section the classification performance of the DensNet 201 architecture is compared to other popular CNN models namely GoogleNet, ALexNet, SqueezeNet and Residual networks (ResNet). In the case of ResNet, three different networks

**Table 2.5.** Comparison of different T–F methods for ensemble channel cross-subject emotion recognition.

| Database | STFT | CWT | Standard ST | Proposed $ST_M$ |
|---|---|---|---|---|
| SEED | $69.35 \pm 1.6$ | $84.67 \pm 0.61$ | $85.12 \pm 0.3$ | $93.26 \pm 0.25$ |
| SEED-IV | $61.67 \pm 0.43$ | $68.61 \pm 0.77$ | $73.6 \pm 0.32$ | $78.6 \pm 0.48$ |

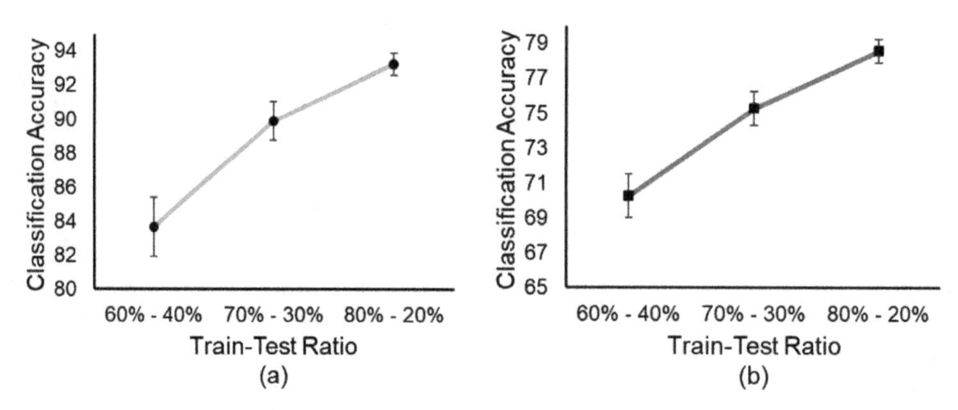


**Figure 2.14.** Variation in classification accuracy for cross-subject emotion recognition with different training to testing ratios for (a) SEED and (b) SEED-IV.

**Table 2.6.** Input image dimensions for different CNN architectures.

| CNN model | Input image size |
|---|---|
| GoogleNet | 224 × 224 × 3 |
| AlexNet | 227 × 227 × 3 |
| SqueezeNet | 227 × 227 × 3 |
| ResNet18 | 224 × 224 × 3 |
| ResNet50 | 224 × 224 × 3 |
| ResNet101 | 224 × 224 × 3 |

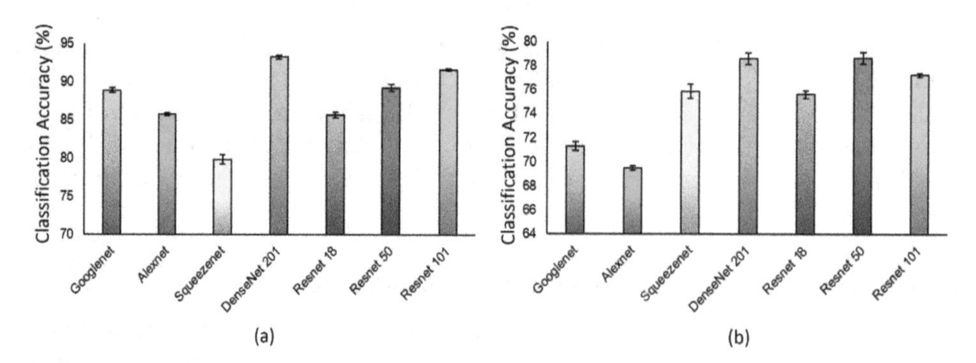


**Figure 2.15.** Cross-subject emotion classification using different CNN architectures for the (a) SEED and (b) SEED-IV databases.

(18, 54 and 101) were used to analyse the performance of the proposed method. For this purpose, the input emotion EEG T–F images obtained using modified ST ($ST_M$), were initially augmented separately for each CNN model. In table 2.6 the input image sizes to the different CNN models are reported.

The classification accuracies achieved using various CNN architectures for cross-subject emotion recognition are displayed in figure 2.15. From figures 2.15 (a)–(b) it can clearly be observed that the DenseNet201 architecture returned the highest classification accuracy compared to other CNN models. It can also be noted that the performance of the other deep CNN models were also found to be satisfactory. Among the other CNN models, the performance of ResNet101 is superior followed by ResNet50 and the GoogleNet model. The performance of SqueezeNet is found to be inferior among the different CNN models. Nevertheless, we can observe that the performance is more or less consistent for different CNN models for the two databases.

### 2.4.7 Comparative study with the existing techniques

In this section we compare the proposed methodology to several existing literature studies which used the same datasets but employed different techniques. As most of the recent literature performed cross-subject emotion recognition, in this comparative study only those studies were considered where cross-subject EEG based emotion recognition has been performed. The comparative study is shown in tables 2.7 and 2.8 for the SEED and SEED-IV datasets, respectively. It is evident from the comparative study presented in the above tables that the proposed method employing the modified ST with DenseNet CNN has delivered better results than the existing studies for both the SEED and SEED-IV databases. From the comparative study, it can be inferred that the proposed methodology can detect different emotional states of the human brain more accurately than the existing methods.

Thus, it can be said that better recognition of human emotion EEG signals can be achieved on the basis of the precise time–frequency representation and deep learning approach compared to traditional techniques. Another important point is that since

**Table 2.7.** Comparative study of the classification performance of cross-subject classification for the SEED database.

| Paper | Method | Accuracy (%) |
|---|---|---|
| [4] | Combined features from time, frequency and non-linear dynamical domain analysis | 83.33 |
| [23] | Dynamical graph convolutional neural networks | 79.95 |
| [18] | Flexible analytic wavelet transform | 93.46 |
| [60] | Multisource transfer learning | 88.92 |
| [61] | Graph regularized sparse linear regression | 88.41 |
| [19] | Domain adaptation techniques | 72.47 |
| [22] | WGAN domain adaptation | 87.07 |
| [20] | Deep adaptation network | 83.81 |
| [24] | Spatial–temporal recurrent neural network | 89.5 |
| [62] | Functional brain connectivity network + local activations | 88 |
| Proposed method | Modified ST ($ST_M$) + CNN (DenseNet201) | 93.26 |

**Table 2.8.** Comparative study of classification performance of cross-subject classification for the SEED-IV database.

| Paper | Method | Accuracy (%) |
|---|---|---|
| [20] | Deep adaptation network | 58.87 |
| [63] | Multimodal deep neural network | 70.58 |
| [64] | Regularized graph neural network | 73.84 |
| [65] | Bi-hemispheric discrepancy model | 69.03 |
| Proposed method | Modified ST ($ST_M$) + CNN (DenseNet201) | 78.6 |

cross-subject emotion classification is performed in this chapter along with subject-specific classification, the generalization of EEG features across different subjects, which is an important issue in emotion recognition, is also addressed here and therefore the proposed system is more robust than the existing techniques.

## 2.5 Conclusion

In this chapter a novel emotion recognition technique from EEG signals employing the Stockwell transform (ST) and a convolution neural network has been reported. We briefly summarize the main outcomes of the present study in the following.

In this study EEG signals representing different human emotions were initially procured from two online datasets. The acquired non-stationary EEG signals were analyzed in the joint time–frequency plane using the ST. In order to achieve better resolution in the time–frequency frame, the conventional GW of the ST has been replaced with a frequency dependent adaptive window function. The parameters of the proposed GW were further optimized using a differential evolution algorithm. Using the optimally parameterized GW parameters, the EEG signals were converted to the time–frequency frame. Investigations showed that the modification of the GW parameters led to a significant improvement in the resolution of the time–frequency images compared to using the conventional GW. The time–frequency images of the EEG signals were subsequently used as inputs to the convolutional neural network for the classification of EEG signals. A predefined CNN architecture, namely 'DenseNet201', was used for the purpose of feature extraction and classification of the emotion EEG signals from the input time–frequency images. In order to reduce the training time of the DenseNet201 CNN model, transfer learning was performed. Both subject-specific as well as cross-subject emotion classification was performed in this study to address the problem of a lack of generalized features across different subjects during the change in emotional behaviour. It was observed that the proposed method yielded high accuracy in discerning different categories of EEG signals for both the SEED and SEED-IV datasets. The performance of the subject-specific classification was found to be superior compared to the cross-subject classification for the recognition of emotion EEG signals. This was observed for both datasets used in this study. Further, the performance of the SEED database was found to be superior compared to the SEED-IV database. The performance of the proposed method when compared to existing methods was found to deliver comparable and even better accuracy for both datasets. Thus it can be inferred that the proposed model can be used to develop an automated human–machine interface system for automated emotion recognition from EEG signals.

However, the present work has certain limitations. In this study the number of emotion states considered are limited in number. The SEED database contains EEG signals corresponding to three emotional states. The SEED-IV database contains EEG data for four emotions. In order to have a better understanding of how the time–frequency distribution of EEG signals changes with different emotional drives, it is necessary to consider a large database containing EEG signals recorded for different emotional drives. In addition, the number of subjects used in this study is also limited. In order to develop an efficient human–machine interface system for automated recognition from EEG signals, the proposed algorithm needs to be verified on a large number of subjects. These aforesaid issues will be taken up in the future as an extension of the present research work. Also, the modified GW of the ST proposed in this chapter can be further applied in the future to study the time–frequency distribution of other types of EEG signals for the classification of motor imagery movements, and neurological diseases such epilepsy, schizophrenia, etc.

## References


[1] Mao X and Li Z 2009 Implementing emotion-based user-aware e-learning *CHI'09 Extended Abstracts on Human Factors in Computing Systems* (Boston MA: ACM) pp 3787–92

[2] Chanel G, Rebetez C, Bétrancourt M and Pun T 2011 Emotion assessment from physiological signals for adaptation of game difficulty *IEEE Trans. Syst. Man Cyber.* **41** 1052–63

[3] Moshfeghi Y 2012 Role of emotion in information retrieval *PhD Thesis* University of Glasgow

[4] Li X, Song D, Zhang P, Zhang Y, Hou Y and Hu B 2018 Exploring EEG features in cross-subject emotion recognition *Front. Neurosci.* **12** 162

[5] Le Bihan D 2003 Looking into the functional architecture of the brain with diffusion MRI *Nat. Rev. Neurosci.* **4** 469–80

[6] Cox D D and Savoy R L 2003 Functional magnetic resonance imaging (fMRI) 'brain reading': detecting and classifying distributed patterns of fMRI activity in human visual cortex *Neuroimage* **19** 261–70

[7] Acharya U R, Sree S V, Swapna G, Martis R J and Suri J S 2013 Automated EEG analysis of epilepsy: a review *Knowl.-Based Syst.* **45** 147–65

[8] Boutros N N, Arfken C, Galderisi S, Warrick J, Pratt G and Iacono W 2008 The status of spectral EEG abnormality as a diagnostic test for schizophrenia *Schizophr. Res.* **99** 225–37

[9] Jeong J 2004 EEG dynamics in patients with Alzheimer's disease *Clin. Neurophysiol.* **115** 1490–505

[10] Oberman L M, Hubbard E M, McCleery J P, Altschuler E L, Ramachandran V S and Pineda J A 2005 EEG evidence for mirror neuron dysfunction in autism spectrum disorders *Cogn. Brain Res.* **24** 190–8

[11] Siuly S and Li Y 2012 Improving the separability of motor imagery EEG signals using a cross correlation-based least square support vector machine for brain–computer interface *IEEE Trans. Neural Syst. Rehabil. Eng.* **20** 526–38

[12] Bose R, Samanta K, Chatterjee S, Bhattacharyya S and Khasnobish A 2019 *Bioelectronics and Medical Devices: From Materials to Devices—Fabrication, Applications and Reliability* ed K Pal, A Khasnobish, I Banerjee, H B Kraatz, S Bag and U Kuruganti (Amsterdam: Elsevier) pp 793–817

[13] Samanta K, Chatterjee S and Bose R 2019 Cross-subject motor imagery tasks EEG signal classification employing multiplex weighted visibility graph and deep feature extraction *IEEE Sens. Lett.* **4** 1–4

[14] Belkacem A N, Hirose H, Yoshimura N, Shin D and Koike Y 2014 Classification of four eye directions from EEG signals for eye-movement-based communication systems *J. Med. Biol. Eng.* **34** 581–8

[15] Acharya U R, Sree S V, Chattopadhyay S and Suri J S 2012 Automated diagnosis of normal and alcoholic EEG signals *Int. J. Neural Syst.* **22** 1250011

[16] Patidar S, Pachori R B, Upadhyay A and Acharya U R 2017 An integrated alcoholic index using tunable-Q wavelet transform based features extracted from EEG signals for diagnosis of alcoholism *Appl. Soft Comput.* **50** 71–8

[17] Zheng W-L and Lu B-L 2015 Investigating critical frequency bands and channels for EEG-based emotion recognition with deep neural networks *IEEE Trans. Auton. Ment. Dev.* **7** 162–75

[18] Gupta V, Chopda M D and Pachori R B 2018 Cross-subject emotion recognition using flexible analytic wavelet transform from EEG signals *IEEE Sensors J.* **19** 2266–74

[19] Lan Z, Sourina O, Wang L, Scherer R and Müller-Putz G R 2018 Domain adaptation techniques for EEG-based emotion recognition: a comparative study on two public datasets *IEEE Trans. Cogn. Develop. Syst.* **11** 85–94

[20] Li H, Jin Y-M, Zheng W L and Lu B L 2018 Cross-subject emotion recognition using deep adaptation networks *Int. Conf. on Neural Information Processing (Sydney)* (Berlin: Springer) pp 403–13

[21] Li Y, Zheng W, Zong Y, Cui Z, Zhang T and Zhou X 2018 A bi-hemisphere domain adversarial neural network model for EEG emotion recognition *IEEE Trans. Affect. Comput.* https://doi.org/10.1109/TAFFC.2018.2885474

[22] Luo Y, Zhang S-Y, Zheng W-L and Lu B-L 2018 WGAN domain adaptation for EEG-based emotion recognition *Int. Conf. on Neural Information Processing (Siem Reap, Cambodia)* (Berlin: Springer) pp 275–86

[23] Song T, Zheng W, Song P and Cui Z 2018 EEG emotion recognition using dynamical graph convolutional neural networks *IEEE Trans. Affect. Comput.* **11** 532–41

[24] Zhang T, Zheng W, Cui Z, Zong Y and Li Y 2018 Spatial–temporal recurrent neural network for emotion recognition *IEEE Trans. Cybernet.* **49** 839–47

[25] Gupta V, Chopda M D and Pachori R B 2019 Cross-subject emotion recognition using flexible analytic wavelet transform from EEG signals *IEEE Sensors J.* **19** 2266–74

[26] Bajaj V, Taran S and Sengur A 2018 Emotion classification using flexible analytic wavelet transform for electroencephalogram signals *Health Inform. Sci. Syst.* **6** 12

[27] Krishna A H, Sri A B, Priyanka K Y V S, Taran S and Bajaj V 2018 Emotion classification using EEG signals based on tunable-Q wavelet transform *IET Sci. Meas. Technol.* **13** 375–80

[28] Khare S K, Bajaj V and Sinha G 2020 Adaptive tunable Q wavelet transform based emotion identification *IEEE Trans. Instrum. Meas.* **69** 9609–17

[29] Taran S and Bajaj V 2019 Emotion recognition from single-channel EEG signals using a two-stage correlation and instantaneous frequency-based filtering method *Comput. Methods Programs Biomed.* **173** 157–65

[30] Lan Z, Sourina O, Wang L, Scherer R and Müller-Putz G R 2019 Domain adaptation techniques for EEG-based emotion recognition: a comparative study on two public datasets *IEEE Trans. Cogn. Develop. Syst.* **11** 85–94

[31] Khare S K and Bajaj V 2020 Time–frequency representation and convolutional neural network-based emotion recognition *IEEE Trans. Neural Netw. Learn. Syst.* https://doi.org/10.1109/TNNLS.2020.3008938

[32] Stockwell R G, Mansinha L and Lowe R 1996 Localization of the complex spectrum: the S transform *IEEE Trans. Signal Process.* **44** 998–1001

[33] Chu P C 1996 The S-transform for obtaining localized spectra *Marine Technol. Soc. J.* **29** 28–38

[34] Mansinha L, Stockwell R, Lowe R, Eramian M and Schincariol R 1997 Local S-spectrum analysis of 1-D and 2-D data *Phys. Earth Planet. Inter.* **103** 329–36

[35] Stockwell R G 1999 S-transform analysis of gravity wave activity from a small scale network of airglow imagers

[36] Stockwell R G, Lowe R P and Mansinha L 1996 Localized cross-spectral analysis with phase-corrected wavelets *Aerospace/Defense Sensing and Controls* (Orlando, FL: International Society for Optics and Photonics) pp 557–64

[37] Stockwell R G, Lowe R P and Mansinha L 1997 Instantaneous wave-vector analysis *AeroSense'97* (Orlando, FL: International Society for Optics and Photonics) pp 349–58

[38] Varanini M, De Paolis G, Emdin M, Macerata A, Pola S, Cipriani M and Marchesi C 1997 Spectral analysis of cardiovascular time series by the S-transform *Computers in Cardiology 1997 (Lund, Sweden)* (Piscataway, NJ: IEEE) pp 383–86

[39] Livanos G, Ranganathan N and Jiang J 2000 Heart sound analysis using the S transform *Computers in Cardiology 2000 (Cambridge, MA)* (Piscataway, NJ: IEEE) pp 587–90

[40] Rakovic P, Sejdic E, Stankovic L and Jiang J 2006 Time-frequency signal processing approaches with applications to heart sound analysis *Computers in Cardiology, 2006 (Valencia, Spain)* (Piscataway, NJ: IEEE) pp 197–200

[41] Dash P, Panigrahi B and Panda G 2003 Power quality analysis using S-transform *IEEE Trans. Power Delivery* **18** 406–11

[42] Dash P, Panigrahi B, Sahoo D and Panda G 2003 Power quality disturbance data compression, detection, and classification using integrated spline wavelet and S-transform *IEEE Trans. Power Delivery* **18** 595–600

[43] Samantaray S, Panigrahi B, Dash P and Panda G 2007 Power transformer protection using S-transform with complex window and pattern recognition approach *IET Gener. Transm. Distrib.* **1** 278–86

[44] McFadden P, Cook J and Forster L 1999 Decomposition of gear vibration signals by the generalised S transform *Mech. Syst. Sig. Process.* **13** 691–707

[45] Eramian M, Schincariol R, Mansinha L and Stockwell R 1999 Generation of aquifer heterogeneity maps using two-dimensional spectral texture segmentation techniques *Math. Geol.* **31** 327–48

[46] Chatterjee S, Choudhury N R and Bose R 2017 Detection of epileptic seizure and seizure-free EEG signals employing generalised S-transform *IET Sci. Meas. Technol.* **11** 847–55

[47] Moukadem A, Bouguila Z, Abdeslam D O and Dieterlen A 2015 A new optimized Stockwell transform applied on synthetic and real non-stationary signals *Digital Signal Process.* **46** 226–38
[48] Reddy M V and Sodhi R 2017 A modified S-transform and random forests-based power quality assessment framework *IEEE Trans. Instrum. Meas.* **67** 78–89
[49] Chatterjee S, Samanta K, Choudhury N R and Bose R 2019 Detection of myopathy and ALS electromyograms employing modified window Stockwell transform *IEEE Sens. Lett.* **3** 1–4
[50] Moukadem A, Bouguila Z, Abdeslam D O and Dieterlen A 2014 Stockwell transform optimization applied on the detection of split in heart sounds *2014 22nd European Signal Processing Conf. (EUSIPCO) (Lisbon, Portugal)* (Piscataway, NJ: IEEE) pp 2015–9
[51] Storn R and Price K 1997 Differential evolution—a simple and efficient heuristic for global optimization over continuous spaces *J. Global Optim.* **11** 341–59
[52] Qin A K, Huang V L and Suganthan P N 2008 Differential evolution algorithm with strategy adaptation for global numerical optimization *IEEE Trans. Evol. Comput.* **13** 398–417
[53] Rogalsky T, Kocabiyik S and Derksen R 2000 Differential evolution in aerodynamic optimization *Can. Aeronaut. Space J.* **46** 183–90
[54] Joshi R and Sanderson A C 1999 Minimal representation multisensor fusion using differential evolution *IEEE Trans. Syst. Man Cyber* **29** 63–76
[55] Storn R 1996 On the usage of differential evolution for function optimization *Proc. of North American Fuzzy Information Processing (Berkeley, CA)* (Piscataway, NJ: IEEE) pp 519–23
[56] Ilonen J, Kamarainen J K and Lampinen J 2003 Differential evolution training algorithm for feed-forward neural networks *Neural Process. Lett.* **17** 93–105
[57] Chaudhary S, Taran S, Bajaj V and Sengur A 2019 Convolutional neural network based approach towards motor imagery tasks EEG signals classification *IEEE Sensors J.* **9** 4494–500
[58] Huang G, Liu Z, Van Der Maaten L and Weinberger K Q 2017 Densely connected convolutional networks *Proc. of the IEEE Conf. on Computer Vision and Pattern Recognition (Honolulu, HI)* pp 4700–08
[59] He K, Zhang X, Ren S and Sun J 2016 Deep residual learning for image recognition *Proc. of the IEEE Conf. on Computer Vision and Pattern Recognition (Las Vegas, NV)* (Piscataway, NJ: IEEE) pp 770–8
[60] Li J, Qiu S, Shen Y Y, Liu C-L and He H 2019 Multisource transfer learning for cross-subject EEG emotion recognition *IEEE Trans. Cybernet.* **50** 3281–93
[61] Li Y, Zheng W, Cui Z, Zong Y and Ge S 2019 EEG emotion recognition based on graph regularized sparse linear regression *Neural Process. Lett.* **49** 555–71
[62] Li P, Liu H, Si Y, Li C, Li F, Zhu X, Huang X, Zeng Y, Yao D and Zhang Y 2019 EEG based emotion recognition by combining functional connectivity network and local activations *IEEE Trans. Biomed. Eng.* **66** 2869–81
[63] Zheng W L, Liu W, Lu Y, Lu B-L and Cichocki A 2018 Emotionmeter: a multimodal framework for recognizing human emotions *IEEE Trans. Cybernet.* **49** 1110–22
[64] Zhong P, Wang D and Miao C 2020 EEG-based emotion recognition using regularized graph neural networks *IEEE Trans. Affective Comput.* https://doi.org/10.1109/TAFFC.2020.2994159
[65] Li Y, Zheng W, Wang L, Zong Y, Qi L, Cui Z, Zhang T and Song T 2020 A novel bi-hemispheric discrepancy model for EEG emotion recognition *IEEE Trans. Cogn. Develop. Syst.* https://doi.org/10.1109/TCDS.2020.2999337

**IOP** Publishing

Modelling and Analysis of Active Biopotential Signals in Healthcare, Volume 2

**Varun Bajaj and G R Sinha**

# Chapter 3

# Multivariate phase synchrony based on fuzzy statistics: application to PTSD EEG signals

**Zahra Ghanbari and Mohammad Hassan Moradi**

*Biomedical Department, Amirkabir University of Technology, Tehran, Iran*

Phase synchronization can be detected in many sophisticated systems, including the human brain. Generalized phase synchrony (GePS) has been proposed as a phase synchrony index for multivariate signals. It uses the linear relationships among instantaneous frequency laws of signals. GePS benefits from multi-channel empirical mode decomposition (MEMD) in addition to applying the Johansen test to intrinsic mode functions extracted from signals. In the current chapter a new generalized phase synchrony, called fuzzy generalized phase synchrony (FGePS), is proposed. FGePS benefits from a fuzzy statistics approach, in addition to the GePS properties. A fuzzy hypothesis testing method, in contrast to the classic method, considers the test statistic and critical values as fuzzy numbers. Such fuzzification enriches the original approach with a more powerful decision making tool which provides strength for the decision made. The results of applying FGePS to the synthetic data demonstrate that it outperforms GePS. Our proposed measure is applied to resting state eyes closed and eyes open EEG datasets. Signals are recorded from three groups of participants including combat related post-traumatic stress disorder (PTSD) subjects, trauma exposed non-PTSD subjects and healthy controls. ANOVA at a confidence level of 99% is used to assess the FGePS performance in detecting significant differences among the three groups. The *p*-values indicate that our proposed measure is more powerful compared to the original one. To our knowledge this is the first study which investigates generalized phase synchrony in PTSD resting state EEG data.

## 3.1 Introduction

The concept of synchronization is defined as the coordination of coupled systems. Coupled systems possess some inter-dependencies. Therefore, they share similarities in their dynamics [1]. Synchronization in chaotic systems has found various applications [2]. Synchronization can be quantified using either linear or nonlinear indices [3]. Recently, however, much attention has been paid to nonlinear approaches, and particularly phase synchrony [4]. Phase synchrony was introduced in relation to the behavior of chaotic oscillators in 1996 [5]. Phase synchronization can be found in various complex systems as a result of the interactions of several sub-systems. The brain is known as a complex system composed of many dynamically interacting subsystems. Phase synchrony has been applied to EEG signals to detect disorders, or their influences, and in cognition and sleep studies [6–13].

Although bivariate phase synchrony is widely applied to brain signals, it cannot necessarily provide a thorough representation of the global interactions occurring within a multivariate signal [14]. Therefore, in addition to bivariate methods, multivariate approaches to the assessment of phase synchrony have been proposed. Some of the best known bivariate and multivariate methods include: phase locking value (PLV) [15, 16], mutual prediction [17], evaluation map [18], general field synchronization [19], instantaneous period [17], frequency flow analysis [20], multivariate phase synchrony (MPS) [21], robust generalized synchrony [22], hyperspherical phase synchrony [23] and empirical mode decomposition (EMD) based approaches [24]. In 2013 Omidvarnia *et al* proposed an approach called generalized phase synchrony (GePS) [25]. GePS was proposed since all previous approaches had been restricted by the assumption of the rationality of the phase locking ratios between signals. GePS tries to circumvent the above restriction by benefiting from the concept of co-integration. In other words, the phase locking ratio can possess an irrational value. It is obvious that the classical definition is embedded in this form. GePS is a method of computing phase synchrony within non-stationary multivariate signals based on a linear relationship between instantaneous frequency laws. In this chapter a modified version of GePS is proposed which is called fuzzy generalized phase synchrony (FGePS). FGePS enriched GePS benefits from fuzzy statistics. Fuzzy logic was introduced by Zadeh at 1965 [26]. Fuzzy approaches have been used in statistics recently [27, 28].

The proposed approach is applied to synthetic data as well as real EEG data. The EEG signals are recorded from post-traumatic stress disorder (PTSD) participants. According to the American Psychiatric Association (APA), post-traumatic stress disorder is defined as a psychiatric disorder which may follow a traumatic experience [29]. PTSD is known as a chronic debilitating anxiety condition characterized by unremitting distressing repetition of the traumatic event, hyper-vigilance, hyper-arousal, dissociation, avoidance, emotional numbing and negative cognitive changes [30].

EEG is an easily available low-cost imaging modality which provides high temporal resolution and optimal observation of brain activities [31–33]. The EEG signals of PTSD patients have been investigated in a few studies. Resting state EEG signals can be recorded with the eyes either closed or open. The former state is called resting state eyes closed (REC) and the latter is called resting state eyes open (REO). REC and ROC possess basically different properties. The alpha band, for example, is defined as being synchronized in REC but desynchronized in REO [34]. Moreover, a comparison of REC and REO in the delta, theta, alpha and beta frequency bands reveals differences in both children and adults [35]. There was previously a hypothesis which inferred the differences of REC and REO as being an impression of visual stimulus on the brain's activity. A study examined this hypothesis by recording REC and REO EEG signals in a completely dark environment and the spectral power and coherence displayed significant differences in the delta, theta, alpha1, alpha2, beta1, beta2 and gamma frequency bands. Therefore, it can be concluded that the differences between REC and REO are independent of any external visual stimulus. Thus, it has been proposed that the differences reflect internally directed attention which is specific to REC and externally directed attention which is specific to REO [36].

Due to the inherent differences existing between REO and REC, this chapter will consider both REC and REO EEG signals. The signals are recorded from combat related PTSD participants. In addition to this group, two control groups are studied, including combat trauma exposed non-PTSD subjects and healthy controls who have not experienced any serious trauma including war.

Our proposed generalized synchrony measure is applied to the synthetic and real EEG signals. The results are reported using a statistical analysis at a 99% significance level for both the proposed method and the original method.

The rest of this chapter is organized as follows. Section 3.2 is dedicated to describing the method. The data are introduced in section 3.3. The results are reported in section 3.4. The last section is dedicated to our conclusions.

## 3.2 Method

Since our proposed method, FGePS, is composed of two underlying parts, these two parts will be described in the following. Generalized phase synchrony is first explained and then the fuzzy statistics are discussed. The last part of this section is dedicated to our proposed approach, which benefits from both concepts.

### 3.2.1 Generalized phase synchrony (GePS)

EEG signals are known as non-stationary signals with varying frequencies. Time–frequency distributions have proven to be a proper choice in the analysis of such signals, since they provide two-dimensional representations. The time–frequency planes provided by time–frequency distributions display the time varying spectra of a non-stationary signal, as well as the number of signal components. They also reveal the pattern of distributed energy. Quadratic time–frequency distributions are the most applicable, which are actually the smoothed version of the Winger–Ville distribution (WVD) [37]. A quadratic time–frequency distribution in its discrete form is as follows:

$$\begin{aligned} \rho_z[n, k] &= \underset{m \to k}{2\mathrm{DFT}}\{G[n, m] *_n (z_s[n + m] z_s^*[n - m])\} \\ k &= 1, \ldots, M \\ n &= 0, \ldots, N - 1, \end{aligned} \tag{3.1}$$

where * stands for the complex conjugate, $\underset{n}{*}$ represents discrete convolution, $G[n, m]$ is the time-lag kernel and $z_s[n]$ denotes the analytic associate of $s[n]$. $\rho_z[n, k]$ is a matrix with $M \times N$ dimensions, where $M$ represents the number of frequency bins. For a given real signal, $s[n]$, the corresponding phase can be obtained as [25]

$$z_s[n] = s[n] + jH(s[n]) = a_z[n]e^{j\varphi_z[n]}, \tag{3.2}$$

where $z_s[n]$ stands for the analytic associate of $s[n]$, $n = 0, 1, \ldots, N - 1$, and $H(\cdot)$ represents the Hilbert transform. $a_z[n]$ is the instantaneous amplitude and $\varphi_z[n]$ is the instantaneous phase of the signal.

The instantaneous frequency of a non-stationary signal indicates its frequency changes in time. Instantaneous frequency, $fz[n]$, is defined based on the derivative of the instantaneous phase:

$$f_z[n] = f_s \frac{\varphi_z[n] - \varphi_z[n-1]}{2\pi}, \tag{3.3}$$

where $f_s$ is the sampling frequency. Several approaches have been proposed for estimating the instantaneous frequency of non-stationary signals. Some of these methods include a new form of the Fourier transform in addition to the local polynomial periodogram, the intersection of confidence intervals (ICI) based approach [38]. Moreover, time–frequency distribution based methods are used for such estimations, including the adaptive short time Fourier transform [39], the T-class of time–frequency distributions, quadratic time–frequency distributions [40], polynomial Winger–Ville distributions and complex time distributions. Reduced interface time–frequency distributions have proved more reliable in EEG signal processing [41]. Therefore, time–frequency distribution instantaneous frequency estimators are used in [25] to determine the best approach. The spectrogram (SPEC), Choi–Williams distribution (CWD) and modified-B distribution (MBD) were examined by Omidvarnia *et al.*

Time–frequency distribution based instantaneous frequency estimation proceeds as follows. The first-order moment corresponding to a digitized quadratic time–frequency distribution, $\rho_z[n, k]$, with respect to frequency, which is actually the weighted frequency of the signal, is defined as follows:

$$f_z[n] = \frac{f_s \sum_{k=0}^{M-1} k\rho_z[n, k]}{2M \sum_{k=0}^{M-1} \rho_z[n, k]}. \tag{3.4}$$

Among the assessed estimators, the CWD with the discrete form of the time-lag kernel as follows is selected:

$$\frac{\sqrt{\pi\sigma}}{2\,|\,m\,|} e^{-\frac{\pi^2\sigma n^2}{4m2}}, \tag{3.5}$$

where $\sigma = 10$ and $L_{\text{lag}} = N/4$.

This assumption does not hold since most real signals are multicomponent rather than monocomponent. Therefore, $\varphi_z[n]$ is no longer the signal phase but represents an ambiguous or meaningless weighted square average of phases associated with the different components of the signal instead. To address the above issue several approaches are used, including filtering signals into frequency bands, applying a wavelet transform to signals and using empirical mode decomposition (EMD). GePS benefits from some features of EMD. EMD is an adaptive data driven approach. It needs no *a priori* knowledge about the signal. Given a non-stationary signal, EMD decomposes it into its intrinsic mode functions (IMFs). The obtained IMF is a monocomponent signal. Furthermore, each IMF contains a limited frequency content of the original signal, because EMD behaves similarly to a dyadic filter bank. Consequently, the obtained IMFs imply simple oscillatory modes which have time varying amplitude and frequency. A real multicomponent signal, $s[n]$, in terms of its IMFs is

$$s[n] = \sum_{k=1}^{M} \mathrm{IMF}_k[n] + r[n], \tag{3.6}$$

where $M$ is the number of components and $r[n]$ stands for the residual. By applying the Hilbert transform to the above equation we have

$$z_s[n] = \sum_{k=1}^{M} a_z^{(k)}[n] \mathrm{e}^{j\varphi_z^{(k)}[n]}, \tag{3.7}$$

where $a_z^{(k)}$ represents the instantaneous amplitudes and $\varphi_z^{(k)}$ stands for the instantaneous phase associated with the $k$th IMF.

The instantaneous frequency can be easily extracted by taking the derivate of the instantaneous phase. Therefore, the Hilbert transform is used to convert the signal into its analytic associate. To obtain smoother instantaneous phase traces, phase angles are corrected by applying an unwrapping approach. In the GePS method the MATLAB script is applied, which adds $\pm 2k\pi$ to the absolute jumps between consecutive points which are $\geqslant \pi$, where $k = 1, 2, \ldots$ [25].

*3.2.1.1 Quantifying phase synchrony*

Given one-dimensional stochastic real signals, $x_1[n]$ and $x_2[n]$, their analytic associates will be

$$z_{x1}[n] = x_1[n] + j\tilde{x}_1[n] = a_{x1}[n] \mathrm{e}^{j\varphi_{x_1}[n]} \tag{3.8}$$

$$z_{x2}[n] = x_2[n] + j\tilde{x}_2[n] = a_{x2}[n] \mathrm{e}^{j\varphi_{x_2}[n]}, \tag{3.9}$$

where $\tilde{x}_1[n]$ stands for the Hilbert transform of $x_1[n]$ and similarly $\tilde{x}_2[n]$ stands for the Hilbert transform of $x_2[n]$. $x_1[n]$ and $x_2[n]$ are considered phase locked of order $P_{x_1} : P_{x_2}$ if we have

$$\Delta\varphi_{x_1,x_2}[n] = P_{x_1}\varphi_{x_1}[n] - P_{x_2}\varphi_{x_2}[n] \approx \text{constant}. \tag{3.10}$$

This condition, however, is rarely satisfied. Therefore, the following condition is usually used, which is a more relaxed version:

$$\left| P_{x1}\varphi_{x1}[n] - P_{x2}\varphi_{x2}[n] \right| < \text{constant}. \tag{3.11}$$

This results in

$$\begin{aligned}\frac{1}{2\pi}\mathrm{diff}(\Delta\varphi_{x_1,x_2}[n]) &= \frac{P_{x_1}}{2\pi}\mathrm{diff}(\Delta\varphi_{x_1}[n]) - \frac{P_{x_2}}{2\pi}\mathrm{diff}(\Delta\varphi_{x_2}[n]) \\ &= P_{x_1} f_{x_1}[n] - P_{x_2} f_{x_2}[n] \approx 0.\end{aligned} \tag{3.12}$$

Therefore, we have

$$f_{x_1}[n] \approx \frac{P_{x_2}}{P_{x_1}} f_{x_2}[n], \tag{3.13}$$

which indicates that the shape of the instantaneous frequency is similar in two signals. In the case of $P_{x_1} = P_{x_2}$, the phase synchrony and instantaneous frequency concepts are connected.

For discrete signals with $P_{x1} = P_{x2} = 1$ PLV is defined as

$$\mathrm{PLV} = \left| \frac{1}{N} \sum_{k=1}^{N-1} \mathrm{e}^{j(\varphi_{x1}[n] - \varphi_{x2}[n])} \right|, \tag{3.14}$$

where $N$ represents the signal length. The resulting value is in [0,1], where 0 reflects the asynchronous case and 1 displays perfect synchronization. Given two signals with similar instantaneous frequency laws on a particular time interval, these signals are said to be phase locked of order 1:1 [42]. This relaxed condition is called phase entertainment and $P_{x1}/P_{x2}$ should be a rational value [43]. However, when the linear relationships between the phase signals can be irrational such a definition can no longer explain the GePS.

We can extend the above definition of bivariate phase synchrony to the multivariate case, benefiting from the co-integration concept [44]. For a multivariate non-stationary signal, they have general phase synchrony if on a reasonably long time period there is a linear relationship between the instantaneous frequencies of a subgroup of channels.

*3.2.1.2 . The co-integration concept*

Integration of order $d$ which is represented by $I(d)$ for a one-dimensional stochastic process exists when the reverse characteristic polynomial of its fitted MVAR model has roots on the unit circle in the $z$-plane [45]. Although $I(d)$ is an unstable process, it can be converted to a stable process $I(0)$ by differentiating $d$ times. Two or more signals are called co-integrated if a linear combination of these signals exists which yields a stationary process. The co-integration rank, $r$, indicates the number of co-integration relationships among the signals [45]. For calculating the co-integration rank and co-integrating coefficients a multivariate Johansen test is used [46].

*3.2.1.3 The multivariate Johansen test*

As mentioned above, the Johansen test is used to test the co-integrating relationships of some zeroth- or first-order integrated processes [47]. Here, we have an $N$-dimensional instantaneous frequency, $f_z[n] \in \mathbb{R}^{N\times 1}$, which is an MVAR model of order $p$ [48]:

$$f_z[n] = \sum_{r=1}^{p} A_r f_z[n-r] + \mu + \varepsilon[n], \tag{3.15}$$

where $A_r \in \mathbb{R}^{N\times N}$ represents the MVAR coefficient matrix corresponding to delay $r$, $\mu \in \mathbb{R}^{N\times 1}$ stands for a constant vector and $\varepsilon[n] \in \mathbb{R}^{N\times 1}$ represents the model white noise. According to the assumption that $f_z[n]$ variables are zero order integrated processes (i.e. a stationary process) or order 1, the above model can be re-written as below, which is called the vector error correlation model (VECM):

$$\Delta f_z[n] = -\Pi f_z[n-1] + \sum_{r=1}^{p-1} \Gamma_r \Delta f_z[n-r] + \varepsilon[n], \tag{3.16}$$

where $\Pi$ and $\Gamma_r$ are as follows:

$$\Pi = 1 - (A_1 + A_2 + \ldots + A_p) \tag{3.17}$$

and

$$\Gamma_r = -\sum_{j=r+1}^{p} A_j. \tag{3.18}$$

In fact, it is represented in terms of its successive delayed values differences, i.e.

$$\Delta f_z[n-i] = f_z[n-i-1]. \tag{3.19}$$

Given $r$, ($r < N$) as the rank of the coefficient matrix, $\Pi$, full rank matrices $\alpha$ and $\beta$ ($\alpha, \beta \in \mathbb{R}^{r\times N}$) exist so that $\Pi = \alpha\beta^T$ and $\beta^T f_z[n]$ is stationary. Therefore, the rank $r$ determines the number of coefficient relationships between $f_z[n]$ dimensions. The process co-integrating vectors are extracted from columns of the matrix $\beta$. For a given $r$, the Johansen test performs two likelihood ratio tests, based on the maximum likelihood estimation of $\beta$:

1. *The trace test*, which tests a null hypothesis, $H_0$, against an alternative hypothesis, $H_1$. $H_0$ and $H_1$ are defined as follows:
   - $H_0$: $r$ co-integrating relationships exist among variables.
   - $H_1$: $k$ co-integrating relationships exist among variables.
2. *The maximum eigenvalue test*, for which there are similarly two hypotheses:
   - $H_0$: $r$ co-integrating relationships exist among variables.
   - $H_1$: $r + 1$ co-integrating relationships exist among variables.

There is a special case of being full rank for $\Pi$ ($r = N$), in which all of the process variables are stationary. Moreover, no co-integrating relationship exists between them [48].

#### *3.2.1.4 Computing phase synchrony*

The given signals $s_1[n]$ and $s_2[n]$ are generally phase synchronized if their corresponding phases satisfy the following condition:

$$\exists\, c_1, c_2 : c_1\varphi_1[n] + c_2\varphi_2[n] = e[n], \tag{3.20}$$

where $c_1$ and $c_2$ are real numbers, $\varphi_1[n]$ and $\varphi_2[n]$ are phase signals, and $e[n]$ represents a normally distributed stationary process with a finite second-order moment. Equation (3.20) indicates a co-integrating relationship between $\varphi_1[n]$ and $\varphi_2[n]$. It can be generalized to a multivariate co-integrating relationship among $K$ phase signals:

$$\exists\, c_1, \dots, c_K : c_1\varphi_1[n] + c_2\varphi_2[n] + \dots + c_K\varphi_K[n] = e[n], \tag{3.21}$$

where $\varphi_s[n] = [\varphi_1[n], \dots, \varphi_K[n]]$ is the multivariate phase signal corresponding to a multivariate real signal with $K$ variables, $s[n] = [s_1[n], \dots, s_K[n]]^T$. GePS uses the Hilbert transform in order to calculate the phase of signal components, $s_i[n]$, separately. If the multivariate phase signal, $\varphi_s[n]$, is integrated of order $r$, there are $r$ stationary linear relationships within $\varphi_s[n]$. In this case, the signal $s[n]$ is required to have generalized phase synchrony of rank $r$ [45, 46]. A higher rank indicates involving a larger number of phase signals in relationships and higher synchronization among channels as a result. The co-integration rank, $r$, and co-integrating coefficients, $c_1, \dots, c_K$, are estimated using the multivariate Johansen test [46]. It is obvious that $0 \leqslant r \leqslant K$.

*3.2.1.5 The implementation of GePS*

Given a non-stationary $K$-dimensional signal, $s[n]$, GePS is calculated based on the following steps:

i. Each channel $s_i[n], i = 1, \dots, K$ is decomposed into $Q$ IMFs $g_{si}^{(q)}[n], q = 1, \dots, Q$ using EMD. A unique value of $Q$ is applied to all channels. This parameter is estimated using the EMD stoppage criteria as

$$Q = \min_i Q_i,\ i = 1, \dots, K, \tag{3.22}$$

where $Q_i$ represents the number of IMFs within the $i$th channel [49].

ii. Corresponding to each IMF the analytic associate for each channel is calculated using the Hilbert transform:

$$z_i^{(q)}[n] = g_{s_i}^{(q)}[n] + jH(g_{s_i}^{(q)}[n]), \tag{3.23}$$

where $i$ indicates the $i$th channel. The instantaneous phase is extracted as

$$\varphi_{z_i}^{(q)}[n] = \text{angle}\left\{\frac{H\left(g_{s_i}^{(q)}[n]\right)}{g_{s_i}^{(q)}[n]}\right\}. \tag{3.24}$$

For the purpose of omitting phase angle jumps between consecutive elements and making phase traces smoother, an unwrapping approach is used to correct phases $\varphi_{z_i}^{(q)}[n]$ [50]. The derivative of unwrapped $\varphi_{z_i}^{(q)}[n]$ provides the instantaneous frequency, $f_{z_i}^{(q)}[n]$.

iii. Corresponding to all channels, at each decomposition level instantaneous frequencies (i.e. $f_z^{(q)} = [f_{z_i}^{(q)}, \dots, f_{z_K}^{(q)}]$) are divided into non-overlapping time segments. The minimum window length is determined according to the requirement of the MVAR parameter estimation. Considering $p$ as the MVAR model order in the Johansen test, then the proper length should be significantly larger than $K^2p$ [51].

iv. The maximum eigenvalue test (Johansen test) is applied to each multivariate segment at a 99% confidence level. The linear relationships among instantaneous frequencies can be extracted as follows:

$$CF = E, \tag{3.25}$$

where $C$ is the coefficient matrix, $F$ is the vector of instantaneous frequencies and $E$ stands for the error vector, which are as follows:

$$C = \begin{bmatrix} c_{11} & \cdots & c_{1K} \\ \vdots & \ddots & \vdots \\ c_{r^{(q)}1} & \cdots & c_{r^{(q)}K} \end{bmatrix}, \tag{3.26}$$

where $c_{ik}$, $k = 1, \ldots, r^{(q)}$ is the $k$th coefficient and $r^{(q)}$, $0 \leqslant r^{(q)} \leqslant K$ represents the number of existing co-integrating relationships within the multivariate segment

$$F = \begin{bmatrix} f_{z1}^{(q)}[n] \\ \vdots \\ f_{zK}^{(q)}[n] \end{bmatrix}, \tag{3.27}$$

where $f_{z_i}^{(q)}$, $i = 1, \ldots, K$ is the $i$th segmented instantaneous frequency corresponding to the $q$th IMF, $q = 1, \ldots, Q$,

$$E = \begin{bmatrix} e_1^{(q)}[n] \\ \vdots \\ e_{r^{(q)}}^{(q)}[n] \end{bmatrix}, \tag{3.28}$$

where $e_k^{(q)}[n]$ represents the stationary residual of the $k$th co-integrating relationship associated with the $q$th IMF.

v. Corresponding to each segment the phase synchronization measure is computed as follows:

$$\eta^{\text{seg}} = \frac{1}{Q \cdot K} \sum\nolimits_{q=1}^{Q} r^{(q)}, \tag{3.29}$$

which is the normalized number of co-integrating relationships over the IMF components. We have

$$0 \leqslant \eta^{\text{seg}} \leqslant 1, \tag{3.30}$$

where 1 represents perfect phase co-integration within the multivariate segment, and 0 indicates the absence of co-integrating relationships within phase signals.

### 3.2.2 Fuzzy hypothesis testing

In 2005 Buckley proposed an approach called fuzzy statistics hypothesis testing. In Buckley's method, despite the classical statistics, the statistics and the confidence interval are both considered fuzzy numbers [52]. Assume $X$ as a random variable with probability density function $f(x;\theta)$ for a single parameter $\theta$. Assume the problem of estimating $\theta$ as an unknown parameter from a random sample $X_1, X_2, \ldots, X_n$. Let $Y = u(X_1, X_2, \ldots, X_n)$ be a statistic for estimating $\theta$. Give the values of theses random variables as

$$X_i = x_i,\ 1 \leqslant i \leqslant n. \tag{3.31}$$

A point estimate for $\theta$ is

$$\theta^* = y = u(x_1, x_2, \ldots, x_n). \tag{3.32}$$

The point estimation is never expected to be exactly equal to $\theta$. Consequently, a $(1 - \alpha)$ confidence interval is often calculated for $\theta$. $\alpha$ is commonly set to 0.01, 0.05 or 0.1. We would like to find the $(1 - \alpha)100\%$ confidence interval for all $\alpha$ values where $\epsilon \leqslant \alpha < 1$. $\epsilon$ is an arbitrary value. Assume $\epsilon = 0.01$, for $0.01 \leqslant \alpha < 1$ these confidence intervals are denoted as

$$[\theta_1(\alpha), \theta_2(\alpha)]. \tag{3.33}$$

$[\theta^*, \theta^*]$, which represents the 0% confidence interval, will be added to the above intervals. By putting these confidence intervals on each other, a triangular shaped fuzzy number, $\bar{\theta}$, will be generated, where the $\beta$-cuts of this fuzzy number are the confidence intervals. Thus, for $0.01 \leqslant \beta \leqslant 1$, we have

$$\bar{\theta}[\beta] = [\theta_1(\beta), \theta_2(\beta)]. \tag{3.34}$$

To have a fuzzy number, the bottom of $\theta$ should be completed. The graph of $\bar{\theta}$ will be dropped straight down to complete its $\beta$-cuts. $\bar{\theta}$ provides more information compared to a single interval estimate.

Let us consider the problem of the mean of a normal with known variance. In the classical situation, a sample of size $n$ is obtained from a $N(\mu, \sigma^2)$ distribution, with known variance $\sigma^2$, to perform the following hypothesis testing:

$$H_0 : \mu = \mu_0 \text{ versus } H_1 : \mu > \mu_0. \tag{3.35}$$

The alternative hypothesis, $H_1$, is considered one-sided in the following. However, the method can be easily generalized to the two-sided case. First of all, the mean of the random sample, $\bar{x}$, is calculated. $\bar{x}$ is a real number, rather than a fuzzy set. The statistic is determined as

$$z_0 = \frac{\bar{x} - \mu_0}{\sigma / \sqrt{n}}. \tag{3.36}$$

The usual values for the significance level of the test, $\alpha$, are 0.01, 0.05 and 0.1. Under the $H_0$ hypothesis $z_0$ is $N(0,1)$. Therefore, the decision rule is as follows:

1. If $z_0 \geqslant z_\alpha$ reject $H_0$.
2. Otherwise, do not reject $H_0$.

In fuzzy hypothesis testing, the estimation of $\mu$ is a triangular shaped fuzzy number, $\bar{X}$, with the following $\beta$-cuts:

$$\bar{X}[\beta] = \left[\bar{x} - z_{\frac{\beta}{2}}\sigma/\sqrt{n},\ \bar{x} + z_{\frac{\beta}{2}}\sigma/\sqrt{n}\right], \tag{3.37}$$

where $0.01 \leqslant \beta \leqslant 1$, and $z_{\frac{\beta}{2}}$ represents the $z$ value so that the probability of exceeding $z$ is $\beta/2$ for a random variable with the $N(0,1)$ probability density. Choosing $a \in \bar{X}[\beta]$, where $a = \bar{x} + \delta$, for some $\delta \in [-z_{\frac{\beta}{2}}\sigma/\sqrt{n},\ z_{\frac{\beta}{2}}\sigma/\sqrt{n}]$ we have the following formulation:

$$z(\delta) = \frac{a - \mu_0}{\sigma/\sqrt{n}} = \frac{\delta}{\sigma/\sqrt{n}} + z_0. \tag{3.38}$$

Therefore, under the $H_0$ hypothesis, $z(\delta)$ is $N\left(\frac{\delta}{\sigma/\sqrt{n}},\ 1\right)$. The $\beta$-cuts of $\bar{Z} = \frac{\bar{X} - \mu_0}{\sigma/\sqrt{n}}$ are

$$\bar{Z}[\beta] = \left[z_0 - z_{\frac{\beta}{2}},\ z_0 + z_{\frac{\beta}{2}}\right]. \tag{3.39}$$

The critical value corresponding to each $z(\delta) = z_0 + \frac{\delta}{\sigma/\sqrt{n}}$ is

$$z_\alpha + \frac{\delta}{\sigma/\sqrt{n}}. \tag{3.40}$$

Actually, $H_0$ will be rejected if $z(\delta) \geqslant z_\alpha + \frac{\delta}{\sigma/\sqrt{n}}$. As the next step, a fuzzy critical value, $\overline{\mathrm{CV}}$, will determine the critical region. The $\beta$-cuts of it are

$$\overline{\mathrm{CV}}[\beta] = \left[z_\alpha - z_{\frac{\beta}{2}},\ z_\alpha + z_{\frac{\beta}{2}}\right]. \tag{3.41}$$

As a result of the above procedure, there are two fuzzy sets corresponding to the test statistic and the critical value, which are called $\bar{Z}$ and $\overline{\mathrm{CV}}$, respectively. The decision on the hypothesis is made based on the relationship between these fuzzy numbers. Figure 3.1 displays these fuzzy numbers. It helps to explain the procedure

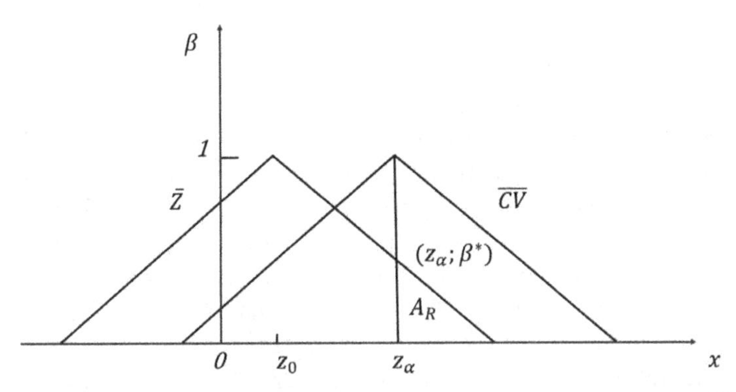


**Figure 3.1.** Fuzzy numbers corresponding to the test statistic ($\bar{Z}$) and critical value ($\overline{\mathrm{CV}}$) [53].

of making decisions (reject the null hypothesis or do not reject it). The vertices of $\overline{\mathrm{CV}}$ and $\bar{Z}$ are located at $x = z_\alpha$ and $x = z_0$, respectively. $A_T$ denotes the total area under the graph of $\bar{Z}$. $A_R$ represents the area under the graph of $\bar{Z}$ to the right side of the vertical line passing through $z_\alpha$. The decision rule proposed by Buckley is

1. If $\frac{A_R}{A_T} \geqslant \phi$ then reject $H_0$.
2. Otherwise do not reject $H_0$.

$\phi$ is a value in (0,1) and is proposed to be set at 0.3.

It should be emphasized that in the fuzzy hypothesis testing two numbers are needed: $\alpha$ as the test significance level and $\phi$ to make the above judgment. If we would like to make use of much more information from the random sample, another $\phi$ should be used. Considering $0 < \phi_1 < \phi_2 < 1$, the decision rule is

1. If $\frac{A_R}{A_T} \geqslant \phi_2$ then reject $H_0$.
2. If $\phi_1 < A_R/A_T < \phi_2$ then make no decision.
3. If $\frac{A_R}{A_T} \leqslant \phi_1$ then do not reject $H_0$.

Here we use the former case. As the last step, $A_R$ and $A_T$ should be calculated. As can be seen in figure 3.1, the horizontal axis is $\alpha$ and the vertical axis is $\beta$. The right side of the membership function corresponding to $\bar{Z}$ is formulated as $x = z_0 + Z_{\frac{\beta}{2}}$, which expresses $x$ as a function of $\beta$. Therefore, we have

$$\alpha = \Phi(x) = 2\left(1 - \int_{-\infty}^{v(x)} \mathcal{N}(0,\, 1)\mathrm{d}x\right), \tag{3.42}$$

where $v(x) = x - z_0$. Thus $A_T$ and $A_R$ are obtained as follows:

$$A_T = 2\int_{z_0}^{z_0+z_{0.005}} \Phi(x)\mathrm{d}x \tag{3.43}$$

$$A_R = \int_{z_\alpha}^{z_0+z_{0.005}} \Phi(x)\mathrm{d}x. \tag{3.44}$$

### 3.2.3 Fuzzy generalized phase synchrony

Our proposed approach is introduced in this section. Our main contribution, as mentioned before, is proposing a fuzzy version of GePS which is called FGePS. This fuzzification is achieved by using fuzzy hypothesis testing instead of the crisp form in the Johansen test. In fact, we first propose a fuzzy version of the Johansen test. Then, we use it to propose the FGePS. As mentioned before, GePS is proposed to benefit from the co-integrating concept. The procedure for calculating the number of co-integrating relationships is expressed above. The Johansen test works via performing a statistical hypothesis test and its results, which are 1 corresponding to the state of rejection of the null hypothesis and 0 corresponding to the case of not rejecting the null hypothesis. On the other hand, fuzzy statistical hypothesis testing, which is explained above, can provide values in [0,1]. Remembering that decision making on a hypothesis is based on the ratio of $\phi = A_R/A_T$, we can use this proportion as a result. This result not only tells us whether the null hypothesis is rejected or not, can provide a quantitative strength for it. In other words, due to the nature of the decision making in fuzzy hypothesis testing, the information of the confidence interval is embedded in our result.

As discussed above, Buckley's approach covers some classical problems. Therefore, some modifications should be applied to make them appropriate for our problem. $\bar{Z}$ and $\overline{CV}$ were defined as the test statistic and critical value, respectively. we need to generate $\bar{Z}$. It is supposed to be an equilateral triangle. The basis of this triangle is considered equal to the basis of the triangle associated with $\overline{CV}$ multiplying by $\sigma/\sqrt{n}$, where $\sigma$ is the standard deviation of the statistic and $n$ is the number of samples.

It should be emphasized that due to the nature of our problem we should use Buckley's method in the right-sided form.

## 3.3 Data

The synthetic and real EEG data used to evaluate the proposed method are introduced in this section.

### 3.3.1 Synthetic data

First of all, the proposed approach is applied to surrogate data which is generated as follows.

Multivariate non-stationary signals with a sampling rate of 100 Hz, a length of 1000 s and unit amplitude are simulated. $2 \leqslant n \leqslant 12$ is considered as the number of channels. For generating asynchronous signal, an integrated process of order 2 ($I(2)$) is considered as the corresponding phase of each channel, $\varphi_i[n]$, where $i$ is the number of channels. Thus $\varphi_i[n]$ will be computed as the output of a linear shift invariant system with $H(z)$ as its impulse response:

$$H(z) = \frac{1}{(1 - z^{-1})^2}. \tag{3.45}$$

$H(z)$ possesses two poles on the unit circle of the $z$ complex plane. It is driven by a white noise process, $w[n]$. Consequently, the corresponding discrete form of the process is as follows:

$$\begin{gathered} \varphi_i[n] = 2\varphi_i[n-1] - \varphi_i[n-2] + w[n] \\ 2 \leqslant i \leqslant 16,\ n = 1, \ldots, L, \end{gathered} \tag{3.46}$$

where $\varphi_i[1] = \varphi_i[2] = 0$ are considered as the initial values and there are $L = 10^5$ samples. This demonstrates that there is no co-integrating relationship among the instantaneous frequencies since instantaneous frequency laws are integrated processes of order one. Therefore, the asynchronous signal is defined as

$$s_{\text{asynch}}[n] = \text{real}\{e^{j\varphi[n]}\}, \tag{3.47}$$

where $\varphi[n] = [\varphi_1[n], \ldots, \varphi_4[n]]$.

The above formulation can be used for generating signals with perfect synchronization as well with random walk phase signals:

$$\begin{aligned} \varphi_i[n] &= \varphi_i[n-1] + w[n] \\ 1 \leqslant i &\leqslant 4,\ n = 1, \ldots, L, \end{aligned} \tag{3.48}$$

where $\varphi_i[1] = 0$ is considered as the initial value. Therefore, instantaneous frequencies with stationary trends and four co-integrating relationships will be achieved. Generated signals are then divided into 4000 ms length segments. As a result, corresponding to each of the synchronous and asynchronous signals, we have 100 segments. Slow drifts of the mean phase are magnified using a moving average process, with a span of 1000 ms. This would artificially slow all phase signals down. Since synthetic signals are composed of multiple random frequency components, generalized phase synchrony can be detected in different IMFs. The signals of each channel are decomposed into five IMFs by applying an EMD sifting approach. Then, for each segment the proposed measure is extracted from all of IMFs. The ultimate index for each segment is computed by averaging over values associated with each IMF. The MVAR model order used in the Johansen test is set to 10. It should be mentioned that the synthetic data used are the same as those used by Omidvarnia *et al* [25] to make the comparison easier.

### 3.3.2 Resting state EEG signals

This part is dedicated to a brief explanation of the real EEG signals used in this study. Forty-five right-handed Iranian were involved in the study. Fifteen participants were diagnosed with chronic combat related PTSD. Fifteen other participants have been exposed to the same trauma, but diagnosed as non-PTSD. In the following the non-trauma exposed healthy control, combat trauma exposed non-PTSD and the combat related PTSD groups will be called control, non-PTSD and PTSD, respectively. All of the participants of the PTSD and non-PTSD groups had participated the Iran–Iraq war (1980–88). The remaining 15 participants were non-trauma exposed healthy controls.

The exclusion criteria were defined as follows: current substance dependence or abuse, a history or current symptoms of a diagnosed neurological disorder, and psychosis. For the two control groups the participants were recruited from the local community. The participants of the control group expressed neither a history of a major trauma, such as a car accident, sexual assault, serious disease, surgery or physical injury, nor combat experience. In addition, the control and non-PTSD participants were not taking medications which could cause psychoactive effects.

The participants of three groups were labeled as PTSD or non-PTSD, based on an examination made by a psychiatrist and a psychologist. All of the participants were asked to answer the structured clinical interview for the *Diagnostic and Statistical Manual of Mental Disorder* fifth edition (DSM$^{5\text{th}}$). In addition, they were asked to

**Table 3.1.** Demographic and clinical information of participants.

| Characteristics | Control mean ± STD | Non-PTSD mean ± STD | PTSD mean ± STD |
|---|---|---|---|
| Age (years) | 50.867 ± 3.870 | 53.1 ± 2.0 | 52.8 ± 4.3 |
| Marital state (1: married, 0: divorced) | 1 ± 0 | 1 ± 0 | 0.8 ± 0.4 |
| Number of children | 1.8 ± 0.8 | 2.4 ± 0.7 | 2.5 ± 1.0 |
| Education (years) | 11.9 ± 2.5 | 14.8 ± 1.9 | 9.0 ± 3.5 |
| Age of trauma | — | 17.6 ± 1.2 | 18.6 ± 3.2 |
| Depression and anxiety | 9.1 ± 3.8 | 8.7 ± 6.8 | 50.0 ± 6.9 |
| PTSD | — | 7.7 ± 4.0 | 55.5 ± 12.7 |

answer the PTSD check-list (PCL) questionnaire as well as the depression, anxiety, stress scales (DASS-21) questionnaire. The PCL questionnaire monitors and measures the severity of avoidance, hyper-arousal and post-traumatic intrusion symptoms in the previous week associated with a trauma. The DASS-21 questionnaire measures anxiety, stress and depressive severity over the previous two weeks. More details are available in [54, 55]. Table 3.1 illustrates the clinical and demographic information of the participants.

The participants were first informed completely about the study procedure. Then they signed a consent form. Finally, at the end of the session, all the participants received financial compensation for their participation. This study was performed under the ethical approval of Shahid Beheshti University on behalf of Sadr Hospital, and Amirkabir University of Technology, Tehran, Iran.

EEG data acquisition was performed in one session, in a quiet room with normal lighting. The signals were recorded using an active electrode g.Tec system, with a sampling rate of 1200 Hz. The data were recorded in 16 channels located at the following sites, according to the standard 10/20 system: AF-3, AF-4, F-3, Pz, F-4, F-8, FC-3, FC-4, T-8, CP-3, P-5, P-6, PO-3, PO-4, O-1 and O-2. The left ear was used as the reference and Fz was the ground. The online filters applied to the signals had 0.1 Hz and 100 Hz cut-off frequencies. Moreover, a 50 Hz notch filter was applied. The electrode impedance was kept below 5 kΩ during data acquisition. The EEG signals were recorded in resting state eyes closed and resting state eyes open, each one for 5 minutes.

Re-referencing was applied to prevent potential bias which may be caused as a result of selecting the left ear as the reference. After removing gross movement artifacts, independent component analysis (ICA) was used for denoising signals. Then the signals were filtered to the delta, theta, alpha, beta and gamma frequency bands, which are defined as 1–4 Hz, 4–8 Hz, 8–13 Hz, 13–25 Hz and 25–40 Hz, respectively. In the following they will be denoted as the $\delta$-band, $\theta$-band, $\alpha$-band, $\beta$-band and $\gamma$-band, respectively. The filtered signals, as well as the whole band signal, are segmented into epochs of 1000 ms length. More details about the dataset can be found in [54].

## 3.4 Results

The results of applying the proposed approach are reported in this section. The results of the synthetic data are reported in the first section. The second section is dedicated to the PTSD EEG data in resting states with eyes closed and eyes open.

### 3.4.1 Simulated data

The simulated signals are generated as explained in section 3.3.1. The synthetic data contain two classes: lack of synchronization and perfect synchronization.

Figure 3.2 presents the area under curve (AUC) for GePS and FGePS. AUC is a measure of the accuracy of a quantitative test. As can be seen, the four-channel synthetic data FGePs outperforms GePS.

Now, the influence of the number of channels will be studied. Figure 3.3 displays the AUC values while the number of channels are increasing from 2 to 9. It can be inferred that FGePS provides higher AUC values compared to GePS, in particular with fewer channels.

### 3.4.2 PTSD EEG data

As mentioned above, the PTSD data were recorded in the resting state with the eyes closed and open. The signals were investigated in five frequency bands including the delta, theta, alpha, beta and gamma bands, as well as the whole band. In this section the results of applying the proposed method will be reported, as well as the results using GePS. Statistical analysis is performed using analysis of variance (ANOVA) at

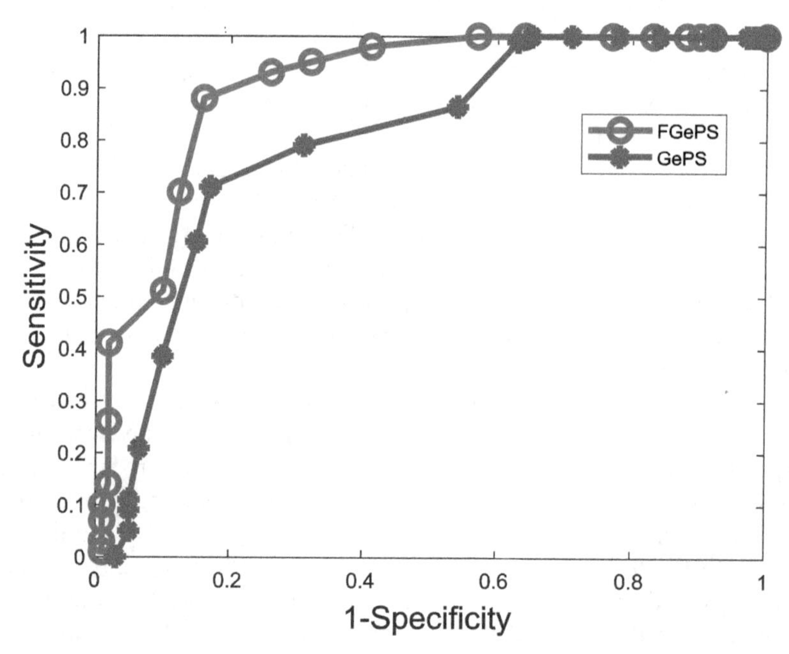


**Figure 3.2.** The area under curve (AUC) associated with GePS and FGePS.

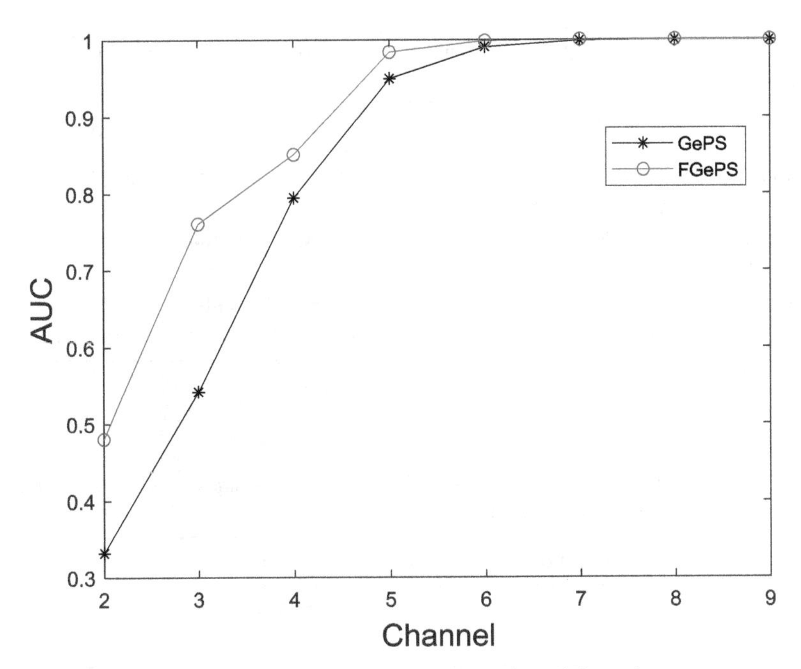


**Figure 3.3.** AUC corresponding to number of channels.

**Table 3.2.** Separability of pairwise groups in resting state eyes closed, based on GePS and FGePS. $p$-values are reported. $p$-values <0.01 are in bold.

| Classes | Method | $\delta$-band | $\theta$-band | $\alpha$-band | $\beta$-band | $\gamma$-band | Whole band |
|---|---|---|---|---|---|---|---|
| **Control–non-PTSD** | FGePS | **<0.0001** | **0.002** | **<0.0001** | **<0.0001** | **<0.0001** | **<0.0001** |
| | GePS | 0.216 | 0.813 | 0.352 | 0.059 | 0.096 | 0.475 |
| **Control–PTSD** | FGePS | 0.047 | **0.004** | **0.007** | **<0.0001** | **<0.0001** | 0.037 |
| | GePS | 1 | 0.917 | 1 | 0.959 | 0.959 | 1 |
| **Non-PTSD–PTSD** | FGePS | **<0.0001** | **<0.0001** | **<0.0001** | **<0.0001** | 0.280 | **<0.0001** |
| | GePS | 0.103 | 1 | 0.578 | 0.724 | 1 | 0.607 |

a 99% significance level. Table 3.2 indicates the separability of pairwise groups in resting state eyes closed, based on GePS and FGePS. $p$-values are reported corresponding to the $\delta$-, $\theta$-, $\alpha$-, $\beta$- and $\gamma$-frequency bands and the whole band signal. Due to the significance level, $p$-values $< 0.01$, which are in bold, illustrate that the pairwise groups are separable using GePS or FGePS. As can be seen, FGePS presents a better performance in telling the difference between pairwise groups compared to GePS. Similarly, table 3.3 summarizes the $p$-values for the resting state eyes open. In this state, similar to the previous results, it can be seen that FGePS outperforms GePS in separating pairwise groups at the 99% significance level. This performance can be found in all frequency bands, as well as the whole

**Table 3.3.** Separability of pairwise groups in resting state eyes open, based on GePS and FGePS. $p$-values are reported. $p$-values <0.01 are in bold.

| Classes | $\delta$-band | $\theta$-band | $\alpha$-band | $\beta$-band | $\gamma$-band | $\delta$-band | Whole band |
|---|---|---|---|---|---|---|---|
| **Control–non-PTSD** | FGePS | **<0.0001** | 0.114 | **0.002** | **<0.0001** | **<0.0001** | **<0.0001** |
| | GePS | 0.155 | 0.391 | 1 | 0.045 | 1 | 0.398 |
| **Control–PTSD** | FGePS | **0.004** | **<0.0001** | **<0.0001** | **<0.0001** | **<0.0001** | **<0.0001** |
| | GePS | 0.630 | 1 | 0.106 | **0.009** | 1 | 0.561 |
| **Non-PTSD–PTSD** | FGePS | **<0.0001** | **0.001** | **<0.0001** | **<0.0001** | **<0.0001** | **<0.0001** |
| | GePS | 1 | 1 | 0.281 | 1 | 0.419 | 1 |

**Table 3.4.** The separability of the three groups in resting state eyes closed. $p$-values corresponding to FGePS and GePS are reported. $p$-values <0.01 are in bold.

| Classes | $\delta$-band | $\theta$-band | $\alpha$-band | $\beta$-band | $\gamma$-band | Whole band |
|---|---|---|---|---|---|---|
| FGePS | **0.002** | 0.438 | **0.007** | **<0.0001** | 0.047 | 0.308 |
| GePS | 1 | 1 | 0.374 | 0.629 | 1 | 1 |

**Table 3.5.** The separability of the three groups in resting state eyes open. $p$-values corresponding to FGePS and GePS are reported. $p$-values <0.01 are in bold.

| Classes | $\delta$-band | $\theta$-band | $\alpha$-band | $\beta$-band | $\gamma$-band | Whole band |
|---|---|---|---|---|---|---|
| FGePS | **<0.0001** | 0.183 | **<0.0001** | **0.007** | **<0.0001** | 0.735 |
| GePS | 1 | 0.628 | 1 | 0.726 | 1 | 1 |

band, except in the $\theta$-band between the control and non-PTSD groups. GePS, however, does not present an appropriate performance.

Table 3.4 summarizes the $p$-values corresponding to the separability of three groups in resting state eyes closed. As can be seen, FGePS can successfully tell the difference between three groups at the 99% confidence level in the $\delta$-, $\alpha$- and $\beta$-bands. However, it cannot detect such differences for the other frequency bands, nor the whole band. GePS does not present an appropriate performance in the frequency bands or the whole band. In a similar manner, the $p$-values associated with the ANOVA test between the three groups in resting state eyes open are summarized in table 3.5.

As can be inferred, FGePS presents an appropriate performance in the delta, alpha, beta and gamma frequency bands at the 99% confidence level. However, GePS was not successful in revealing differences.

## 3.5 Conclusion

A novel generalized phase synchrony measure is proposed in the present chapter, which is called fuzzy generalized phase synchrony (FGePS). It can be used for computing the global phase synchrony within multivariate non-stationary signals, in particular EEG. FGePS is a modified version of generalized phase synchrony (GePS), first proposed by Omidvarnia *et al.* GePS estimates the phase synchrony based on the linear relationship between the instantaneous frequency laws of signals. It uses the Johansen test to assess the phase synchrony. Our main contribution is proposing a fuzzy version for the Johansen test based on the fuzzy hypothesis testing concept. Then we use this fuzzified Johansen test to propose a modified version of GePS.

FGePS is applied to synthetic data. Comparing the results with GePS reveals the superiority of our proposed method. It is also applied to real EEG data. Our database contains EEG signals in resting state eyes closed and eyes open, from three groups of participants, including combat related PTSD as the study group, and combat trauma exposed non-PTSD and healthy non-trauma exposed controls as the control groups. To our knowledge this is the first time that PTSD EEG signals are assessed using a global synchrony measure. The signals are studied in the delta, theta, alpha, beta and gamma bands, as well as the whole band. Statistical analysis is used. The ANOVA test at the 99% confidence level is applied. The resulting $p$-values reflect that FGeP performs more powerfully in making a difference between pairwise groups, compared to GePS, which confirms that our proposed measure, FGePS, outperforms its predecessor.

## Acknowledgments

We wish to express our appreciation to Professor Amir Omidvarnia for his kind and valuable advice. Moreover, we are grateful for the valuable support of Professor Soltani, Mrs Dehghan (Amirkabir University of Technology, Tehran, Iran), Professor Alireza Moradi (Institute for Cognitive Science Studies and Kharazmi University, Tehran, Iran), Dr Jafar Mirazaei, Dr Daneshmand, Dr Nasiri and the staff of Sadr Psychiatric Hospital (Tehran, Iran).

## References

[1] Ahmadlou M, Adeli H and Adeli A 2012 Fuzzy synchronization likelihood-wavelet methodology for diagnosis of autism spectrum disorder *J. Neurosci. Methods* **211** 203–9

[2] Ghanbari Z and Moradi M H 2020a FSIFT-PLV: an emerging phase synchrony index *Biomed. Signal Process. Control* **57** 101764

[3] Dauwels J, Vialatte F, Musha T and Cichocki A 2010 A comparative study of synchrony measures for the early diagnosis of Alzheimer's disease based on EEG *NeuroImage* **49** 668–93

[4] Aviyente S and Mutlu A Y 2011 A time-frequency-based approach to phase and phase synchrony estimation *IEEE Trans. Signal Process.* **59** 3086–98

[5] Rosenblum M G, Pikovsky A S and Kurths J 1996 Phase synchronization of chaotic oscillators *Phys. Rev. Lett.* **76** 1804–7

[6] Raeisi K, Mohebbi M, Khazaei M, Seraji M and Yoonessi A 2020 Phase-synchrony evaluation of EEG signals for multiple sclerosis diagnosis based on bivariate empirical mode decomposition during a visual task *Comput. Biol. Med.* **117** 103596

[7] Li Y, Kang C, Qu X, Zhou Y, Wang W and Hu Y 2016 Depression-related brain connectivity analyzed by EEG event-related phase synchrony measure *Front. Human Neurosci.* **10** 477

[8] Yan T *et al* 2017 Increased local connectivity of brain functional networks during facial processing in schizophrenia: evidence from EEG data *Oncotarget* **8** 107312–22

[9] Aviyente S, Tootell A and Bernat E M 2017 Time–frequency phase-synchrony approaches with ERPs *Int. J. Psychophysiol.* **111** 88–97

[10] Kawano T *et al* 2017 Large-scale phase synchrony reflects clinical status after stroke: an EEG study *Neurorehab. Neural Repair* **31** 561–70

[11] Mezeiová K and Paluš M 2012 Comparison of coherence and phase synchronization of the human sleep electroencephalogram *Clin. Neurophysiol.* **123** 1821–30

[12] Fell J and Axmacher N 2011 The role of phase synchronization in memory processes *Nat. Rev. Neurosci.* **12** 105

[13] Kong W, Wang L, Xu S, Babiloni F and Chen H 2019 EEG fingerprints: phase synchronization of EEG signals as biomarker for subject identification *IEEE Access* **7** 121165–73

[14] Blinowska K J, Kuś R and Kamiński M 2004 Granger causality and information flow in multivariate processes *Phys. Rev.* E **70** 050902

[15] Mormann F, Lehnertz K, David P and Elger C E 2000 Mean phase coherence as a measure for phase synchronization and its application to the EEG of epilepsy patients *Physica* D **144** 358–69

[16] Lachaux J P, Rodriguez E, Martinerie J and Varela F J 1999 Measuring phase synchrony in brain signals *Hum. Brain Mapp.* **8** 194–208

[17] Rosenblum M G, Cimponeriu L, Bezerianos A, Patzak A and Mrowka R 2002 Identification of coupling direction: application to cardiorespiratory interaction *Phys. Rev.* E **65** 041909

[18] Rosenblum M G and Pikovsky A S 2001 Detecting direction of coupling in interacting oscillators *Phys. Rev.* E **64** 045202

[19] Koenig T, Lehmann D, Saito N, Kuginuki T, Kinoshita T and Koukkou M 2001 Decreased functional connectivity of EEG theta-frequency activity in first-episode, neuroleptic-naïve patients with schizophrenia: preliminary results *Schizophr. Res.* **50** 55–60

[20] Rudrauf D, Douiri A, Kovach C, Lachaux J-P, Cosmelli D, Chavez M, Adam C, Renault B, Martinerie J and Le Van Quyen M 2006 Frequency flows and the time-frequency dynamics of multivariate phase synchronization in brain signals *NeuroImage* **31** 209–27

[21] Jalili M, Barzegaran E and Knyazeva M G 2014 Synchronization of EEG: bivariate and multivariate measures *IEEE Trans. Neural Syst. Rehabil. Eng.* **22** 212–21

[22] Li S, Li D, Deng B, Wei X, Wang J and Wai-Loc C 2013 A novel feature extraction method for epilepsy EEG signals based on robust generalized synchrony analysis *25th Chinese Control and Decision Conf. (CCDC) (25–27 May 2013)* pp 5144–7

[23] Mutlu A Y and Aviyente S 2012 Hyperspherical phase synchrony for quantifying multivariate phase synchronization *Statistical Signal Processing Workshop (SSP) (5–8 August 2012)* (Piscataway, NJ: IEEE) pp 888–91

[24] Mutlu A Y and Aviyente S 2010 Multivariate empirical mode decomposition for quantifying multivariate phase synchronization *EURASIP J. Adv. Signal Process.* **2011** 615717

[25] Omidvarnia A, Azemi G, Colditz P B and Boashash B 2013 A time–frequency based approach for generalized phase synchrony assessment in nonstationary multivariate signals *Digital Signal Process.* **23** 780–90

[26] Zadeh L A 1965 Fuzzy sets *Inform. Control* **8** 338–53

[27] Chachi J, Taheri S M and Viertl R 2016 Testing statistical hypotheses based on fuzzy confidence intervals *Austrian J. Stat.* **41** 267–86

[28] Filzmoser P and Viertl R 2004 Testing hypotheses with fuzzy data: the fuzzy *p*-value *Metrika* **59** 21–9

[29] Vahia V N 2013 Diagnostic and Statistical Manual of Mental Disorders 5: a quick glance *Indian J. Psych.* **55** 220–3

[30] Kessler R C 2000 Posttraumatic stress disorder: the burden to the individual and to society *J. Clin. Psych.* **61** 4–12

[31] Taran S and Bajaj V 2018 Drowsiness detection using instantaneous frequency based rhythms separation for EEG signals *2018 Conf. on Information and Communication Technology (CICT) (26–28 October 2018)* pp 1–6

[32] Bajaj V, Rai K, Kumar A, Sharma D and Singh G K 2017 Rhythm-based features for classification of focal and non-focal EEG signals *IET Signal Process.* **11** 743–8

[33] Taran S and Bajaj V 2018b Rhythm-based identification of alcohol EEG signals *IET Sci. Meas. Technol.* **12** 343–9

[34] Adrian E D and Matthews B H C 1934 The Berger rhythm: potential changes from the occipital lobes in man *Brain* **57** 355–85

[35] Barry R J and De Blasio F M 2017 EEG differences between eyes-closed and eyes-open resting remain in healthy ageing *Biol. Psychol.* **129** 293–304

[36] Boytsova Y A and Danko S G 2010 EEG differences between resting states with eyes open and closed in darkness *Human Physiol.* **36** 367–9

[37] Boashash B (ed) 2016 Advanced implementation and realization of TFDs *Time-Frequency Signal Analysis and Processing* 2nd edn (Oxford: Academic) ch 6

[38] Lerga J and Sucic V 2009 Nonlinear IF estimation based on the pseudo WVD adapted using the improved sliding pairwise ICI rule *IEEE Signal Process. Lett.* **16** 953–6

[39] Kwok H K and Jones D L 2000 Improved instantaneous frequency estimation using an adaptive short-time Fourier transform *IEEE Trans. Signal Process.* **48** 2964–72

[40] Hussain Z M and Boashash B 2002 Adaptive instantaneous frequency estimation of multicomponent FM signals using quadratic time–frequency distributions *IEEE Trans. Signal Process.* **50** 1866–76

[41] Khlif M S, Mesbah M, Boashash B and Colditz P 2010 Detection of neonatal seizure using multiple filters *10th Int. Conf. on Information Science, Signal Processing and their Applications (ISSPA 2010) (10–13 May 2010)* pp 284–7

[42] Rudrauf *et al* 2006b Frequency flows and the time-frequency dynamics of multivariate phase synchronization in brain signals *Neuroimage* **31** 209–27

[43] Wacker M and Witte H 2011 On the stability of the n:m phase synchronization index *IEEE Trans. Biomed. Eng.* **58** 332–8

[44] Murray M P 1994 A drunk and her dog: an illustration of cointegration and error correction *Am. Stat.* **48** 37–9

[45] Kammerdiner A R and Pardalos P M 2010 Analysis of multichannel EEG recordings based on generalized phase synchronization and cointegrated VAR *Computational*

*Neuroscience Springer Optimization and Its Applications* ed W Chaovalitwongse, P Pardalos and P Xanthopoulos vol 38 (New York: Springer)
[46] Johansen S 1991 Estimation and hypothesis testing of cointegration vectors in Gaussian vector autoregressive models *Econometrica* **59** 1551–80
[47] Stevenson N J, Mesbah M and Boashash B 2010 Multiple-view time-frequency distribution based on the empirical mode decomposition *IET Signal Processing* **4** 447–56
[48] Österholm P and Hjalmarsson E 2007 *Testing for Cointegration using the Johansen Methodology When Variables are Near-integrated* (Washington, DC: International Monetary Fund)
[49] Huang N E Introduction to the Hilbert–Huang transform and its related mathematical problems *Hilbert-Huang Transform and Its Applications* (Singapore: World Scientific) pp 1–26
[50] Karam Z N and Oppenheim A V 2007 Computation of the one-dimensional unwrapped phase *2007 15th Int. Conf. on Digital Signal Processing (1–4 July 2007)* pp 304–7
[51] Hytti H, Takalo R and Ihalainen H 2006 Tutorial on multivariate autoregressive modelling *J. Clin. Monit. Comput.* **20** 101–8
[52] Buckley J J 2005 Fuzzy statistics: hypothesis testing *Soft Comput.* **9** 512–8
[53] Ghanbari Z and Moradi M H 2017 Generalized phase synchrony using fuzzified Johansen test *2017 Artificial Intelligence and Signal Processing Conf. (AISP) (25–27 October 2017)* pp 294–9
[54] Ghanbari Z, Moradi M H, Moradi A and Mirzaei J 2020 Resting state functional connectivity in PTSD veterans: an EEG study *J. Med. Biol. Eng.* **40** 505–16
[55] Ghanbari Z and Moradi M H 2020b Fuzzy scale invariant feature transform phase locking value and its application to PTSD EEG data *Modelling and Analysis of Active Biopotential Signals in Healthcare* vol 1 (Bristol: IOP Publishing) https://iopscience.iop.org/book/978-0-7503-3279-8

**IOP** Publishing

Modelling and Analysis of Active Biopotential Signals in Healthcare, Volume 2

**Varun Bajaj and G R Sinha**

# Chapter 4

# A study of the influence of meditation and music therapy on the vital parameters of the human body through EEG signal analysis: a review

**Anurag Shrivastava, Bikesh Kumar Singh and Neelamshobha Nirala**
*Department of Biomedical Engineering, National Institute of Technology Raipur, India (CG)-492010*

Meditation and music therapy can have significant positive effects on both the human body and emotions [1, 2]. According to various studies, different forms of meditation may have different effects on the body's vital parameters. Among all the meditation techniques, mindfulness meditation has proven to be one of the most popular globally accepted meditational practices and helps to reduce distress among patients suffering from critical diseases (such as breast cancer [3]) and disorders. Along with meditation, music therapy has also proven to be effective in treating various disorders and diseases [4] and also in pain management [5]. Music therapy also finds a role in clinical applications [4], for example, fast music can make one feel more alert resulting in improved concentration, while slower music can relax the brain and muscles and can be an effective stress reducing tool. The music can be as simple as the sound of rippling water or as complex as classical music. EEG is considered to be the most convenient non-invasive and powerful tool for the analysis of the electrical activity of the brain [6, 7]. Traditionally, the EEG power spectral density is categorized into four spectral bands: delta (0–4 Hz), theta (4–7 Hz), alpha (8–13 Hz) and beta (13–30 Hz). Each band reflects different activities of the brain. The current chapter focuses on the research, theory, experimental methods and findings based on the application of meditation and music in disease management, task performance, etc, using EEG. In the conclusion, it is reported that meditation and listening to music have a positive influence on releasing stress, performing tasks and in some cases can be used as supplementary tools for curing diseases. These studies have implications in neuroscience, neurasthenics and music therapy.

doi:10.1088/978-0-7503-3411-2ch4

## 4.1 Introduction

Meditation is basically a mind practice that can be used to increase calmness and relaxation of the mind [8, 9]. Various forms of meditation are found in different countries. Currently, meditation plays a significant role as an alternative medication in various countries and also in improving the efficiency of performance of various cognitive tasks [10]. However, to ensure proper effectiveness and to allow it to be recommended, a significant amount of scientific study is required so as to validate a particular meditation method and its effectiveness. Different forms of meditation are known across the world, of which the most popular types are mindfulness meditation, transcendental meditation, religious or spiritual meditation, focused meditation, mantra meditation, movement meditation, and loving and kindness meditation (metta meditation) [8]. Along with meditation, music therapy has also proven beneficial in curing various disorders and diseases alongside medication. Several studies have shown that listening to music can have a positive effect on patients with different diseases such as cancer, Alzheimer's, behavioral and emotional disorders, and other psychological problems [2]. Furthermore, listening to music can improve a person's ability to perform tasks associated with reinforcement learning. There are other physical and vital parameters that can be improved while listening to music. This chapter concentrates specifically on mindfulness meditation and music therapy and their effects on the vital parameters of the human body using EEG signal analysis.

## 4.2 Meditation

Meditation is the exercise of the mind and is used to keep the mind fit, similarly to how exercise keeps the body fit and in a healthy condition. The most important part of your body to exercise is the brain. Meditation in combination with more traditional therapies has been found to be effective for various medical and psychiatric conditions. There are various meditation centers in different countries offering courses on different forms of meditation.

### 4.2.1 The origin of meditation

According to archaeologists, meditation was first discovered from 5000 to 3500 BCE in the Indus valley; this is based on evidence found by dating wall art. Meditation is practiced in different religions and the earliest recorded meditation technique is found in the Hindu tradition. In the Hindu scriptures there are descriptions of meditation dating to around 3000 years ago. The spread of meditation globally began through the Silk Route around the fifth to sixth centuries BCE. However, it was only around the twentieth century that meditation moved beyond the realm of specific religions. Meditation practice is part of different religious beliefs around the world, but most significantly in Hinduism, ancient Egypt and China, Judaism, Jainism and Buddhism.

According to the Hindu literature, the earliest references to meditation are available in the Upanishads in the Mahabharata and Bhagavad Gita. The

Patañjali's Yoga sutras (circa 400 CE), which outlines eight branches: discipline referred to as yamas, rules referred to as niyamas, physical postures referred to as āsanas, breath control referred to as prāṇāyama, withdrawal from the senses referred to as pratyāhāra, singular-focus of mind referred to as dhāraṇā, meditation referred to as dhyāna and finally samādhi, which is one of the most ancient and influential Hindu yogas. Buddhist meditation arose around 2600 years ago, when the Buddha founded an experiential path involving sitting calmly in the mode of mindful awareness and breathing as a means to long-lasting peace. Some of the most popular meditation practices, such as mindfulness meditation, Zen meditation and Vipasana arose from the Buddhist tradition. Modern meditation is mainly associated with spiritual traditions belonging to Asian countries and a few of these practices are Theravada, Tibetan, Buddhist and Zen meditation. Meditation in western countries became popular by the middle of the twentieth century [11]. Dr Jon Kabat Zinn, a scholar of Zen Buddhist meditation practices, established the Mindfulness Meditation Center in 1979 at the Medical School of Massachusetts University. The mindfulness-based stress reduction (MBSR) program developed by Dr Jon Kabat Zinn is offered to the public and researchers by different medical centers and health organizations worldwide, without any religious overtones [12].

### 4.2.2 Types of meditation

Different forms of meditation can be found in different parts of the world.

There are popularly seven types of meditation practices [8]:

- Mindfulness meditation.
- Transcendental meditation.
- Spiritual or religious meditation.
- Mantra meditation.
- Focused meditation.
- Movement meditation.
- Loving and kindness meditation (metta meditation).

### 4.2.3 Mindfulness meditation

There is an alarming situation in today's world of an increasing incidence of mental health issues and a primary factor responsible for these increasing mental health problems is stress. Mindfulness meditation, although rooted in Buddhist traditions, is a non-religious mental training regime, involving breathing practices and awareness of the mind and body followed by relaxation of the muscles and body, that teaches you how to calm your body and mind, reducing negativity. Mindfulness meditation can be performed anywhere and any time, it does not require any props or preparation [13]. Devised by Dr Jon Kabat Zinn, the MBSR program is proving beneficial in treating chronic pain and countering stress and other ailments, and mindfulness courses can be found in many institutions ranging from schools to prisons and sports teams. Even the US Army recently adopted an MBSR program to improve military resilience [12].

Meditation is playing a vital role in the current COVID-19 situation. For example, in April 2020 the Indian Ministry of AYUSH provided guidelines on immunity-boosting measures during the COVID-19 crisis for the self-care of individuals, which suggested the regular practice of meditation and yoga for at least half an hour a day, and the United Nations has organized workshops on mindfulness meditation every week for all staff at its Headquarters and in the field. It has been suggested by UNESCO that mindfulness meditation can help to reduce anxiety and stress when incorporated by an individual into their daily routine.

The major benefits of mindfulness meditation include:

- reduced stress,
- increased focus,
- improved physical well-being,
- improved creativity,
- a stronger immune system,
- reduced depression,
- increased collaboration, etc.

## 4.3 Music therapy and its application in medical science

Music therapy (MT) involves the application of clinical and evidence-based musical interventions with the help of professionals who have undergone a certified music therapy training to accomplish individualized goals. After assessing the requirements of a client, the music therapist applies treatment which includes singing and/or listening to music. Music therapy involves meditation, listening to music, playing an instrument and singing. Different types of music therapy are known across the world, some of which are neurologic music therapy, guided meditation, tuning fork therapy and many more. Music therapy can be used to treat the symptoms of conditions such as cancer, anxiety disorders, learning difficulties, behavioral and psychiatric disorders, autism spectrum disorder and post-traumatic stress disorder [14].

The American Music Therapy Association (AMTA) considers music therapy as an evidence-based and clinical tool to accomplish individualized goals with the help of a certified music therapist [15].

Music therapy has wide pathological applications in the following areas [16]:

1. Disorders of consciousness.
2. Psychiatry.
3. Rehabilitation.
4. Chronic conditions.

Studies show that listening to music can have an activating effect on the brain of people suffering from disorders of consciousness, and also show that long term music therapy helps to reduce depression and anxiety in psychiatric patients. Furthermore, listening to music is widely used to reduce anxiety and stress in almost all areas of medicine. Patients with critical and chronic illnesses (such as cancer and those undergoing chemotherapy) and those admitted to intensive care often suffer stress and anxiety, which can lead to prolonged recovery times and other adverse effects. For these types of patients, listening to music leads to a reduction in stress levels and anxiety which leads to reliable and effective treatment [16].

Listening to music also helps to improve sleep quality and induce overall relaxation, which can lower the cardiac workload and reduce oxygen consumption, therefore providing better and more effective ventilation for patients undergoing mechanical ventilation.

Another study has shown significant differences in the cardiovascular parameters (diastolic and systolic blood pressure) and stress levels between experimental and control groups of subjects.

Music interventions have proven to be an effective way to obtain relief from acute and chronic pain. Music therapy also reduces the risk of stroke, boosts immunity, lowers blood pressure and has many more positive effects. Many hospitals are using music therapy to help in the management of acute and chronic pain for the treatment of other pathological conditions.

According to an article published in the *Deccan Herald* (April 2020) frontline workers against COVID-19, which include doctors and police personnel, who have an exhausting schedule, find music therapy to be an effective tool to reduce stress levels. Some professionals are giving free therapy to these frontline personnel. According to the same report, the National Institute of Mental Health and Science (NIMHANS), Bangalore, is studying the same field and obtaining positive results.

### 4.3.1 Meditation and music therapy

Meditation and music therapy are related to a certain extent. A form of meditation known as guided meditation uses a combination of meditation, music and mindfulness. Guided meditation uses sound healing in which a person can meditate to voiced instruction, using a video or app or directly through a trained instructor. Guided meditation can involve chanting or repeating mantras or prayers [14].

## 4.4 Measurement of neurological response

The above-discussed meditation and music therapy have a significant effect on the brain parameters which can be analysed by recording the neurological responses generated from the brain. The brain is divided into four sections called lobes. The frontal lobe is responsible for cognitive function. The second lobe, known as the parietal lobe, helps to process information regarding movement, taste, touch and temperature. Vision is taken care of by the third lobe, known as the occipital lobe. The fourth lobe is the temporal lobe, which is responsible for memory processing and is also responsible for the sensations of sound, touch, light and taste. Therefore, each lobe will be generating signals according to the task being performed and it is necessary to measure these signals [17]. There are two broad categories of methods for the measurement of neurological response: invasive methods and non-invasive methods.

### 4.4.1 Invasive methods

In these methods electrodes are implanted under the skull. Invasive methods measure the neural activity of the brain on the cortical surface or intra-cortically from within the motor cortex. The greatest advantage of this method is that it provides high spatial and temporal resolution and therefore increases the quality of the obtained signal and enables the precise pattern of the occurrence of a certain activity to be determined. However, the major drawback from the researcher's point of view is the requirement for very expensive equipment and a complex set-up which involves surgery and therefore these methods are less popular among researchers.

Invasive methods include cortical surface electrocartiography (ECoG) and intra-cortical electrocartiography.

### 4.4.2 Non-invasive methods

Non-invasive signal acquisition does not require penetrating external objects or surgical procedures on the subject's brain. This method is safe and is widely used by researchers compared to the invasive methods.

Non-invasive methods include:

- Electroencephalogram (EEG).
- Magnetic resonance imaging (MRI).
- Functional magnetic resonance imaging (fMRI).
- Functional near-infrared spectroscopy (fNIRS).

*4.4.2.1 Electroencephalogram*

EEG is a non-invasive and powerful electrophysiological tool used for the analysis of the brain's activity through electrical signals generated inside the brain. Traditionally the EEG power spectral density is categorized into four spectral bands: delta (0–4 Hz), theta (4–7 Hz), alpha (8–13 Hz) and beta (13–30 Hz). Each band reflects different activities of the brain. EEG is mainly used to determine disorders such as sleep disorders, epilepsy, stroke and other brain disorders. It is also used to diagnose the depth of anesthesia, state of meditation, brain death, coma and other conditions. EEG is low cost and portable and therefore popular among researchers.

*4.4.2.2 Magnetic resonance imaging*

The MRI technique creates a brain image by using a set of radio waves and a magnetic field to create an image of the brain. MRI plots brain maps at a particular time moment. The image produced by an MRI scan provides structural information on organs/tissues. The information we obtain from MRI is very effective for identifying abnormalities in the brain (a tumor for example).

*4.4.2.3 Functional magnetic resonance imaging*

fMRI is used for the measurement of brain activity which is directly related to changes in the blood flow inside the brain. fMRI helps to examine the active regions of the brain along with structural imaging. As fMRI depends on oxygen levels in the blood it is called blood oxygen level dependent (BOLD). The primary energy source for the brain is glucose, which cannot be stored inside the brain. Therefore when a particular area of the brain requires energy to perform any task, the amount of blood flow increases to carry more glucose into that active area of the brain, thus more oxygen-rich blood enters an active area of the brain. fMRI images metabolic function whereas MRI scans the anatomical brain image structure. fMRI is task oriented.

#### *4.4.2.4 Functional near-infrared spectroscopy*

fNIRS is an optical imaging technique used for the analysis of the brain's neural activity by measuring changes in Hb concentrations in the tissues of the brain. It is based on the principle that near-infrared light easily passes through the tissues of the brain. Both the fMRI and fNIRS methods are applied for the analysis of the hemodynamic response to the neural activity of the brain. While fMRI depends on the paramagnetic properties of Hb, fNIRS depends on the absorption properties of biological chromophores. fNIRS results in the measurement of oxy-Hb and deoxy-Hb.

#### *4.4.2.5 Why EEG?*

The studies in [6] and [7] are comparisons between popular neuromeasurement techniques on the basis of various components and the results suggest that advantages exist for EEG over the other techniques. It is non-invasive, portable and relatively cheaper, which makes EEG ideal for researchers in recording the electrical activities of the brain. EEG is best suited when considering various parameters such as effectiveness, application cost, artifacts during body movements in the context of music therapy, brain controlled interfaces (BCIs) and meditation.

## 4.5 EEG signal processing

The block diagram shown in figure 4.1 represents the different stages of EEG signal processing.

### 4.5.1 Data acquisition

An EEG recording is obtained by proper placement of electrodes on the surface of the scalp. The EEG signal amplitude ranges from 10 $\mu$V to 100 $\mu$V. Electrode can be made of gold-plated silver and Ag/AgCl (silver, silver-chloride) and range in number from 4 to 256, but for most clinical and research applications the internationally recognized 10–20 electrode system is used. The '10' and '20' in the name of the electrode system refer to the distances between the adjacent electrodes which can be either 10% or 20% of the right to left or front to back distance of the total skull. Each

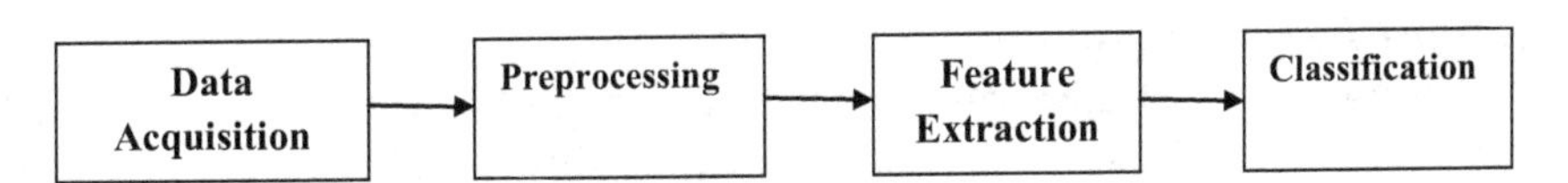


**Figure 4.1.** Block diagram of EEG signal processing.

electrode has a label that will identify the position of the lobe of the brain. The notations are as follows: F = frontal lobe, A = temporal lobe, O = occipital lobe, P = parietal lobe, Fp = frontal polar, C = central and Pg = nasopharyngeal. A 'Z' (zero) indicates electrodes which are placed on the skull's sagittal plane midline. Reference points are represented by Cz, FpZ and Fz.

### 4.5.2 Preprocessing

The next step includes removal of artifacts from the recorded signals. The main sources of artifacts in the EEG signals are blinking of the eyes, muscular activities and power line interference electrical noise. There are various methods available to eliminate these unwanted signals.

Principal component analysis (PCA) is used to reduce the environmental noise and for physiological and cardiac artifacts independent component analysis (ICA) can be used [18].

#### *4.5.2.1 Principal component analysis*

PCA is one of the factor analysis methods. It is used to reduce a wide range of random variables to a smaller set. The PCA method separates components which contain artifacts. It also decreases the size of an EEG signal. As a result, an artifact-free EEG signal is obtained [19].

#### *4.5.2.2 Independent component analysis*

The second method that can be used in preprocessing is ICA, which is based on the principle of 'blind source separation' (BSS) and can be used for signal data which are a mixture of more than one independent signal. If the undesired component which represents noise is eliminated then an artifact-free EEG signal will be obtained [19].

Furthermore, preprocessing includes averaging the signal, thresholding the output and signal enhancement, followed by edge detection. A high pass filter (HPF) can be used to eliminate direct current components, whereas a low pass filter (LPF) can be applied for the removal of high frequency components [20].

### 4.5.3 Feature extraction

Feature extraction is a method that is used for deriving the most useful and relevant descriptors from the signals. The commonly used feature extraction methods for EEG signals are the fast Fourier transform (FFT), wavelet transform (WT), autoregressive method (AR), time–frequency distribution and eigenvectors [21]. Here is a brief description of all the feature extraction techniques.

#### *4.5.3.1 Fast Fourier transform*

The FFT is used for the transformation of time-domain EEG signals into frequency-based signals and is followed by feature extraction through filtering the EEG signal and using the median, mean and standard deviations of each EEG signal. The results obtained from the feature extraction are used for classification purposes.

#### *4.5.3.2 Wavelet transform*

The FFT method is a transformation method suited to stationary signals. For the transformation of non-stationary signals where frequency varies over time, such as EEG signals, a linear transformation method called WT is used. One study demonstrated tunable Q factor wavelet transform (TQWT) based feature extraction and various classification techniques for distinction between alertness and drowsiness using EEG signals and obtained optimum results [22]. Analysis performed using WT produced more precise information over certain classes of signal compared to other techniques [23]. Another study showed that the wavelet method is most suitable for unstable signals, as the varying window size will ensure suitable time–frequency resolution [21]. Applications of WT include audio, video and image processing, biomedical image and signal processing, etc. Another research study proposed adaptive TQWT and a multiclass least-squares support vector machine (SVM) for feature extraction and classification, respectively, for recognition of human emotions using EEG signals [24]. The optimal flexible analytic wavelet transform (OFAWT) uses the genetic algorithm for the minimization of decomposition error. OFAWT decomposition is helpful for the extraction of information related to sleep stages using EEG signals [25].

#### *4.5.3.3 The autoregressive method*

An AR method is a linear regression model whose current observation of a series is calculated from more than one past observation of the series. It is most commonly employed for extracting the features of EEG signals. [26]. AR methods can be used for EEG signal analysis corrupted by white noise. The advantages of AR include better spectral resolution and low computation cost. AR models offer better smooth power spectra and frequency resolution in comparison to the FFT [27]. Autoregression is not suited to real-time analysis due to speed limitations.

#### *4.5.3.4 Time–frequency distribution*

The TFD method is used for feature extraction of non-stationary signals. TFD analysis is considered to be one of the most powerful tools in processing EEG signals [28]. This method is only suitable for noiseless signals, therefore preprocessing methods should be employed carefully to so that all the artifacts are removed [21].

#### *4.5.3.5 Eigenvectors*

Eigenvectors are applied to the estimate frequency and power of the signal from a signal distorted by artifacts [21]. Eigenvectors are used to evaluate the power and frequencies when the signals are corrupted by noise. Each of the methods suffers from certain advantages and disadvantages. The accuracy and speed of the feature extraction stages are crucial. The time-domain method lacks detailed information whereas the frequency domain method lacks high quality performance.

### 4.5.4 Classification

EEG classification can be used to detect tumors, seizures and drowsiness. There are basically five machine learning techniques available for EEG classification: logistic regression (LR), SVMs, *k*-nearest neighbors (*k*NN), artificial neural networks (ANNs), random forest (RF) and the multiple layer perceptron (MLP).

#### *4.5.4.1 Logistic regression*

LR is the most common statistical model used in machine learning and pattern recognition. Regression analysis (RA) is employed to determine the relationship between the dependent variable and independent variable. If two or more variables are used it is called multiple regression. LR comes under the category of supervised learning and this method is mainly used for brainwave classification.

#### *4.5.4.2 Support vector machine*

The SVM is a machine learning method that is widely used for EEG signal classification. The SVM uses the concept of a discriminate hyperplane for class identification. The process includes the selection of the hyperplane and then mapping the data onto high dimensional space. The SVM can be available as both a linear and non-linear SVM. A linear SVM is used when the dataset can be separated linearly. Otherwise, for non-separable data the non-linear SVM can be used. It is proved that the multiple kernel SVM provides more accurate results compared to the single kernel SVM [29]. The SVM classifier has proven to be an accurate tool for predicting meditation compared to the control state in addition to respiration in EEG data. It can also be used for the analysis of real-time EEG signals [30]. Bajaj [31] proposed the classification of focal and non-focal EEG using a least-squares SVM and the results showed improvement compared to other existing methods.

#### *4.5.4.3* k*-nearest neighbors*

*k*NN is a multiclass classifier. *k* refers to the nearest neighbors taken into account, with $k = 1$ referring to the nearest neighbor and larger values referring to two or more neighbors. The author of [29] claimed that *k*NN out-performed SVM and ANN. *k*NN uses a data-mining approach and can be used for emotion recognition, with an RF improved decision tree method for EEG classification.

#### *4.5.4.4 Artificial neural network*

An ANN is a type of machine learning tool which models the brain using a set of algorithms. It receives a complex pattern as the input and provides simple decisions as the output. ANN has found various applications in the medical and scientific fields. The architecture of ANN consists of a minimum of three layers. The three mandatory layers for ANN are the input layer, output layer and hidden layer. The input layer simply passes the input information to the hidden layers, the hidden layer receives the information and performs multilevel processing on it, and the final decision that results is passed on to the output layer. Research has proposed the employment of deep learning algorithms to study human cognitive ability [32].

#### *4.5.4.5 Random forest classifier*

RF is an ensemble tree-based learning algorithm. It combines more than one algorithm of its kind. One advantage is the ability to handle a large number of variables with high accuracy. Therefore it is suited to large datasets.

#### *4.5.4.6 Multiple layer perceptron*

The MLP is a kind of ANN which is often used for the classification of EEG signals. The MLP consists of more than three layers (input layer, output layer and one or more hidden layers). The hidden layers are the heart of the neural network and consist of maximum numbers of neurons to perform the task. The neurons basically work as a processor producing a sequence of activations provided by the weights which are updated according to the learning procedure applied to perform the task. The MLP comes under the category of feed-forward ANNs and is used to perform more complex operations compared to single perceptrons. A stress detection system based on the real-time analysis of EEG signals in meditation has been developed [33]. The FFT was used for feature extraction and *k*NN was used for classification, and the functionality testing result ranged from 80%–100%. One study used a random forest based classification method to classify the state of mind of real-time subjects, whether the state of mind was one of concentration or meditation [34].

## 4.6 Studies conducted globally in the field of meditation and music therapy using EEG signal processing

There are various studies going back decades in the area of EEG of meditation and music therapy. Travis [35] suggested the importance of the spatial and temporal characteristics of EEG signals recorded during meditative practices and discussed the issue that a power derived measure misses the coherence analysis. The author's suggestion was to calculate alpha1, alpha2, theta1 and theta2 for different cognitive processes and apply both power and coherence analysis for more accurate results. Harne *et al* [36] employed an SVM for classification of EEG signals for Om based meditation and the results showed high accuracy and less complexity out of a total of 23 samples. The features of 17 subjects were used to train the SVM and six subject samples were used for testing. The study showed that Om meditation reduces the electrical activity in the brain, reflected in the delta waves. Hata *et al* [37] reported the effect of short-term meditation on patients with post-traumatic residual disability (PTRD). Their method included ten meditators trained by the Japan Yoga Therapy Society. The statistical mapping was performed on eLORETA software. The results indicate that there is increased gamma activity in the subjects suffering from PTRD compared to normal controls. The overall results show increased activity of the psychological system. Vasquez *et al* [38] made a comparison of the neuro-physiological changes observed between expert and novice meditators on the basis of spectral entropy analysis, and the alpha and theta frequency bands.

From the results it can be concluded that there is a change in the alpha and theta frequency bands depending on the experience of the meditator. It has also been observed that spectral entropy plays an important role as an indicator of long term therapy progression

Braboszcz *et al* [39] compared changes in the EEG activity of three different types of meditation (Vipassana, Himalayan and Isha Shoonya yoga). Their method included 10 min of meditation and 10 min of instructed mind wandering (IMW) followed by a questionnaire. The EEG data were recorded and a post-hoc test was performed by EEGLAB. All of the meditators showed an increase in gamma amplitude and in addition increased alpha activity was observed in the Vipassana group of meditators compared to the other groups. Sharma *et al* [40] highlighted the effect of long term Rajyoga meditation on EEG brain dynamics. Their method included the recording of semi-high-density EEG which was recorded before and during meditation in the long term meditation group and resting in the control group. The spectral properties of alpha1 (8–9 Hz) and alpha2 (10–13 Hz) were calculated separately. The results indicate significant difference in the alpha1 power of the meditative and control subject groups, and high spectral alpha power is also observed, which indicates less engagement of the brain. There is also an increase in theta power and the alpha and theta amplitudes were also found to be greater during meditation. This study has the limitation that the beta and other faster brain waves were not considered. Also, only male meditators were considered in the study, so there is a scope for further study. Also, there is scope to apply the study to a particular population of a specific age group. Harne [41] studied the positive effect of Om mantra based meditation on the brain using EEG spectral analysis. The process involved the chanting of Om by 23 naïve meditators with closed eyes. The EEG signals were recorded before and after meditation. The results indicate an increase in the theta power after meditation. Sik *et al* [13] tried to determine the relation between the heart and brain under different experimental conditions using the wavelet entropy method. The method involved recruiting 11 adults, who were given training on an MBSR meditation technique, and recording the EEG and ECG signals during MBSR mindful breathing and the closed eye resting state. EEGLAB software was used for the EEG analysis. The results show there is significant correlation between the entropies of the brain and heart, but the entropies are not correlated during the resting state. It can be concluded that entropy helps to monitor the state of the brain during meditation and can be used in the mental training field quite effectively. Fingelkurts *et al* [42] highlighted the positive and negative effects of meditation for an individual, which means that each form of meditation might not be suitable for everyone. Therefore, to ensure the effectiveness of meditation given as a treatment, the individual's neuro-physiological characteristics have to be identified. The author proposed quantitative EEG (qEEG) as an effective tool for the identification of the neuro-physiological type of an individual and on the basis of that a training program can be designed to benefit the individual. However, this was a review paper based on research performed previously, so there remains the possibility to validate the study through experimental methods. Ahani *et al* [30] analysed the EEG and respiratory signals during mindfulness meditation. Their methodology included the recording of EEG data signals using 32-channel EEG, respiration signals and ECG signals using Bio-Semi (Amsterdam, The Netherlands). A total of 28 men and six women with a mean age of 61 participated in the experiment. They were given training on MBSR and mindfulness-based cognitive therapy (MBCT) and the analysis was performed during two different conditions: the first included listening to a National Public Radio broadcast for 15 min with the eyes closed and the second included simply sitting in mindfulness meditation. SVM classification was used for the EEG phase classification analysis of the recorded data. The results support the evidence that theta and alpha power are increased during meditation. Jadhav *et al* [43] studied the effect of meditation on the cognitive workload of the brain. In this work 11 participants were selected all of whom were students. Focused attention meditation was used and the EEG signal was recorded at the beginning and at the end of four weeks of meditation training. Seven-level arithmetic single- and three-digit addition were used to induce the cognitive workload. The power spectral densities were calculated and compared pre- and post-meditation and the author concluded from the result that after four weeks of meditation practice the stress level was reduced quite significantly. Kaur *et al* [44] investigated the variation of different EEG bands in different forms of meditation. The conclusion of this paper reported that there was an increase in independent brain processes compared to task-free resting during meditation. On the basis of the studies done thus far the author suggested that there is a still requirement to categorize EEG signatures which corresponds to different meditation practices. Also, there is a need for neurobiological studies related to various types of meditation such as Buddhist yoga and others.

In his review paper Kucikiene [4] highlighted the studies conducted to evaluate the effect of music therapy on healthy subjects, and patients suffering from conditions such as psychiatric diseases, disorders of consciousness and chronic conditions, using brainwave spectral analysis. The author concluded that brainwave spectral analysis is a powerful tool for examining the effect of music therapy. Lee [5] conducted a study on music therapy (MT) and music medicine (MM) with a total of 97 trials. The results suggest that both MT and MM have overall beneficial effects on the pain intensity, use of anesthetics, distress from pain, heart rate, blood pressure and respiration rate. Compared to MM, MT showed a greater clinical impact on pain intensity, whereas MM shows significant reduction in analgesic use whereas MT did not. The author concluded that both types of music intervention have proved effective in the relief from acute, procedural and cancer/chronic pain in a medical setting. Nakamura [45] studied the receptive aspect of music and the effect of music was simultaneously measured using positron emission tomography (PET), regional cerebral blood flow (rCBF) and EEG. Eight volunteers were selected for the study. The results suggest that, compared to the rest condition, listening to music caused a significant increase in the EEG beta power spectrum. The beta power spectrum positively correlates with regional cerebral blood flow (rCBF). Sun [46] conducted a study on 40 cases of traumatic brain injury coma patients who were divided into groups of 20 each. One group of patients was given music therapy along with regular treatment and the patients in the other group were not given music therapy. The Glasgow Coma Scale (GCS) value increased for the group of patients who were given music therapy compared to the other group. Taran *et al* [47] proposed an extreme learning machine (ELM) approach for the classification of normal and alcohol EEGs. ELM proved to be effective with 98% accuracy.

## 4.7 Limitations and future scope

Meditation and music therapy are basically mind practices that can be used to increase calmness and relaxation of the mind. There are various forms of meditation and music therapy found in different countries, and currently meditation and music therapy are playing a significant role as alternative medications in various countries. However, to ensure proper effectiveness and to obtain appropriate recommendations, a significant amount of scientific study is required to validate a particular meditation method and determine its effectiveness. Negligible research related to meditation has been performed on children. One of the most untouched areas in the current field is identifying the signature for a specific type of meditation, which is very important for it to be recommended as treatment. Much less technically verified research has been conducted on music therapy. Also, there is a lot of scope to perform studies on patients who are suffering from particular disorders or diseases as to which type of meditation or therapy proves to be effective. The major limitations encountered when performing studies in research areas related to music therapy and meditation, in particular in hospitals, is a lack of proper feedback from patients who are critically ill, and maintaining coordination with hospital staff is also very difficult.

## 4.8 Conclusion

From the above studies it can be concluded that meditation and music therapy are effective stress reducing tools which can also be helpful in maintaining proper coordination between the heart and the brain. They can also be used to reduce cognitive workload. Meditation and music therapy play vital roles in clinical applications for the treatment of various disorders and critical diseases all over the globe, and various programs related to stress reduction using meditation and music therapy have been developed and have already been implemented in various fields (corporate, medical, defense, etc). There is a lot of scope for researchers in both the field of meditation and music therapy, or a combination of both, as discussed in the section on future scope. Even in our current situation of the COVID-19 pandemic, meditation and music therapy have provided effective results for the reduction in the stress levels of COVID-19 frontline workers. To analyse the effects of meditation and music therapy one has to perform analysis of brain signals. As per the discussion above, among the several techniques that are available, EEG has proven to be the most economical and portable in addition to other benefits. Thus it is beneficial for researchers to use EEG based analysis in the field of meditation and music therapy.

## References


[1] Arya N K, Singh K, Malik A and Mehrotra R 2018 Effect of heartfulness cleaning and meditation on heart rate variability *Indian Heart J.* **70** S50-S55

[2] Ikonomidou E, Rehnström A and Naesh O 2004 Effect of music on vital signs and postoperative pain *AORN J.* **80** 269–78

[3] Lesiuk T 2016 The development of a mindfulness-based music therapy (MBMT) program for women receiving adjuvant chemotherapy for breast cancer *Healthcare* **4** 53

[4] Kučikienė D and Praninskienė R 2018 The impact of music on the bioelectrical oscillations of the brain *Acta Med. Litu.* **25** 101–6

[5] Lee D J, Kulubya E, Goldin P, Goodarzi A and Girgis F 2018 Review of the neural oscillations underlying meditation *Front. Neurosci.* **12** 178

[6] Castermans T, Duvinage M, Cheron G and Dutoit T 2013 Towards effective non-invasive brain–computer interfaces dedicated to gait rehabilitation systems *Brain Sci.* **4** 1–48

[7] Fachner J and Stegemann T 2013 Electroencephalography (EEG) and music therapy: on the same wavelength? *Music Med.* **5** 217–22

[8] Bertone H 2017 Which type of meditation is right for me? *Healthline* https://www.healthline.com/health/mental-health/types-of-meditation

[9] Thomas J W and Cohen M 2014 A methodological review of meditation research *Front. Psych.* **5** 74

[10] Edwards M K and Loprinzi P D 2018 Experimental effects of acute exercise and meditation on parameters of cognitive function *J. Clin. Med.* **7** 125

[11] Wikipedia 2020 Meditation *Wikipedia* https://en.wikipedia.org/w/index.php?title=Meditation&oldid=956361264

[12] Wikipedia 2020 Jon Kabat-Zinn *Wikipedia* https://en.wikipedia.org/w/index.php?title=Jon_Kabat-Zinn&oldid=948347548

[13] Sik H H, Gao J, Fan J, Wu B W Y, Leung H K and Hung Y S 2017 Using wavelet entropy to demonstrate how mindfulness practice increases coordination between irregular cerebral and cardiac activities *J. Visual. Exp.* **123** e55455

[14] Healthline 2020 The uses and benefits of music therapy *Healthline* https://www.healthline.com/health/sound-healing

[15] Rafieyan R and Ries R 2007 A description of the use of music therapy in consultation-liaison psychiatry *Psychiatry* **4** 47–52

[16] Kučikienė D and Praninskienė R 2018 The impact of music on the bioelectrical oscillations of the brain *Acta Med. Lit.* **25** 101–6

[17] Wikipedia 2020 Lobes of the brain *Wikipedia* (Accessed: 29 July 2020)

[18] Garcés M A, Lorena L and Orosco 2008 EEG signal processing in brain–computer interface *Smart Wheelchairs and Brain–Computer Interfaces* pp 95–110

[19] Kaczorowska M, Plechawska-Wojcik M, Tokovarov M and Dmytruk R 2017 Comparison of the ICA and PCA methods in correction of EEG signal artefacts *10th Int. Symp. on Advanced Topics in Electrical Engineering (ATEE) (Bucharest, 23–25 March 2017)* 262–7

[20] Riera A 2020 EEG signal processing for dummies—neuroelectrics *Neuroelectrics Blog* https://www.neuroelectrics.com/blog/2014/12/18/eeg-signal-processing-for-dummies/

[21] Al F and Al F 2014 Methods of EEG signal features extraction using linear analysis in frequency and time–frequency domains *ISRN Neurosci.* **2014** 1–7

[22] Bajaj V, Taran S, Khare S K and Sengur A 2020 Feature extraction method for classification of alertness and drowsiness states EEG signals *Appl. Acoust.* **163** 107224

[23] Rhif M, Ben Abbes A, Farah I R, Martinez B and Sang Y 2019 Wavelet transform application for/in non-stationary time-series analysis: a review *Appl. Sci.* **9** 1347
[24] Khare K, Bajaj V and Sinha G R 2020 Adaptive tunable Q wavelet transform based emotion identification *IEEE Trans. on Instrume. Meas.* **69** 9607–17
[25] Taran S, Sharma P C and Bajaj V 2020 Automatic sleep stages classification using optimize flexible analytic wavelet transform *Knowl.-Based Syst.* **192** 105367
[26] Zhang Y, Liu B, Ji X and Huang D 2016 Classification of EEG signals based on autoregressive model and wavelet packet decomposition *Neural Process. Lett.* **45** 365–78
[27] Atyabi A, Shic F and Naples A 2016 Mixture of autoregressive modeling orders and its implication on single trial EEG classification *Expert Syst. Appl.* **65** 164–80
[28] Ridouh A, Boutana D and Bourennane S 2017 EEG signals classification based on time frequency analysis *J. Circuits Syst. Comput.* **26** 1750198
[29] Shaabani M N A H, Fuad N, Jamal N and Ismail M F 2020 *k*NN and SVM classification for EEG: a review *InECCE2019* ed A Kasruddin Nasir *et al* (Lecture Notes in Electrical Engineering vol 632) (Singapore: Springer)
[30] Ahani A, Wahbeh H, Nezamfar H, Miller M, Erdogmus D and Oken B 2014 Quantitative change of EEG and respiration signals during mindfulness meditation *J. NeuroEng. Rehab.* **11** 87
[31] Bajaj V, Rai, Kumar K, Sharma A, Singh D and Kumar G 2017 Rhythm-based features for classification of focal and non-focal EEG signals *IET Signal Proc.* **11** 743–8
[32] Sinha G R, Raju K S, Patra R K, Aye D W and Khin D T 2018 Research studies on human cognitive ability *Int. J. Intell. Defence Support Syst.* **5** 298
[33] Purnamasari P and Fernandya A 2019 Real time EEG-based stress detection and meditation application with *k*-nearest neighbor *IEEE R10 Humanitarian Technology Conf. (R10-HTC) (47129) (Depok, West Java, Indonesia)* pp 49–54
[34] Edla D R, Mangalorekar K, Dhavalikar G and Dodia S 2018 Classification of EEG data for human mental state analysis using random forest classifier *Proc. Comp. Sci.* **132** 1523–32
[35] Travis F 2020 Temporal and spatial characteristics of meditation EEG. Psychological trauma: theory *Res. Pract. Policy* **12** 111–5
[36] Harne B P and Hiwale A S 2018 EEG spectral analysis on OM mantra meditation: a pilot study *Appl. Psychophysiol. Biofeedback* **43** 123–9
[37] Hata M, Hayashi N, Ishii R, Canuet L, Pascual-Marqui R D, Aoki Y and Ito T 2019 Short-term meditation modulates EEG activity in subjects with post-traumatic residual disabilities *Clin. Neurophysiol. Pract.* **4** 30–6
[38] Vasquez J J S and Estelles E D G 2019 EEG feature extraction as markers for states and traits associated with mindfulness meditation practice *2019 IEEE 39th Central America and Panama Convention*
[39] Braboszcz C, Cahn B R, Levy J, Fernandez M and Delorme A 2017 Increased gamma brainwave amplitude compared to control in three different meditation traditions *PLoS One* **12** e0170647
[40] Sharma K, Chandra S and Dubey A K 2018 Exploration of lower frequency EEG dynamics and cortical alpha asymmetry in long-term Rajyoga meditators *Int. J. Yoga* **11** 30–6
[41] Harne B P, Bobade Y, Dhekekar D R S and Hiwale A 2019 SVM classification of EEG signal to analyze the effect of Om mantra meditation on the brain *2019 IEEE 16th India Council Int. Conf. (INDICON) (Rajkot, India)* pp 1–4

[42] Fingelkurts A A, Fingelkurts A A and Kallio-Tamminen T 2015 EEG-guided meditation: a personalized approach *J. Physiol.-Paris* **109** 180–90

[43] Jadhav N, Manthalkar R and Joshi Y 2017 Assessing effect of meditation on cognitive workload using EEG signals *Second Int. Workshop on Pattern Recognition*

[44] Kaur C and Singh P 2015 EEG derived neuronal dynamics during meditation: progress and challenges *Adv. Prevent. Med.* **2015** 1–10

[45] Nakamura S, Sadato N, Oohashi T, Nishina E, Fuwamoto Y N D and Yonekura Y 999 Analysis of music–brain interaction with simultaneous measurement of regional cerebral blood flow and electroencephalogram beta rhythm in human subjects *Neurosci. Lett.* **275** 222–6

[46] Sun J and Chen W 2015 Music therapy for coma patients: preliminary results *Eur. Rev. Med. Pharmacol. Sci.* **19** 1209–121

[47] Taran S and Bajaj V 2018 Rhythm-based identification of alcohol EEG signals in IET science *Meas. Technol.* **12** 343–9

**IOP** Publishing

Modelling and Analysis of Active Biopotential Signals in Healthcare, Volume 2

**Varun Bajaj and G R Sinha**

# Chapter 5

# Cross-wavelet transform aided focal and non-focal electroencephalography signal classification employing deep feature extraction

**Sayanjit Singha Roy[1], Sudip Modak[1], Somshubhra Roy[2] and Soumya Chatterjee[1]**

*[1]Department of Electrical Engineering Techno India University Kolkata, India*

*[2]Department of Electrical Engineering Institute of Engineering and Management Kolkata, India*

The accurate recognition of focal electroencephalography (EEG) signals is important for identifying the precise location of the epileptogenic focal points within the human brain. With this in mind, in this chapter a novel focal and non-focal epileptic EEG signal detection and classification framework implementing cross-wavelet transform based extraction of deep neural features is presented. In this study focal and non-focal EEG signals were obtained from an online data archive. A sample EEG signal from each class (focal and non-focal) was selected as a reference and a cross-wavelet transform of the remaining EEG signals belonging to both the focal and non-focal categories was performed with their respective chosen reference EEG signals. From the magnitude of the cross-wavelet spectrum matrix, a deep feature extraction technique employing an autoencoder was implemented. Along with the sparse autoencoder, deep feature extraction employing a convolutional neural network was also performed for comparing the performance of the proposed framework. After extracting deep features from the autoencoder and CNN based models, segregation of the focal epileptic EEG signals from non-focal ones was carried out using three benchmark machine-learning classifiers. The proposed methodology was observed to deliver satisfactorily high classification accuracy in discriminating the focal and non-focal EEG signals. Therefore, the proposed disease detection framework can be potentially applied for real-time identification of focal epilepsy using EEG signals.

doi:10.1088/978-0-7503-3411-2ch5

## 5.1 Introduction

Electrical signals generated by the neurons inside the human brain are a type of biopotential known as electroencephalography (EEG) signals. EEG recordings contain significant information regarding various brain related activities. The analysis of EEG signals is very common in the context of the diagnosis of various neurodegenerative disorders, the reason for which is that the abnormal discharging patterns of the neurons can be traced using EEG recordings [1]. One of the most commonly occurring neurological disorders, epilepsy, affects people of all ages. Epilepsy can be identified by symptoms such as a rapid decline in mental and physical performance, resulting in recurrent seizure activity. Epileptic seizures can be manifested in the intermittent spikes observed in EEG signals. Hence, expert neurologists and doctors in clinics use EEG recordings extensively for the detection and diagnosis of epilepsy [2]. Based on the affected region of the brain, two distinct categories of epilepsy can be recognized, i.e. generalized epilepsy and partial or focal epilepsy [3, 4]. In the case of generalized epilepsy, the affected region is distributed throughout almost the entire human brain, whereas partial epilepsy stems from certain localized regions identified as focal epileptogenic points. Further, EEG signals corresponding to partial or focal epilepsy can be distinguished into two classes based on the place of occurrence of the seizure onsets. EEG signals acquired from specific regions of the human brain where the first seizure inception takes place are defined as focal EEG (FL) signals. Conversely, non-focal EEG (NFL) signals are procured from the regions of the brain that have no contribution to the seizure onsets. Therefore, the focal epileptogenic regions can be identified accurately by differentiating the FL and NFL EEG signals [1]. In addition, it is noted in the existing literature that a large portion of epileptic patients, i.e. roughly 60% of total cases, become resistant to standard drugs and medication procedures [3, 5]. Hence, locating and isolating the epileptogenic regions becomes extremely important before attempting to remove them surgically. For this purpose proper discrimination of FL and NFL EEG signals is mandatory. Conventional methods of detecting focal epilepsy include human observation of FL and NFL EEG recordings, which is obviously error-prone and also time consuming. Another sophisticated method of diagnosing focal epilepsy incorporates the interpretation of functional magnetic resonance images (fMRI) and diffusion magnetic resonance images (dMRI) [6]. However, as reported in the existing literature, not only are these methods expensive and complex but they also yield a poorer time-resolution compared to the EEG signal analysis techniques [7]. Considering the above-mentioned issues, developing an automated and computerized disease detection scheme for the diagnosis of focal epilepsy still remains an important area of research.

Several studies have been reported in the literature in which various sophisticated signal-processing and machine-learning techniques are reported for identifying focal epileptic EEG signals. Multi-resolution analysis of FL and NFL epileptic EEG signals employing the discrete wavelet transform (DWT) was performed in [5] to extract entropy features for the identification of focal epilepsy. The empirical mode decomposition (EMD) technique was implemented in [8] for classifying FL and

NFL EEG signals. In [9] the EMD technique was employed to decompose the EEG signals into several intrinsic mode functions (IMFs) and the entropy features extracted from the IMFs were used as inputs for the least-squares support vector machine (LS-SVM) classifier to segregate EEG signals into FL and NFL classes. A similar experiment based on the extraction of IMFs using the EMD technique was carried out in [10], where two distinct entropy features, namely the average Renyi entropy and the average negentropy, were computed from the IMFs and fed to a neural network classifier for discriminating FL and NFL EEG signals. Implementation of the EMD technique was also reported in [11] to compute the average bandwidth ratio from the extracted IMFs for classifying EEG signals into focal and non-focal classes. The application of joint DWT-EMD domain based analysis was reported in [12] to extract entropy features and recognition of the focal epileptic EEG signals was performed with the help of SVM and *k*-nearest neighbour (*k*NN) classifiers. The implementation of a bivariate empirical mode decomposition (BEMD) technique was also reported in [13] to separate FL and NFL EEG signals. The application of a clustering variational mode decomposition (CVMD) technique was reported in [14] to identify focal epileptic EEG signals. A hybrid technique employing the EMD method and Teager–Kaiser energy operator was implemented in [2] to segregate EEG signals into the FL and NFL categories using statistical features extracted from the obtained IMFs. The detection of focal epilepsy was reported in [15], using delay permutation entropy features and an SVM classifier. The extraction of different entropy features using a flexible analytic wavelet transform (FAWT) for classifying EEG signals into FL and NFL classes was reported in [16]. Multivariate sub-band entropy analysis utilizing the tunable Q-factor wavelet transform (TQWT) was also reported in [17] for the identification of focal epileptic EEG signals. In [18] EEG signals were separated into different rhythms according to their instantaneous frequencies using a hybrid implementation of the EMD technique and Hilbert transform and, finally, recognition of focal epileptic EEG signals was performed using different kernel functions of the LS-SVM classifier. The empirical wavelet transform (EWT) was applied in [19] for rhythmic separation of the focal epileptic EEG signals and extensive analysis based on phase-plane reconstruction of the EEG rhythms was carried out for classifying the EEG signals into the FL and NFL categories. Joint application of a dual tree complex wavelet transform (DT-CWT) and neuro-fuzzy interference system (ANIFS) was reported in [20] for the recognition of focal epileptic EEG signals. The application of orthogonal localized filter banks to discriminate FL and NFL EEG signals was reported in [7]. The implementation of continuous wavelet transform (CWT) and a combination of texture and higher-order spectra features [21], cumulants [22], entropy features [23], etc, for discrimination of FL and NFL signals were also reported in the existing literature. Various non-linear methods for analysing the focal epileptic EEG signals, including recurrence quantification analysis [24], were reported for classifying FL and NFL EEG signals. The application of fractal mathematics employing a multifractal detrended fluctuation analysis (MFDFA) based non-linear feature extraction framework was proposed in [1] to investigate the chaotic behaviour of the focal epileptic EEG signals in the fractal domain. In several previous studies, features extracted from time–frequency images obtained using various signal-processing techniques, including the short time Fourier transform (STFT) [25], smoothed pseudo-Wigner–Ville distribution (SPWVD) [26] technique, etc, were used to recognize focal epileptic EEG signals. Recently, the application of graph theory employing visibility graph (VG) analysis was reported in [27] for the detection of focal regions within the human brain, where several network theory features extracted from the VG of the EEG signals were used to classify the FL and NFL EEG signals.

Although several techniques have been implemented in the literature to identify focal epileptic EEG signals, one of the major shortcomings of existing studies is that the majority of them are dependent on manual feature extraction procedures, which can be considered unsophisticated in the context of developing a fully automated and computer-aided disease diagnosis system. In addition, the selection of suitable features is also important to detect focal epilepsy more accurately and for ensuring reliability of the disease diagnostic scheme as well. However, the risk of selecting redundant features always imperils the classification performance in the case of extracting features manually, which might lead to erroneous and inaccurate diagnosis. Another important issue is signal adulteration. During the recording of EEG signals, the presence of random and uncorrelated noise may corrupt the recorded signal. One way to minimize the influence of random noise is by using suitable filters, but there is always the risk of losing signal information. To contend with the above-mentioned issues, in this chapter we are proposing a cross-wavelet transform (XWT) based deep learning aided fully automated feature extraction and classification framework for the detection of focal epileptic EEG signals.

The XWT is an efficient joint time–frequency domain based signal-processing tool, which has been reported previously for various feature extraction purposes [29–36]. The distinct advantage of using the cross-wavelet transform is that it is capable of obtaining the degree of correlation between two time signals in the time scale and in the joint time–frequency frame as well. Thus, using the XWT, extraction of more meaningful information is possible from a signal than the existing conventional time–frequency analysis techniques. Moreover, the correlation technique has an added advantage in that it can suppress the influence of random uncorrelated noise present in any two cross-correlated time series. This implies that if two time signals are perturbed with random uncorrelated noise, then the resultant cross-correlogram of these two signals will be free of the influence of that noise as the cross correlation coefficient for random uncorrelated noise is small [28]. This eliminates the need to design an additional filtering framework during recording of EEG signals. The application of the XWT has been reported in [29] for the analysis of non-linear geophysical time series. The application of XWT based feature extraction schemes has been extensively reported in past studies for various condition monitoring tasks in electrical equipment, such as the classification of multiple power quality disturbances [30], detecting impulse faults in transformers [31] and partial discharge recognition purposes [32–34]. In [35] a recent implementation of the XWT for the discrimination of healthy, myopathic and amyotrophic lateral sclerosis EMG signals was reported using deep neural features. Analysis of non-stationary EMG signals using XWT for classifying left and right hand movements has also been reported in [36]. In this work, we investigated the feasibility of implementing the XWT for developing a fully automated disease detection scheme to identify focal epilepsy by discriminating FL and NFL EEG signals. The novelty of the proposed XWT based deep learning aided disease detection scheme is that it does not require the implementation of any additional manual feature extraction procedure. The proposed method not only discards the influence of redundant features but also reduces the computational complexity.

For this purpose, we initially procured EEG signals of the FL and NFL categories from a benchmark data archive available online. After that, we randomly chose a reference EEG signal from both the FL and NFL categories and performed XWT on the remaining EEG signals, with respect to the selected reference signal for either class. As a result, from the XWT operation cross-wavelet spectrum (XWS) images were acquired for the corresponding categories of the EEG signals. For the automated feature extraction purposes, we implemented a sparse autoencoder based deep neural network model in this study. The performance of the feature extraction scheme was also investigated using several other well-known convolutional neural network (CNN) architectures, namely AlexNet and GoogleNet. The acquired XWS images were fed to the above-mentioned deep learning models for feature extraction. For training the CNN architectures, transfer learning was implemented in this work, which enabled us to retain the pre-trained layer information including the respective layer weights and biases. The high dimensional feature vectors extracted from the sparse autoencoder model and from the CNN based deep learning models were finally served as inputs to three highly efficient machine-learning algorithms, including random forest (RF), support vector machines (SVMs) and *k*-nearest neighbour (*k*NN). From the classification results, it was observed that very high recognition accuracies were achieved in discriminating FL and NFL EEG signals, which points towards the potential implementation of the employed technique for developing a computer-aided and automated disease detection system. The proposed methodology is presented using a flowchart in figure 5.1.

The rest of the chapter is arranged as follows. A brief section of the EEG database used in this study is provided in section 5.2. Section 5.3, i.e. the methodology section, deals with the theoretical background of the cross-wavelet transform, sparse autoencoder and convolutional neural network. The penultimate portion of this study, which includes the results and discussion, is section 5.4 and finally section 5.5 concludes the chapter.

## 5.2 Dataset description

In this study, FL and NFL epileptic EEG signals were procured from an open access publicly available benchmark data archive. A brief description of the dataset used in this work is presented below.

The FL and NFL epileptic signals utilized in this work were procured from the publicly available Bern–Barcelona benchmark data archive [4]. The dataset contains EEG signals acquired from five drug-resistant temporal lobe epilepsy affected

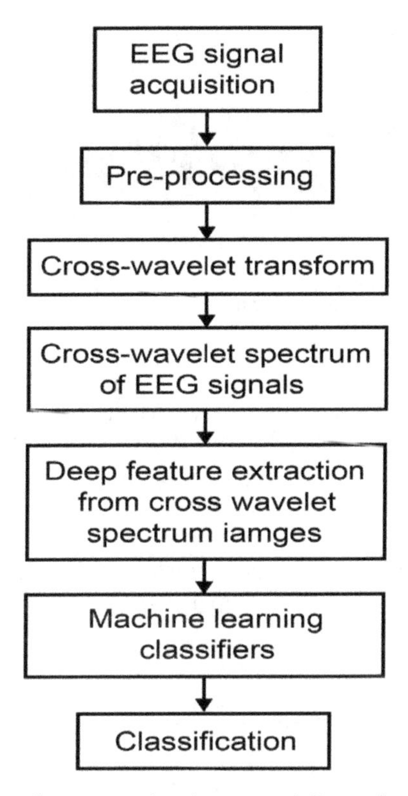


**Figure 5.1.** Brief outline representing the proposed disease detection framework.

patients. Monitored by expert neurologists, recording of the EEG signals was carried out by placing the electrodes intracranially. Between the Fz and Pz locations, an extracranial reference electrode was positioned as described for the 10/20 electrode system. According to [4], the electrodes where the first ictal changes were observed are denoted as FL EEG channels and the rest of the electrodes are indicated as NFL EEG channels. Additionally, in the case of the FL and NFL EEG recordings, a random channel was selected as $x$, and the neighbourhood channel was termed as $y$. In this way, two pairs of EEG signals were obtained for either class of the EEG signals. Each of the EEG signals was further digitized at a sampling frequency of 512 Hz and contains 10 240 of sample points, since the duration of the EEG recording process was 20 s. In this work a total of 7500 EEG signals for both FL and NFL categories (3750 pairs belonging to each of the $x$ and $y$ channels) has been acquired from the database for the purpose of analysis. Sample EEG signals of the FL and NFL categories recorded from channels $x$ and $y$, respectively, are shown below in figure 5.2.

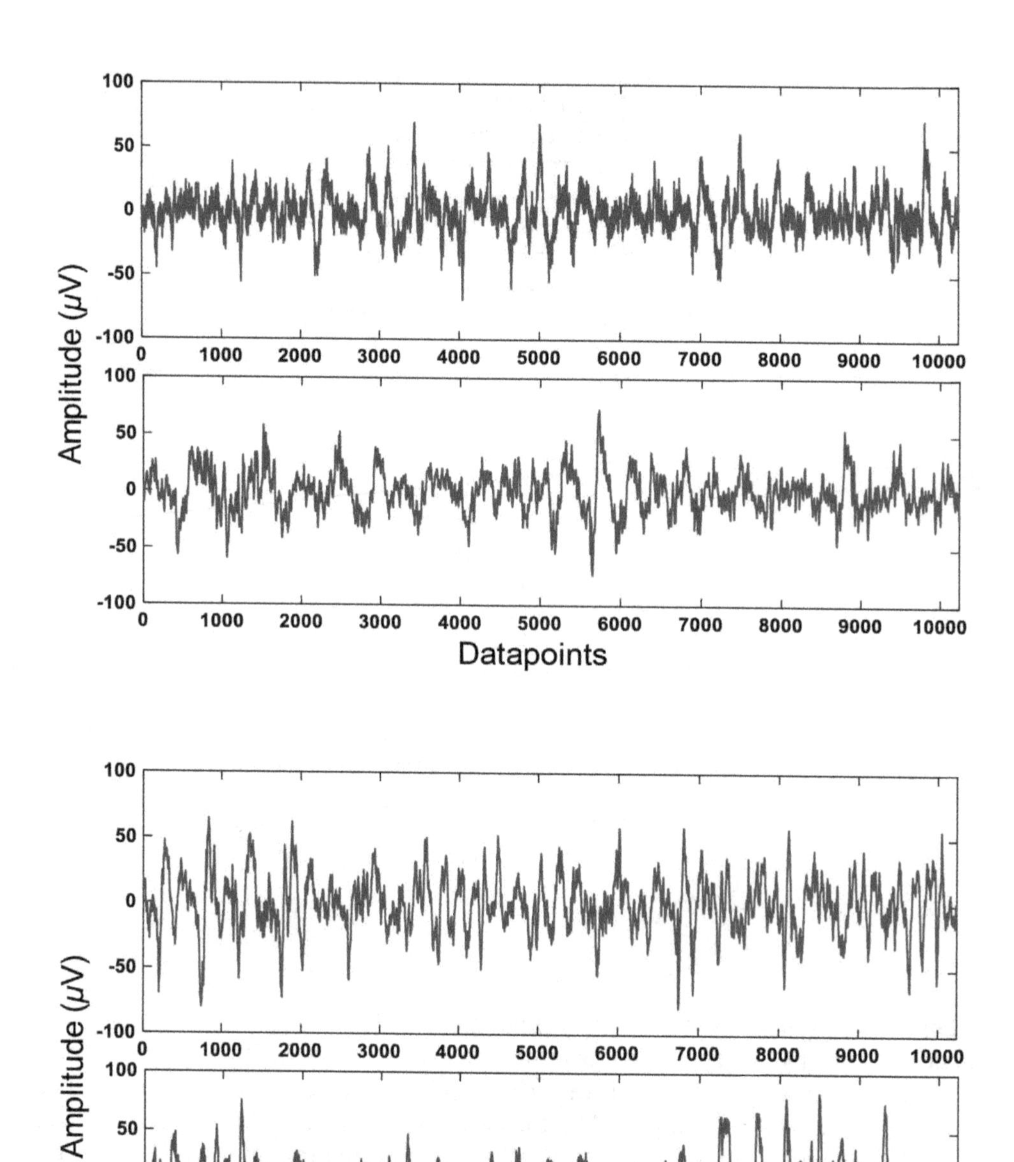


**Figure 5.2.** Time domain FL and NFL epileptic EEG signals. (a) Focal EEG signal. Top: channel $x$. Bottom: channel $y$. (b) Non-focal EEG signal. Top: channel $x$. Bottom: channel $y$.

## 5.3 Methodology

### 5.3.1 Cross-wavelet transform

The cross-wavelet transform (XWT) is a modified form of the conventional continuous wavelet transform (CWT) method. Detailed formulation of the XWT has been reported in the literature [30]. Mathematically, we can express the XWT between two time series $f(t)$ and $g(t)$ and using the following equation:

$$\mathrm{XWT}^{fg}(p, \phi) = \frac{1}{m_\xi} \int_{-\infty}^{+\infty} \int_{-\infty}^{+\infty} W^f(x, y).\ W^{g*}\left(\frac{x}{p}, \frac{y-\phi}{p}\right) \frac{\mathrm{d}x\mathrm{d}y}{x^2}, \tag{5.1}$$

where the CWT of the two time functions $f(t)$ and $g(t)$ are represented by $W^f$ and $W^g$, respectively, and the complex conjugate operator is represented by *. The notations, $p$ and $\phi$ indicate the 'dilation' and the 'translation' parameters, respectively. It should be mentioned here that in XWT based analysis, the selection of the mother wavelet is a crucial choice and it is dependent upon the characteristics of the time series being analyzed. In the initial portion of this study, different mother wavelets such as Meyer, Morlet, Mexican hat, Haar, etc, were used to compute the XWT, but it was observed that the Morlet wavelet provided the best information compared to other mother wavelets. Hence, in this work we have chosen the Morlet mother wavelet to compute the XWT of the EEG signals. In (5.1) $\psi$ indicates the mother wavelet, which can be expressed as

$$m_\psi = \int_{-\infty}^{+\infty} \frac{|\psi(\omega)|^2}{|\omega|} \mathrm{d}\omega < \infty. \tag{5.2}$$

Although the XWT is a modified version of the CWT algorithm, it has a distinct advantage over CWT, as it can measure the degree of (dis)similarity between two random signals in both the time scale and joint time–frequency frame, which the CWT cannot. Thus, the XWT can extract more meaningful information from any non-stationary signal compared to the CWT. The XWT of any two random time signals yields cross-wavelet spectrum (XWS) images as outputs, which indicates the dissimilarity between the signals in the time scale and in the joint time–frequency plane as well. The XWS is a complex matrix, usually obtained by plotting the magnitude $|\mathrm{XWT}^{fg}|$ and phase angle $\theta$ as

$$|\mathrm{XWT}^{fg}| = \sqrt{(\mathrm{Re}\{\mathrm{XWT}^{fg}\})^2 + (\mathrm{Im}\{\mathrm{XWT}^{fg}\})^2} \tag{5.3}$$

$$\theta = \arctan \frac{(\mathrm{Im}\{\mathrm{XWT}^{fg}\})}{(\mathrm{Re}\{\mathrm{XWT}^{fg}\})}, \tag{5.4}$$

where the real and imaginary parts of the complex $\mathrm{XWT}^{fg}$ matrix are indicated by $\mathrm{Re}\{\mathrm{XWT}^{fg}\}$ and $\mathrm{Im}\{\mathrm{XWT}^{fg}\}$, respectively.

The XWS images signify the regions where the two respective time series are highly correlated, i.e. where they exhibit the highest common power [31]. Further, it also points to the region in the time scale and in the joint time–frequency frame where time signals are poorly correlated [37]. In addition, as mentioned earlier, the XWT has the added advantage of suppressing the influence of random and uncorrelated noise present in the corresponding cross-correlated time series. Taking into account the above-mentioned advantages, we have used the XWT in this study to obtain the time–frequency representation of the FL and NFL signals.

### 5.3.2 Deep feature extraction

A deep feature extraction technique has been implemented in this chapter for discriminating the EEG signals of the FL and NFL categories. It is well known that feature extraction is an integral component of any classification task and the selection of redundant features always imposes the risk of incorrect analysis in the case of manual feature extraction procedures, which harms the performance of the classification tasks. In addition to that, the whole process becomes very tedious and consumes a great amount of time. Thus to resolve the aforesaid problem, an automated feature extraction framework using deep learning has been proposed in this work, which obviates the burden of implementing any additional manual feature extraction step, thus making it more convenient in developing an automated disease diagnosis system. Two deep learning models, i.e. a sparse autoencoder and convolutional neural network, have been used for automated feature extraction from the XWT of EEG signals. The detailed theoretical background of these two aforesaid deep neural network models is provided below.

#### *5.3.2.1 The sparse autoencoder*

The sparse autoencoder is formed by combining two fundamental units, namely the encoder and decoder. The encoder encrypts high quality feature data from the given inputs into the hidden representations and the decoder tries to recreate an approximated form of the original input while only preserving the relevant properties of the input data. Figure 5.3 shows the basic structure of an autoencoder model representing its component elements.

The mapping of input data into the hidden layers inside the encoder and decoder units of an autoencoder model is achieved with the help of unique activation functions, namely sigmoid functions. Mathematically, the sigmoid functions used by the encoder and the decoder units can be expressed as follows:

$$S_e = f(x_{in}) = \sigma(W_e x_{in} + b_e) \tag{5.5}$$

$$S_d = \sigma(W_d x_{in} + b_d), \tag{5.6}$$

where $S_e$ and $S_d$ represent the sigmoid activation functions used by the encoder and the decoder units, respectively, to map any input data $x_{in} \in M^C$ into the hidden representations. In (1.5) and (1.6), the corresponding weight matrices of the encoder and decoder are denoted by $W_e$ and $W_d$, respectively, where $W_e$, $W_d \in M^{C \times N}$ and the size of the input data is denoted by $N$. Additionally, the bias vectors of the same are indicated by $b_e$ and $b_d$, respectively. Sparse auto-encoders possess the advantage of constraining sparsity in the hidden layers, which ensures elimination of redundant

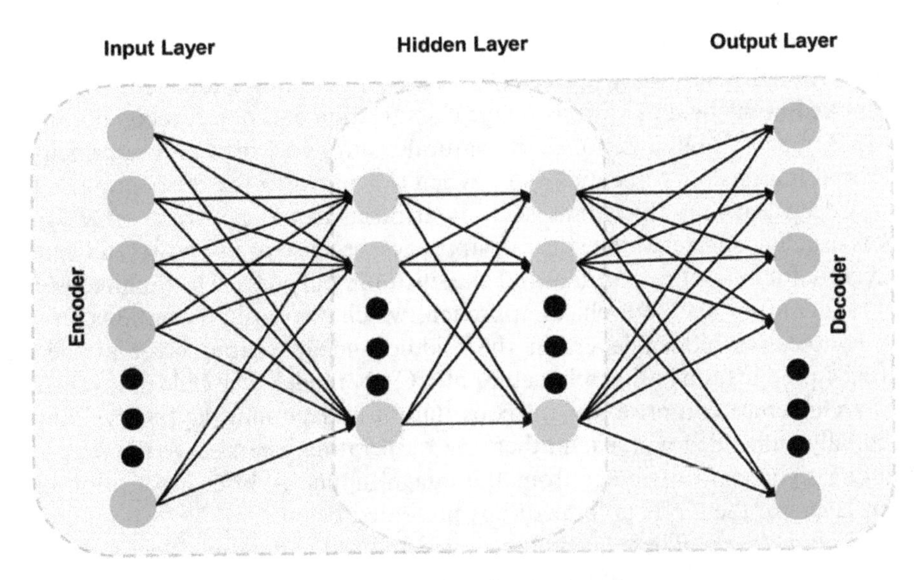


**Figure 5.3.** Structure of an autoencoder.

feature information during the training of the input data. The control parameter termed as a 'sparsity regularizer' serves to maintain the sparsity in the hidden layers to produce the output data [38]. The impact of the sparsity regularization parameter ($\rho_s$) in the hidden layers can be realized using the following expression:

$$\zeta_{\text{sparsity}} = \text{KL}(\rho_s \| \theta_s) = \rho_s \log \frac{\rho_s}{\theta_s} = (1 - \rho_s)\log\left(\frac{1-\rho_s}{1-\theta_s}\right), \tag{5.7}$$

where the average activation value of the sparsity regularizer is expressed as $\theta_s = \frac{1}{m}\sum_1^m y_{\text{out}}$. It is worth mentioning here that the KL function in equation (5.7) corresponds to the Kullback–Leibler divergence model, which we have implemented as the sparsity regularization technique in this work. A detailed description of the sparse autoencoder along with various sparsity and learning rate controlling parameters is given in [39]. The cost function used in this work for the autoencoder for the approximation of input data can be expressed as

$$C(W, b) = \gamma = \frac{1}{M}\sum_{i=1}^{M}\sum_{j=1}^{N}(a_{ij} - \hat{a}_{ij})^2 + \alpha \times \zeta_{\text{weights}} + \beta \times \zeta_{\text{sparsity}}, \tag{5.8}$$

where the mean squared error is written as $\gamma$ and the sparsity regularizer is denoted by $\zeta_{\text{sparsity}}$. In addition, we have adopted the L2 regularization method to restrain the sparse autoencoder model from overfitting. In equation (5.8), $\alpha$ and $\beta$ represent the weights of the L2 regularization coefficient and sparsity regularization coefficient, respectively. The latter part of the sparse autoencoder including the decoder layer has been removed for extracting deep neural features in this study.

*5.3.2.2 Convolutional neural network*

A convolutional neural network (CNN) is a deep learning framework which has been applied in the field of various image classification and object recognition tasks [40–44]. A CNN closely resembles the natural brain and works on the principle of regularized multi-layer perception [45]. When compared to the other popular feed-forward deep learning algorithms such as artificial neural networks (ANNs), the CNN is usually considered the superior approach because of its end-to-end learning architecture for extracting features and classification purposes. The architecture of a CNN model follows a hierarchical approach, which compositely contains an input layer, consecutive hidden layers in the middle and an output layer at the end. Figure 5.4 presents the basic architecture of a CNN model. The hidden layers in the CNN architecture comprise several convolution and pooling layers. As they are sequentially embedded with each other, these operating layers serve the purpose of extracting significant attributes from the image inputs. A brief description of the hidden layers of the CNN architecture is presented below.

*Convolution layers.* These layers are considered as the building blocks of a CNN model as it relies on heavy computations to produce the initial high dimensional feature maps from the image input. Convolution layers consist of a fixed number of independent filters, most commonly known as kernels or edge detectors [39]. These filters are generally two-dimensional arrays of weights which are sequentially processed to perform convolution on the input data. The choice of several convolution layer parameters such as filter size, filter count, stride, padding, etc, are very important in the training process since they heavily affect the performance of the CNN model. The filters are convoluted transversely through the entire input image on a feed-forward basis to construct the high-level feature maps.

*Rectified linear units (ReLU).* ReLU blocks are the commonly used activation functions in most deep learning architectures. These activation units remain integrated in the convolution layers and provide non-linearity to the CNN architecture for a more robust analysis [46].

*Pooling layers.* In the CNN architecture, pooling layers are merged in succession with the convolution layers to reduce the dimension and computation complexity of the network and also to avoid the model overfitting and the loss of useful

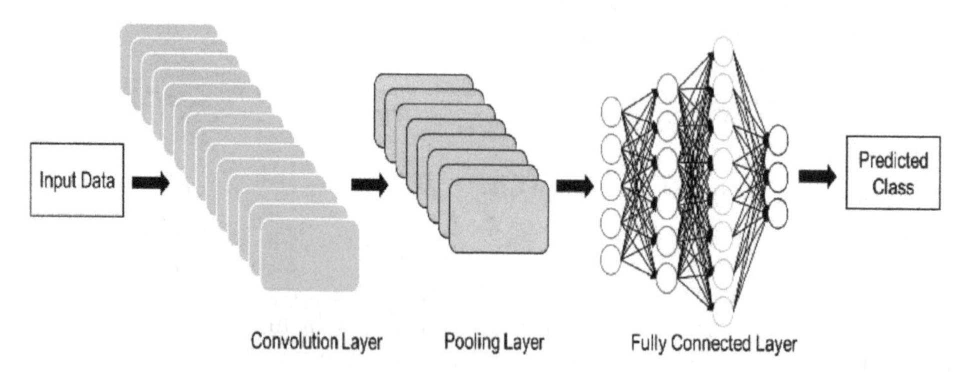


**Figure 5.4.** The basic architecture of a CNN model.

information [47]. The dimension reduction is performed with the help of applying non-linear down sampling on the previously generated high dimensional feature output. Usually, a series of non-intersecting rectangular sub-regions are generated from the regions covered by the feature maps and from these sub-regions. Pooling operations are performed with help of various operators such as maximum pooling, average pooling, global pooling, etc. Hence, the spatial volume of the output is reduced and also the training time of the CNN model is shortened by reducing the network parameters.

*Fully connected layers.* The final structure in the CNN architecture is termed as the fully connected layers. The purpose of this layer is to produce attributes corresponding to the feature output of the previous pooling layers and differentiate between different labels of the input data. The fully connected layer flattens the two-dimensional feature outputs of the preceding convolution and pooling layers into one-dimensional feature vectors. A softmax classification layer is also included in the fully connected layer, which allocates class labels to each category of the input data depending on the scores calculated by the previous operating layers [48].

In training the CNN architecture, errors are calculated using the conventional back propagation algorithm. In the existing literature, different CNN architectures have been reported for feature extraction and classification of input image data. In this chapter, we explore two such well-known CNN architectures, namely AlexNet and GoogleNet for the purpose of automated feature extraction from XWT spectrum images for FL and NFL epileptic EEG signal classification.

#### *5.3.2.3 AlexNet*

AlexNet, proposed in [49], is a popular CNN architecture which was the winner of the ImageNet Large-Scale Visual Recognition Challenge (ILSVRC), 2012. The network was trained on the ILSVRC database, containing over 1.2 million images corresponding to 1000 different class labels. The AlexNet architecture contains roughly 650 000 neurons and 60 million parameters. The initial hidden layers of the AlexNet architecture have a depth of eight layers, containing five convolutional layers and three max-pooling layers, respectively. The first two convolution layers make use of 96 and 256 kernels with sizes of $11 \times 11 \times 3$ and $48 \times 5 \times 5$, respectively. A max-pooling layer, using a filter of size $3 \times 3$, follows each of these two layers. The rest of the convolution layers, i.e. the third, fourth and the fifth layers, are directly connected with each other and use 384, 384 and 256 kernels with the size of the filters being $256 \times 3 \times 3$, $192 \times 3 \times 3$ and $192 \times 3 \times 3$, respectively. Another max-pooling layer, using a filter of size $3 \times 3$, follows the fifth convolution layer and links the succeeding two fully connected layers, namely Fc6 and Fc7, containing 4096 neurons each. The fully connected layers produce a high dimensional feature vector, which is utilized as the input to the terminating classification layer with softmax activation that is capable of differentiating between images of 1000 data labels. The AlexNet architecture employs the dropout regularization method in the classification layer and ReLU activation functions in the convolution layers. For this study, the last fully connected layer, i.e. Fc7, was utilized to extract deep neural features from the AlexNet CNN model. The basic structure of AlexNet is shown in figure 5.5.

*5.3.2.4 GoogleNet*

GoogleNet is another highly efficient CNN architecture and was the winner of ILSVRC 2014 [50]. Similarly to AlexNet, GoogleNet was also trained on the ILSVRC database, containing over 1.2 million images corresponding to 1000 different class labels. The CNN model incorporates a novel 'inception module' that allows simultaneous processing of input data through several convolutional layers, which increases the depth of the network while avoiding computation complexity [51]. The inception modules are comprised of two convolutional layers, four dimension reducing convolution layers and one max-pooling layer. The basic structure of GoogleNet consists of nine such inception modules, two convolution layers, four max-pooling layers, one average-pooling layer and one linear classification layer with softmax activation. GoogleNet uses a global average-pooling layer at the bottom of the architecture instead of a fully connected layer to lower the error rate. It employs the dropout regularization method to avoid overfitting in the classification layer and ReLU activation functions in the convolution layers. GoogleNet also contains 12 times fewer parameters when compared to the above-mentioned AlexNet CNN architecture [52]. For feature extraction purposes, the softmax classification layer was discharged in this study and the corresponding high dimensional feature vector was acquired as an output from the average-pooling layer. The basic structure of GoogleNet is shown in figure 5.6.

### 5.3.3 Transfer learning

The term transfer learning (TL) signifies the flow of knowledge gathered from one domain to another domain. In the context of deep learning, transfer leaning (TL) embodies the application of a finely tuned CNN architecture for feature extraction and classification purposes, which was previously trained on a larger dataset. Recently, TL has gained the attention of many researchers because in practical scenarios, training the CNN architecture through the TL process is much more advantageous than training it from scratch using random weights. The pre-trained

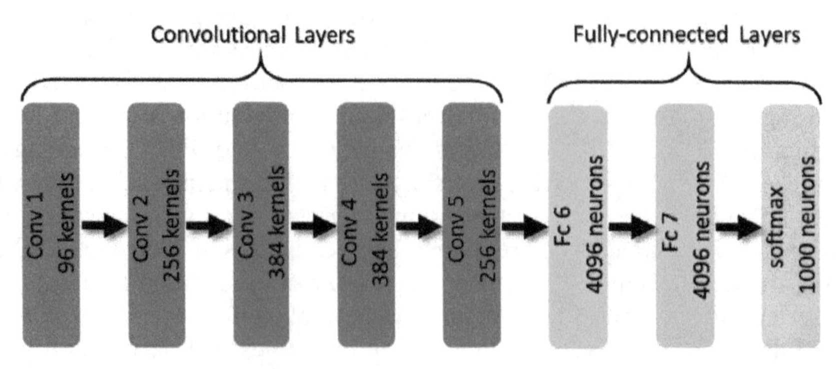


**Figure 5.5.** The structure of AlexNet.

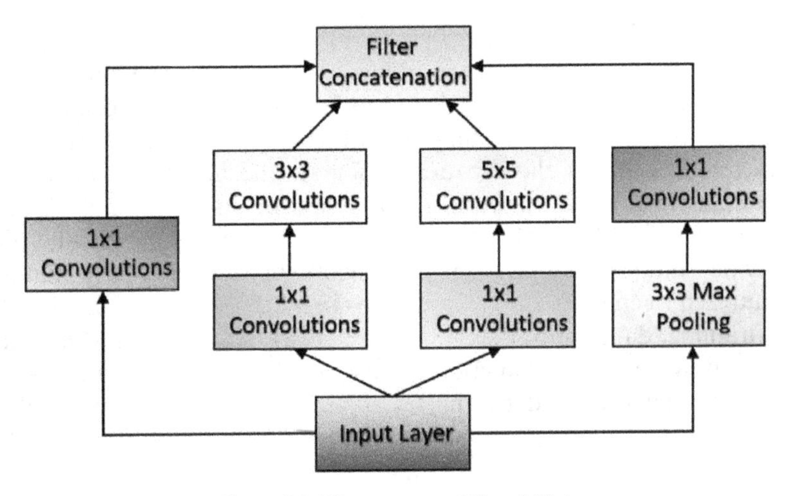


**Figure 5.6.** The structure of GoogleNet.

CNN architecture contains weights and biases which typify the features learned from the previous datasets and these features are in fact transferable to different datasets [53, 54]. Through TL, the last couple of operating layers in the CNN architecture, namely the fully connected layers and the softmax classification layer, can be discharged, and the rest of the CNN model can be trained on different image datasets for detecting and extracting meaningful features as well as to perform various classification tasks. In other words, TL allows fine-tuning of a pre-trained CNN architecture to operate on a fresh classification task. In this chapter, TL is implemented to train two pre-trained CNN architectures, namely AlexNet and GoogleNet, for the purposes of automated feature extraction.

### 5.3.4 Machine-learning classifiers

#### *5.3.4.1 Random forest*

Random forest (RF) is an efficient machine-learning algorithm which is generally formed by combining a bunch of standard decision tree classifiers together. RF makes use of a unique bootstrap-bagging technique for allocating training inputs to each of the tree classifiers. In this process, the number of training samples assigned to the nodes of each tree corresponds to square root times the original feature dimension [55]. Also, random splitting of the training samples amongst the nodes of tree classifiers makes the RF algorithm robust against noise. The output of the RF classifier is dependent on the cumulative decision of all the tree classifiers in the forest, based on a majority voting technique. In [56] a detailed formulation of the RF algorithm has been reported. The selection of the number of trees is an important issue in order to optimize the performance of the RF classifier. Hence, to achieve the best classification performance, we have varied the number of optimal trees in the wide range of 50–150 and observed that the best classification results were obtained using 80 decision trees.

*5.3.4.2 The support vector machine*

We have employed another efficient machine-learning classifier, namely the support vector machine (SVM), in our study. A detailed description of the formulation of the SVM can be found in the existing literatures [57]. The SVM was primarily configured to solve binary classification problems and hence we implemented the SVM based classification framework in our study for discriminating FL and NFL EEG signals. The SVM establishes an optimum separating hyperplane (OSH) to map the input training samples and the theory of structural risk minimization (SRM) is utilized to enhance the class margins between different data labels [58]. To map the training data onto the OSH, the SVM incorporates different fitting techniques, known as kernel functions, satisfying Mercer's theorem. Some of the various fitting functions used by the SVM algorithm include liner, polynomial, radial basis function (RBF), etc. Mathematically, they are formulated as the following, respectively:

$$f(a, b) = a^T b \tag{5.9}$$

$$f(a, b) = (1 + a^T b)^m \tag{5.10}$$

$$f(a, b) = \mathrm{e}^{\frac{-(\|a-b\|^2)}{2\sigma^2}}, \tag{5.11}$$

where $a$ and $b$ represent the random input training samples. In (5.10) and (5.11) $m$ denotes the order of the polynomial kernel function and $\sigma$ signifies the width of the RBF kernel, respectively. It is important to acknowledge that varying these kernel parameters significantly affects the classification performance. Hence, we have selected the polynomial index and varied the RBF kernel width within a range of 0.5–100 with a step size of 0.1 for investigating the performance of the SVM classifier.

*5.3.4.3* k*-nearest neighbours*

In this work, we have implemented another well-known machine-learning classifier, namely $k$-nearest neighbours ($k$NN) for segregating EEG signals between the FL and NFL categories. This particular machine-learning algorithm has found popularity in solving different classification problems because of its relatively simple implementations and robustness towards ambiguous data samples. Depending on the most frequently occurring class labels inside a certain dimensional cluster of training samples, $k$NN makes data annotations based on a majority voting technique [59]. Hence assigning appropriate values to the two operating parameters of $k$NN is important to optimize the performance of the classifier. The distance parameter is the first one, which corresponds to defining the class labels and the latter is the choice of $k$-value, which decides the dimension of the clusters formed with the training samples. In this study, we have selected the Euclidean distance function and varied the value of $k$ within a range of 2–15 with a step size of 1 for evaluating the performance of the $k$NN classifier.

## 5.4 Results and discussion

### 5.4.1 EEG signal analysis using the XWT

For this study, we procured 3750 sets of EEG signals belonging to the FL and NFL categories. For the XWT operation we randomly chose an EEG signal from both the FL and NFL categories and, using equation (5.1), the XWT of the rest of the EEG signals of both categories was performed with respect to their respective selected reference signals. In this way, the XWTs of the FL and NFL epileptic EEG signals were obtained for channels $x$ and $y$. The computation of the XWT results is a complex matrix, whose magnitude and phase are computed using equations (5.3) and (5.4), respectively. The magnitude of the XWT is known as the cross-wavelet spectrum (XWS), which shows the degree of similarity or dissimilarity between two cross-correlated signals in the time frame and in the joint time–frequency space. Figures 5.7 (a)–(b) shows the obtained XWS images for a sample FL and NFL category for channels $x$ and $y$, respectively. The XWS images were obtained by taking the magnitude of the XWT complex matrix obtained for the FL and NFL EEG signals, for both channels $x$ and $y$, respectively. The $Y$-axis of the XWS images corresponds to the scale that represents the inverse of the frequency components and

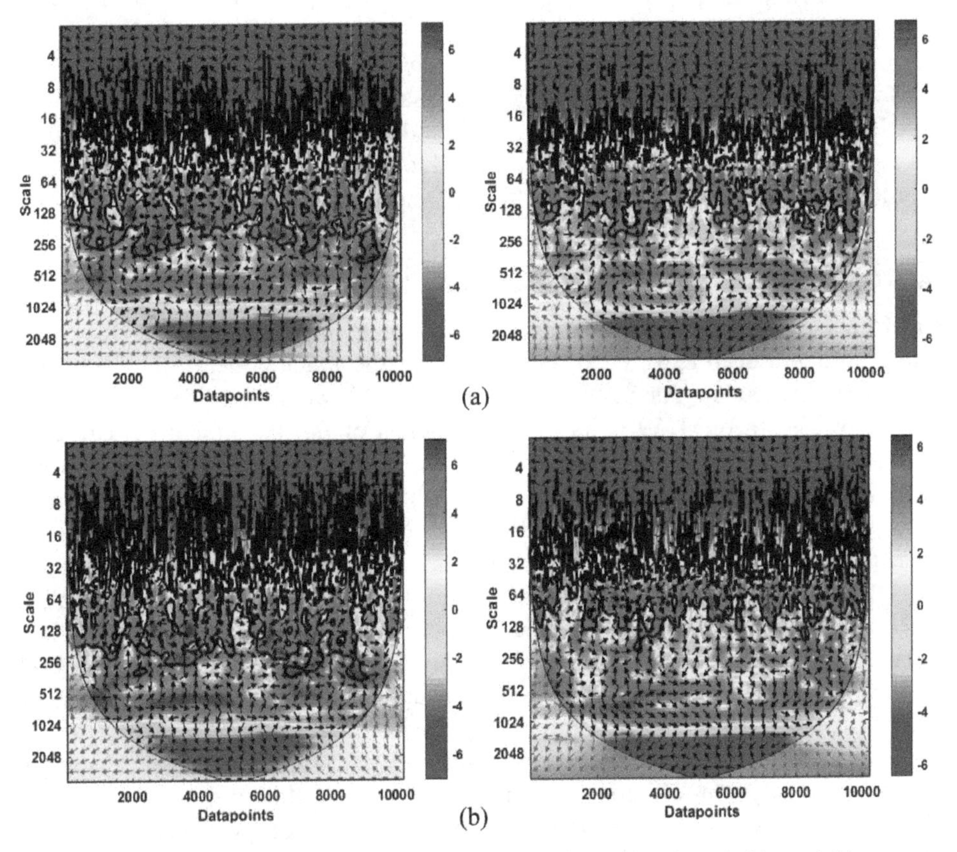


**Figure 5.7.** XWS images for a sample FL and NFL category for channels (a) $x$ and (b) $y$.

the number of data points of the time series is indicated on the $X$-axis. The colorbar at any particular point in the XWS image signifies the strength of the magnitude of the XWT at that point. Since XWT measures correlate in the joint time–frequency scale, higher strength indicates a higher degree of correlation between the two time signals, which corresponds to the point in the joint time–frequency frame where the time signals exhibits maximum common power. In addition, the phase differences between the two cross-correlated time series are represented by black arrows in the XWS images, where the arrows pointing in the rightward direction signify the components which have similar phases and the arrows pointing towards the leftward direction indicate anti-phase, respectively. The black U-shape portion in the CWS images signifies the region of interest, commonly known as the cone of influence (COI) [35]. The COI signifies regions in the XWS spectrum where the edge effects are prominent. Due to the finite length of the two cross-correlated signals, edge effects occur at the beginning or at the end of the waveform. To eradicate this problem, zero padding was performed, which in turn introduces discontinuities. The amplitude of the CWT coefficients decreases with the addition of zeros, in particular at larger scales. These effects become pronounced in the region of the COI of the XWS. This is denoted by the e-folding time of the autocorrelation of the wavelet power at each scale, i.e. beyond this point the discontinuity drops by a factor of $e^{-2}$ ($e = 2.7182$) resulting in minimal edge effects.

From figures 5.7 (a)–(b) we can observe significant discrimination between the XWS images of the FL and NFL categories. In addition, the XWS images of the FL and NFL epileptic EEG signals were also found to be different for either channel $x$ and $y$, respectively. The XWS images were fed to the employed deep learning models for automated feature extraction and for subsequent classification purposes.

### 5.4.2 Performance analysis of the proposed method

Performance evaluation of the proposed methodology implementing a cross-wavelet transform based deep feature extraction scheme and machine-learning classifiers is reported in this section. As mentioned in section 5.4.1, we have performed the cross-wavelet transform on 3750 pairs of EEG signals of the FL and NFL categories corresponding to channels $x$ and $y$, respectively. Since one EEG signal belonging to either class is chosen as the reference, we obtained 3749 pairs of XWS images belonging to either category, i.e. FL and NFL for the respective channels. Subsequently, those images were utilized as inputs to the employed deep learning architectures for automated feature extraction purposes. Finally, the extracted high dimensional deep features were utilized to differentiate the FL and NFL EEG signals employing three machine-learning classifiers, namely RF, SVM and $k$NN. We have computed four performance indices, namely accuracy, sensitivity, specificity and precision, respectively, from the respective confusion matrices to assess the performance of the machine-learning classifiers. For training and testing purposes, we have adopted a ten-fold cross validation technique in this study to avoid the pitfalls of overfitting and have split the input data in proportions of 9:1. The classification performance of the employed machine-learning classifiers is reported in this study with respect to the percentage of their mean and standard deviation values after performing each iteration ten times for each fold.

*5.4.2.1 Performance evaluation using sparse autoencoder based deep features*

In this study the XWS images of FL and NFL epileptic EEG signals had initial dimensions of 656 × 875 × 3. These obtained RGB images were converted to grayscale and the dimensions were further reduced to 224 × 224 before feeding them as inputs to the autoencoder model. The control parameters of the sparse autoencoder model were properly fine-tuned to train the inside hidden layers properly based on the characteristics of the obtained XWS images. For training the autoencoder, the values assigned to different control variables, namely the number of hidden layers, sparsity proportion, sparsity regularization, learning rate and L2 weight regularization parameter were taken as 50, 0.01, 4, 0.005 and 0.005, respectively. The performance of the sparse autoencoder model is shown in figure 5.8 for reconstructing the corresponding feature representations from the image inputs, based on training of the hidden layers. From figure 1.8 it can be seen that the mean squared error between the reconstructed feature output and the input data is less than 0.005, which indicates satisfactory performance of the sparse autoencoder model. Moreover, from figure 5.8 it can be noted that the best training performance has been achieved at the 150th epoch value. Hence the number of training epochs has been chosen as 150 in this work. In this study, 50 features were extracted using the autoencoder model. The extracted features were utilized as inputs of the three machine-learning classifiers for EEG signal discrimination. The performance of the three machine-learning classifiers in classifying FL and NFL epileptic EEG signals is reported in tables 5.1 and 5.2 for channels $x$ and $y$, respectively. Table 5.3 reports the mean classification performance in discriminating FL and NFL epileptic EEG signals that is obtained by computing the arithmetic mean of the performance

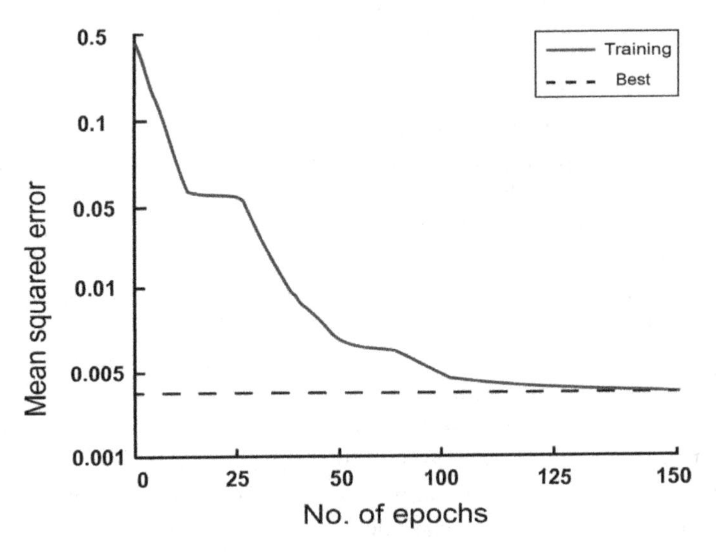


**Figure 5.8.** Performance of the autoencoder.

**Table 5.1.** Classification performance for channel $x$ using sparse autoencoder features.

| Classifier | Accuracy (%) | Sensitivity (%) | Specificity (%) | Precision (%) |
|---|---|---|---|---|
| RF | 92.52 ± 0.75 | 92.08 ± 0.66 | 92.97 ± 0.71 | 93.07 ± 0.63 |
| SVM-linear | 93.46 ± 0.62 | 93.12 ± 0.71 | 93.80 ± 0.58 | 93.87 ± 0.70 |
| SVM-polynomial | 92.92 ± 0.29 | 92.89 ± 0.51 | 93.26 ± 0.34 | 93.33 ± 0.39 |
| SVM-RBF ($\sigma$ = 1.9) | 94.27 ± 0.36 | 95.6 ± 0.49 | 93.01 ± 0.35 | 92.8 ± 0.42 |
| *k*NN | 94.00 ± 0.38 | 93.67 ± 0.46 | 94.59 ± 0.29 | 94.67 ± 0.33 |

**Table 5.2.** Classification performance for channel $y$ using sparse autoencoder features.

| Classifier | Accuracy (%) | Sensitivity (%) | Specificity (%) | Precision (%) |
|---|---|---|---|---|
| RF | 93.06 ± 0.69 | 92.61 ± 0.64 | 93.51 ± 0.70 | 93.6 ± 0.66 |
| SVM-linear | 93.99 ± 0.58 | 93.65 ± 0.67 | 94.34 ± 0.55 | 94.4 ± 0.62 |
| SVM-polynomial | 93.19 ± 0.35 | 92.63 ± 0.47 | 93.77 ± 0.30 | 93.87 ± 0.43 |
| SVM-RBF ($\sigma$ = 1.4) | 94.67 ± 0.30 | 95.15 ± 0.39 | 94.2 ± 0.32 | 94.13 ± 0.38 |
| *k*NN | 94.12 ± 0.28 | 93.88 ± 0.32 | 94.10 ± 0.25 | 94.13 ± 0.25 |

**Table 5.3.** Mean classification performance using sparse autoencoder features.

| Classifier | Accuracy (%) | Sensitivity (%) | Specificity (%) | Precision (%) |
|---|---|---|---|---|
| RF | 92.79 ± 0.72 | 92.35 ± 0.58 | 93.21 ± 0.65 | 93.34 ± 0.59 |
| SVM-linear | 93.73 ± 0.6 | 93.39 ± 0.65 | 94.07 ± 0.53 | 94.14 ± 0.66 |
| SVM-polynomial | 93.05 ± 0.32 | 92.77 ± 0.48 | 93.51 ± 0.28 | 93.6 ± 0.4 |
| SVM-RBF ($\sigma$ = 1.6) | 94.47 ± 0.35 | 94.29 ± 0.43 | 94.63 ± 0.37 | 94.66 ± 0.35 |
| *k*NN | 94.06 ± 0.36 | 93.77 ± 0.39 | 94.34 ± 0.22 | 95.4 ± 0.27 |

indices achieved for both channels in tables 5.1 and 5.2, respectively. From the classification results reported in tables 5.1 and 5.2, we can observe that all three machine-learning classifiers have delivered satisfactorily high degrees of recognition accuracy in segregating the FL and NFL epileptic EEG signals. From tables 5.1–5.2 it is also seen that among the different classifiers used the SVM classifier employing the RBF kernel function has yielded the best classification accuracies for both the $x$ and $y$ channels, followed by the *k*NN classifier. The performance of the RF classifier was found to be slightly inferior to the other two classifiers. Further, for all three classifiers, the performance parameters obtained for channel $y$ were perceived to be marginally improved compared to those of channel $x$. The performance parameters reported in the above tables were computed from the confusion matrix for both the FL and NFL signals and also for channels $x$ and $y$. The confusion matrix for the FL and NFL epileptic EEG signal classification is shown below for an SVM-RBF kernel corresponding to channels $x$ and $y$, respectively (figure 5.9).

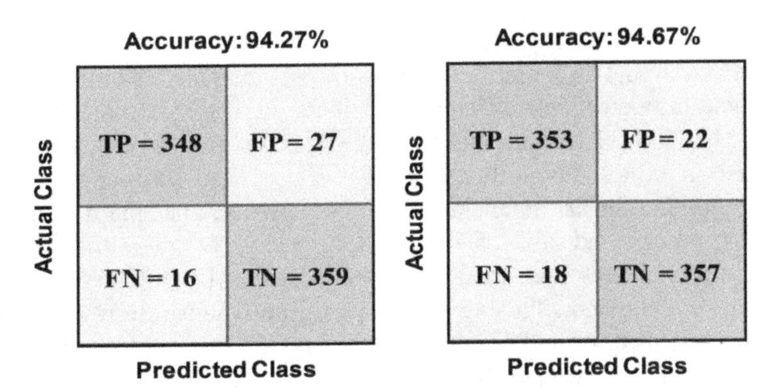


**Figure 5.9.** Confusion matrix of SVM-RBF: (a) channel $x$ and (b) channel $y$.

**Table 5.4.** Classification performance for channel $x$ using AlexNet CNN features.

| Classifier | Accuracy (%) | Sensitivity (%) | Specificity (%) | Precision (%) |
|---|---|---|---|---|
| RF | 97.06 ± 0.58 | 97.07 ± 0.49 | 97.06 ± 0.62 | 97.07 ± 0.65 |
| SVM-linear | 97.33 ± 0.46 | 96.83 ± 0.56 | 97.84 ± 0.45 | 97.87 ± 0.51 |
| SVM-polynomial | 96.53 ± 0.4 | 96.29 ± 0.37 | 96.77 ± 0.35 | 96.8 ± 0.52 |
| SVM-RBF ($\sigma = 2.9$) | 98.13 ± 0.37 | 97.88 ± 0.26 | 98.39 ± 0.35 | 98.4 ± 0.42 |
| $k$NN ($k = 7$) | 97.73 ± 0.46 | 97.11 ± 0.39 | 98.37 ± 0.33 | 98.4 ± 0.35 |

*5.4.2.2 Performance evaluation using CNN based deep features*

The performance of the three machine-learning classifiers evaluated using deep features extracted from the AlexNet and the GoogleNet CNN model is analysed in this section. Using the TL technique, the weights and biases of the initial hidden layers of the aforesaid CNN architectures were retained and rest of the operating layers were discharged for deep feature extraction. In addition, the acquired XWS images were further augmented into lesser dimensions as per the image input requirements of the CNN architectures before the feature extraction procedure. For the AlexNet architecture, the XWS images were resized into dimensions of 227 × 227 × 3 and for the GoogleNet model, resizing was performed to reduce the dimensions to 224 × 224 × 3. For training the above-mentioned CNN models, the learning rate was kept at 0.0001 and a minimum batch size of 20 was considered in this study. The maximum epoch size was kept fixed as 50. We also implemented the L2 regularization technique to prevent the CNN models from overfitting in the training stage. In this work, the output of the AlexNet model yielded a 4096-dimensional feature vector. Similarly, from the output of the GoogleNet model, a 1024-dimensional vector was obtained. Finally, the high dimensional feature vectors were fed to three machine-learning classifiers as inputs for discriminating the EEG signals. The performance of the three machine-learning classifiers assessed using deep features extracted from the AlexNet model are presented in tables 5.4 and 5.5 for channels $x$ and $y$, respectively. Similarly, in tables 5.7 and 5.8 the mean classification performace parameters obtained using deep features extracted from the GoogleNet model are reported for two corresponding channels. The mean performance parameters of different classifiers in distinguishing EEG signals between the FL and NFL classes utilizing the AlexNet and the GoogleNet features are reported in tables 5.6 and 5.9, respectively. These mean parameters were computed by taking the arithmetic average of the performance parameters obtained in tables 5.4 and 5.5, and tables 5.7 and 5.8, respectively.

From the results presented in tables 5.4–5.9, it can be observed that all three machine-learning classifiers have delivered significantly accurate results in discriminating EEG signals into the FL and NFL categories. The SVM classifier using the RBF kernel was found to yield better results than the remaining classifiers for FL and NFL classification for both channels $x$ and $y$ using GoogleNet features, while in

**Table 5.5.** Classification performance for channel $y$ using AlexNet CNN features.

| Classifier | Accuracy (%) | Sensitivity (%) | Specificity (%) | Precision (%) |
|---|---|---|---|---|
| RF | 97.2 ± 0.46 | 96.58 ± 0.35 | 97.83 ± 0.54 | 97.87 ± 0.39 |
| SVM-Linear | 97.46 ± 0.53 | 96.84 ± 0.45 | 98.1 ± 0.47 | 98.13 ± 0.37 |
| SVM-polynomial | 95.99 ± 0.68 | 95.51 ± 0.75 | 96.49 ± 0.65 | 96.53 ± 0.53 |
| SVM-RBF ($\sigma = 2.2$) | 98.26 ± 0.21 | 97.63 ± 0.32 | 98.92 ± 0.15 | 98.93 ± 0.25 |
| $k$NN ($k = 7$) | 98.67 ± 0.36 | 99.19 ± 0.24 | 98.15 ± 0.3 | 98.13 ± 0.29 |

**Table 5.6.** Mean classification performance using AlexNet CNN features.

| Classifier | Accuracy (%) | Sensitivity (%) | Specificity (%) | Precision (%) |
|---|---|---|---|---|
| RF | 97.13 ± 0.52 | 96.82 ± 0.45 | 97.44 ± 0.58 | 97.47 ± 0.53 |
| SVM-linear | 97.38 ± 0.46 | 96.71 ± 0.49 | 97.97 ± 0.42 | 98.2 ± 0.32 |
| SVM-polynomial | 96.26 ± 0.54 | 95.9 ± 0.49 | 96.63 ± 0.42 | 96.67 ± 0.61 |
| SVM-RBF ($\sigma = 2.5$) | 98.19 ± 0.25 | 97.59 ± 0.22 | 98.73 ± 0.29 | 98.8 ± 0.32 |
| $k$NN ($k = 7$) | 98.19 ± 0.37 | 97.63 ± 0.3 | 98.78 ± 0.26 | 98.8 ± 0.34 |

**Table 5.7.** Classification performance for channel $x$ using GoogleNet CNN features.

| Classifier | Accuracy (%) | Sensitivity (%) | Specificity (%) | Precision (%) |
|---|---|---|---|---|
| RF | 98 ± 0.45 | 97.37 ± 0.57 | 98.64 ± 0.36 | 98.67 ± 0.35 |
| SVM-linear | 97.46 ± 0.65 | 96.84 ± 0.59 | 98.1 ± 0.71 | 98.13 ± 0.55 |
| SVM-polynomial | 97.2 ± 0.76 | 96.34 ± 1.16 | 98.09 ± 0.64 | 98.13 ± 0.84 |
| SVM-RBF ($\sigma = 3.2$) | 99.07 ± 0.24 | 99.46 ± 0.19 | 98.68 ± 0.15 | 98.67 ± 0.23 |
| $k$NN ($k = 3$) | 98.4 ± 0.34 | 97.89 ± 0.42 | 98.92 ± 0.45 | 98.93 ± 0.29 |

**Table 5.8.** Classification performance for channel $y$ using GoogleNet CNN features.

| Classifier | Accuracy (%) | Sensitivity (%) | Specificity (%) | Precision (%) |
|---|---|---|---|---|
| RF | 98.53 ± 0.42 | 97.89 ± 0.44 | 99.19 ± 0.27 | 99.2 ± 0.16 |
| SVM-linear | 98.4 ± 0.52 | 97.64 ± 0.34 | 99.18 ± 0.25 | 99.2 ± 0.37 |
| SVM-polynomial | 97.86 ± 0.57 | 97.11 ± 0.65 | 98.64 ± 0.42 | 98.67 ± 0.48 |
| SVM-RBF ($\sigma = 2.4$) | 99.33 ± 0.25 | 99.73 ± 0.16 | 98.94 ± 0.19 | 98.93 ± 0.26 |
| $k$NN ($k = 5$) | 99.2 ± 0.14 | 98.68 ± 0.46 | 99.73 ± 0.16 | 99.65 ± 0.27 |

**Table 5.9.** Mean classification performance using GoogleNet CNN features.

| Classifier | Accuracy (%) | Sensitivity (%) | Specificity (%) | Precision (%) |
|---|---|---|---|---|
| RF | 98.26 ± 0.36 | 97.65 ± 0.51 | 98.92 ± 0.32 | 98.94 ± 0.3 |
| SVM-linear | 97.93 ± 0.58 | 97.24 ± 0.45 | 98.64 ± 0.54 | 98.67 ± 0.46 |
| SVM-polynomial | 97.58 ± 0.64 | 96.73 ± 0.74 | 98.37 ± 0.52 | 98.35 ± 0.66 |
| SVM-RBF ($\sigma = 2.8$) | 99.15 ± 0.12 | 98.94 ± 0.22 | 99.52 ± 0.35 | 99.4 ± 0.19 |
| $k$NN ($k = 5$) | 98.8 ± 0.23 | 98.29 ± 0.33 | 99.33 ± 0.25 | 99.3 ± 0.16 |

the case of AlexNet, SVM-RBF delivered the best performance for channel $x$ and $k$NN delivered better performance for channel $y$. Nevertheless, all three classifiers were found to deliver satisfactory performance using the deep features extracted from the AlexNet and the GoogleNet CNN models. Between AlexNet and GoogleNet, the features extracted from the GoogleNet model were found to perform marginally better at delivering higher mean classification accuracy, sensitivity, specificity and precision values for all three classifiers. In addition, the standard deviation values (indicated in brackets) obtained for each classifier were low, indicating the robustness of the proposed technique. Another interesting observation is that the performance of all the three machine-learning classifiers was found to be better using the features extracted from the CNN models compared to the autoencoder model. The confusion matrices obtained for the FL and NFL EEG signal classification is shown in figures 5.10 and 5.11 for both channels $x$ and $y$, respectively.

### 5.4.3 Comparison to the existing literature

In this section, the performance of the employed methodology was measured against some of the existing techniques proposed for FL and NFL epileptic EEG signal classification and studied using the same database. Moreover, this comparative study has been performed considering only those studies in the literature which have used 3750 pairs of EEG signals and used a ten-fold cross validation method. A comparative performance assessment of the proposed scheme with the existing literatures is shown in table 5.10, with the percentage values of accuracy, sensitivity and specificity.

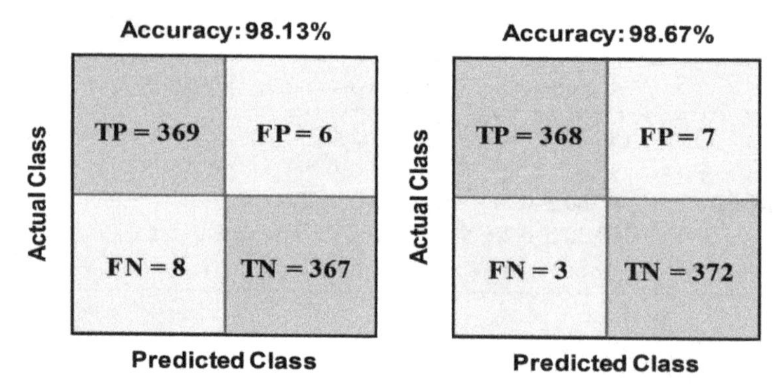


**Figure 5.10.** Confusion matrices: (a) SVM-RBF channel $x$ and (b) $k$NN channel $y$.

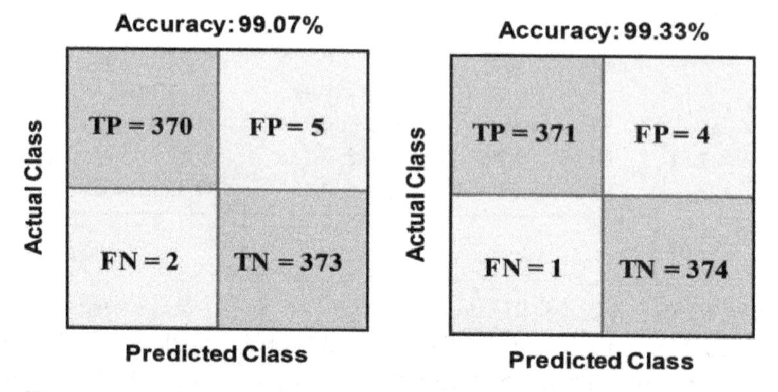


**Figure 5.11.** Confusion matrices of SVM-RBF: (a) channel $x$ and (b) channel $y$.

From table 5.10 we can observe that the proposed XWT aided deep feature extraction technique has delivered better performance in detecting the focal epileptic EEG signals, in comparison to existing studies. Another significant observation made from the comparative study was that most existing studies have used manually selected features for FL and NFL epileptic signal classification. Unlike the existing methods, the proposed method has the advantage of automated feature extraction. Thus, the problem of manual feature extraction that may lead to inaccurate classification can be eliminated using the proposed approach. Therefore, it can be said that the proposed cross spectrum aided deep feature extraction method can be successfully implemented for the detection and diagnosis of focal seizures, and is capable of achieving a high degree of recognition accuracy.

## 5.5 Conclusion

In this chapter, a novel FL and NFL epileptic EEG signal recognition scheme implementing a cross-wavelet transform and deep neural feature extraction is proposed. The EEG signals of FL and NFL epileptic patients were acquired from an online data archive. A sample EEG signal from both categories (FL and NFL) was selected as a reference and the XWT of the remaining EEG signals was

**Table 5.10.** Comparison to some previous studies.

| References | Methods | Accuracy (%) | Sensitivity (%) | Specificity (%) |
|---|---|---|---|---|
| [60] | Entropy features + EMD + LS-SVM | 83.18 | 85.78 | 80.45 |
| [61] | Bivariate EMD + LS-SVM | 84.01 | 83.47 | 84.56 |
| [17] | Fuzzy entropy + TQWT + LS-SVM | 84.67 | 83.86 | 85.46 |
| [62] | Non-linear features + LS-SVM | 87.93 | 89.97 | 85.89 |
| [2] | Statistical moments + EMD + SVM | 92.65 | 90.70 | 93.15 |
| [7] | Entropy features + time–frequency filter bank + LS-SVM | 94.25 | 91.95 | 96.56 |
| [16] | Entropy features + FAWT + LS-SVM | 94.41 | 93.25 | 95.57 |
| [63] | Statistical features + EMD + RF | 95.94 | 95.80 | 96.10 |
| [64] | Non-linear features + TQWT + LS-SVM | 94.92 | 96.37 | 93.47 |
| [65] | Neighbourhood component analysis + SVM | 96.10 | 92.60 | 94.41 |
| [66] | Entropy features + multiple classifiers | 98 | 100 | 96 |
| [67] | Sliding mode singular spectrum analysis + SAE-RBFN | 99.11 | 98.52 | 99.70 |
| This work | XWT + GoogleNet + SVM | 99.15 | 98.94 | 99.52 |

performed with their respective chosen reference signals. The process was repeated for both channels $x$ and $y$. The XWT operation yielded a complex cross-wavelet matrix for both categories of EEG signals. The magnitude of the complex cross-wavelet matrix was computed to obtain the cross-wavelet spectra (XWS) of the FL and NFL EEG signals, which were subsequently utilized as inputs of the deep neural networks for the purpose of automated feature extraction. Two well-known deep learning models, namely a stacked sparse autoencoder and convolutional neural network, were used for the purpose of feature extraction from the input XWS images. In the case of the CNN, two well-known networks, namely AlexNet and GoogleNet, were utilized and the training of these two deep neural network models was carried out using a transfer learning (TL) procedure. The extracted deep features from the autoencoder as well as from the CNN models were used to segregate the FL and NFL EEG signals, employing the RF, SVM and *k*NN classifiers. The main outcomes of the investigation are summarized briefly as follows.

We observed that the proposed deep feature extraction framework delivered reasonably accurate results in segregating EEG signals into the FL and NFL categories. From comparing the performance of the autoencoder model and the CNN architectures, the employed machine-learning classifiers were found to perform better using the deep CNN features, delivering better recognition accuracies. Further, between AlexNet and GoogleNet, the performance of the machine-learning classifiers were found to be superior for the GoogleNet features. Comparative analysis to the existing literature revealed that the proposed method delivered analogous performance with respect to the existing disease detection methods and was found to perform even better in some cases. Thus, it is plausible to infer that the proposed cross-wavelet transform aided deep feature extraction framework to discriminate FL and NFL EEG signals can be potentially applied for the detection and diagnosis of focal seizures.

However, the present study has certain limitations. We have validated our proposed scheme by procuring two-channel EEG data from five subjects. Prior to clinical application, further experiments are required to evaluate the disease detection scheme on EEG data procured from a large number of patients and also using multiple channels. This could be performed as part of a future work. Also, the proposed XWT aided deep feature extraction technique will be implemented in the future to diagnose several other neurological disorders such as schizophrenia, autism, Alzheimer's disease, etc.

## References


[1] Chatterjee S, Pratiher S and Bose R 2017 Multifractal detrended fluctuation analysis based novel feature extraction technique for automated detection of focal and non-focal electroencephalogram signals *IET Sci. Meas. Technol.* **11** 1014–21

[2] Chatterjee S 2019 Detection of focal electroencephalogram signals using higher order moments in EMD-TKEO domain *Healthc. Technol. Lett.* **6** 64–9

[3] Pati S and Alexopoulos A V 2010 Pharmacoresistant epilepsy: from pathogenesis to current and emerging therapies *Cleve. Clin. J. Med.* **77** 457–567

[4] Andrzejak R G, Schindler K and Rummel C 2012 Nonrandomness, nonlinear dependence, and nonstationarity of electroencephalographic recordings from epilepsy patients *Phys. Rev.* E **86** 046206

[5] Sharma R, Pachori R B and Acharya U R 2015 An integrated index for the identification of focal electroencephalogram signals using discrete wavelet transform and entropy measures *Entropy* **17** 5218–40

[6] Seeck M *et al* 1998 Non-invasive epileptic focus localization using EEG-triggered functional MRI and electromagnetic tomography *Electroencephalogr. Clin. Neurophysiol.* **106** 508–12

[7] Sharma M, Dhere A, Pachori R B and Acharya U R 2017 An automatic detection of focal EEG signals using new class of time–frequency localized orthogonal wavelet filter banks *Knowl.-Based Syst.* **118** 217–27

[8] Sharma R, Pachori R B and Gautam S 2014 Empirical mode decomposition based classification of focal and non-focal seizure EEG signals *Proc. Int. Conf. on Medical Biometrics (China, May–June 2014)* pp 135–40

[9] Sharma R, Pachori R B and Acharya U R 2015 Application of entropy measures on intrinsic mode functions for the automated identification of focal electroencephalogram signals *Entropy* **17** 669–91

[10] Rai K, Bajaj V and Kumar A 2015 Features extraction for classification of focal and non-focal EEG signals *Inform. Sci. Appl. Lect. Notes Electr. Eng.* **339** 599–605

[11] Rai K, Bajaj V and Kumar 2015 A novel feature for identification of focal EEG signals with *K*-means and fuzzy *C*-means algorithms *Proc. 20th IEEE Int. Conf. on Digital Signal Processing (DSP) (Singapore, July 2015)* pp 412–6

[12] Das A B and Bhuiyan M I H 2016 Discrimination and classification of focal and non-focal EEG signals using entropy-based features in the EMD-DWT domain *Biomed. Signal Process. Control* **29** 11–21

[13] Sharma R and Pachori R B 2018 Automated classification of focal and non-focal EEG signals based on bivariate empirical mode decomposition *Biomedical Signal and Image Processing in Patient Care* ed M H Kolekar and V Kumar (Hershey, PA: IGI Global) pp 13–33

[14] Taran S and Bajaj V 2018 Clustering variational mode decomposition for identification of focal EEG signals *IEEE Sens. Lett.* **2** 7001304

[15] Zhu G, Li Y, Wen P P, Wang S and Xi M 2013 Epileptogenic focus detection in intracranial EEG based on delay permutation entropy *Proc. AIP Conf. Proc.* 1559 pp 31–6

[16] Gupta V, Priya T, Yadav A K, Pachori R B and Acharya U R 2017 Automated detection of focal EEG signals using features extracted from flexible analytic wavelet transform *Pattern Recognit. Lett.* **94** 180–8

[17] Bhattacharyya A, Pachori R B and Acharya U R 2017 Tunable-Q wavelet transform based multivariate sub-band fuzzy entropy with application to focal EEG signal analysis *Entropy* **19** 99

[18] Bajaj V, Rai K, Kumar A, Sharma D and Singh G K 2017 Rhythm-based features for classification of focal and non-focal EEG signals *IET Signal Process* **11** 743–8

[19] Bhattacharyya A, Sharma M, Pachori R B, Sircar P and Acharya U R 2018 A novel approach for automated detection of focal EEG signals using empirical wavelet transform *Neural Comput. Appl.* **29** 47–57

[20] Deivasigamani S, Senthilpari C and Yong W H 2016 Classification of focal and nonfocal EEG signals using ANFIS classifier for epilepsy detection *Int. J. Imag. Syst. Technol.* **26** 277–83

[21] Acharya U R, Yanti R, Zheng J W, Krishnan M M R, Tan J H, Martis R J and Lim C M 2013 Automated diagnosis of epilepsy using CWT, HOS and texture parameters *Int. J. Neural Syst.* **23** 1350009

[22] Acharya U R, Sree S V and Suri J S 2011 Automatic detection of epileptic EEG signals using higher order cumulant features *Int. J. Neural Syst.* **21** 403–14

[23] Acharya U R, Molinari F, Sree S V, Chattopadhyay S, Ng K H and Suri J S 2012 Automated diagnosis of epileptic EEG using entropies *Biomed. Signal Process. Control* **7** 401–8

[24] Acharya U R, Sree S V, Chattopadhyay S, Yu W and Ang P C A 2011 Application of recurrence quantification analysis for the automated identification of epileptic EEG signals *Int. J. Neural Syst.* **21** 199–211

[25] Bajaj V, Taran S, Tanyildizi E and Sengur A 2019 Robust approach based on convolutional neural networks for identification of focal EEG signals *IEEE Sens. Lett.* **3** 1–4

[26] Walde S, Rani P, Bajaj V and Sharma D 2016 Time frequency image based features for detection of focal EEG signals *Proc. Int. Conf. on Signal Processing and Communication (ICSC) (Noida, Uttar Pradesh, India, December 2016)* pp 358–62

[27] Modak S, Roy S S, Samanta K, Chatterjee S, Dey S, Bhowmik R and Bose R 2020 Detection of focal EEG signals employing weighted visibility graph *Proc. 4th IEEE Int. Conf. on Int. Conf. on Computer, Electrical and Communication Engineering (ICCECE) (Kolkata, West Bengal, India, January 2020)*

[28] Chatterjee S, Dalai S, Chakravorti S and Chatterjee B 2018 Use of chirp excitations for frequency domain spectroscopy measurement of oil-paper insulation *IEEE Trans. Dielect. Elect. Insul.* **25** 1103–11

[29] Grinsted A, Moore J C and Jevrejeva S 2004 Application of the cross wavelet transform and wavelet coherence to geophysical time series *Nonlinear Process. Geophys.* **11** 561–6

[30] Dalai S, Dey D, Chatterjee B, Chakravorti S and Bhattacharya K 2014 Cross-spectrum analysis-based scheme for multiple power quality disturbance sensing device *IEEE Sensors J.* **15** 3989–97

[31] Dey D, Chatterjee B, Chakravorti S and Munshi S 2008 Rough-granular approach for impulse fault classification of transformers using cross-wavelet transform *IEEE Trans. Dielect. Elect. Insul.* **15** 1297–304

[32] Dey D, Chatterjee B, Chakravorti S and Munshi S 2010 Cross-wavelet transform as a new paradigm for feature extraction from noisy partial discharge pulses *IEEE Trans. Dielect. Elect. Insul.* **17** 157–66

[33] Muñoz-Muñoz F and Rodrigo-Mor A 2020 Partial discharges and noise discrimination using magnetic antennas, the cross wavelet transform and support vector machines *Sensors* **20** 3180

[34] Biswas S, Dey D, Chatterjee B and Chakravorti S 2016 Cross-spectrum analysis based methodology for discrimination and localization of partial discharge sources using acoustic sensors *IEEE Trans. Dielect. Elect. Insul.* **23** 3556–65

[35] Roy S S, Samanta K, Modak S, Chatterjee S and Bose R 2020 Cross spectrum aided deep feature extraction based neuromuscular disease detection framework *IEEE Sens. Lett.* **4**

[36] Roy S S, Samanta K, Chatterjee S, Dey S, Nandi A, Bhowmik R and Mondal S 2020 Hand movement recognition using cross spectrum image analysis of EMG signals—a deep learning approach *Proc. National Conf. on Emerging Trends on Sustainable Technology and Engineering Applications (NCETSTEA) (Durgapur, West Bengal, India, February 2020)* pp 1–5

[37] Das S, Purkait P, Dey D and Chakravorti S 2011 Monitoring of inter-turn insulation failure in induction motor using advanced signal and data processing tools *IEEE Trans. Dielect. Elect. Insul.* **18** 1599–608

[38] Samanta K, Chatterjee S and Bose R 2019 Cross-subject motor imagery tasks EEG signal classification employing multiplex weighted visibility graph and deep feature extraction *IEEE Sens. Lett.* **4** 1–4

[39] Zhang C, Cheng X, Liu J, He J and Liu G 2018 Deep sparse autoencoder for feature extraction and diagnosis of locomotive adhesion status *J. Control Sci. Eng.* **2018** 1–9

[40] Lawrence S, Giles C L, Tsoi A C and Back A D 1997 Face recognition: a convolutional neural-network approach *IEEE Trans. Neural Netw.* **8** 98–113

[41] Pu Y, Apel D B, Szmigiel A and Chen J 2019 Image recognition of coal and coal gangue using a convolutional neural network and transfer learning *Energies* **12** 1735

[42] LeCun Y, Bottou L, Bengio Y and Haffner P 1998 Gradient-based learning applied to document recognition *Proc. IEEE* **86** 2278–324

[43] Nguyen D T, Nguyen T N, Kim H and Lee H J 2019 A high-throughput and power-efficient FPGA implementation of YOLO CNN for object detection *IEEE Trans. Very Large Scale Integr. (VLSI) Syst.* **27** 1861–18732

[44] Shi B, Bai X and Yao C 2016 An end-to-end trainable neural network for image-based sequence recognition and its application to scene text recognition *IEEE Trans. Pattern Anal. Mach. Intell.* **39** 2298–304

[45] Chaudhary S, Taran S, Bajaj V and Sengur A 2019 Convolutional neural network based approach towards motor imagery tasks EEG signals classification *IEEE Sensors J.* **19** 4494–500

[46] Nair V and Hinton G E 2010 Rectified linear units improve restricted Boltzmann machines *Proc. 27th Int. Conf. on Machine Learning (ICML) (Haifa, Israel, January 2010)* pp 807–14
[47] Sengur A, Akbulut Y, Guo Y and Bajaj V 2017 Classification of amyotrophic lateral sclerosis disease based on convolutional neural network and reinforcement sample learning algorithm *Health Inf. Sci. Syst.* **5** 1–7
[48] Sengur A, Gedikpinar M, Akbulut Y, Deniz E, Bajaj V and Guo Y 2017 DeepEMGNet: an application for efficient discrimination of ALS and normal EMG signals *Proc. Int. Conf. Mechatronics (Australia, February 2017)* (Cham: Springer) pp 619–25
[49] Krizhevsky A, Sutskever I and Hinton G E 2012 ImageNet classification with deep convolutional neural networks *Proc. Adv. in Neural Information Processing Systems (Lake Tahoe, CA, USA, December 2012)* pp 1097–105
[50] Szegedy C, Liu W, Jia Y, Sermanet P, Reed S, Anguelov D, Erhan D, Vanhoucke V and Rabinovich A 2015 Going deeper with convolutions *Proc. IEEE Computer Society Conf. on Computer Vision and Pattern Recognition (Boston, MA, USA, June 2015)* pp 1–9
[51] Ghazi M M, Yanikoglu B and Aptoula E 2017 Plant identification using deep neural networks via optimization of transfer learning parameters *Neurocomputing* **235** 228–35
[52] Singla A, Yuan L and Ebrahimi T 2016 Food/non-food image classification and food categorization using pre-trained GoogleNet model *Proc. 2nd Int. Workshop on Multimedia Assisted Dietary Management (Amsterdam, Netherlands, October 2016)* pp 3–11
[53] Akçay S, Kundegorski M E, Devereux M and Breckon T P 2016 Transfer learning using convolutional neural networks for object classification within x-ray baggage security imagery *Proc. IEEE Int. Conf. on Image Processing (ICIP) (Arizona, USA, September 2018)* pp 1057–61
[54] Shin H C, Roth H R, Gao M, Lu L, Xu Z, Nogues I, Yao J, Mollura D and Summers R M 2016 Deep convolutional neural networks for computer-aided detection: CNN architectures, dataset characteristics and transfer learning *IEEE Trans. Med. Imag.* **35** 1285–98
[55] Roy S S, Dey S and Chatterjee S 2020 Autocorrelation aided random forest classifier based bearing fault detection framework *IEEE Sensors J.* **20** 10792–800
[56] Zhang T, Chen W and Li M 2017 AR based quadratic feature extraction in the VMD domain for the automated seizure detection of EEG using random forest classifier *Biomed. Signal Process. Control* **31** 550–9
[57] Vapnik V N 1995 *The Nature of Statistical Learning Theory* (New York: Springer)
[58] Roy S S, Jana R, Barman R, Modak S, Halder P, Bose R and Chatterjee S 2018 Detection of healthy and neuropathy electromyograms employing Stockwell transform *Proc. 1st IEEE Int. Conf. on Applied Signal Processing (ASPCON) (Kolkata, West Bengal, India, December 2018)*
[59] Chatterjee S, Roy S S, Bose R and Pratiher S 2020 Feature extraction from multifractal spectrum of electromyograms for diagnosis of neuromuscular disorders *IET Sci. Meas. Technol.* **14** 817–24
[60] Gupta V and Pachori R B 2019 A new method for classification of focal and non-focal EEG signals *Machine Intelligence and Signal Analysis* ed M Tanveer and R B Pachori (Advances in Intelligent Systems and Computing vol 748) (Singapore: Springer) 235–46 pp
[61] Sharma R R, Varshney P, Pachori R B and Vishvakarma S K 2018 Automated system for epileptic EEG detection using iterative filtering *IEEE Sens. Lett.* **2** 1–4
[62] Acharya U R, Hagiwara Y, Deshpande S N, Suren S, Koh J E W, Oh S L, Arunkumar N, Ciaccio E J and Lim C M 2019 Characterization of focal EEG signals: a review *Future Gener. Comput. Syst.* **91** 290–9

[63] Subasi A, Jukic S and Kevric J 2019 Comparison of EMD, DWT and WPD for the localization of epileptogenic foci using random forest classifier *Measurement* **146** 846–55
[64] Sharma R, Kumar M, Pachori R B and Acharya U R 2017 Decision support system for focal EEG signals using tunable-Q wavelet transform *J. Comput. Sci.* **20** 52–60
[65] Raghu S and Sriraam N 2018 Classification of focal and non-focal EEG signals using neighborhood component analysis and machine learning algorithms *Expert Syst. Appl.* **113** 18–32
[66] Arunkumar N, Kumar K R and Venkataraman V 2018 Entropy features for focal EEG and non focal EEG *J. Comput. Sci.* **27** 440–4
[67] Siddharth T, Tripathy R K and Pachori R B 2019 Discrimination of focal and non-focal seizures from EEG signals using sliding mode singular spectrum analysis *IEEE Sensors J.* **19** 12286–96

**IOP** Publishing

Modelling and Analysis of Active Biopotential Signals in Healthcare, Volume 2

**Varun Bajaj and G R Sinha**

# Chapter 6

# Local binary pattern based feature extraction and machine learning for epileptic seizure prediction and detection

**Abdulhamit Subasi[1,2], Turker Tuncer[3], Sengul Dogan[3] and Dahiru Tanko[3]**

[1]*Institute of Biomedicine, Faculty of Medicine, University of Turku, Kiinanmyllynkatu 10, 20520, Turku, Finland*

[2]*Effat University, College of Engineering, Department of Computer Science, Jeddah, 21478, Saudi Arabia*

[3]*Firat University, Technology Faculty, Department of Digital Forensics Engineering, Elazig, Turkey*

Electroencephalogram (EEG) signals have been extensively utilized to identify brain disorders such as epilepsy. Local descriptor based feature extraction methods have been utilized to generate features from EEG signals in order to understand them. The best known local descriptor is the local binary pattern (LBP). The LBP was presented for images and, at the same time, the one-dimensional form was presented for signals to use the advantages of the LBP. The advantages of the LBP are as follows: (i) it has a low computational complexity, (ii) it generates valuable features and (iii) it is easily implemented. In order to benefit from these advantages, many LBP-like local descriptors have been presented for both images and signals. In this chapter we discuss the effect of one-dimensional LBP and LBP-like descriptors on EEG signal recognition. We also explain the multilevel feature extraction model using LBP-like descriptors for EEG signals. In this chapter an LBP based feature extraction framework for epileptic seizure prediction and detection is presented. Feature selection/reduction is the crucial stage of EEG signal classification. Neighborhood component analysis (NCA) and ReliefF based feature reduction methods are explained. Finally, different machine learning techniques will be compared for epileptic seizure prediction and detection.

doi:10.1088/978-0-7503-3411-2ch6  

## 6.1 Introduction

Brain cells communicate electrically. Electroencephalography (EEG) is a technique that is used to measure and evaluate the brain activities using these electrical signals [1, 2]. EEG signals without a constant amplitude and frequency were presented by Hans Berger in 1930 [3]. Microvolts measure the signal intensity produced by the dendrites of adjacent neurons in the cortical surface. EEG signals have low amplitude and high noise [4, 5]. These signals are gathered with the aid of electrodes attached to the cortical surface or scalp. Measuring the electrical activity of neurons by placing electrodes on the cortical surface is an invasive technique. A method that obtains the activity of cerebral neurons through electrodes placed directly on the scalp is called non-invasive [1, 6], and the alpha, beta, delta, theta and gamma bands of the electroencephalographic signals of the brain can thus be obtained. The different types of brain waves control different states of consciousness, from sleep level to active thinking. These brain waves work simultaneously. Each of these waves works at different frequencies in the brain and can be at different intensities in the body in our daily lives for reasons such as stress, exercise, trauma and pollution. The different types of brain waves can be described as follows [1, 7].

*Delta brain waves* contain frequencies in the range of 1–4 Hz. They occur in the third phase of sleep and increases in the fourth phase. In these cases, healing functions are stimulated. The brain produces delta waves intensively in relation to brain injuries and learning difficulties. Too few delta brain waves causes poor sleep.

*Theta brain waves* are frequencies in the range of 4–7 Hz. These waves occur when a person is sleepy. These waves are related to people's emotional states. If the theta level rises too high while a person is awake, it causes a dream-like feeling and carelessness. Too few of these waves cause anxiety and stress in humans.

*Alpha brain waves* occur in the range of 8–12 Hz. Alpha waves increase when the eyes are closed, in particular when falling asleep and disappear when mental activity increases. Anxiety increases in humans when alpha waves are intense. When they are low, anxiety and insomnia appear.

*Beta brain waves* are in the range of 13–38 Hz. While beta waves are are quite intense a person is vigilant, careful or anxious. When these waves are low, depression and learning disabilities can be seen.

*Gamma brain waves* are in the 30–100 Hz range. These waves have the highest frequency of any brain wave. They are associated with high levels of concentration and cognition. High gamma wave intensity causes increased memory function and a high sense of happiness. When they are low, learning disability and memory weakness occur.

To evaluate EEG signals effectively, the quality of the measured signal must be high. The quality of the device used is essential for this purpose [8–10]. In addition, in non-invasive methods, the scalp should be thoroughly degreased during the registration of the device, and otherwise appropriately prepared [11]. However, many signal restoration techniques have been implemented on EEG signals to increase their quality. The signal processing techniques used go through stages of preprocessing, feature extraction, feature selection and classification to automatically process large amounts of EEG signals. In this way, it can be ensured that the

EEG signals are interpreted accurately and completely. These methods are in demand to guide doctors and speed up procedures.

Epilepsy is a neurological disorder characterized by a frequent tendency of the brain to produce abrupt electrical activity bursts [12]. Such events are called seizures and they occur spontaneously. High and synchronized neuronal activity causes epileptic seizures [13, 14]. Epilepsy is one of the most common neurological diseases, along with strokes, and affects more than 1% of the world's population [15]. For the diagnosis and detection of epileptic seizures, the EEG of the patient must usually be monitored for several days. The monitoring process is tedious, time-consuming and burdensome. The neurologist needs to identify chronic epileptic activity from the recorded EEG data to determine whether or not the medication used is working. A computer-aided epileptic seizure detection system is therefore of vital importance [16–20].

For several decades electroencephalograms (EEGs) have been used for the clinical diagnosis of epilepsy. The EEG is a safe and non-invasive tool for detecting brain activity compared to other methods such as the electrocorticogram (ECoG). The clinical examination of EEG traces is well known in detecting seizures. However, the efficiency of automated EEG based approaches depends on the types of features being evaluated and how the signal is handled. Epilepsy patients suffer from frequent attacks that manifest as physical or behavioral changes requiring treatment or surgery to intervene [21]. A lot of channels are used for EEG signal recording and it is a time-consuming to examine such a large number of channels. Therefore, parallel processing is an essential feature of EEG signal processing to reduce the processing time.

## 6.2 Literature review

Through EEG signals, the evaluation of brain activity can be provided for diseases such as schizophrenia, head injury, motor cortex dysfunctions, headache, sleep disorders, dizziness and memory problems [22–27]. In the literature, various methods have been developed for these and similar diseases. Bakhshali *et al* [28] proposed an imagined speech recognition method. This method used non-invasive EEG signals. In this study, a correntropy spectral density matrix was obtained from these signals and the Riemannian distances of this matrix were calculated. Finally, accuracy rates of 90.25% were presented by classifying the signals using *k*-nearest neighbors (*k*NN). In this study the Kara One dataset [29] was used. George *et al* [30] suggested a study on the classification of epilepsy. This study was based on a tunable-Q wavelet transform (TQWT). The EEG signals were divided into sub-bands via a TQWT. Entropy-based feature generation was applied and the most informative features were selected using particle swarm optimization (PSO). They selected ANN as a classifier. In this study the KITS [31, 32] and TUH [33] databases were used. The accuracy rates were calculated as 100.0% and 97.4% for the KITS and TUH databases, respectively. Aydemir *et al* [34] suggested a model for diagnosing epilepsy using EEG. In this study they used TQWT and a local feature generator for generating multilevel features. The effective features were selected using neighborhood component analysis (NCA). A 98.4% accuracy rate was achieved using the *k*NN classifier. This study used the Bonn [35] dataset. Bajaj *et al* [36] proposed an automated and accurate EEG signal classification approach. Their approach used TQWT as a preprocessor and feature generator. The features obtained were evaluated by the Kruskal–Wallis (KW) test. A 91.84% accuracy rate was obtained using the MIT/BIH dataset [37]. Rahman *et al* [38] suggested a hybrid method to extract features. In their hybrid method, the *t*-statistics method and principal component analysis were used to reduce the signal dimensionality and select the effective features. In this study the SEED dataset [39] was used and the accuracy rate was calculated as 86.57%. Tuncer *et al* [40] presented a local graph structure method based approach for EEG signal classification. NCA and the ReliefF method were used together for feature reduction. The Bonn [35] and Indian [41–43] epilepsy datasets were used. The accuracy rate for the Bonn dataset was calculated as 97.20%. Similarly, 98.67% was reached for the Indian epilepsy dataset. Sharma *et al* [44] developed an approach to detect the abnormality of EEG signals using the stop-band energy method and wavelet decomposition. EEG signals of 2130 subjects collected from Temple University Hospital were used to present the success of the method. The study achieved a success rate of 79.34%. Jaiswal and Banka [45] proposed a study to detect epilepsy. The main purpose of this study was to perform an automatic classification of epilepsy from EEG signals. Two different patterns, the one-dimensional local gradient pattern and local neighbor descriptive pattern were used in feature extraction. The Bonn [35] dataset was used for experimental results. The accuracy rate was calculated as 99.82% according to the results obtained. Kumar *et al* [46] presented a local binary pattern (LBP) based study for detecting epilepsy. EEG signals consisting of epileptic patients and healthy individuals were pretreated using Gabor filters. In the study, histograms were extracted using the local binary pattern and this was used as a feature set. The feature set obtained was classified and a 98.33% accuracy rate was obtained. Kocadagli and Langari [47] proposed a classification method for the early detection of epileptic seizures. A hybrid method was used during feature extraction. The fuzzy relations were preferred in the feature reduction phase. The results obtained in the classification stage showed that the proposed method performed effective analysis of EEG signals. Ghaemi *et al* [48] suggested an approach to determine the most appropriate channel in brain–computer interface applications. An improved binary gravitation search algorithm was used to provide left or right hand classification. The accuracy rate obtained in the classification stage was 80.0%. Ubeyli [49] suggested an model for EEG classification. In their study, three types of EEG signals were trained (healthy individuals with eyes open, epilepsy patients and patients during epileptic seizures). This method was based on a discrete wavelet transform and Levenberg–Marquardt algorithm. In this study, the accuracy rate was obtained at 94.83%.

EEG plays a significant role in controlling the brain functions of epileptic patients and has been widely used for diagnosing epilepsy. Clinical interpretation of EEG signals is a very time-consuming and laborious process. Thus, automated seizure detection is vital, and many automated EEG signal classification/clustering models have been developed/presented to detect seizures simply. Zhang *et al* [50] proposed three convolutionary neural networks (CNN) by using transfer learning. In the transfer learning applications, the CNNs were trained using the ImageNet dataset. They used pre-trained VGG16, VGG19 and ResNet50 for automatic cross-subject seizure detection. The initial dataset was the EEG dataset on the CHB-MIT scalp. Spectrogram images were generated using the Fourier transform, and the obtained spectrogram images were utilized as the input of these CNNs, since CNNs have been used for computer vision. The accuracies calculated by the VGG16, VGG19 and ResNet50-based deep transfer CNNs were 97.75%, 98.26% and 96.17%, respectively. Our system may prove to be a successful tool for the identification of cross-subject seizures in such experimental tests.

EEG signals provide essential information about the brain's electrical activity and are commonly used to facilitate the diagnosis of epilepsy. A daunting aspect in the treatment of epilepsy, the exact definition of different epileptic disorders, is of particular importance and has been thoroughly studied. Gao *et al* [51] suggested a new form of deep learning classification, namely the classification of epileptic EEG signals (EESC). This model transformed EEG signals into power spectrum density energy diagrams in the preprocessing phase. The generated spectrogram images were forwarded to the convolutional neural network (CNN). They selected a transfer learning method on a CNN. A pre-trained CNN (transfer learning) was used for classification and it classified epileptic states into four types. This model was compared to previously presented EEG classification models and it yielded higher classification accuracy (greater than 90%).

Among neurological disorders, epilepsy is the most unpredictable and chronic disorder and early detection of seizures is crucial for patients. Variable machine learning-based seizure classification models have been proposed to detect seizures. Abbasi *et al* [52] proposed a deep learning-based EEG classification model and they used an EEG dataset. This dataset has three classes and these classes are called ictal, pre-ictal and inter-ictal. They preferred a recurrent neural network (RNN) as a classification model and they selected long short term memory (LSTM) models for classification. They reached >95% classification accuracy. The EEG signal is not defined as a mathematical function. Therefore, it is a real-world signal. Hurst exponent and autoregressive moving average (ARMA) functions were utilized as feature extractors for EEG signals. Two LSTMs were used by Abbasi *et al* [52] and single-layer memory units and double-layered memory units were both modeled. The generated features were standardized and then forwarded to the double-layered LSTM deep model. The comparative results demonstrated that this model outperformed other studies.

Alickovic *et al* [53] presented a framework for detecting abnormal EEG signals. They employed two publicly published EEG datasets (Freiburg, CHB-MIT) to evaluate their model. Their model has four fundamental components: (i) noise reduction, (ii) signal decomposition, (iii) feature generation with statistical moments and (iv) classification using shallow classifiers. 100% accuracy was reached for ictal versus inter-ictal EEG for both datasets. This model reached 99.77% accuracy using the inter-ictal, ictal and pre-ictal classes.

Subasi *et al* [54] introduced a metaheuristic optimization-based automated seizure detection method using the genetic algorithm (GA) and PSO to assign optimum parameters of SVM classifiers. Kernel parameters were set using optimization algorithms. The used optimized SVMs achieved 99.38% accuracy for seizure detection.

To detect seizures, Savadkoohi and Oladduni [55] studied the features of brain electrical activity from various monitoring regions and physiological states. They used the Bonn EEG dataset and this dataset has five clusters, and it is one of the widely preferred EEG signal datasets. They extracted features using a Butterworth filter, Fourier transform and wavelet transform, respectively. To select the relevant/informative features statistical analyses, namely the *T*-test and sequential forward floating selection (SFFS) were used. They selected the *k*NN and SVM classifiers to obtain results.

The fact that the seizures occur at random times decreases the quality of life of the patients. Liu *et al* [56] implemented a machine learning-based seizure predictor for alerting patients on impending seizures. They used both the frequency domain and time domain to generate features from EEG signals. They also used multi-view convolutional neural network architecture to achieve high classification accuracy. 82% and 89% AUC values were reached using the CHB-MIT dataset and collected dataset, respectively. This work demonstrated the effective application of deep learning in critical healthcare.

Abiyev *et al* [57] presented a CNN based EEG classification model for reaching high performance on epilepsy datasets. The cross-validation model used produced robust results. Other classifiers and the CNN were compared for performance evaluation and the obtained results obviously demonstrated the effectiveness of the CNN.

## 6.3 Theoretical background

### 6.3.1 Feature extraction

The local binary pattern (LBP) is one of the most widely preferred manual feature generators in the literature. The primary philosophy of the LBP is to generate local features from micro-structures to achieve optimal global features such as metaheuristic optimization methods. Therefore, LBP divides $3 \times 3$ sized overlapping blocks into an image (figure 6.1). The $3 \times 3$ sized block is the main neighborhood block since all neighborhood relations are defined in the $3 \times 3$ sized block as below [58].

The prime aim of the LBP is to find neighborhood relations using the signum function. The signum function is a primary function and it is considered as the kernel

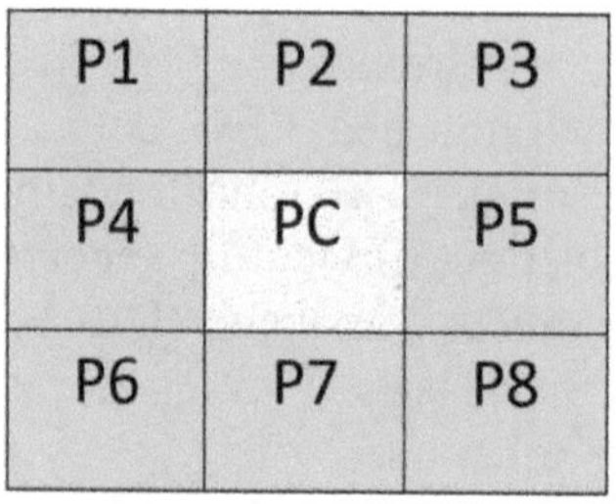

| P1 | P2 | P3 |
|---|---|---|
| P4 | PC | P5 |
| P6 | P7 | P8 |

**Figure 6.1.** The basic neighborhood matrix used. PC is the center pixel/value and the other pixels are neighborhood pixels.

function of the LBP. Equation (1.1). denotes the mathematical expression of the bit extraction (binary feature generation):

$$\text{signum}(t, d) = \begin{cases} 0, & t - d < 0 \\ 1, & t - d \geqslant 0 \end{cases}. \tag{6.1}$$

As seen in equation (1.1), the signum function is a comparison function. Here, variable kernels can be used to extract binary features. To better explain the LBP function, the steps are as follows.

**Step 1.** Divide the used gray-level image into $3 \times 3$ sized overlapping blocks. Here, $W - 2 \times H - 2$ blocks are obtained from a $W \times H$ sized image where $W$ and $H$ are the width and height of the image.

**Step 2.** Generate binary features from a $3 \times 3$ sized overlapping block:

$$\text{bit}(j) = \text{signum}(p_j, \text{PC}),\ j = \{1,2, \dots, 8\}. \tag{6.2}$$

As seen from equation (6.2), eight bits are generated from a block.

**Step 3.** Convert binary features to the decimal value

$$\text{map}_{k,\,l} = \sum_{j=1}^{8} \text{bit}_{j*} 2^{j-1},\ k = \{1,2, \dots, W - 2\},\ l = \{1,2, \dots, H - 2\}. \tag{6.3}$$

**Step 4.** Generate a histogram of the map value. The size of the generated histogram is calculated as $2^8 = 256$. Histogram extraction of the map value is defined mathematically as

$$\text{hstgrm}(\text{map}_{k,\,l}) = \text{hstgrm}(\text{map}_{k,\,l}) + 1, \tag{6.4}$$

where hstgrm is the histogram.

The extracted histograms are utilized as a feature vector in the LBP.

After this model, there are many LBP-like feature generators have been presented because of the effectiveness of the LBP. The benefits of the LBP are as follows [59]:

- It has an informative/distinctive feature generation capability.
- The mathematical structure of the LBP is simple. Therefore, the implementation of it is straightforward.
- Broad implementation areas.
- Low computational complexity.

A graphical example of the LBP is shown in figure 6.2.

The one-dimensional version of the LBP is the binary pattern (BP), and details of the BP are given below [60].

The *binary pattern* (BP) is the one-dimensional version of the LBP. To use the benefits of the LBP on the one-dimensional signals, the BP was presented. The loaded or used signal is divided into *nine* overlapping blocks. The fifth value of the overlapping block is assigned as the center value since the LBP uses $3 \times 3$ sized blocks [60]. The steps of the BP are as follows:

**Step 0.** Load a one-dimensional signal.

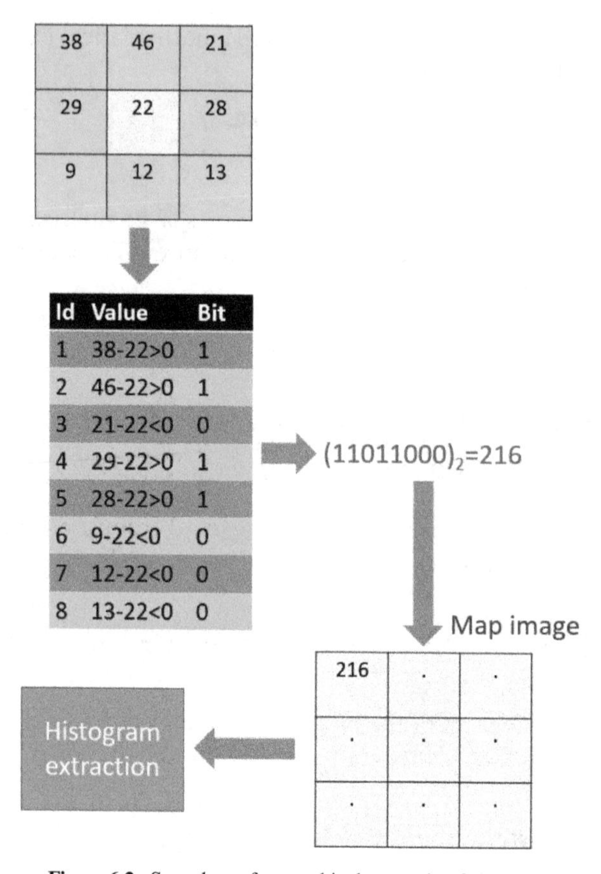


**Figure 6.2.** Snapshot of a graphical example of the LBP.

**Step 1.** Divide the loaded signal into a block with a length of 9.
**Step 2.** Determine the center value of the block. Generally, the fifth value is determined as the center value.
**Step 3.** Extract bits using the signum function:

$$\text{bit}(j) = \text{signum}(v_j, \text{center}), \ j = \{1,2, \dots, 8\}. \tag{6.5}$$

Use steps 3–4 of the LBP and extract the features.

A summary of the BP is shown in figure 6.3.

In this example, the center value is 22 and the background of it is yellow. Others (neighbors) are shown using a black background color. The center value (22) is compared to the others. Eight bits are generated and these bits are converted to a decimal value. A map signal is created using the generated decimal values. In the final phase, histogram extraction is processed.

The *ternary pattern* (TP) is a one-dimensional feature generator like BP. The main difference between the TP and BP is the kernel function used. The ternary function is utilized as a kernel function in the TP [61]. Therefore, TP generates two map signals and they are called the upper and lower features. These two map values are coded by

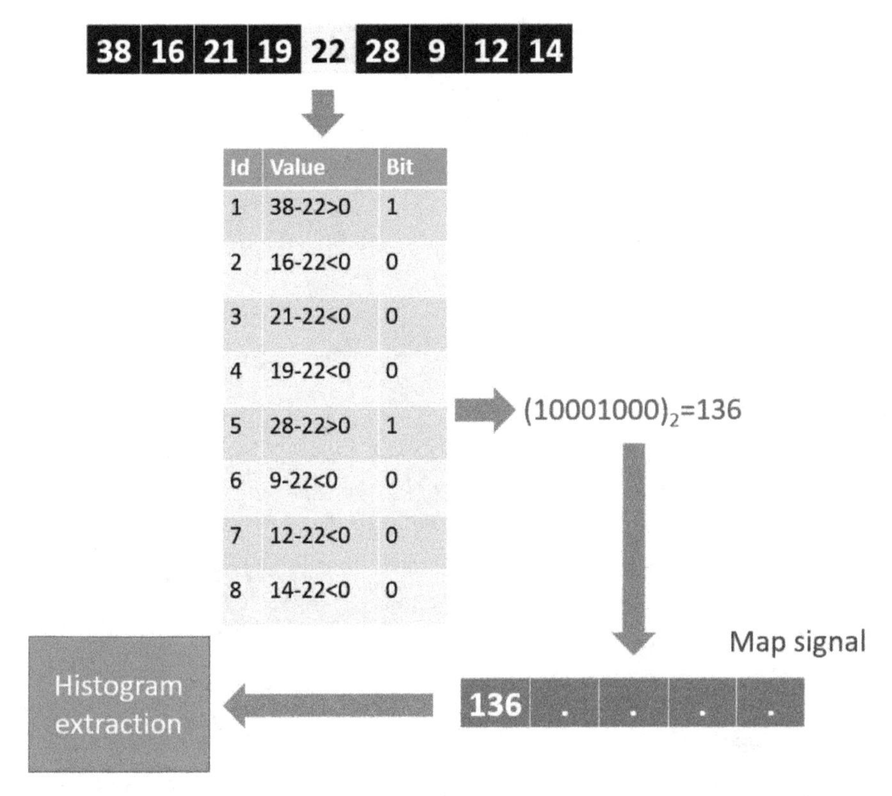


**Figure 6.3.** Summary of the BP feature extraction process using a numerical example.

eight bits. Thus, 256 features are extracted from each map signal. Thus 512 features (256 features are lower and 256 features are upper) are finally obtained. The steps of TP are as follows [62]:

**Step 1.** Apply steps 1–2 of the BP.

**Step 2.** Use the ternary function to generate ternary values:

$$\mathrm{ter}(j) = \mathrm{ternary}(v_j, \mathrm{center}), \; j = \{1,2, \ldots, 8\}, \tag{6.6}$$

$$\mathrm{ternary}(v_j, \mathrm{center}, \mathrm{tresh}) = \begin{cases} -1, & v_j - \mathrm{center} < -\mathrm{tresh} \\ 0, & -\mathrm{tresh} \leqslant v_j - \mathrm{center} \leqslant \mathrm{tresh} \\ 1, & v_j - \mathrm{center} > \mathrm{tresh} \end{cases} \tag{6.7}$$

where ter is the ternary value and tresh represents the threshold value.

**Step 3.** Calculate the lower and upper bits using ternary values:

$$\mathrm{bit}_j^{\mathrm{lower}} = \begin{cases} 0, & \mathrm{ter}(j) > -1 \\ 1, & \mathrm{ter}(j) = -1 \end{cases}, \tag{6.8}$$

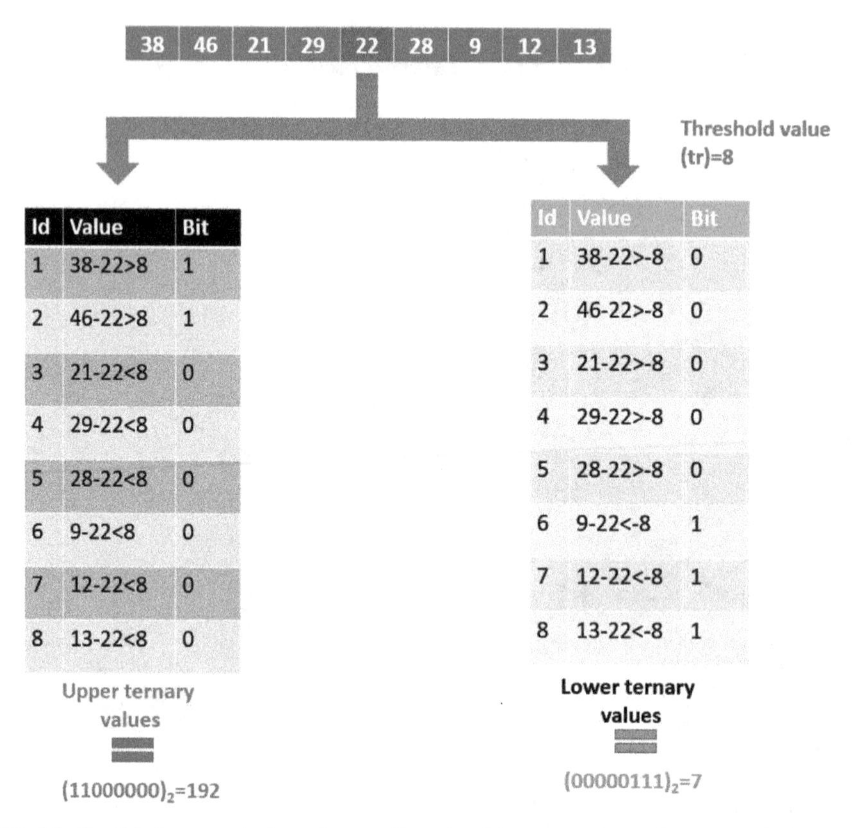


**Figure 6.4.** Graphical summarization of a one-dimensional ternary pattern using a numerical example.

$$\mathrm{bit}_j^{\mathrm{lower}} = \begin{cases} 0, & \mathrm{ter}(j) < 1 \\ 1, & \mathrm{ter}(j) = 1 \end{cases}. \tag{6.9}$$

**Step 4.** Generate lower and upper maps:

$$\mathrm{map}_i^{\mathrm{lower}} = \sum_{j=1}^{8} \mathrm{bit}_j^{\mathrm{lower}} * 2^{j-1},\ i = \{1,2, \ldots, L-8\}, \tag{6.10}$$

$$\mathrm{map}_i^{\mathrm{upper}} = \sum_{j=1}^{8} \mathrm{bit}_j^{\mathrm{upper}} * 2^{j-1}. \tag{6.11}$$

**Step 5.** Extract histograms of the lower and upper map values.
**Step 6.** Apply histogram concatenation (figure 6.4).

### 6.3.2 Classification techniques

*k-nearest neighbors* ($k$NN) groups the dataset into categories of related objects and then assigns the test set to a group whose objects are dominantly closer to it. $k$NN uses labeled observations, distance metrics and *k*-values. Some variants of the *k*NN use the voting method. The *k*NN has many variants. The users and developers can use variable *k*-values, distance metrics and voting/non-voting methods. The similarity is measured by using variable distance metrics, for example Euclidean, cosine, Manhattan, Spearman and Minkowsky metrics. It is a fast classifier for small datasets. However, it is not effective for large datasets. Since it measures the distance of all observations, it is a widely preferred shallow and conventional classifier for lightweight machine learning applications [63].

The *support vector machine* (SVM) is one of the most accurate and robust shallow/conventional classifiers among all well-known classifiers. In a binary classification task, the purpose of the SVM is to find the best classification function (kernel) in the training data to distinguish the members of the two classes. A linear classification function defines a hyperplane that passes through the middle of the two classes and separates them for a linear classification task. Since there are many such linear hyperplanes, what SVM additionally guarantees is that it finds the best function by maximizing the margin between the two classes. It aims to optimize the parameters for separation. Therefore, it uses variable kernels, for example linear, polynomial (quadratic, cubic) and Gaussian kernels. It can be used on both small and large datasets [63].

*Decision tree* (DT). Tree-based learning algorithms are one of the most used supervised learning algorithms. The general problem can be adapted to resolve all the problems addressed. Algorithms such as decision trees, random forests and gradient reinforcement (gradient augmentation) work widely in all types of data science problems. A DT is used to divide a dataset containing a large number of records into smaller groups by applying simple decision steps.

Advantages of DTs:

1. They are easy to understand and interpret. The tree structures used can be visualized.
2. They require a small amount of data preparation. However, it should not be forgotten that this model does not support lost values.
3. The time cost of a DT is logarithmic with the number of data points used to train the tree.
4. They can process both numerical and categorical data.
5. They can deal with multiple output problems.
6. It is possible to verify a model using statistical tests.

Decision trees can be considered as a nonparametric method. In other words, it does not have an approach to space distribution and classification structure.

Disadvantages of DTs:

1. Extremely complex trees can be produced that do not describe the data well. In this case, tree branching may not be followed.
2. Over-fitting is one of the most commonly seen problems in DT. Methods such as restrictions and pruning can be used on model parameters to solve this problem. Pruning refers to the removal of leaf nodes containing a small number of objects from the decision tree graph [64].

*Ensemble methods.* Ensemble theory is widely used to improve the performance of machine learning algorithms. They use more than one classifier to reach high accuracies. New subsets are obtained from the data. Each subset is trained with a basic learner, i.e. a machine learning algorithm. The results are combined. Success in ensemble learning should be based on two basic constructs: the basic learners should have high performance (accuracy) and the more successful learners are used, the higher the collective success. The decisions of the basic learners should be different and diverse [65, 66].

*Bagging/bootstrap aggregating* is one of the most preferred ensemble models in the literature for increasing performance. The samples are held in bootstrap samples. They are sampled uniformly. Each bootstrap sample uses a learning algorithm and these are forwarded to variable classifiers. A combined classifier is used to select the optimal classifier for each bootstrap to increase their performance. This selection process is processed by using the voting model. The widely used bagged model is a bagged tree and it is available in the MATLAB classification learner (MCL) [65, 66].

The *random subspace method* is one of the most used ensemble methods in the literature. It is similar to the bagged model. Here, observations are divided into subspaces randomly, and they are utilized as input data of the used classifiers. The predicted values of each classifier are calculated. Finally, by employing a combination algorithm, the final result is obtained. In the MCL, there are the subspace discriminant and subspace *k*NN. These methods are ensemble versions of the discriminant and *k*NN classifiers [67].

## 6.4 The proposed approach

### 6.4.1 Implementation of the BP on EEG signals

In this section, the BP implementation of the EEG signals is given. The presented BP and TP are utilized as a feature generation method. They extract 256 and 512 features from an EEG signal. The proposed model must be used for feature extraction, feature selection and classification to create an EEG classification model. Variable models can be presented by using only these feature generators. These variants are given below.

### 6.4.2 BP feature generation based EEG recognition

The first implementation is called BP based EEG classification. We only employ the BP for the EGG signals and the features generated are forwarded to the classifiers. The defined datasets given above are used. In this model, the BP is utilized as a feature generator and the classification is processed using BP features. The graphical summarization of this model is shown in figure 6.5.

First, we coded the BP function (table 6.1).

The primary function is used to extract features from each EEG signal. The primary function used is shown below (table 6.2).

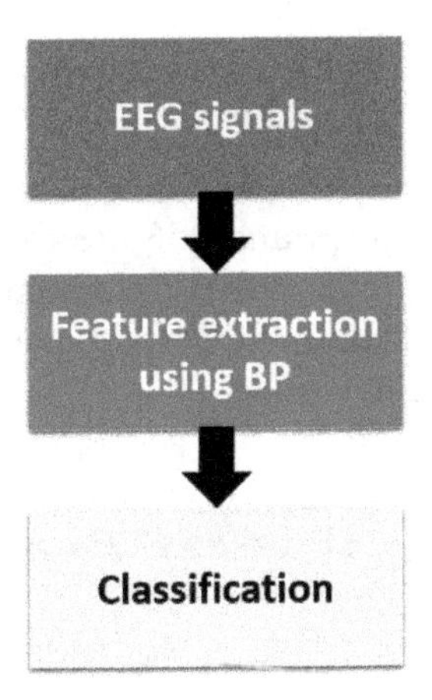


**Figure 6.5.** Graphical summarization of the BP based EEG classification model.

**Table 6.1.** MATLAB code of the BP function.

Input: EEG signal (signal).
Output: BP features (histo) with a length of 256.

```
function histo=bp(signal)
histo=zeros(1,256);
for i=1:length(signal)-8
  block=signal(i:i+8);
  say=0; u(i)=0; l(i)=0;
  for j=1:9
    if j~=5
      if block(j)-block(5)>0
        l(i)=l(i)+2^say;
      end
      say=say+1;
    end
  end
  histo(l(i)+1)=histo(l(i)+1)+1;
end
```

**Table 6.2.** The MATLAB code of the presented BP based EEG classification model.

**Input:** EEG signals
**Output:** Feature vector (feature_vector) with size of 500 x 257. The 257th column belongs to the target.

```
clc,clear all,close all
Fl=dir('C:\Users\tunce\Desktop\EEG\*.txt'); % Read txt files
counter=1; % This counter is defined to assign target values.
for k=1:500
  signal=textread(Fl(k).name); % Read each EEG signal
  X(k,1:256)=bp(signal); % Feature extraction
  y(k)=counter; % Target assigning.
  if mod(k,100)==0 % Each classs has 100 EEG signals.
    counter=counter+1;
  end
end
feature_vector=[X y'];
```

### 6.4.3 TP feature generation based EEG recognition

The second presented feature generator is the TP. The TP is the most preferred variant of the BP. It uses a ternary function as a kernel. It generates 512 features, of which 256 are called lower features and 256 are called upper features. The main problem of the TP is to find an optimal threshold point. In this section, the details of the TP based EEG based classification model are given. A flowchart of the TP based EEG classification is shown in figure 6.6.

The MATLAB code of the TP feature generator is given in table 6.3 to better understand the TP based EEG classification model.

The primary function of the TP based EEG classification model is shown in table 6.4.

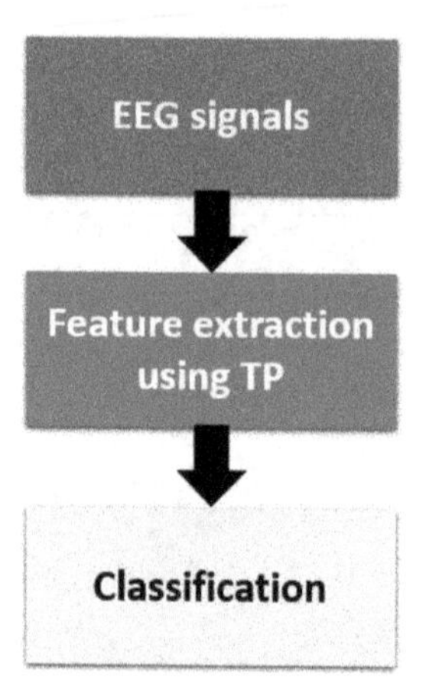


**Figure 6.6.** The TP based EEG classification model.

**Table 6.3.** The MATLAB code of the TP feature generator.

**Input:** EEG signal (signal) and the multiplier of the threshold value.
**Output:** Feature vector (histo) with a length of 512.

```
function histo=tp(signal,tt)
t=std(signal)*tt/10;
h1=zeros(1,256); h2=zeros(1,256);
for i=1:length(signal)-8
  block=signal(i:i+8);
  say=0; u(i)=0; l(i)=0;
  for j=1:9
    if j~=5
      if block(j)-block(5)<-t
        l(i)=l(i)+2^say;
      elseif block(j)-block(5)>t
        u(i)=u(i)+2^say;
      end
      say=say+1;
    end
  end
  h1(l(i)+1)=h1(l(i)+1)+1; h2(u(i)+1)=h2(u(i)+1)+1;
end
histo=[h1 h2];
```

**Table 6.4.** The TP based EEG classification model.

**Input:** EEG signals
**Output:** Feature vector (feature_vector) with size of 500 x 513. The 513th column belongs to the target.

```
clc,clear all,close all
Fl=dir('C:\Users\tunce\Desktop\EEG\*.txt'); % Read txt files
counter=1; % This counter is defined to assign target values.
for k=1:500
  signal=textread(Fl(k).name); % Read each EEG signal
  X(k,1:512)=tp(signal,5); % Feature extraction. Here, half of the standard deviation of the signal is used as threshold value
  y(k)=counter; % Target assigning.
  if mod(k,100)==0 % Each classs has 100 EEG signals.
    counter=counter+1;
  end
end
feature_vector=[X y'];
```

### 6.4.4 Multilevel models

Multilevel models have been used to generate multileveled features (low-level, mid-level and high-level). By using this feature generation strategy higher classification rates can be achieved. A decomposition model must be used to create a multilevel feature generation model. Here, we used discrete wavelet transform [68] and TQWT [69–71] as decomposition methods. These methods are two of the most effective decomposition models for signal decomposition. In the multileveled feature generation models huge feature vectors can be generated. Therefore, a feature selector must be used. In this chapter, ReliefF [72] and neighborhood component analysis [73] were used to select informative features. The general schematic diagram of the multilevel model is shown in figure 6.7.

### 6.4.5 DWT based feature generation

The DWT is one of the most widely used decomposition methods in the literature. It has variable filters for decomposition. Haar, Symlets, Daubechies are widely used filters. The primary objective of the wavelet decomposition is to generate wavelet filters to generate features comprehensively. The wavelet decomposition generates two sub-bands and these are called the high-pass filter sub-band (high) and low-pass filter sub-band (low). The multileveled feature generation used is defined as [68]

$$[\text{low}^1,\ \text{high}^1] = \text{DWT}(\text{signal, filter}) \tag{6.12}$$

$$[\text{low}^k,\ \text{high}^k] = \text{DWT}(\text{low}^{k-1}, \text{filter}),\ \ k > 1. \tag{6.13}$$

Low sub-bands are generally used to extract features. Here, the BP or TP is employed on these signals for feature generation:

$$\text{feat}^1 = \text{FG}(\text{signal}) \tag{6.14}$$

$$\text{feat}^k = \text{FG}(\text{low}^{k-1}), \tag{6.15}$$

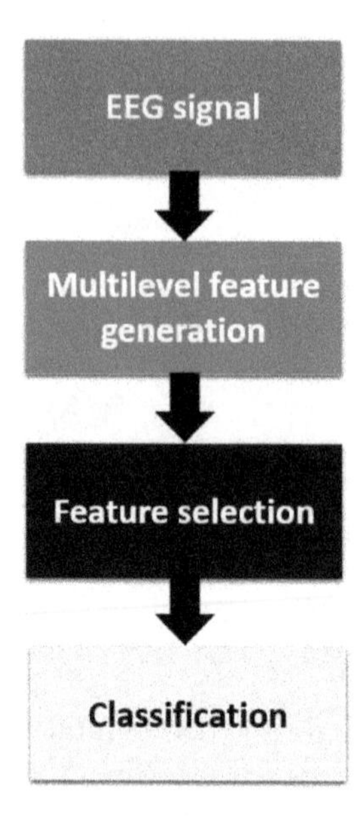


**Figure 6.7.** The schematic explanation of a multilevel EEG signal classification model.

where FG(·) represents the used feature generation function and features are extracted in each phase using this function (the BP or TP). After this step, the features are concatenated to obtain the final feature vector:

$$\text{ff} = \text{conc}(\text{feat}^1, \text{feat}^2, \ldots, \text{feat}^k), \tag{6.16}$$

where ff represents the final feature vector.

Generally, the sizes of the final features are enormous. Therefore, a feature selector must be used to yield high classification accuracy. In this chapter we used NCA [73] and ReliefF to select the discriminative features. The selected features are forwarded to MCL and classification results were obtained. The sub-phases of the multilevel feature generation are given in the subsections.

### 6.4.6 TQWT based feature generation

TQWT is a new generation decomposition model [74, 75]. It is a parametrical model and these parameters are called the Q-factor ($Q$), redundancy value ($r$) and the number of levels ($J$). Variable wavelet decomposition methods have been proposed by employing these parameters. TQWT has a quick response and abnormal signals can be detected easily using the TQWT [76]. The MATLAB tool for the TQWT can be downloaded using http://eeweb.poly.edu/iselesni/TQWT/. A MATLAB implementation of the TQWT is shown in table 6.5.

### 6.4.7 Feature generation

In this study, the TP and BP are utilized as feature generators, respectively. We used five levels of wavelet decomposition using the 'sym4' filter. The sym4 filter has been

**Table 6.5.** TQWT decomposition-based feature generation.

```
Input: The one-dimensional signal:
Output: Feature vector.
01: Load a one-dimensional signal.
02: R = TQWT(signal,Q,r,J);
03: for i = 1 to J + 1
04:       feat^i = FG(R{i});
05: end for i
```

**Table 6.6.** Feature generation of the multilevel feature generation model.

```
Input: The one-dimensional signal:
Output: Feature vector.
01: Load a one-dimensional signal.
02: for i = 1 to iteration
03:      feat^i = FG(signal);
04:      [low, high] = DWT(signal, filter);
05:      signal = low; // Updating signal
06: end for i
```

used for both denoising and decomposition. Therefore, we selected it as the wavelet filter. The MATLAB codes for this model are shown in table 6.6.

### 6.4.8 Feature selection

Two feature selectors have been used to reduce the dimension of the features. These are NCA and ReliefF. These are distance based feature selectors. Both of them used a Manhattan distance based fitness function. ReliefF generates both negative and positive weights. However, NCA generates only positives weights for each feature. Also, NCA uses an optimizer such as stochastic gradient descent (SGD) or ADAM. The most informative features can be selected using the generated weights. The larger weights represent informative features and the smaller weights define the redundant features. In the ReliefF, negative weighted features are expressed as redundant features. Hence, the feature elimination using ReliefF is very simple. However, all features have positive weights using NCA. Therefore, the NCA based redundant feature is a more difficult process than ReliefF. The MATLAB code of the used feature selection model is shown in tables 6.4 and 6.5.

*Hybrid feature selector—RFNCA.* A hybrid feature selector was presented by Tuncer *et al* [77] to use the advantages of the ReliefF and NCA together. As discussed above, ReliefF can generate negative features. Hence, the redundant feature elimination process

is easy using ReliefF. In the first step, ReliefF is deployed to the features and positive weighted features are selected. Then the NCA is employed to the features chosen by ReliefF, and the most discriminative features are selected by NCA. In this feature selection phase, the optimal number of features selected is a significant problem. However, it is a good solution for feature selection. The NCA(·) and ReliefF(·) functions express the NCA and ReliefF feature selectors to express the model easily.
**Step 0.** Load feature vector ($X$) with the size of $k$.
**Step 1.** Deploy ReliefF and calculate the weights of the ReliefF ($w^{\mathrm{RF}}$):

$$w^{\mathrm{RF}} = \mathrm{ReliefF}(X, y). \tag{6.17}$$

**Step 2.** Select features with positive weights ($X^{+}$):

$$X^{+}(\mathrm{count}) = X(i), \;\; w^{\mathrm{RF}}(i) > 0, \;\; \mathrm{count}\ {+}{+}. \tag{6.18}$$

**Step 3.** Deploy NCA to $X^{+}$:

$$w^{\mathrm{NCA}} = \mathrm{NCA}(X^{+}, y). \tag{6.19}$$

**Step 4.** Sort the NCA weights ($w^{\mathrm{NCA}}$) and calculate indices (end):

$$[\mathrm{maximum}, \;\; \mathrm{end}] = \mathrm{sort}(w^{\mathrm{NCA}}, \mathrm{descending}). \tag{6.20}$$

**Step 5.** Select the most informative features as below:

$$\mathrm{feature}(k) = X^{+}(\mathrm{end}(k)). \tag{6.21}$$

*Iterative NCA.* The primary purpose of iterative feature selectors is to choose the optimal number of features without using parameters. Since NCA and ReliefF are parametric feature selectors a trial and error method has been used to select the optimal features. By employing iterative feature selection, a loss value generator should be used. Therefore, a classifier is chosen as a loss value generator. Here, we used shallow classifiers to decrease the computational complexity of the iterative feature selector. This iterative feature selection model is deployed on the NCA in this section, this model is thus called INCA. Tuncer *et al* [78] presented INCA in 2020. INCA aims to select optimal features without using any trial and error method. The pseudo-code of INCA is shown in figure 6.8.

## 6.5 Results and discussion

Implementations of the multileveled feature generation models for EEG classification are given in this section. Two decomposition methods have been used in this section, and these are the DWT and TQWT. Moreover, the TP and BP are deployed as a feature generation model.

### 6.5.1 The used datasets

The first dataset used in this chapter is the Bonn dataset [35]. It has been widely preferred in EEG classification studies. This dataset contains 500 EEG signals in five classes. These classes are called A, B, C, D and E, or Z, O, N, F and S. There are one

| Procedure: $INCA(X, target, C)$ |
|---|
| Input: Extracted features ($X$) with size of d x K, target with size of d and classifier (C).<br>Output: Final feature vector ($last$) with size of j. |
| 00: Load $X$<br>01: for i=1 to $K$ do<br>02: $X(:,i) = \frac{X(:,i)-min(X(:,i))}{max(X(:,i))-min(X(:,i))}$; //Apply min-max normalization to each column of the feature vector. NCA is a distance based feature selector. In order to use NCA effectively, normalization must be performed.<br>03: end for i<br>04: $endex = NCA(X, target)$; // Calculate indices of the features by using NCA. These indices are ordered by descending.<br>05: for k=1 to 900 do // Apply iterative NCA. Select number of features from 100 to 1000. The main aim of this step is to reduce computational complexity of this method.<br>06: for l=1 to 99+k do<br>07: $f^{NCA}(:,l) = X(:, endex(l))$; // Iterative NCA feature selection process.<br>08: end for l<br>09: $error(k) = C(f^{NCA})$; // Calculate error rates of each selected feature vector.<br>10: end for k<br>11: $[minimum, index] = \min(error)$; // Find minimum error and index of it.<br>12: for j=1 to $index + 99$ do // Select optimal features.<br>13: $last(:,j) = X(:, endex(j))$;<br>14: end for j |

**Figure 6.8.** The pseudo-code of the INCA feature selector. The used selector uses distances to calculate the coefficients of the features and target values. Therefore, min-max normalization is deployed. Here, the iteration is initiated from 100 to 1000.

hundred EEG signals in each class. In this view this dataset is homogeneous. By using these classes variable cases can be defined. The number of samples of each EEG signal is 4097. It is a single channeled EEG dataset. Healthy EEG signals are in the Z and O classes. In the Z and O clusters/classes, EEG signals were gathered from a healthy subject when the eyes were open and the eyes were closed, respectively. The N and F classes represent inter-ictal EEG signals. These signals were collected from epilepsy subjects. The N and F clusters define EEG signals from before a seizure and in the epileptic zone, respectively. EEG signals of the E cluster were collected during the seizure. The length of these EEG signals is 23.6 s. The sampling rate of these signals was approximately 170 Hz.

### 6.5.2 DWT and BP based EEG classification

In this implementation a multilevel DWT and BP are utilized to create a multilevel feature generation network. The five-level DWT is employed to the EEG signals for decomposition. Here, four feature selectors are deployed to extracted features and these are NCA, ReliefF, RFNCA and INCA. The steps of this model are given below.

**Step 1.** Apply multilevel DWT with five levels:

$$[\text{low}^1,\ \text{high}^1] = \text{DWT}(\text{EEG, filter}) \tag{6.22}$$

$$[\text{low}^k,\ \text{high}^k] = \text{DWT}(\text{low}^{k-1}, \text{filter}),\ k > 1,\ k = \{2,3, \dots, 5\}. \tag{6.23}$$

**Step 2.** Employ the BP for feature extraction:

$$X(h, 1 : 256) = \text{BP}(\text{EEG}) \tag{6.24}$$

$$X(h, k \times 256{+}1 : (k + 1) \times 256) = \text{BP}(\text{low}^k). \tag{6.25}$$

**Step 3.** Select the most discriminative features using any selector used (FS):

$$\text{last} = \text{FS}(X). \tag{6.26}$$

The general steps of the multilevel feature generation model used are given above. The results of this model are given below. In the first model, the NCA is utilized as feature selector and 512 of the most informative features are selected. The MATLAB code of this implementation is shown in table 6.7.

The selected 512 features are forwarded to the MCL toolbox and the best result is obtained by using a fine *k*NN with a city block distance classifier. Ten-fold cross-validation is chosen as the validation methodology.

The other feature selector is ReliefF. Using ReliefF 512 features are selected to compare these feature selectors. The MATLAB code of the ReliefF based model is shown in table 6.8.

As seen from these codes, 512 features are selected using ReliefF. These features are classified and the best-result classifier is found to be the cubic SVM and it reached 86% accuracy.

The third feature selector is RFNCA. In this model the redundant features are determined using ReliefF and NCA selects 512 the most informative ones. The MATLAB code of this model is shown in table 6.9.

Fine *k*NN reached 90% classification accuracy using an RFNCA based model. ReliefF selected 880 features from the extracted 1536 features, and NCA selected 512 of the selected 880 features. 90% accuracy was achieved using fine *k*NN with a city block distance.

The last feature selector is INCA. The computational complexity of it is greater than others since it uses iterations. INCA has the selection capability of the optimal number of feature. The MATLAB code of the INCA based EEG classification model is shown in table 6.10.

**Table 6.7.** The MATLAB code of the NCA feature selector.

```
clc,clear all,close all
Fl=dir('C:\Users\tunce\Desktop\EEG\*.txt'); % Create file
say=1;
for k=1:500
   signal=textread(Fl(k).name); % Read EEG signals
   for i=1:6 % Multilevel feature extraction and concatenation.
      X(k,(i-1)*256+1:i*256)=bp(signal,5); % Feature extraction and concatenation
      [low,high]=dwt(signal,'sym4'); % Signal decomposition using sym4 filter
      clear signal
      signal=low; % Signal updating
      clear low;
   end
   y(k)=say;
   if mod(k,100)==0
      say=say+1;
   end
end
[mm,nn]=size(X);
for i=1:nn % Min-max normalization
   X(:,i)=(X(:,i)-min(X(:,i)))/(max(X(:,i))-min(X(:,i))+eps);
end
mdl=fscnca(X,y,'Solver','sgd','Verbose',1); % NCA applying
xx=mdl.FeatureWeights; % Assign weights to xx
[aa,ind]=sort(xx,'desc'); % Sort weights and find the most valuable indices of the features
for i=1:512 % The most informative 512 features selection
   last(:,i)=X(:,ind(i));
end
last(:,i+1)=y;
```

**Table 6.8.** The MATLAB code of the ReliefF feature selector.

```
clc,clear all,close all
Fl=dir('C:\Users\tunce\Desktop\EEG\*.txt'); % Create file
say=1;
for k=1:500
   signal=textread(Fl(k).name); % Read EEG signals
   for i=1:6 % Multilevel feature extraction and concatenation.
      X(k,(i-1)*256+1:i*256)=bp(signal,5); % Feature extraction and concatenation
      [low,high]=dwt(signal,'sym4'); % Signal decomposition using sym4 filter
      clear signal
      signal=low; % Signal updating
      clear low;
   end
   y(k)=say;
   if mod(k,100)==0
      say=say+1;
   end
end
[mm,nn]=size(X);
for i=1:nn % Min-max normalization
   X(:,i)=(X(:,i)-min(X(:,i)))/(max(X(:,i))-min(X(:,i))+eps);
end
[ind,weights]=relieff(X,y',10); % ReliefF applying
for i=1:512 % The most informative 512 features selection
   last(:,i)=X(:,ind(i));
end
last(:,i+1)=y;
```

**Table 6.9.** The MATLAB code of the ReliefF and NCA feature selector.

```
clc,clear all,close all
Fl=dir('C:\Users\tunce\Desktop\EEG\*.txt'); % Create file
say=1;
for k=1:500
   signal=textread(Fl(k).name); % Read EEG signals
   for i=1:6 % Multilevel feature extraction and concatenation.
      X(k,(i-1)*256+1:i*256)=bp(signal,5); % Feature extraction and concatenation
      [low,high]=dwt(signal,'sym4'); % Signal decomposition using sym4 filter
      clear signal
      signal=low; % Signal updating
      clear low;
   end
   y(k)=say;
   if mod(k,100)==0
      say=say+1;
   end
end
% RFNCA feature selector
[mm,nn]=size(X);
for i=1:nn % Min-max normalization
   X(:,i)=(X(:,i)-min(X(:,i)))/(max(X(:,i))-min(X(:,i))+eps);
end
[ind,weights]=relieff(X,y',10); % ReliefF applying
counter=1;
for i=1:2304 % Positive weighted feature selection
   if weights(i)>0
      X_plus(:,counter)=X(:,i);
      counter=counter+1;
   end
end
mdl=fscnca(X_plus,y,'Solver','sgd','Verbose',1); % NCA applying
xx=mdl.FeatureWeights; % Assign weights of xx
[aa,ind]=sort(xx,'desc'); % Sort weights and find the most valuable indices of the features
for i=1:512 % The most informative 512 features selection
   last(:,i)=X_plus(:,ind(i));
end
last(:,i+1)=y;
```

The INCA selected 329 features as an optimal number of features. Fine *k*NN classifiers achieved 94.6% classification accuracy.

As stated, the best feature selector is INCA for this dataset and the used feature generation model. However, INCA is more complicated than the other feature selectors. The presented MATLAB codes can be used in other models.

### 6.5.3 DWT and TP based EEG classification

The TP is utilized as a feature extractor. The main difference between the above models is the feature generation. Therefore only results were given in this section (table 6.11).

### 6.5.4 TQWT and BP based EEG classification

The second decomposition method used is the TQWT. The TQWT is an effective decomposition model. However, the parameter setting is very difficult on the TQWT. Therefore, three parameters ($Q$, $r$, $J$) must be set to achieve high

**Table 6.10.** The MATLAB code of the INCA feature selector.

```
clc,clear all,close all
Fl=dir('C:\Users\tunce\Desktop\EEG\*.txt'); % Create file
say=1;
for k=1:500
   signal=textread(Fl(k).name); % Read EEG signals
   for i=1:6 % Multilevel feature extraction and concatenation.
      X(k,(i-1)*256+1:i*256)=bp(signal,5); % Feature extraction and concatenation
      [low,high]=dwt(signal,'sym4'); % Signal decomposition using sym4 filter
      clear signal
      signal=low; % Signal updating
      clear low;
   end
   y(k)=say;
   if mod(k,100)==0
      say=say+1;
   end
end
% INCA feature selector
[mm,nn]=size(X);
for i=1:nn % Min-max normalization
   X(:,i)=(X(:,i)-min(X(:,i)))/(max(X(:,i))-min(X(:,i))+eps);
end
mdl=fscnca(X,y,'Solver','sgd','Verbose',1); % NCA applying
xx=mdl.FeatureWeights; % Assign weights of xx
[aa,ind]=sort(xx,'desc'); % Sort weights and find the most valuable indices of the features
mdl=fscnca(X,y,'Solver','sgd','Verbose',1);
xx=mdl.FeatureWeights;
[aa,ind]=sort(xx,'desc');
for ts=0:900
   for i=1:100+ts
      poz(:,i)=X(:,ind(i));
   end
   mdl = fitcknn(... % Loss value calculation
   poz, ...
   y, ...
   'Distance', 'Cityblock', ...
   'Exponent', [], ...
   'NumNeighbors', 1, ...
   'DistanceWeight', 'Equal', ...
   'Standardize', true, ...
   'ClassNames', [1:5]');
   kk=crossval(mdl,'KFold',10); % Calculate loss values using 10-fold CV
   ll(ts+1) = kfoldLoss(kk);
   clear poz
   end
   [eb,inde]=min(ll);
   for i=1:inde+99 % Select the optimal features
      last(:,i)=X(:,ind(i));
   end
   last(:,i+1)=y;
```

classification accuracy. Here, values of 2, 3 and 7 are defined for the $Q$, $r$ and $J$ parameters, respectively. Hence $256 \times 9 = 2304$ features are extracted in total. The MATLAB code of the TQWT and BP classification model using NCA feature selection is shown in table 6.12.

The best-result classifier is found as a medium Gaussian SVM [79] and it reached 78.6% classification accuracy.

**Table 6.11.** The results of DWT and TP based EEG classification.

| Feature selection method | Classifier | Number of features | The best result (%) |
|---|---|---|---|
| NCA | Fine *k*NN with city block distance | 512 | 95.0 |
| ReliefF | Ensemble subspace *k*NN | 512 | 86.0 |
| RFNCA | Ensemble subspace *k*NN | 512 | 87.8 |
| INCA | Fine *k*NN with city block distance | 139 | 96.6 |

**Table 6.12.** The MATLAB code of the TQWT and BP classification model using NCA feature selection.

```
clc,clear all,close all
Fl=dir('C:\Users\tunce\Desktop\EEG\*.txt'); % Create file
say=1; % Class target
for k=1:500
   signal=textread(Fl(k).name); % Read EEG signals
   R=tqwt(signal(1:4096),2,3,7); % Apply TQWT
   X(k,1:256)=bp(signal);
   for i=1:8 % Multilevel feature extraction and concatenation.
      X(k,(i)*256+1:(i+1)*256)=bp(signal,5); % Feature extraction and concatenation
   end
   y(k)=say;
   if mod(k,100)==0
      say=say+1;
   end
end
[mm,nn]=size(X);
for i=1:nn % Min-max normalization
   X(:,i)=(X(:,i)-min(X(:,i)))/(max(X(:,i))-min(X(:,i))+eps);
end
mdl=fscnca(X,y,'Solver','sgd','Verbose',1); % NCA applying
xx=mdl.FeatureWeights; % Assign weights of xx
[aa,ind]=sort(xx,'desc'); % Sort weights and find the most valuable indices of the features
for i=1:512 % The most informative 512 features selection
   last(:,i)=X(:,ind(i));
end
last(:,i+1)=y;
```

In the second model, ReliefF is utilized as a feature selector and 512 features are selected using ReliefF. The calculated results are given as below. Also, the MATLAB code of this model is shown below (table 6.13).

The best-result classifier is an ensemble bagged tree and it achieved 76.4% classification accuracy.

The third selector used is RFNCA. The MATLAB code of the TQWT + BP and RFNCA based EEG classification model is shown in table 6.14.

Here, we selected 400 feature using RFNCA since ReliefF selected 408 positive weighted features. The best-result classifier is the medium Gaussian SVM classifier. It yielded 74.6% accuracy.

**Table 6.13.** The MATLAB code of the TQWT and BP classification model using ReliefF feature selection.

```
clc,clear all,close all
Fl=dir('C:\Users\tunce\Desktop\EEG\*.txt'); % Create file
say=1;
for k=1:500
   signal=textread(Fl(k).name); % Read EEG signals
   R=tqwt(signal(1:4096),2,3,7);
   X(k,1:256)=bp(signal,5);
   for i=1:8 % Multilevel feature extraction and concatenation.
      X(k,(i)*256+1:(i+1)*256)=bp(signal,5); % Feature extraction and concatenation
   end
   y(k)=say;
   if mod(k,100)==0
      say=say+1;
   end
end
[mm,nn]=size(X);
for i=1:nn % Min-max normalization
   X(:,i)=(X(:,i)-min(X(:,i)))/(max(X(:,i))-min(X(:,i))+eps);
[ind,weights]=relieff(X,y',10); % ReliefF applying
for i=1:512 % The most informative 512 features selection
   last(:,i)=X(:,ind(i));
end
last(:,i+1)=y;
```

**Table 6.14.** The MATLAB code of the TQWT and BP classification model using RFNCA feature selection.

```
% RFNCA feature selector
[mm,nn]=size(X);
for i=1:nn % Min-max normalization
   X(:,i)=(X(:,i)-min(X(:,i)))/(max(X(:,i))-min(X(:,i))+eps);
end
[ind,weights]=relieff(X,y',10); % ReliefF applying
counter=1;
for i=1:1536 % positive weighted feature selection
   if weights(i)>0
      X_plus(:,counter)=X(:,i);
      counter=counter+1;
   end
end
mdl=fscnca(X_plus,y,'Solver','sgd','Verbose',1); % NCA applying
xx=mdl.FeatureWeights; % Assign weights of xx
[aa,ind]=sort(xx,'desc'); % Sort weights and find the most valuable indices of the features
for i=1:400 % The most informative 400 features selection
   last(:,i)=X_plus(:,ind(i));
end
last(:,i+1)=y;
```

The last feature selector is INCA. INCA resulted in the best accuracy in the DWT based model. The MATLAB code of the TQWT + BP + INCA model is shown in table 6.15. INCA selects 285 features.

The best result is calculated as 80.2% by employing the quadratic SVM.

**Table 6.15.** The MATLAB code of the TQWT and BP classification model using INCA feature selection.

```
clc,clear all,close all
Fl=dir('C:\Users\tunce\Desktop\EEG\*.txt'); % Create file
say=1;
for k=1:500
  signal=textread(Fl(k).name); % Read EEG signals
  R=tqwt(signal(1:4096),2,3,7);
  X(k,1:256)=bp(signal,5);
  for i=1:8 % Multilevel feature extraction and concatenation.
    X(k,(i)*256+1:(i+1)*256)=bp(signal,5); % Feature extraction and concatenation
  end
  y(k)=say;
  if mod(k,100)==0
    say=say+1;
  end
end
% INCA feature selector
[mm,nn]=size(X);
for i=1:nn % Min-max normalization
  X(:,i)=(X(:,i)-min(X(:,i)))/(max(X(:,i))-min(X(:,i))+eps);
end
mdl=fscnca(X,y,'Solver','sgd','Verbose',1); % NCA applying
xx=mdl.FeatureWeights; % Assign weights of xx
[aa,ind]=sort(xx,'desc'); % Sort weights and find the most valuable indices of the features
mdl=fscnca(X,y,'Solver','sgd','Verbose',1);
xx=mdl.FeatureWeights;
[aa,ind]=sort(xx,'desc');
for ts=0:900
  for i=1:100+ts
    poz(:,i)=X(:,ind(i));
  end
  mdl = fitcknn(... % Loss value calculation
  poz, ...
  y, ...
  'Distance', 'Cityblock', ...
  'Exponent', [], ...
  'NumNeighbors', 1, ...
  'DistanceWeight', 'Equal', ...
  'Standardize', true, ...
  'ClassNames', [1:5]');
  kk=crossval(mdl,'KFold',10); % Calculate loss values using 10-fold CV
  ll(ts+1) = kfoldLoss(kk);
  clear poz
  end
  [eb,inde]=min(ll);
  for i=1:inde+99 % Select the optimal features
    last(:,i)=X(:,ind(i));
  end
  last(:,i+1)=y;
```

### 6.5.5 TQWT and BP based EEG classification

The TP is utilized as a feature extractor. The main difference between the above models is the feature generation. Therefore, only the results are given in table 6.16.

As seen in the results, the best-result model is the DWT + TP + INCA based model, since it yielded a 96.6% classification accuracy using the fine *k*NN classifier.

**Table 6.16.** The results of the TQWT and BP based EEG classification.

| Feature selection method | Classifier | Number of features | The best result (%) |
|---|---|---|---|
| NCA | Medium Gaussian SVM | 512 | 87.2 |
| ReliefF | Ensemble bagged tree | 512 | 84.6 |
| RFNCA | Cubic SVM | 512 | 86.0 |
| INCA | Fine *k*NN with city block distance | 843 | 88.6 |

## 6.6 Conclusions

In this chapter variable EEG classification models are presented by employing BP and TP feature generators. Using these feature generators, 256 and 512 features are extracted, respectively. These feature extractors are applied to the Bonn EEG dataset. Also, 1D-DWT and TQWT decomposition algorithms are used to generate multilevel feature generators. In the feature selection phase four selectors are used and these are NCA, ReliefF, RFNCA and INCA. MATLAB codes are also provided for the implementation of the presented EEG signal classification model. According to the results, the best classification accuracy was obtained by employing the DWT + TP + INCA method using the fine *k*NN classifier. This method yielded 96.6% classification accuracy.

## References


[1] Paszkiel S 2020 *Analysis and Classification of EEG Signals for Brain-Computer Interfaces* (Cham: Springer)

[2] Khosla A, Khandnor P and Chand T 2020 A comparative analysis of signal processing and classification methods for different applications based on EEG signals *Biocybern. Biomed. Eng.* **40** 649–90

[3] Berger H 1929 Über das Elektroenkephalogramm des Menschen *Archiv. Psychiatr. Nervenkr.* **87** 527–70

[4] Mosher J C, Baillet S and Leahy R M 1999 EEG source localization and imaging using multiple signal classification approaches *J. Clin. Neurophysiol.* **16** 225–38

[5] Subasi A 2019 *Practical Guide for Biomedical Signals Analysis using Machine Learning Techniques: A MATLAB based Approach* (Cambridge, MA: Academic)

[6] de Aguiar Neto F S and Rosa J L G 2019 Depression biomarkers using non-invasive EEG: a review *Neurosci. Biobehav. R.* **105** 83–93

[7] Siuly S, Alcin O F, Bajaj V, Sengur A and Zhang Y 2018 Exploring Hermite transformation in brain signal analysis for the detection of epileptic seizure *IET Sci. Meas. Technol.* **13** 35–41

[8] Ullal A and Pachori R B 2020 EEG signal classification using variational mode decomposition, arXiv: 2003.12690

[9] Taran S, Bajaj V and Siuly S 2017 An optimum allocation sampling based feature extraction scheme for distinguishing seizure and seizure-free EEG signals *Health Inform. Sci. Syst.* **5** 7

[10] Bajaj V, Rai K, Kumar A and Sharma D 2017 Time–frequency image based features for classification of epileptic seizures from EEG signals *Biomed. Phys. Eng. Express* **3** 015012
[11] Subasi A and Gursoy M I 2010 EEG signal classification using PCA, ICA, LDA and support vector machines *Expert Syst. Appl.* **37** 8659–66
[12] Waterhouse E 2003 New horizons in ambulatory electroencephalography *IEEE Eng. Med. Biol.* **22** 74–80
[13] Pieter B, Benjamin S, David V and Dirk S 2008 Real-time epileptic seizure detection on intra-cranial rat data using reservoir computing *Advances in Neuro-Information Processing. ICONIP 2008* ed M Köppen, N Kasabov and G Coghill (Lecture Notes in Computer Science vol 5506) (Berlin: Springer) pp 56–63
[14] Subasi A and Ercelebi E 2005 Classification of EEG signals using neural network and logistic regression *Comput. Methods Prog. Bio.* **78** 87–99
[15] Alkan A, Koklukaya E and Subasi A 2005 Automatic seizure detection in EEG using logistic regression and artificial neural network *J Neurosci. Meth.* **148** 167–76
[16] Ali Shahidi Z, Manouchehr J, Guy A D and Reza T 2010 Automated real-time epileptic seizure detection in scalp EEG recordings using an algorithm based on wavelet packet transform *IEEE T. Bio-Med. Eng.* **57** 1639–51
[17] Subasi A 2007 Application of adaptive neuro-fuzzy inference system for epileptic seizure detection using wavelet feature extraction *Comput. Biol. Med.* **37** 227–44
[18] Peker M, Sen B and Delen D 2016 A novel method for automated diagnosis of epilepsy using complex-valued classifiers *IEEE J. Biomed. Health* **20** 108–18
[19] Khanmohammadi S and Chou C-A 2018 Adaptive seizure onset detection framework using a hybrid PCA–CSP approach *IEEE J. Biomed. Health* **22**
[20] Shafiul Alam S M and Bhuiyan M I H 2013 Detection of seizure and epilepsy using higher order statistics in the EMD domain *IEEE J Biomed. Health* **17** 312–8
[21] Subasi A 2006 Automatic detection of epileptic seizure using dynamic fuzzy neural networks *Expert Syst. Appl.* **31** 320–8
[22] Sabeti M, Katebi S and Boostani R 2009 Entropy and complexity measures for EEG signal classification of schizophrenic and control participants *Artif. Intell. Med.* **47** 263–74
[23] Johnstone J and Thatcher R W 1991 Quantitative EEG analysis and rehabilitation issues in mild traumatic brain injury *J. Insur. Med.* **23** 228–32
[24] Ford J M, Roach B J, Faustman W O and Mathalon D H 2008 Out-of-synch and out-of-sorts: dysfunction of motor-sensory communication in schizophrenia *Biol. Psychiat.* **63** 736–43
[25] Sayyari E, Farzi M, Estakhrooeieh R R, Samiee F and Shamsollahi M B 2012 Migraine analysis through EEG signals with classification approach *2012 11th Int. Con. on Information Science, Signal Processing and their Applications (ISSPA)* (Piscataway, NJ: IEEE) pp 859–63
[26] Estrada E, Nazeran H, Nava P, Behbehani K, Burk J and Lucas E 2004 EEG feature extraction for classification of sleep stages *The 26th Annual Int. Con. of the IEEE Engineering in Medicine and Biology Society* (Piscataway, NJ: IEEE) pp 196–9
[27] Cao C, Tutwiler R L and Slobounov S 2008 Automatic classification of athletes with residual functional deficits following concussion by means of EEG signal using support vector machine *IEEE T. Neur. Sys. Reh.* **16** 327–35
[28] Bakhshali M A, Khademi M, Ebrahimi-Moghadam A and Moghimi S 2020 EEG signal classification of imagined speech based on Riemannian distance of correntropy spectral density *Biomed. Signal Proces* **59** 101899

[29] Zhao S and Rudzicz F 2015 Classifying phonological categories in imagined and articulated speech *2015 IEEE Int. Con. on Acoustics, Speech and Signal Processing (ICASSP)* (Piscataway, NJ: IEEE) pp 992–6

[30] George S T, Subathra M, Sairamya N, Susmitha L and Premkumar M J 2020 Classification of epileptic EEG signals using PSO based artificial neural network and tunable-Q wavelet transform *Biocybern. Biomed. Eng.* **40** 709–28

[31] Selvaraj T G, Ramasamy B, Jeyaraj S J and Suviseshamuthu E S 2014 EEG database of seizure disorders for experts and application developers *Clin. EEG Neurosci.* **45** 304–9

[32] Selvan S E, George S T and Balakrishnan R 2015 Range-based ICA using a nonsmooth quasi-Newton optimizer for electroencephalographic source localization in focal epilepsy *Neural Comput.* **27** 628–71

[33] Obeid I and Picone J 2016 The temple university hospital EEG data corpus *Front. Neurosci-Switz.* **10** 196

[34] Aydemir E, Tuncer T and Dogan S 2020 A tunable-Q wavelet transform and quadruple symmetric pattern based EEG signal classification method *Med. Hypotheses* **134** 109519

[35] Andrzejak R G, Lehnertz K, Mormann F, Rieke C, David P and Elger C E 2001 Indications of nonlinear deterministic and finite-dimensional structures in time series of brain electrical activity: dependence on recording region and brain state *Phys. Rev.* E **64** 061907

[36] Bajaj V, Taran S, Khare S K and Sengur A 2020 Feature extraction method for classification of alertness and drowsiness states EEG signals *Appl. Acoust.* **163** 107224

[37] Goldberger A L, Amaral L A, Glass L, Hausdorff J M, Ivanov P C, Mark R G, Mietus J E, Moody G B, Peng C-K and Stanley H E 2000 PhysioBank, PhysioToolkit, and PhysioNet: components of a new research resource for complex physiologic signals *Circulation* **101** e215–20

[38] Rahman M A, Hossain M F, Hossain M and Ahmmed R 2020 Employing PCA and *t*-statistical approach for feature extraction and classification of emotion from multi-channel EEG signal *Egyptian Informat. J.* **21** 23–35

[39] 2020 SEED Datase http://bcmi.sjtu.edu.cn/home/seed/

[40] Tuncer T, Dogan S, Ertam F and Subasi A 2020 A novel ensemble local graph structure based feature extraction network for EEG signal analysis *Biomed. Signal Proces.* **61** 102006

[41] Gandhi T, Panigrahi B K, Bhatia M and Anand S 2010 Expert model for detection of epileptic activity in EEG signature *Expert Syst. Appl.* **37** 3513–20

[42] Gandhi T K, Chakraborty P, Roy G G and Panigrahi B K 2012 Discrete harmony search based expert model for epileptic seizure detection in electroencephalography *Expert Syst. Appl.* **39** 4055–62

[43] Gandhi T, Panigrahi B K and Anand S 2011 A comparative study of wavelet families for EEG signal classification *Neurocomputing* **74** 3051–7

[44] Sharma M, Patel S and Acharya U R 2020 Automated detection of abnormal EEG signals using localized wavelet filter banks *Pattern Recogn. Lett.* **133** 188–94

[45] Jaiswal A K and Banka H 2017 Local pattern transformation based feature extraction techniques for classification of epileptic EEG signals *Biomed. Signal Proces.* **34** 81–92

[46] Kumar T S, Kanhangad V and Pachori R B 2015 Classification of seizure and seizure-free EEG signals using local binary patterns *Biomed. Signal Proces.* **15** 33–40

[47] Kocadagli O and Langari R 2017 Classification of EEG signals for epileptic seizures using hybrid artificial neural networks based wavelet transforms and fuzzy relations *Expert Syst. Appl.* **88** 419–34

[48] Ghaemi A, Rashedi E, Pourrahimi A M, Kamandar M and Rahdari F 2017 Automatic channel selection in EEG signals for classification of left or right hand movement in brain computer interfaces using improved binary gravitation search algorithm *Biomed. Signal Process.* **33** 109–18

[49] Übeyli E D 2009 Combined neural network model employing wavelet coefficients for EEG signals classification *Digit. Signal Process.* **19** 297–308

[50] Zhang B, Wang W, Xiao Y, Xiao S, Chen S, Chen S, Xu G and Che W 2020 Cross-subject seizure detection in EEGs using deep transfer learning *Comput. Math. Method M.* **2020** 7902072

[51] Gao Y, Gao B, Chen Q, Liu J and Zhang Y 2020 Deep convolutional neural network-based epileptic electroencephalogram (EEG) signal classification *Front. Neurol.* **11** 375

[52] Abbasi M U, Rashad A, Basalamah A and Tariq M 2019 Detection of epilepsy seizures in neo-natal EEG using LSTM architecture *IEEE Access* **7** 179074–85

[53] Alickovic E, Kevric J and Subasi A 2018 Performance evaluation of empirical mode decomposition, discrete wavelet transform, and wavelet packed decomposition for automated epileptic seizure detection and prediction *Biomed. Signal Proces.* **39** 94–102

[54] Subasi A, Kevric J and Canbaz M A 2019 Epileptic seizure detection using hybrid machine learning methods *Neural Comput. Appl.* **31** 317–25

[55] Savadkoohi M and Oladduni T 2020 A machine learning approach to epileptic seizure prediction using electroencephalogram (EEG) signal *Biocybern. Biomed. Eng.* **40** 1328–41

[56] Liu C-L, Xiao B, Hsaio W-H and Tseng V S 2019 Epileptic seizure prediction with multi-view convolutional neural networks *IEEE Access* **7** 170352–61

[57] Abiyev R, Arslan M, Idoko J B, Sekeroglu B and Ilhan A 2020 Identification of epileptic EEG signals using convolutional neural networks *Appl. Sci.-Basel* **10** 4089

[58] Ojala T, Pietikäinen M and Mäenpää T 2001 A generalized local binary pattern operator for multiresolution gray scale and rotation invariant texture classification *Int. Con. on Advances in Pattern Recognition* (Berlin: Springer) pp 399–408

[59] Ojala T, Pietikäinen M and Mäenpää T 2002 Multiresolution gray-scale and rotation invariant texture classification with local binary patterns *IEEE T. Pattern Anal.* **24** 971–87

[60] Kaya Y, Uyar M, Tekin R and Yıldırım S 2014 1D-local binary pattern based feature extraction for classification of epileptic EEG signals *Appl. Math. Comput.* **243** 209–19

[61] Tan X and Triggs B 2010 Enhanced local texture feature sets for face recognition under difficult lighting conditions *IEEE T. Image Process.* **19** 1635–50

[62] Kuncan M, Kaplan K, Minaz M R, Kaya Y and Ertunc H M 2020 A novel feature extraction method for bearing fault classification with one dimensional ternary patterns *ISA T.* **100** 346–57

[63] Wu X, Kumar V, Quinlan J R, Ghosh J, Yang Q, Motoda H, McLachlan G J, Ng A, Liu B and Philip S Y 2008 Top 10 algorithms in data mining *Knowl. Inf. Syst.* **14** 1–37

[64] Grąbczewski K 2014 *Meta-learning in Decision Tree Induction* vol 1 (Berlin: Springer)

[65] Flach P 2012 *Machine Learning: The Art and Science of Algorithms that Make Sense of Data* (Cambridge: Cambridge University Press)

[66] Subasi A 2020 *Practical Machine Learning for Data Analysis Using Python* (New York: Academic)

[67] Skurichina M and Duin R P 2002 Bagging, boosting and the random subspace method for linear classifiers *Pattern Anal. Appl.* **5** 121–35

[68] Lai C-C and Tsai C-C 2010 Digital image watermarking using discrete wavelet transform and singular value decomposition *IEEE T. Instrum. Meas.* **59** 3060–3
[69] Krishna A H, Sri A B, Priyanka K Y V S, Taran S and Bajaj V 2018 Emotion classification using EEG signals based on tunable-Q wavelet transform *IET Sci. Meas. Technol.* **13** 375–80
[70] Taran S and Bajaj V 2019 Motor imagery tasks-based EEG signals classification using tunable-Q wavelet transform *Neural Comput. Appl.* **31** 6925–32
[71] Khare S K, Bajaj V and Sinha G 2020 Adaptive tunable Q wavelet transform based emotion identification *IEEE T. Instrum. Meas.* **69** 9607–17
[72] Reddy T R, Vardhan B V, GopiChand M and Karunakar K 2018 *Intelligent Engineering Informatics* (Berlin: Springer) pp 169–76
[73] Raghu S and Sriraam N 2018 Classification of focal and non-focal EEG signals using neighborhood component analysis and machine learning algorithms *Expert Syst. Appl.* **113** 18–32
[74] Mohdiwale S, Sahu M and Sinha G 2020 LJaya optimisation-based channel selection approach for performance improvement of cognitive workload assessment technique *Electron. Lett.*
[75] Mohdiwale S, Sahu M, Sinha G and Bajaj V 2020 Automated cognitive workload assessment using logical teaching learning based optimization and PROMETHEE multi-criteria decision making approach *IEEE Sens. J.*
[76] Wang H, Chen J and Dong G 2014 Feature extraction of rolling bearing's early weak fault based on EEMD and tunable Q-factor wavelet transform *Mech. Syst. Signal. Pr.* **48** 103–19
[77] Tuncer T, Ertam F, Dogan S and Subasi A 2020 An automated daily sport activities and gender recognition method based on novel multi-kernel local diamond pattern using sensor signals *IEEE T. Instrum. Meas.* **69** 9441–8
[78] Tuncer T, Dogan S, Özyurt F, Belhaouari S B and Bensmail H 2020 Novel multi center and threshold ternary pattern based method for disease detection method using voice *IEEE Access* **8** 84532–40
[79] Ekız S and Erdoğmuş P 2017 Comparative study of heart disease classification *2017 Electric Electronics, Computer Science, Biomedical Engineerings' Meeting (EBBT)* (Piscataway, NJ: IEEE) pp 1–4

**IOP** Publishing

Modelling and Analysis of Active Biopotential Signals in Healthcare, Volume 2

**Varun Bajaj and G R Sinha**

# Chapter 7

# Increasing the usability of the Devanagari script input based P300 speller

**Ghanahshyam B Kshirsagar and Narendra D Londhe**

*Department of Electrical Engineering, National Institute of Technology Raipur, Chhattisgarh, India, 492010*

The existing Devanagari script-based P300 speller (DS-P3S) suffers from (i) an increase in user-related problems during the spelling of a lengthy word using the crowdy row–column (RC) paradigm and (ii) a high mental demand due to the visual periphery task for event-related potential (ERP) generation which affects the quality of the ERP and performance.

To overcome the above-mentioned problems, we proposed (i) a modified display paradigm called the zigzag RC paradigm, (ii) an overt attention based spelling task and (iii) an efficient compact model for classification of the ERPs in a single trial. The experiments are performed on an ERP-EEG dataset of 64 DS characters collected from seven healthy subjects. The experimental evaluation includes the analysis of the classification performance, the error rate in the character detection and the ERP amplitudes, and the evaluation of workload.

There is an approximately 5%–7% increment in character detection using the zigzag paradigm with covert attention compared to the RC paradigm and an 8%–11% increment using the zigzag paradigm with overt attention compared to the RC and zigzag paradigms with covert attention. Moreover, the high character detection rate in just a single trial, the improved ERP amplitude and the low mental demand score in the workload evaluation ensure the high usability of the zigzag paradigm with overt attention for DS-P3S.

## 7.1 Introduction

In most neuromuscular diseases, such as amyotrophic lateral sclerosis or spinal muscular atrophies, the patient partially or completely loses their voluntary control

doi:10.1088/978-0-7503-3411-2ch7  

of muscles [1–5]. Therefore, such patients are unable to communicate naturally. In such cases there is a need for external assistance for communication [1–5]. However, due to the loss of natural motor ability, only one possible mode of communication remains, i.e. cognitive ability. Therefore, to provide such external communication, a device called a brain–computer interface (BCI) system was proposed which acquires, processes and converts human cognitive responses into meaningful commands for communication and control applications [2]. These neurocognitive signals can be collected either invasively from the surface of the cortex, i.e. electrocorticographic (ECoG) signals, or non-invasively from the surface of the scalp, i.e. electro-encephalographic (EEG) signals [3]. The non-invasive EEG based approach is more popular and has been adopted by researchers to implement various BCI systems for communication and control [4]. EEG based BCI systems have been implemented in the design of various applications of control in the medical field, such as the design of neuro-rehabilitation and neuro-ergonomics devices and the control of wheelchairs, and in non-medical applications, such as authentication and security, in the entertainment field, such as gaming, for communication, such as EEG spellers, for neurofeedback based applications, such as neuromarketing, and in the education sector [5]. Of these applications, BCI based EEG spelling systems had been studied the most, in order to set up communication in various languages such as Devanagari script (DS) [6–11], English [12], Japanese [13], Arabic [14] and Chinese [15]. These speller systems work on event-related potential (ERP) or visually evoked potential (VEP) generated by presenting visual, audio or somatosensory based external stimuli to the user [5].

Using the above-mentioned neurofeedback, the P300 based EEG speller has been explored and implemented by many research communities due to its exceptional spatiotemporal characteristics and ease of use, as it requires negligible skills or practical training of the subject [1–15]. In most the P300 spellers, the oddball principle is used to generate an ERP [6–15]. This neurocognitive response to the desired stimulus produces a positive amplitude with latency at 250–500 ms [6–9] and has a maximum value at approximately 300 ms post-stimulus, hence it is referred to as P300 [6–9].

The existing implemented P300 based BCI systems for English, Japanese, Arabic, Chinese and DS utilize several display paradigm designs with different visual presentations, such as a matrix based display paradigm such as row–column (RC) [6–15] and single character (SC) [16], region-based (RB) paradigms [17], the checker board (CB) paradigm [18], the steady-state visual-evoked potential (SSVEP) [19, 20], the rapid serial visual presentation (RSVP)-based paradigm [21], the GIBS block speller [22], the hex-o-spell paradigm [23] and the geometric speller [24, 25].

The matrix based display paradigm was first designed by Farwell and Donchin [12], in which characters are randomly flashed in a row or column, and hence it is also called the RC paradigm. The standard RC paradigm gained popularity for its simplicity and ease of implementation. Thus researchers have implemented various approaches, such as state-of-the-art machine learning and advanced optimization techniques, to improve the classification performance [26–33]. These approaches have achieved accuracy ranging from 70%–96% in 5–15 trials. Later, researchers found that the conventional approaches, even after adopting channel selection, did not provide satisfactory results on reducing the number of trials to spell the characters. Hence, further advancements of existing approaches, such as adopting ensemble learning [10, 32, 34] and other advanced techniques such as particle filters [35], adaptive learning [34], partially observable Markov decision processes (POMDP) [35], dynamic stopping [34], predictive spelling [37–41] and deep learning [6–9, 42, 43], have been proposed to improve the accuracy with fewer trials and achieved accuracies ranging from 88.34%–94.80% in 3–9 trials, respectively. These approaches have managed to achieve a information transfer rate of 32.33–59.39 BPM with the minimum spelling time ranging from 9–14 s (i.e. 3–6 trials). Later, a group of researchers found that RC created some user-related problems, such as adjacency distraction, crowding effects and double flashing. Therefore, advancements of the display paradigm to overcome such problems have been proposed in the last two decades [16–25]. A modified matrix based paradigm such as the SC paradigm was proposed by Guan and Wu [25]. Although the SC approach increases the performance by reducing the user-related problems, it still has a major issue with spelling time as it takes approximately four times longer to spell a character than in the standard RC approach. Further, to overcome the crowding effect of the matrix based paradigm, a region-based display paradigm was implemented [26]. However, the highly complex design of the RB paradigm became the main hurdle for its online implementation. Next, RSVP and its variant [21] based ERP generation were found to be quite efficient and showed promising results in the accuracy of character detection, as this approach only contains a single flashing window, unlike the matrix based approaches which have a whole set of characters in the display paradigm. This particular approach overcomes the problems of the matrix based paradigm. However, RSVP requires similar spelling time as in the SC paradigm, which is four times more than that of the RC paradigm. Another type of evoked potential based EEG speller system, i.e. the steady-state visual-evoked potential (SSVEP), was implemented recently [19, 20]. Unfortunately, this approach is quite risky for patients with epileptic seizures as the flickering nature of the visual paradigm may trigger seizures. Further, other non-matrix based spellers such as GIBS block [22], hex-o-spell [23] and geospell [24, 25] are proposed to overcome the limitations of the matrix based paradigm. However, these paradigms are highly complex to design and again have very poor spelling time compared to the RC approach. Therefore, we have still not addressed the main objective of reducing the trade-off between spelling time and performance.

Based on the above-described literature on the display paradigm, although the RC paradigm has a few user-related issues, it also has several advantages, such as shorter spelling time and ease of implementation. Hence, in previous studies on DS-P3S [6–11] the RC paradigm with a 8×8 matrix of Devanagari letters and other characters was implemented. However, the user-related problems associated with the RC paradigm were found to increase due to greater number of characters in the Devanagari script. As a result, the accuracy of the DS based P300 speller is limited to 94%–95% for 15 trials and is found to be degraded drastically for a reduced number of trials, even after adopting state-of-the-art classifiers.

Even after adopting advanced machine (deep) learning approaches, the performance of the existing DS-P3S suffers from the following problems:

- The existing DS paradigm has 64 characters to spell which are arranged in an 8 × 8 matrix. The increased number of characters makes the paradigm more crowded. This leads to a further increase in user-related problems such as the crowding effect, adjacency distraction, double flashing, task difficulty and fatigue.
- The DCNN and WE-DCNN used for the classification of P300 in the existing DS-P3S and other EEG based applications [44–48] generate numerous trainable parameters. This makes it computationally cumbersome and requires high computational power and a lot of time.
- The standard covert attention based data acquisition protocol involves counting of the flashing of the target row–column. This leads to an increase in the mental demand on the user as it requires remembering the count during the long flashing cycle. This generates an error-related potential (ErrP) corresponding to the flashing of the non-target characters.

These user-related problems require heavy-duty classifiers to provide satisfactory performance. Therefore, user-related problems, computational demand and mental demand restrict the usability of the DS-P3S. On the one hand, to handle the display paradigm related problems, there is a need to optimize the design of the RC paradigm which helps to reduce the crowding effect and thus any adjacency distraction problems. On the other hand, to reduce the mental demand, a few studies have explored the use of an attentional blink task for the spelling of the target character [23], wherein intentional blinking is used to generate the ERP instead of imaginary counting. This has shown proven results to reduce the mental demand and thus the fatigue problems of the user. However, an intentional blink has very similar morphology to a non-intentional or involuntary blink, hence it is quite a challenging task to separate the two. To overcome this problem, we are replacing the single blinking task with a double blink. Therefore, to increase the usability of the DS-P3S we are proposing the following hypotheses.

- Hypothesis 1: Implementation of a zigzag RC paradigm to reduce the crowding effect and adjacency distraction problem.
- Hypothesis 2: Overt attention based data acquisition wherein the user is asked to perform eye moments instead of a counting task. This will help to reduce the mental demand on the subject and will reduce other user-related problems such as task difficulty and fatigue.
- Hypothesis 3: Implementation of a compact convolutional neural network (CNN) for accurate and robust detection of P300 components.

The above-mentioned hypotheses will provide a novel display paradigm, an efficient and compact CNN based classification model, and finally a new data acquisition protocol which reduces the mental demand as well as user-related problems. To evaluate the hypotheses, the experiments are performed on an ERP–EEG dataset of 64 DS characters collected from seven healthy subjects. Moreover, the experimental evaluation, including the analysis of the classification performance, the error rate in the character detection, the ERP amplitudes and the workload are also presented in this work.

## 7.2 Methodology

In this section we present the details of the design of both display paradigms (i.e. conventional RC and the proposed zigzag RC), the proposed data acquisition and preprocessing methods, and the classification of P300 components using a compact CNN.

### 7.2.1 Design of the DS display paradigm

The display paradigm is one of the key components in visual P300 spellers, as it provides the external stimulus to the user. Hence, it also plays an important role in the quality of the ERP in the case of amplitude and latency. Therefore, the efficient and intelligent design of a display paradigm has prime importance to improve the performance of P300 spellers. In the existing DS-P3S, the 8 × 8 standard RC paradigm was implemented, which includes 64 characters, as shown in figure 7.1(a). The details of the display paradigm design are given in [6]. As discussed in the earlier studies on DS-P3S, the increased number of characters and the size of the matrix make the standard RC paradigm more crowded. Therefore, the user may experience distraction during the spelling of the characters [6]. A detailed pictorial representation of user-related problems in the DS display paradigm is presented in figure 7.1 and a detailed description is provided in [6].

Therefore, to increase the inter-character distance in the rows and columns we propose a modified RC paradigm called the zigzag RC DS paradigm. As the name suggests, the characters are arranged in the same 8 × 8 matrix but in a zigzag pattern, as shown in figure 7.2. Due to the shifting of the alternate rows, the display paradigm becomes zigzag and it is found to reduce the number of characters surrounding the targeted spelling character from eight to six as shown in figure 7.3(a). The proposed zigzag paradigm has one change: as two DS characters, i.e. ङ ('ng') and ञ ('ny'), are not frequently used in written Hindi and Marathi text, so these two characters are replaced with special characters for the commands of

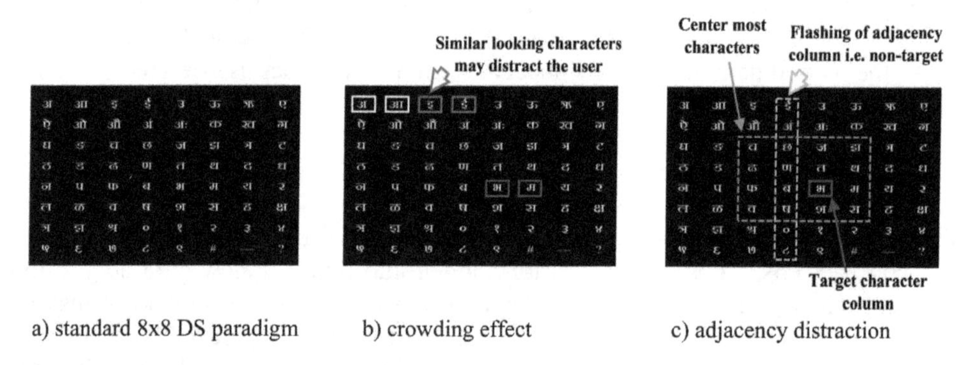


**Figure 7.1.** A pictorial representation of the user-related challenges in the DS paradigm: (a) the standard 8 × 8 DS paradigm, (b) the crowding effect and (c) adjacency distraction.

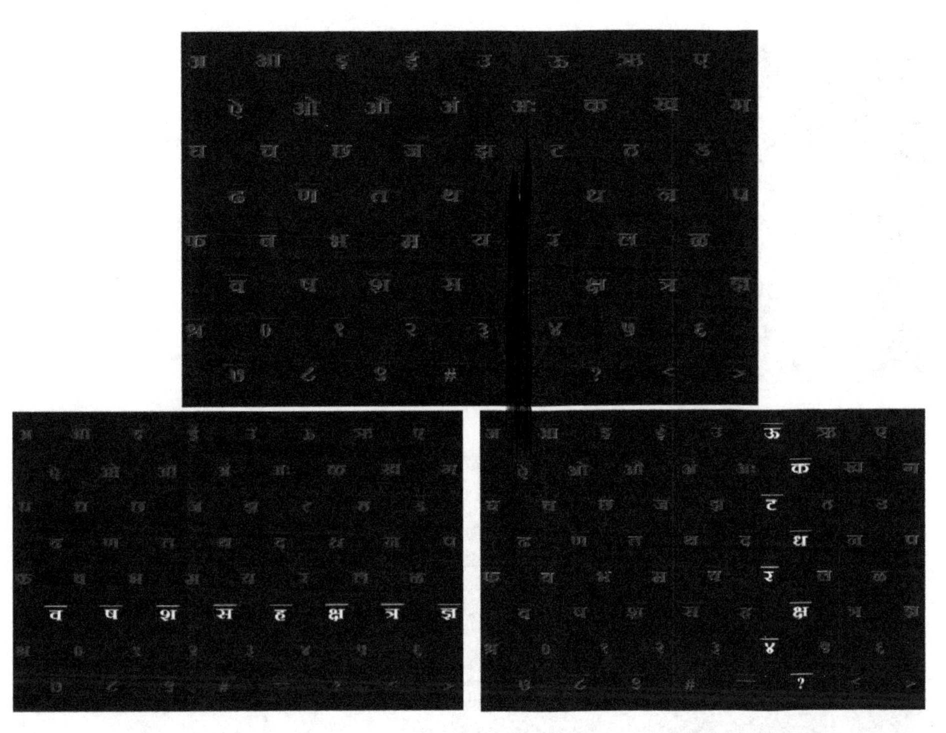

**Figure 7.2.** The proposed 8x8 matrix based zigzag RC paradigm for DS-P3S.

forward shifting the cursor and deleting a character, i.e. '>' and '<', respectively, as shown in figure 7.1. In experiments with both paradigms, i.e. the existing standard RC 8 × 8 DS paradigm and the proposed zigzag RC DS paradigm, a standard display screen of size 22 cm × 42 cm (height × width of the actual display screen) is used. In the existing standard RC 8 × 8 DS paradigm, the distance between the characters in the same row or column is $d_w$ and the distance between the characters in different rows and columns is $d_b = \sqrt{(d_w)^2 + (d_w)^2}$, as shown in figure 7.3(a). However, in the proposed zigzag paradigm, the distance between the characters in the same row and column remain the same, i.e. $d_w$, but the distance between different rows and columns is increased by $d_w/2$ as the width is increased by $(w + d_w/2)$, where $w$ is the original width between the adjacent row and column. Hence, the distance between the adjacent characters in RC is $d_{b(\text{new})} = \sqrt{(d_w)^2 + (w + d_w/2)^2}$, as shown in figure 7.3(b). In addition, we have also visually modified the conventional display, for example designing a cubic character ($h \times w \times d$) (see figure 7.3(c)), and the use of colored characters to increase the attention of the user. Moreover, in earlier studies, it has been found that the visual modifications in the standard RC paradigm, i.e. the use of a blue-green color for the flashing of the row–column, improved the attention of the user. However, the flashing of blue-green feels unpleasant to the user. Therefore, we have modified the flashing to be blue-white to maintain the user's attention without any unpleasant feelings.

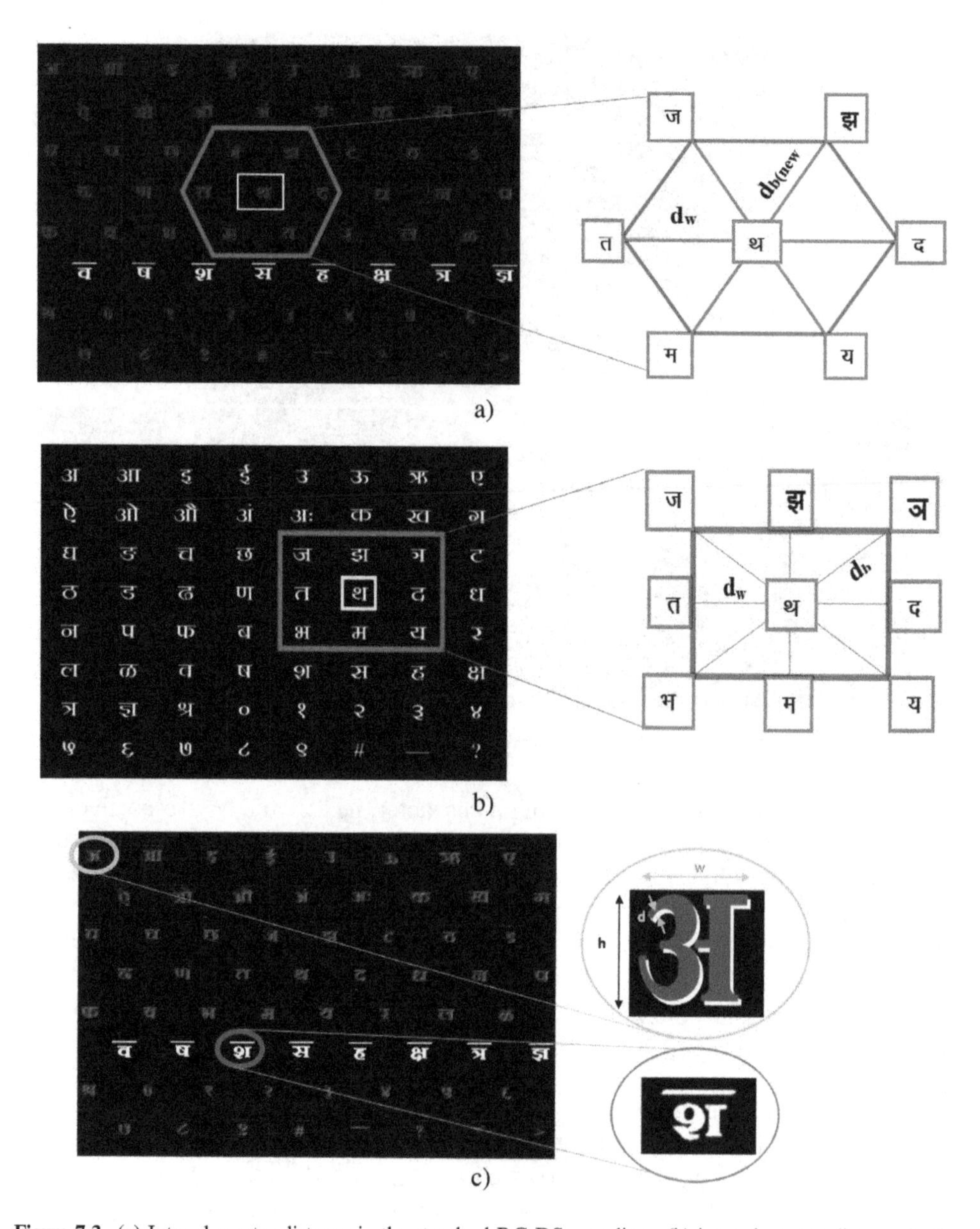


**Figure 7.3.** (a) Inter-character distance in the standard RC DS paradigm, (b) inter-character distance in the proposed zigzag RC DS paradigm and (c) visual modifications in the proposed zigzag paradigm.

### 7.2.2 Data acquisition and preprocessing

The cognitive responses corresponding to desired characters are recorded using a 16 channel EEG recorder at a 500 Hz sampling rate from seven healthy subjects at the National Institute of Technology Raipur, India. The standard 10%–20% EEG arrangement is used to place 16 electrodes (FZ, FC1, FC2, C3, CZ, C4, CP5, CP1, CP2, CP6,

**Table 7.1.** The details of the dataset in both the standard RC and zigzag paradigm using covert and overt attention tasks.

| Data acquisition protocol | Paradigm | P300 ($T \times N_c \times m_1$) | Non-P300 ($T \times N_c \times m_2$) |
|---|---|---|---|
| Covert attention (1D) | Standard RC | 300 × 30 720 | 300 × 1689 600 |
| | Zigzag | 300 × 30 720 | 300 × 1689 600 |
| Overt attention (1D) | Zigzag | 400 × 30 720 | 400 × 1689 600 |
| Covert attention (2D) | Standard RC | 300 × 16 × 13 440 | 300 × 16 × 94 082 |
| | Zigzag | 300 × 16 × 13 440 | 300 × 16 × 94 082 |
| Overt attention (2D) | Zigzag | 400 × 16 × 13 440 | 400 × 16 × 94 082 |

Note: $m$ (total number of ERPs) = $m_1$ (number of P300 components) + $m_2$ (number of non-P300 components)

P3, PZ, P4, O1, OZ and O2) on the different scalp regions, i.e. frontal, parietal, central and occipital. The previous covert attention based ERP recording methods increased the mental demand on the user as they required remembering the count over a long flashing cycle, as in the 8 × 8 DS paradigm. This leads to the generation of an error-related potential (ErrP) during the flashing of non-desired characters. Therefore, in this work we have designed an overt attention based (i.e. double blinking task) EEG recording to reduce the mental demand on the user. Unlike other overt attention based systems that used an extra set of electrodes to record eye movements, i.e. EOG, we use the same set of scalp electrodes to record the overt attention in the response of the user. This reduces implementation costs as well as providing greater comfort to the subject, as it is free of extra sets of electrodes other than on the scalp. To record the overt attention, the user is asked to voluntarily double blink when the desired character is flashed. Here, the purpose of the double blinking is to differentiate the involuntary blinking noise from double blinking. As the paradigm contains an 8 × 8 row–column matrix, it generates a total of 16 flashes in a single trial, with further categories of two targets and the remaining 14 non-targets. Likewise, the dataset for 15 trials is recorded to meet the standard recording protocol. The same process was followed to collect the ERP responses of seven healthy subjects for 64 characters. The details of the flashing duration are: flashing time = 200 ms, inter-flashing interval (pause) = 50 ms and inter-character interval = 2 s. Further, the recorded EEG signals are preprocessed to enhance the signal-to-noise ratio (SNR). For that, bandpass filtering with 0.1–20 Hz of passband frequency is applied and post-flashing components between 1–800 ms were extracted as the voluntary eye closure response requires approximately between 300–400 ms [23]. Moreover, it helps to remove redundant or extra unwanted information from the EEG signal. Therefore, the final dataset becomes $D \in \mathbb{R}^{T \times N_c \times m}$, where $T = 400$, $N_c = 16$ (number of electrodes) and $m = 107\,520$ (64 characters × 15 trials × 16 P300 and non-P300 × seven subjects) as shown in table 7.1. Note that the data are further reshaped for a 1D convolution experiment to compare the performance with the existing 1D DCNN from our previous work. The details of the 1D and 2D shape data are given in table 7.1.

As seen in table 7.1, the P300 and non-P300 components are in a ratio of 1:7. This high imbalance may lead to bias in the classifier. Therefore, we have used a weighted class balanced approach by assigning lower weights to the majority class, i.e. non-P300 components and vice versa.

### 7.2.3 Classification of P300

In the existing DS-P3S, different classification approaches, such as state-of-the-art machine learning [10, 11] and deep learning [6, 8, 9] with ensemble learning [7], are used for accurate detection of P300 components. However, in the case of conventional machine learning, the classification accuracy gradually decreases on reducing the number of trials, even when adopting channel optimization approaches. This trade-off has been addressed by using DCNN [6] and WE-DCNN [7], but it creates the problem of generating numerous trainable parameters and increasing the computational burden. Therefore, there is a need for a compact and efficient model that will handle both the problems of conventional machine learning and deep learning approaches [44]. For that purpose we have implemented efficient shallow and compact CNN models. The main contribution is in the development of a modified shallow CNN [43] and EEGNet [50] with a filter pruning [49] approach to further reduce the trainable parameters. In filter pruning, in both models, we selected the kernel size to be one quarter of the sampling frequency, i.e. 500 Hz/4 = 125, to collect the information above 4 Hz, i.e. the theta band. This also reduces the number of parameters by half compared to the original at the first layer of convolution. Furthermore, we have used a windowing approach to limit the upper bandwidth of the EEG. For that purpose, the overlapping windowing with the size $w = 8$ (25 ms) is selected to extract a frequency below 40 Hz in our proposed model. Hence, using this approach we are collecting information between 4–40 Hz, i.e. the theta to the beta range. The details of the architecture are given below.

#### *7.2.3.1 The shallow CNN*

In the shallow CNN, a five-layered model is implemented, which includes the first temporal layer with an activation and subsampling layer, followed by a spatial convolution with activation and subsampling to reduce the dimensionality. Finally, there is a classification layer which predicts the probability of the P300 and non-P300 components. The details of the architecture are presented below and are pictorially represented in figure 7.4.

**Block 1** ($L_0$): Input

The input ERP–EEG data $D \in \mathbb{R}^{T \times N_c \times m}$ in the form of a tensor with order 3 is provided to the model where $T$ is the number of components in ERP, $N_c$ is the number of channels and $m$ is the number of samples, as shown in table 7.1.

**Block 2** ($L_1$–$L_2$): Temporal information

In $L_1$, to extract the temporal information from the given data, a convolution operation is performed with a kernel of size $N = (n, 1)$ is performed over the length of the ERP ($T$), where $n = F_s/4 = 125$ is selected to extract the frequency above 4 Hz, $D_1 = N_c = 16$ is the depth, the stride = 8 with window size ($w$) = 25 ms or 50 samples to extract the information below 40 Hz. The mathematical representation of the convolution operation at $L_1$ is

$$C^T(j) = f\left(\sum_i^{N_c}, \sum_{j=1}^{j \leqslant T} D_j^T \cdot w_j^T(n, 1) + b_j\right), \tag{7.1}$$

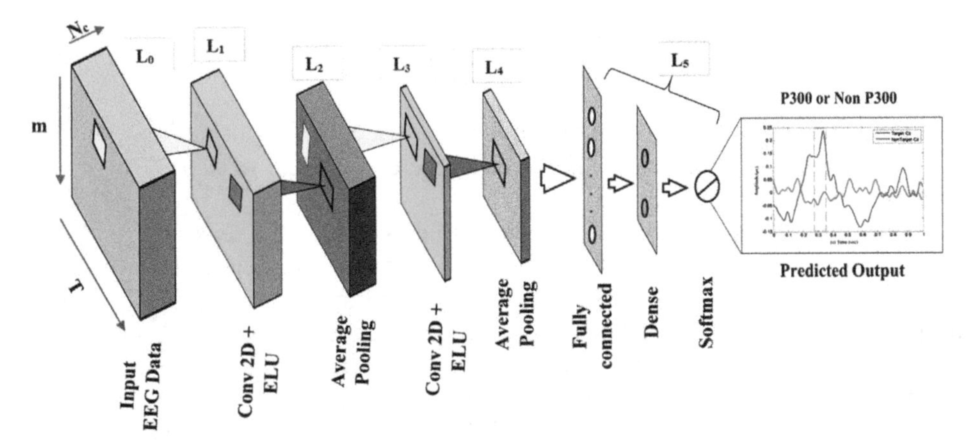


**Figure 7.4.** A pictorial representation of the architecture of the shallow CNN.

where $C^T(j)$ is the output of convolution performed over the temporal domain, i.e. $1 \geqslant j \leqslant T$, followed by a subsampling layer, i.e. average pooling is used at layer $L_2$ with a filter size of (6, 1) to down-sample the sampling frequency to approximately 80 Hz. The main advantage of the average pooling over max-pooling is that it takes the average over the selected size of the kernel. Therefore, other relevant information is also extracted which is neglected using max-pooling.

**Block 3** ($L_3$–$L_4$): Spatial information

At $L_3$ convolution over a spatial domain with the kernel of size (1, $N_c$) is performed to eliminate the redundant electrodes, where $N_c = 16$ is the number of electrodes, depth $D_2 = 2 \times D_1$ and the stride = 1. The mathematical representation of the convolution operation at $L_3$ is

$$C^{N_c}(j) = f\left(\sum\nolimits_{i}^{N_c}, \sum\nolimits_{j=1}^{j \leqslant T} D_j^{N_c}. w_j^{N_c}(n, 1) + b_j\right), \tag{7.2}$$

where $C^{N_c}(j)$ is the output of convolution performed over the temporal domain, i.e. $1 \geqslant j \leqslant N_c$, followed by a subsampling layer, i.e. average pooling is used at layer $L_4$ of filter size (2, 1) to down-sample the sampling frequency to approximately 40 Hz.

**Block 4** ($L_5$): Class prediction

Before the prediction of the probabilities, the extracted temporal and spatial information is mapped at $L_5$, i.e. a dense layer which also reduces the dimensionality of the features. The mathematical expression of the dense operation is given below:

$$d = g\left(\sum_{j=1}^{x}, (C^T \times C^{N_c}) \times W + b\right), \tag{7.3}$$

where $g(\cdot)$ is the activation function, $d$ is the output of the dense layer and $\times$ is the reduced feature map which is equal to the number of classes.

Further, the maximum probability-based class predicted is performed using softmax and is mathematically expressed as

$$Y(X)_j = \frac{e^{x_i}}{\sum_1^k e^{x_k}}, \tag{7.4}$$

where $Y(X)_j$ is the probability obtained for $i$ classes, $x_i$ is the output of the layer $L_6$ and the denominator represents the sum of the exponential of $k$ ($k = 1, 2, \ldots, i$) components of the input $x$.

*Activation and regularization.* In previous studies on DS-P3S, leaky rectified linear units (ReLUs) [6, 7] were adopted to accelerate the training and to avoid the problem of dying ReLUs. However, on differentiation, the leaky ReLU has the propensity to turn out to be a linear function. Also, extra batch normalization layers were used in the existing DS-P3S to provide a covariate shift. However, this also generates the trainable parameters. Therefore, to overcome these two problems, we used an exponential linear unit (ELU) [51] which exhibits both properties, namely accelerating training with a partial non-linear nature even after differentiation. Therefore, we have used ELU activation at each convolution layer in this work.

*Loss function and optimizer.* In this work the binary cross-entropy loss function is used to calculate the deviation between the predicted and actual class, which is expressed as

$$L = -\sum_i O_i \ln(\hat{O}_i), \tag{7.5}$$

where $L$ is the loss, $O_i$ is the target categories and $(\hat{O}_i)$ is a predicted probability. Further, an adaptive learning approach of a gradient-based optimizer called RmsProp [52] is used to minimize the loss function in equation (7.5) by updating the weights. The main advantage of the RmsProp optimizer is that it uses a moving average of the square gradients of the previous gradient to update the next gradient. The weights are updated using the following equations:

$$E[g^2]_t = \gamma E[g^2]_{t-1} + (1 - \gamma)g^2_t \tag{7.6}$$

$$w_{t+1} = w_t - \frac{\eta}{\sqrt{E[g^2]_t + \varepsilon}}, \tag{7.7}$$

where $\gamma$ is the decay parameter, $\eta$ is the learning rate and $E[g^2]_t$ is the moving average of square gradients.

#### *7.2.3.2 Compact CNN*

Although the compact CNN includes the same convolutional process in the temporal feature extraction, it utilizes depthwise and separable layers to extract the spatial information. This is the major difference between the shallow CNN and compact CNN. The details for the temporal feature extraction are similar. Therefore, it is repeated as it is to provide proper flow and connectivity of the explanation. The details of the compact CNN are given below and its pictorial representation is given in figure 7.5.

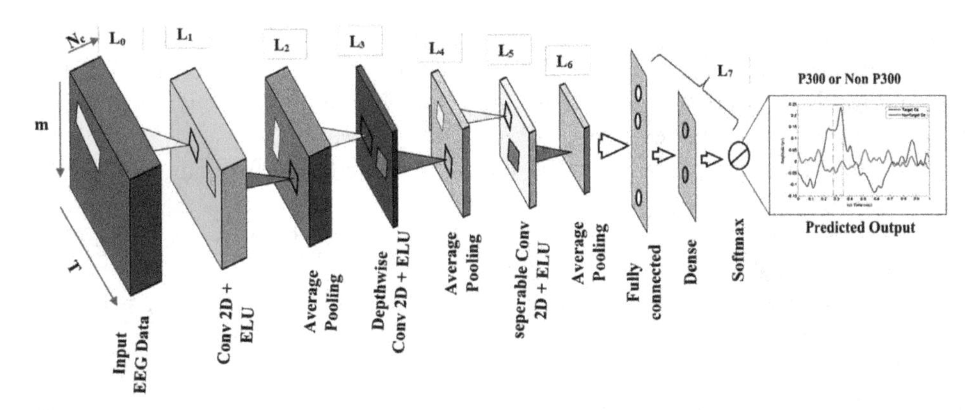


**Figure 7.5.** The pictorial representation of the architecture of the modified EEGNet.

**Block 1** ($L_0$): Input

The input ERP–EEG data $D \in \mathbb{R}^{T \times N_c \times m}$ in the form of a tensor with order three is provided to the model, where $T$ is the number of components in ERP, $N_c$ is the number of channels and $m$ is the number of samples, as shown in figure 7.5 and table 7.1.

**Block 2** ($L_1$–$L_2$): Temporal information

In $L_1$, to extract the temporal information from the given data, a convolution operation is performed, with a kernel of size $N = (n, 1)$, over the length of the ERP ($T$) using equation (7.1), where $n = F_s/4 = 125$ is selected to extract the frequency above 4 Hz, the depth is $D_1 = N_c = 16$, the stride = 8 with window size is $(w) = 25$ ms or 50 samples to extract the information below 40 Hz.

This is followed by a subsampling layer, i.e. average pooling is used at layer $L_2$ of filter size (6, 1) to down-sample the sampling frequency to approximately 80 Hz. The main advantage of the average pooling over max-pooling is it takes the average over the selected size of the kernel. Therefore, other relevant information is also extracted which is neglected using max-pooling.

**Block 3** ($L_3$–$L_5$): Spatial information

For the computation of the spatial information, a depthwise convolution of size $(1, N_c)$ is used at $L_3$. As depthwise convolutions do not provide full connections with their preceding layer, hence it helps to reduce the trainable parameters [50] as well as eliminating the lossy channels. The depth $D_1$ is selected based on the criteria of $D_1 = 2 \times F_1$ [50]. Furthermore, to reduce the trainable parameters, the second subsampling layer, i.e. average pooling of size (2, 1), is added at layer $L_4$ to down-sample the sampling frequency from 80 Hz to approximately 40 Hz. Here we kept the depth parameter $D = 1$ for our experiment to limit the trainable parameters. Followed by that, a separable convolution of size (1, 1) applied at layer $L_6$ is used to provide the pointwise learning [50] of the feature maps by decoupling the relationship within and between the feature maps. In addition, a separable convolution manages to reduce a few parameters.

**Block 4** ($L_6$–$L_7$): Class prediction

Before the prediction of the probabilities, the extracted temporal and spatial information is mapped at a dense layer by using equation (7.3) which also reduces the dimensionality of the features.

At $L_7$ the maximum probability-based class predicted is performed using softmax using equation (7.4).

*Activation and regularization.* A similar ELU is used here as an activation at layers $L_1$, $L_3$ and $L_5$.

*Loss function and optimizer.* Here as well, the binary cross-entropy loss function given in equation (7.5) is used to calculate the deviation between the predicted probabilities.

Further, an adaptive learning approach of a gradient-based optimizer called RmsProp is used to minimize the loss function in equation (7.5) by updating the weights using equations (7.6) and (7.7) (tables 7.2 and 7.3).

## 7.3 Experiments and results

### 7.3.1 Experimental set-up

This section presents the various experiments implemented in this work to evaluate the stated hypotheses. For the first hypothesis we are implementing a spelling task by designing a proposed zigzag paradigm and collecting the ERP–EEG using a covert attention task. For our second hypothesis, we are acquiring an ERP–EEG using overt attention tasks using the same proposed zigzag paradigm. Further, for the third and last hypothesis, we are implementing three different classification approaches, namely deep CNN, shallow CNN and compact CNN for the detection of P300 in both the first two experiments, i.e. overt and covert attention.

For implementing DCNN, shallow CNN and compact CNN, we used Python with the Keras [53] libraries and TensorFlow backend [54]. The selected parameters are as follows: batch size = 50, the initial weights and bias for experiments with 15 trials are initialized using the Glorot uniform method [55], dropout ($p$) = 0.25 [56], loss = binary cross-entropy (see equation (7.5)) and optimizer = RmsProp [52], with the parameters of RmsProp being a learning rate $\eta = 0.001$ and $\varepsilon = 0.9$ with no decay (see equations (7.6) and (7.7)). We have used a ten-fold cross validation to improve the generalization ability of the model.

### 7.3.2 Results of classification of P300

In this section we present the classification performance of the ERP–EEG collected using the covert attention task and overt attention task for the proposed zigzag RC DS paradigm using deep, shallow and compact CNN models. Moreover, we also present within trial and cross-subject analyses for the covert and overt attention tasks.

#### *7.3.2.1 Covert attention task*

Figure 7.6 illustrates the results of the P300 detection in the zigzag RC paradigm with covert attention using DCNN for classification. In cross-trial and cross-subject analysis, it is observed that the range of accuracy for 15, 12, 9, 6, 3 and 1 trials is

**Table 7.2.** The details of the architecture of the shallow CNN and modified shallow CNN.

| | | Shallow CNN | | | Modified shallow CNN | | |
|---|---|---|---|---|---|---|---|
| Block no. | Layers | Filters | Output size | No. of parameters | Filters | Output size | No. of parameters |
| 1 | Input | | $400 \times 16 \times 1$ | | | $400 \times 16 \times 1$ | |
| 2 | Conv2D + ELU | $F_1 = 16$, $N = (250, 1)$, $S = (1, 1)$ | $151 \times 16 \times 16$ | 4016 | $F_1 = 16$, $N = (125, 1)$, $S = (8, 1)$ | $32 \times 16 \times 16$ | 2016 |
| | Avg. pooling | $P_1 = (6, 1)$ | $25 \times 16 \times 16$ | 0 | $P_1 = (6, 1)$ | $7 \times 16 \times 16$ | 0 |
| 3 | Conv2D + ELU | $D_1 = 2$, $N = (1, 16)$, $S = (1, 1)$ | $25 \times 1 \times 32$ | 8224 | $D_1 = 2$, $N = (1, 16)$, $S = (1, 1)$ | $7 \times 1 \times 32$ | 8224 |
| | Avg. pooling | $P_2 = (2, 1)$ | $13 \times 1 \times 32$ | 0 | $P_2 = (2, 1)$ | $3 \times 1 \times 32$ | 0 |
| | Dropout | $p = 0.25$ | | 0 | $p = 0.25$ | | 0 |
| 4 | Fully connected | | $1 \times 416$ | 0 | | $1 \times 96$ | |
| | Dense | $C = 2$ | $1 \times 2$ | 832 | | $1 \times 2$ | 192 |
| | Total number of parameters | | | 13 072 | | 10 432 | |

**Table 7.3.** The details of the architecture of EEGNet and the modified EEGNet.

| | | EEGNet | | | Modified EEGNet | | |
|---|---|---|---|---|---|---|---|
| Block no. | Layers | Filters | Output size | No. of parameters | Filters | Output size | No. of parameters |
| 1 | Input | | $400 \times 16 \times 1$ | | | $400 \times 16 \times 1$ | |
| 2 | Conv2D + ELU | $F_1 = 16$, $N = (250, 1)$, $S = (1, 1)$ | $151 \times 16 \times 16$ | 4016 | $F_1 = 16$, $N = (125, 1)$, $S = (8, 1)$ | $46 \times 16 \times 16$ | 2016 |
| | Avg. pooling | $P_1 = (6, 1)$ | $25 \times 16 \times 16$ | 0 | $P_1 = (6, 1)$ | $6 \times 16 \times 16$ | 0 |
| 3 | Depthwise Conv2D + ELU | $D_1 = 2$, $N = (1, 16)$, $S = (1, 1)$ | $25 \times 1 \times 32$ | 544 | $D_1 = 2$, $N = (1, 16)$, $S = (1, 1)$ | $6 \times 1 \times 32$ | 544 |
| | Avg. pooling | $P_2 = (2, 1)$ | $13 \times 1 \times 32$ | 0 | $P_2 = (2, 1)$ | $3 \times 1 \times 32$ | 0 |
| 4 | Separable Conv2D + ELU | $D_2 = 1$, $N = (1, 1)$, $S = (1, 1)$ | $13 \times 1 \times 32$ | 2144 | $D_2 = 1$, $N = (1, 1)$, $S = (1, 1)$ | $3 \times 1 \times 1$ | 129 |
| | Dropout | $p = 0.25$ | | 0 | $p = 0.25$ | | |
| 5 | Fully connected | | $1 \times 640$ | 0 | | $1 \times 3$ | 0 |
| | Dense | $C = 2$ | $1 \times 2$ | 130 | | $1 \times 2$ | 6 |
| | Total number of parameters | | | 6834 | | 2695 | |

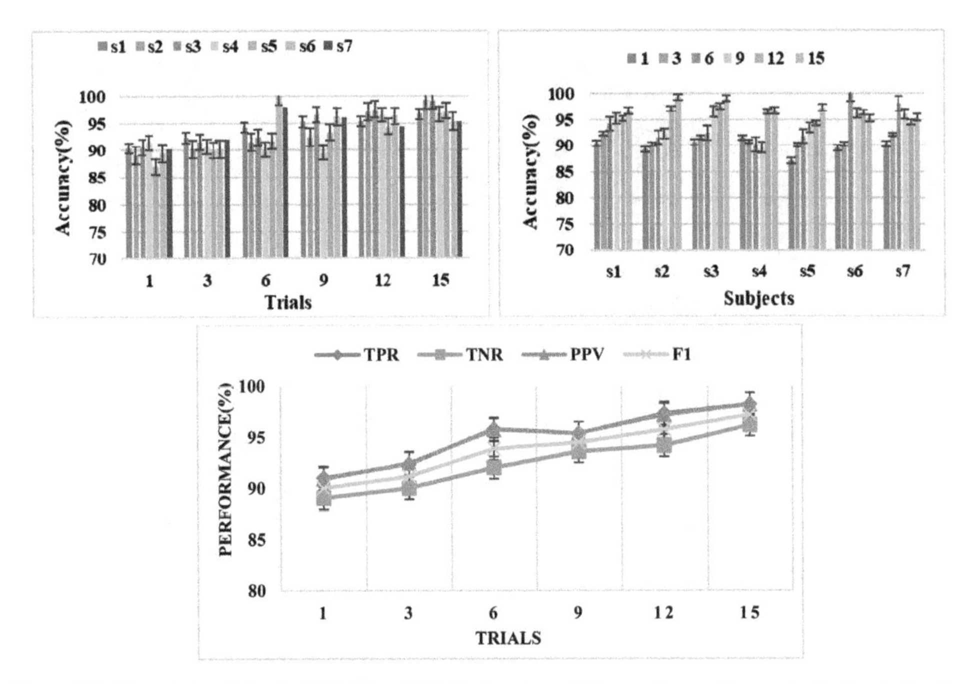


**Figure 7.6.** The results of the classification of P300 in the zigzag RC paradigm with covert attention task using DCNN: (a) cross-trial analysis and (b) cross-subject analysis. (c) The performance metrics calculated for P300 and non-P300 detection in the zigzag paradigm using DCNN.

95%–99% with an average of 97.05% ± 1.59, 94%–97% with an average of 95.93% ± 1.29, 89%–96% with an average of 94.18% ± 2.56, 91%–98% with an average of 93.97% ± 3.65, 89%–92% with an average of 91.00% ± 0.95, and 89%–90% with an average of 89.79% ± 1.04, respectively. From the cross-trial analysis in figure 7.6(a), it is found that the data with trial numbers of 6 and 9 have the highest variation among the subjects. The average drop in accuracy among the trials for all subjects is ~9%. Among all the subjects, it is observed that the performance of each subject varies from trial to trial. Subjects 2 and 5 have the highest variation, i.e. 10%–11% (SD ~ 4%), in the performance within the lowest to highest trials. For the highest trials, i.e. 15, DCNN is able to discriminate the ERP and non-ERP components very efficiently even without a channel selection approach.

Further, we have also analysed the other performance metrics such as true positive rate (TPR), true negative rate (TNR), positive predictive value (PPV) and F1-score. The ranges of the TPR, TNR, PPV and F1-score are 91%–98%, 89%–96%, 90%–98% and 90%–97% for 15–1 trials, respectively, as shown in figure 7.6(c).

Figure 7.7 presents the results of the P300 detection in covert attention using shallow CNN. The range of accuracies for 15, 12, 9, 6, 3 and 1 trials are 94%–98% with an average of 96.34% ± 1.27, 93%–96% with an average of 94.77% ± 1.49, 90%–96% with an average of 93.41% ± 2.27, 90%–97% with an average of 92.44% ± 3.19, 90%–92% with an average of 91.11% ± 0.85, and 84%–89% with an average of 87.34% ± 1.74, respectively. From the cross-trial and cross-subject analysis in

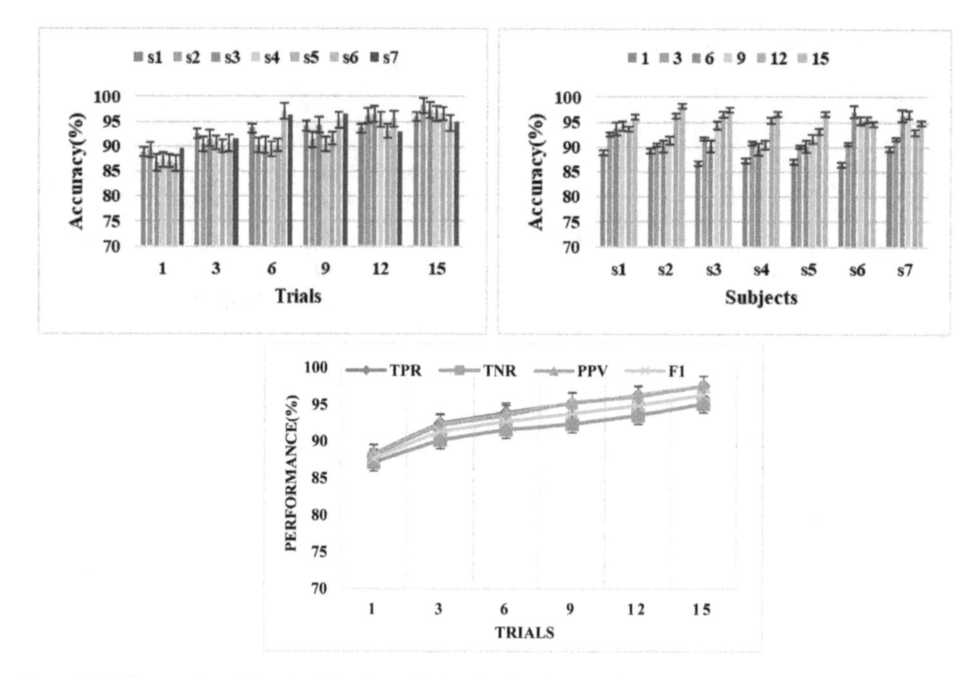


**Figure 7.7.** The results of the classification of P300 in the zigzag RC paradigm with a covert attention task using shallow CNN: (a) cross-trial analysis and (b) cross-subject analysis. (c) The performance metrics calculated for P300 and non-P300 detection in the zigzag paradigm using the shallow CNN.

figures 7.7(a) and (b), it is found that the variation, i.e. SD across trials and across subjects, is reduced by 1%–2% when adopting the spatial information based channel selection. This shows that the generalization of the model across trials and across subjects has improved. Moreover, the average drop in accuracy among the trials for all subjects is 8%. Among the subjects, it is observed that the performance of each subject varies from trial to trial. Subject 2 has the highest variation, i.e. 11% (SD ~ 4%) in performance between the lowest to highest trials. The calculated TPR, TNR, PPV and F1-score are 87.25%–97.54%, 87%–95%, 88%–97% and 88%–96% for 15–1 trials, respectively, as shown in figure 7.7(c).

Figure 7.8 presents the results of the P300 detection in covert attention using the compact CNN. The range of accuracies for 15, 12, 9, 6, 3 and 1 trials are 94%–97% with an average of 95.83% ± 1.31, 89%–96% with an average of 93.61% ± 2.42, 89%–95% with an average of 92.07% ± 2.65, 87%–95% with an average of 90.05% ± 3.32, 85%–89% with an average of 87.89% ± 1.23 and 83%–87% with an average of 85.61% ± 1.39, respectively. From cross-trial and cross-subject analysis in figure 7.8(a) and (b), it is found that the average drop in accuracy among the trials for all subjects is 10%. Among the subjects, it is observed that the performance of each subject varies from trial to trial. Subject 3 has the highest variation, i.e. 12% (SD ~ 4.6%), in the performance within the lowest to highest trials. The calculated TPR, TNR, PPV and F1-score are 87.5%–97.66%, 83.87%–95.11%, 87.51%–97.25% and 85.68%–96.37% for 15–1 trials, respectively, as shown in figure 7.8(c).

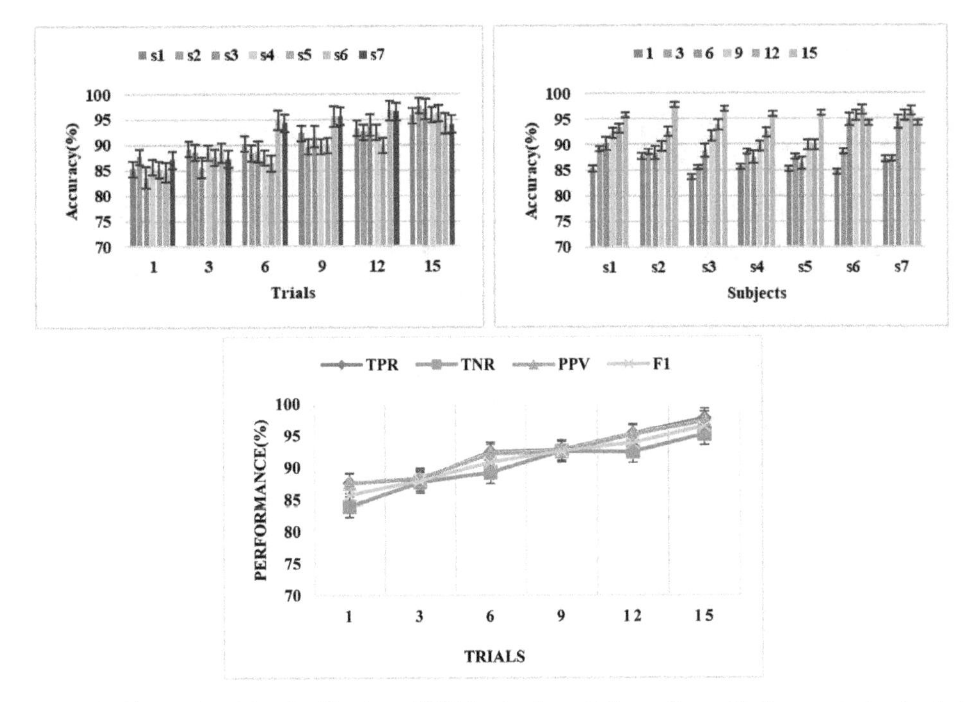


**Figure 7.8.** The results of the classification of P300 in the zigzag RC paradigm with the covert attention task using compact CNN: (a) cross-trial analysis and (b) cross-subject analysis. (c) The performance metrics calculated for P300 and non-P300 detection in the zigzag paradigm using compact CNN.

*7.3.2.2 Overt attention task*

Figure 7.9 illustrates the results of the P300 detection in the zigzag RC paradigm with overt attention using the DCNN. In cross-trial and cross-subject analysis, it is observed that the range of accuracy for 15, 12, 9, 6, 3 and 1 trials are 97%–99% with an average of 98.64% ± 0.94, 96%–98% with an average of 97.17% ± 0.69, 95%–97% with an average of 96.24% ± 1.01, 94%–96% with an average of 95.38% ± 0.92, 83%–95% with an average of 94.37% ± 0.80 and 90%–94% with an average of 92.96% ± 1.36, respectively. From the cross-trial analysis in figure 7.9(a), it is found that the data with trials 1 and 9 having high variation among the subjects. The average drop in accuracy among the trials for all subjects is ~6%. Among the subjects, it is observed that the performance of each subject varies from trial to trial. Subjects 2 and 5 have the highest variation, i.e. 7%–8% (SD ~ 2%), in the performance within the lowest to highest trials. The highest accuracy for P300 classification is obtained with 15 trials using the DCNN.

The calculated TPR, TNR, PPV and F1-score are 93.25%–99.35%, 92.45%–98.12%, 93.12%–99.12% and 92.86%–98.73% for 15–1 trials, respectively, as shown in figure 7.9(c).

Figure 7.10 illustrates the results of the P300 detection in the zigzag RC paradigm with overt attention using the shallow CNN. In cross-trial and cross-subject analysis, it is observed that the range of accuracy for 15, 12, 9, 6, 3 and 1 trials is

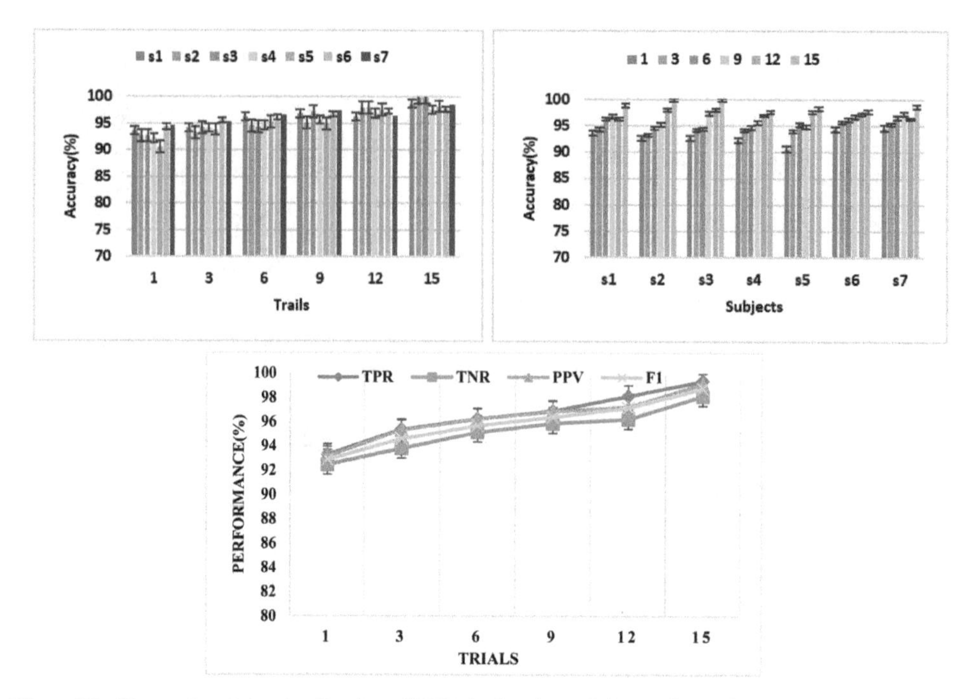


**Figure 7.9.** The results of the classification of P300 in the zigzag RC paradigm with the overt attention task using DCNN: (a) cross-trial analysis and (b) cross-subject analysis. (c) The performance metrics calculated for P300 and non-P300 detection in the zigzag paradigm with the overt attention task using the DCNN.

97%–99% with an average of 98.21% ± 0.77, 96%–97% with an average of 97.11% ± 0.52, 95%–97% with an average of 96.49% ± 1.00, 93%–96% with an average of 94.94% ± 1.46, 92%–95% with an average of 93.9% ± 1.23, and 91%–92% with an average of 92.05% ± 0.63, respectively. From the cross-trial analysis in figure 7.10(a) it is found that the data with trial numbers of 3 and 6 have high variation among the subjects. The average drop in accuracy among the trials for all subjects is ~6%. Among the subjects, it is observed that the performance of each subject varies from trial to trial. Subjects 2 and 5 have the highest variation, 7% (SD ~ 3%), in the performance within the lowest to highest result trials. Unlike the performance of the covert attention task, the shallow CNN obtained equal accuracy as the DCNN, even with a huge drop in the number of parameters.

The calculated TPR, TNR, PPV and F1-score are 93.12%–99.33%, 91.25%–97.45%, 93.10%–99.23% and 92.18%–98.38% for 15–1 trials, respectively, as shown in figure 7.10(c).

Figure 7.11 illustrates the results of the P300 detection in the zigzag RC paradigm with overt attention using the compact CNN. In the cross-trial and cross-subject analyses, it is observed that the range of accuracy for 15, 12, 9, 6, 3 and 1 trials is 95%–97% with an average of 97.00% ± 0.75, 94%–96% with an average of 95.90% ± 0.84, 93%–95% with an average of 94.82% ± 1.04, 92%–94% with an average of 93.60% ± 1.25, 91%–93% with an average of 92.09% ± 1.40, and 88%–90% with an average of 90.06% ± 1.05, respectively. From the cross-trial analysis in figure 7.11

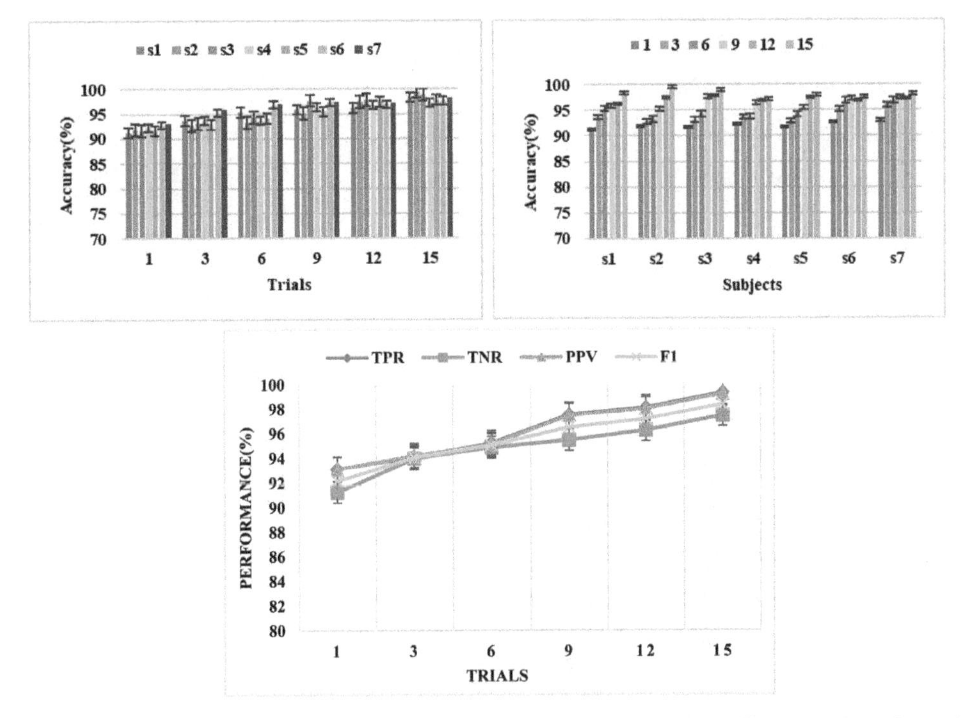


**Figure 7.10.** The results of the classification of P300 in the zigzag RC paradigm with an overt attention task using the shallow CNN: (a) cross-trial analysis and (b) cross-subject analysis. (c) The performance metrics calculated for P300 and non-P300 detection in the zigzag paradigm with an overt attention task using shallow CNN.

(a) it is found that the data with trial numbers of 3 and 6 have high variation among the subjects. The average drop in accuracy among the trials for all the subjects is ~7%. Among the subjects, it is observed that the performance of each subject varies from trial to trial. Subjects 2 and 5 have the highest variation, i.e. 8%–97% (SD = 3%–4%), in performance between the lowest to highest trials. The compact CNN with overt attention obtained a higher accuracy than the DCNN with covert attention in a single trial with fewer parameters.

The calculated TPR, TNR, PPV and F1-score are 90.12%–98.28%, 89.89%–96.01%, 93.12%–98.28% and 90.00%–97.10% for 15–1 trials, respectively, as shown in figure 7.11(c).

## 7.4 Discussion

In this section comparative analyses of the performance of the standard RC and proposed zigzag classification approaches, and the covert and overt attention based DS-P3S are presented. Workload evaluation is then presented to analyse the usability of the DS-P3S.

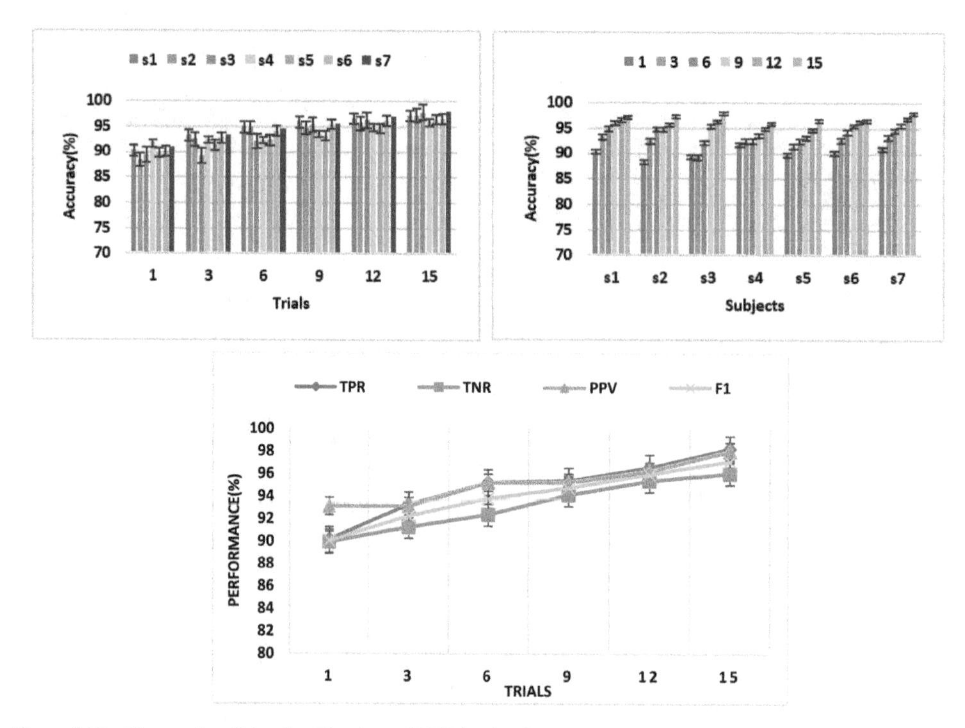


**Figure 7.11.** The results of the classification of P300 in the zigzag RC paradigm with overt attention task using compact CNN: (a) cross-trial analysis and (b) cross-subject analysis. (c) The performance metrics calculated for P300 and non-P300 detection in the zigzag paradigm with overt attention tasks using compact CNN.

## 7.4.1 Comparative analysis

### *7.4.1.1 Hypothesis 1: RC versus zigzag RC*

To evaluate our first hypothesis, a comparative analysis between the standard RC paradigm and proposed zigzag paradigm for DS and other languages based on detection rate and spelling time are presented in table 7.4. It can be seen from the table that the detection of P300 in the standard RC paradigm-based DS-P3S using machine learning and channel optimization achieved an accuracy of 79%–94% with 5–15 trials, which required a minimum spelling time of 16 s. To reduce the spelling time and false detection rate, the DCNN and WE-DCNN were adopted, which achieved an accuracy of 82.30%–95.07% with spelling time ranging from 3.2–48 s (see table 7.4). The accuracy for 1–15 trials degrades further on adopting shallow and compact CNN.

The trend of the trade-off between the number of trials and performance persists even after adopting the DCNN. As we can observe from figure 7.3(a), the DS display is highly crowded and has very little spacing between two consecutive characters, which leads to the generation of ErrPs, overlapping and poor quality of P300. These must be the main reasons for the increase in the false detection rate in the existing DS-P3S.

**Table 7.4.** Comparative analysis of the RC and zigzag RC for DS language with covert and overt attention using different machine and deep learning techniques.

| Author, Year | #Parameters | #Trials | Acc (%) | Time (s) |
|---|---|---|---|---|
| Chaurasiya *et al*, 2016 [11] | | 5–15 | 79.2–94. | 16 |
| Kshirsagar G *et al*, 2018 [6] | 153 730 | 3–15 | 88.20–95.79 | 9.6 |
| Kshirsagar G *et al*, 2019 [7] | 153 730 | 1 | 92.6 | 3.2 |
| DCNN (covert-RC) | 153 730 | 1–15 | 82.30–95.07 | 3.2–48 |
| Shallow CNN (covert-RC) | 10 432 | 1–15 | 80.44–93.74 | 3.2–48 |
| Compact CNN (covert-RC) | 2693 | 1–15 | 79.59–91.80 | 3.2–48 |
| DCNN (covert-zigzag) | 153 730 | 1–15 | 89.79–97.05 | 3.2–48 |
| Shallow CNN(covert-zigzag) | 10 432 | 1–15 | 87.34–96.34 | 3.2–48 |
| Compact CNN (covert-zigzag) | 2693 | 1–15 | 85.61–95.83 | 3.2–48 |
| DCNN (overt-zigzag) | | 1–15 | 92.96–99.64 | 4.0–60 |
| Shallow CNN (overt-zigzag) | 11 138 | 1–15 | 92.05–98.21 | 4.0–60 |
| Compact CNN (overt-zigzag) | 2695 | 1–15 | 90.00–97.01 | 4.0–60 |

On the other hand, the classification accuracy with zigzag RC was improved by 5%–7% ($p < 0.05$, paired $T$-test) for a single trial (see table 7.4). In other words, the false detection rate was reduced by 5%–7% due to modifications in the DS paradigm as shown in figure 7.3(b). The increase in the spacing between the characters reduced the incorrectly generated ERPs. Moreover, there is an improvement in the average amplitude of ~3 $\mu$V ($p < 0.05$, paired $T$-test) (see table 7.5) and the reduction in the variations in the average latency are also noted, as the SD among the latencies of the RC is 60.33 ms whereas the SD among the latencies of the zigzag RC is 22.61. This shows that the modification in the distance and visual representation of characters presented in figures 7.3(b) and (c) helped to improve the attention and the quality of P300.

*7.4.1.2 Hypothesis 2: covert attention task versus overt attention task*

Further, to validate our second hypothesis, we present the subject-wise analysis of the amplitude and latency of the ERPs generated during the spelling of characters using covert and overt attention tasks, as shown in table 7.5. For the covert attention task the range of peak amplitudes for the standard RC and zigzag RC paradigms is 2.89–5.92 $\mu$V with average 3.93 ± 0.99 $\mu$V and 6.68–8.81 $\mu$V with an average of 7.36 ± 1.08 $\mu$V, and the corresponding latencies are 261–382 ms with an average of 319.65 ± 60.33 and 323.23 ± 22.61 ms, respectively. On the other hand, the amplitudes and latencies of the zigzag RC with the overt attention task are in the range of 13.21–17.21 $\mu$V with an average of 14.83 ± 1.82 $\mu$V and the corresponding latencies range from 455–522 ms with an average of 488.43 ± 26.66 ms. The improvement in the average amplitudes of ~7–11 $\mu$V ($p < 0.01$, paired $T$-test) shows that the overt attention task may be the best candidate for the design of EEG based spelling systems. Moreover, the double blinking task can handle the overlapping problem of the RC paradigm as the ERPs with a double blink include the two peaks

**Table 7.5.** The subject-wise analysis of the amplitude and latency of ERPs generated using covert attention and overt attention tasks.

| | Standard RC (covert) | | Zigzag RC (covert) | | Zigzag RC (overt) | |
|---|---|---|---|---|---|---|
| Subjects | Amplitude ($\mu$V) | Latency ( ms) | Amplitude ($\mu$V) | Latency ( ms) | Amplitude ($\mu$V) | Latency ( ms) |
| S1 | 5.92 ± 3.64 | 382.35 ± 21.65 | 8.61 ± 1.07 | 341.20 ± 15.02 | 17.21 ± 3.22 | 485.21 ± 10.25 |
| S2 | 3.78 ± 1.67 | 379.89 ± 27.21 | 7.41 ± 2.69 | 322.12 ± 10.06 | 14.10 ± 4.11 | 501.55 ± 20.25 |
| S3 | 3.21 ± 1.89 | 358.13 ± 19.22 | 7.06 ± 3.22 | 315.99 ± 14.21 | 14.74 ± 3.58 | 455.10 ± 15.22 |
| S4 | 4.09 ± 1.71 | 261.81 ± 41.33 | 8.11 ± 2.69 | 302.87 ± 22.14 | 16.45 ± 5.21 | 470.21 ± 13.33 |
| S5 | 4.13 ± 1.66 | 223.09 ± 40.05 | 7.89 ± 3.10 | 289.54 ± 18.78 | 15.88 ± 4.63 | 519.21 ± 26.54 |
| S6 | 2.89 ± 3.20 | 308.92 ± 10.33 | 6.68 ± 4.53 | 336.65 ± 17.35 | 13.21 ± 4.25 | 522.44 ± 27.25 |
| S7 | 3.49 ± 2.48 | 323.39 ± 16.66 | 7.82 ± 3.45 | 354.25 ± 13.32 | 14.24 ± 3.11 | 465.33 ± 18.58 |
| Mean ± SD | 3.93 ± 0.99 | 319.65 ± 60.33 | 7.36 ± 1.08 | 323.23 ± 22.61 | 14.83 ± 1.82 | 488.43 ± 26.66 |

in its morphology. Therefore, the morphologies of overlapping ERPs and double blinks show similarity in their structures which further helps the model to learn this kind of complex pattern and hence it further increases the performance by 8%–11% ($p < 0.01$) and 3%–5% ($p < 0.05$) compared to the covert RC and covert zigzag RC, respectively, as shown in table 7.4. The pictorial representation of the ERPs generated for the spelling task in subject 1 for the covert and overt attention tasks are presented in figure 7.12.

*7.4.1.3 Hypothesis 3: bias versus variance analysis*

Table 7.6 presents the bias and variance analysis of the classification of P300 for the RC and zigzag paradigm with covert and overt attention tasks using the DCNN, shallow CNN and compact CNN. For the RC paradigm with an overt attention task, all three models, i.e. the DCNN, shallow CNN and compact CNN, have both high a bias ($E$) of 17%–20% and a high variance ($\sigma^2$) of 4%–8% for the classification of P300. Further, the bias and variance are reduced significantly on adopting the proposed zigzag paradigm and an overt attention task for spelling. Although the DCNN achieved the highest accuracy for classification in all three classification experiments of P300, it still generated approximately 153 730 parameters in the case of the covert attention based task and even more for overt attention. This increases the computational burden and hence the parameters have been pruned further using

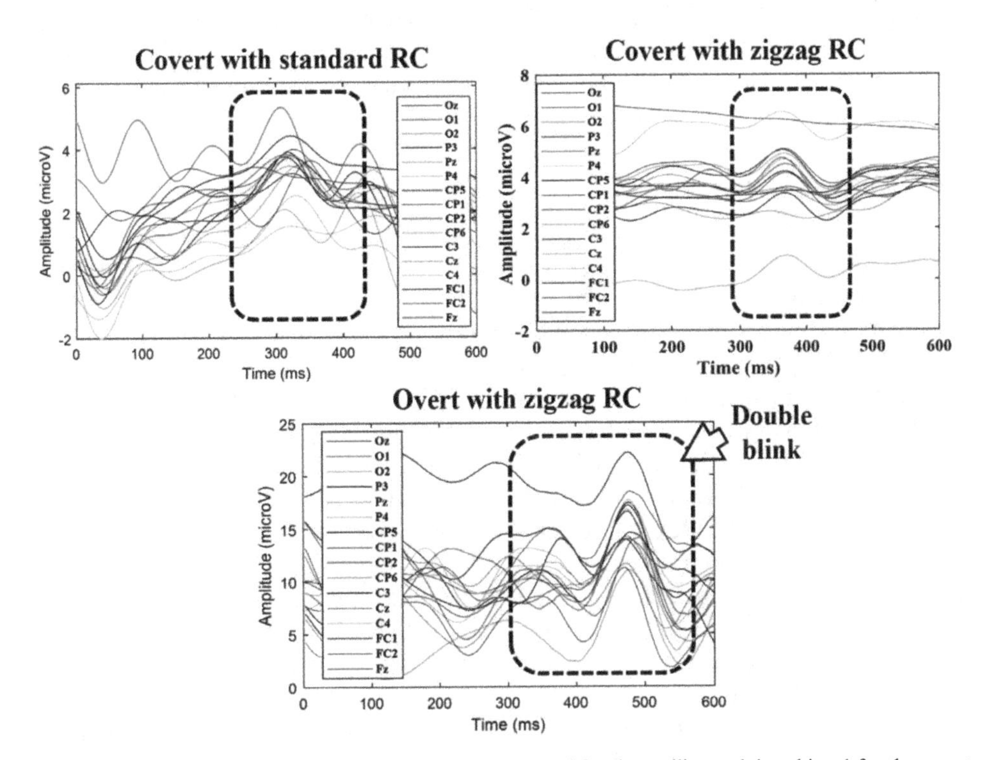


**Figure 7.12.** The pictorial representation of the ERPs generated for the spelling task in subject 1 for the covert and overt attention tasks.

**Table 7.6.** The bias and variance analysis of the classification of P300 for the RC and zigzag paradigms with covert and overt attention tasks the using DCNN, shallow CNN and compact CNN.

| | 1 trial | | 15 trials | |
|---|---|---|---|---|
| Methods | $E$ | $\sigma^2$ | $E$ | $\sigma^2$ |
| DCNN (covert-RC) | 17.70 | 7.61 | 4.93 | 6.88 |
| Shallow CNN (covert-RC) | 19.56 | 4.35 | 6.26 | 3.78 |
| Compact CNN (covert-RC) | 20.41 | 5.41 | 8.20 | 4.89 |
| DCNN (covert-zigzag) | 10.21 | 1.92 | 2.95 | 2.52 |
| Shallow CNN(covert-zigzag) | 12.66 | 1.66 | 3.66 | 1.63 |
| Compact CNN (covert-zigzag) | 14.39 | 1.23 | 4.17 | 1.72 |
| DCNN (overt-zigzag) | 7.04 | 1.86 | 0.36 | 0.88 |
| Shallow CNN (overt-zigzag) | 7.95 | 0.40 | 1.79 | 0.58 |
| Compact CNN (overt-zigzag) | 10 | 1.12 | 2.99 | 0.56 |

modified shallow and compact CNNs, as shown in tables 7.2 and 7.3, respectively. The parameter pruning approach with windowing helped to not only reduce the parameters by (1/14) to (1/57) times, but it also boosted the training speed of both models. Although reducing the trainable parameters also costs accuracy, it is just 2%–4% compared to the DCNN. If we consider the huge drop in trainable parameters then a 90%–92% accuracy is comparatively satisfactory. These promising results with only a few trainable parameters will help to implement the DS-P3S on mobile devices in the future.

### 7.4.2 Workload evaluation

Finally, we have also analysed the usability of all three approaches using the NASA Task Load Index (NASA-TLX) questionnaire [57] based workload evaluation and this is presented in table 7.7. The overall workload shows the sum of all seven parameters, namely mental demand (MD), physical demand (PD), temporal demand (TD), performance (Perf), efforts (E), frustration (F) and fatigue (FG). It can be seen from table 7.7 that the zigzag RC with overt attention (Z(O)) has a smaller overall workload, i.e. $30 \pm 5.65$, whereas the zigzag RC (Z(C)) and standard RC (RC(C)) with covert attention have $35 \pm 6.44$ and $39 \pm 14.33$ overall workloads, respectively. Further, we have thoroughly investigated the individual parameters of the NASA task load index questionnaire. It is observed that there is a trade-off between the mental and physical demands for the covert and overt tasks. For covert attention based spelling tasks, the mental demand increases with a reduction in physical demand, whereas for overt attention based spelling tasks the physical demand increases with a reduction in mental demand. The increase in the mental demand has a direct effect on the quality of the ERPs whereas the physical demand may create the problem of fatigue. However, there is no significant difference found as $p > 0.05$ for fatigue (see table 7.7) between the covert and overt attention tasks. Moreover, the parameter Perf has a direct correlation, and E and TD have an

**Table 7.7.** NASA task load index (NASA-TLX) questionnaire based workload evaluation [57] of the RC and zigzag paradigms with covert and overt attention spelling tasks.

| Paradigm | Overall workload (0–100) | MD (0–12) | PD (0–12) | TD (0–12) | Perf (0–12) | E (0–12) | F (0–12) | FG (0–12) | An (1–6) | Dz (1–6) |
|---|---|---|---|---|---|---|---|---|---|---|
| Z(O) | 30 | 2 | 6 | 2 | 8 | 4 | 2.5 | 5.5 | 2 | 4 |
| | (5.65) | (0.87) | (2.54) | (1.36) | (1.65) | (1.11) | (0.56) | (1.04) | (1.0) | (0.57) |
| R/C(C) | 39 | 8 | 2 | 4 | 4 | 7 | 7 | 7 | 4 | 3 |
| | (14.33) | (1.49) | (0.90) | (1.09) | (1.65) | (1.31) | (0.82) | (3.54) | (0.75) | (0.66) |
| Z(C) | 35 | 8 | 2 | 4 | 6 | 6 | 3 | 6 | 3 | 2 |
| | (6.44) | (1.01) | (0.79) | (1.22) | (0.89) | (0.97) | (0.76) | (2.45) | (0.86) | (0.47) |

MD—mental demand; PD—physical demand; TD—temporal demand; Perf—performance; E—effort; F—frustration; FG—fatigue; An—annoyance; Dz—dizzy inverse correlation with the task difficulty, and lower values of E and TD with a high Perf ($p < 0.01$) in Z(O) ensures that the zigzag with overt attention manages to reduce the task difficulty in the RC paradigm.

## 7.5 Conclusion

In this work, we provided three novel contributions to improve the usability of the DS-P3S. In the first experiment, i.e. the design of a zigzag RC paradigm, the positive increment of 5%–7% ($p < 0.05$, paired $T$-test) for a single trial in the performance of detection of P300 and the improvement in the average amplitude by ~3 $\mu$V ($p < 0.05$, paired $T$-test) (see table 7.5), as well as the reduction in the variations in the average latency, show that the modification in the distance and visual representation of characters in the RC paradigm (see figure 7.3 (b) and (c)) help to improve the attention of the user, the quality of P300, and to reduce the effect of crowding and adjacency distraction. In the second experiment, i.e. the overt attention based spelling task, improvements of 8%–11% and 3%–5% in the performance of P300 detection were noted compared to the covert RC and covert zigzag RC based spelling tasks, respectively. Moreover, the reduction in the mental demand improved the quality of the ERP by ~7–11 $\mu$V ($p < 0.01$, paired $T$-test). In addition, the lower overall workload for Z(O) shows that the overt attention task may be the best candidate for the design of an EEG based spelling system. In the last experiment, the filter pruning and windowing approach in the design of the shallow and compact CNNs leads to a significant reduction in the trainable parameters, i.e. by 1/14 to 1/52, compared to the number of trainable parameters in the DCNN. Although the accuracy is reduce by 2%–4% compared to the DCNN for a single trial, it still produces high performance for such a compact model. This compact CNN with high performance will be useful for BCI research as it will further be adopted to design an efficient BCI system for mobile devices. The only limitation of the overt attention task is that it cannot be used for patients with severe neuromuscular disability who are not able to perform eye moments.

## Bibliography


[1] Wolpaw J R, Birbaumer N, McFarland D J, Pfurtscheller G and Vaughan T M 2002 Brain–computer interfaces for communication and control *Clin. Neurophysiol.* **113** 767–91

[2] Ramadan Rabie A and Vasilakos A V 2017 Brain computer interface: control signals review *Neurocomput.* **223** 26–44

[3] Rezeika A, Benda M, Stawicki P, Gembler F, Saboor A and Volosyak I 2018 Brain–computer interface spellers: a review *Brain Sci.* **8** 57

[4] Abiri R, Borhani S, Sellers E W, Jiang Y and Zhao X 2019 A comprehensive review of EEG-based brain–computer interface paradigms *J. Neural Eng.* **16** 011001

[5] Abdulkader S N, Atia A and Mostafa M S M 2015 Brain computer interfacing: applications and challenges *Egypt. Informat. J.* **16** 213–30

[6] Kshirsagar G B and Londhe N D 2018 Improving performance of Devanagari script input-based P300 speller using deep learning *IEEE Trans. Biomed. Eng* **66** 2992–3005

[7] Kshirsagar G B and Londhe N D 2019 Weighted ensemble of deep convolution neural networks for single-trial character detection in Devanagari-script-based P300 speller *IEEE Trans. Cogn. Developm. Syst* **12** 551–60
[8] Kshirsagar G B and Londhe N D 2017 Deep convolutional neural network based character detection in Devanagari script input based P300 speller *Proc. IEEE Int. Conf. Electrical, Electronics, Communication, Computer, and Optimization Techniques (ICEECCOT)* (Piscataway, NJ: IEEE) pp 507–11
[9] Kshirsagar G B and Londhe N D 2018 Performance improvement for Devanagari script input based P300 speller *Proc. IEEE 5th Int. Conf. Signal Processing and Integrated Networks (SPIN)* (Piscataway, NJ: IEEE) pp 142–7
[10] Chaurasiya R K, Londhe N D and Gosh S 2016 Binary DE-based channel selection and weighted ensemble of SVM classification for novel brain–computer interface using Devanagari script-based P300 speller paradigm *Int. J. Hum. Comput. Interact* **32** 861–77
[11] Chaurasiya R K, Londhe N D and Gosh S 2017 Multi-objective binary DE algorithm for optimizing the performance of Devanagari script-based P300 speller *Biocybern. Biomed. Eng.* **37** 422–31
[12] Farwell L A and Donchin E 1988 Talking off the top of your head: toward a mental prosthesis utilizing event-related brain potentials *Electroencephalogr. Clin. Neurophysiol.* **70** 510–23
[13] Yamamoto Y *et al* 2015 Improvement of performance of Japanese P300 speller by using second display *J. Artif. Intell. Soft Comput. Res.* **5** 221–6
[14] Kabbara M *et al* 2015 An efficient P300-speller for Arabic letters *Int. Conf. Adv. Biomed. Eng. ICABME 2015* pp 142–5
[15] Minett J W *et al* 2012 A Chinese text input brain–computer interface based on the P300 speller *Int. J. Hum. Comput. Interact* **28** 472–83
[16] Guan C, Thulasidas M and Wu J 2004 High performance P300 speller for brain–computer interface *IEEE Int. Workshop on Biomedical Circuits and Systems* pp S3–5
[17] Oralhan Z 2019 A new paradigm for region-based P300 speller in brain computer interface *IEEE Access* **7** 106618–27
[18] Townsend G *et al* 2010 A novel P300-based brain–computer interface stimulus presentation paradigm: moving beyond rows and columns *Clin. Neurophysiol.* **121** 1109–20
[19] Yin E *et al* 2013 A novel hybrid BCI speller based on the incorporation of SSVEP into the P300 paradigm *J. Neural Eng.* **10** 026012
[20] Chen X, Chen Z, Gao S and Gao X 2014 A high-ITR SSVEP-based BCI speller *Brain-Comp. Interf.* **1** 181–91
[21] Lin Z, Zhang C, Zeng Y, Tong L and Yan B 2018 A novel P300 BCI speller based on the triple RSVP paradigm *Sci. Rep.* **8** 1–9
[22] Pires G, Nunes U and Castelo-Branco M 2011 GIBS block speller: toward a gaze-independent P300-based BCI *2011 Ann. Int. Conf. of the IEEE Engineering in Medicine and Biology Society* pp 6360–4
[23] Treder M S and Blankertz B 2010 (C)overt attention and visual speller design in an ERP-based brain–computer interface *Behav. Brain Funct.* **6** 28
[24] Aloise F *et al* 2012 A covert attention P300-based brain–computer interface: Geospell *Ergonomics* **55** 538–51
[25] Sato H and Washizawa Y 2016 An N100-P300 spelling brain–computer interface with detection of intentional control *Computers* **5** 31

[26] Garduño E *et al* 2016 P300 detection based on EEG shape features *Comput. Math. Methods Med.* **2016** 1–14
[27] Turnip A *et al* 2013 P300 detection using a multilayer neural network classifier based on adaptive feature extraction *Int. J. Brain Cognit. Sci.* **2** 63–75
[28] Xu N *et al* 2004 BCI competition 2003-data set IIb: enhancing P300 wave detection using ICA-based subspace projections for BCI applications *IEEE Trans. Biomed. Eng* **51** 1067–72
[29] Li K *et al* 2009 Single trial independent component analysis for P300 BCI system *Proc. 31st Annu. Int. Conf.* (Minneapolis, MN: IEEE EMBS) pp 4035–8
[30] Kaper M, Meinicke P, Grossekathoefer U, Lingner T and Ritter H 2004 BCI competition 2003-data set IIb: support vector machines for the P300 speller paradigm *IEEE Trans. Biomed. Eng.* **51** 1073–6
[31] Thulasidas M *et al* 2006 Robust classification of EEG signal for brain computer interface *IEEE Trans. Neural Syst. Rehab. Eng.* **14** 24–9
[32] Rakotomamonjy A and Guigue V 2008 BCI competition III: dataset II ensemble of SVMs for BCI P300 speller *IEEE Trans. Biomed. Eng.* **55** 1147–54
[33] Lotte F *et al* 2018 A review of classification algorithms for EEG-based brain–computer interfaces: a 10-year update *J. Neural Eng.* **15** 031005
[34] Vo K, Nguyen D N, Kha H H and Dutkiewicz E 2017 Subject-independent P300 BCI using ensemble classifier, dynamic stopping and adaptive learning *GLOBECOM 2017-2017 IEEE Global Communications Conf.* pp 1–7
[35] Speier W, Arnold C W, Deshpande A, Knall J and Pouratian N 2015 Incorporating advanced language models into the P300 speller using particle filtering *J. Neural Eng.* **12** 046018
[36] Park J and Kim K E 2012 A POMDP approach to optimizing P300 speller BCI paradigm *IEEE Trans. Neural Syst. Rehabil. Eng.* **20** 584–94
[37] Speier W *et al* 2014 Integrating language information with a hidden Markov model to improve communication rate in the P300 speller *IEEE Trans. Neural Syst. Rehabil. Eng.* **22** 678–84
[38] Mainsah B O, Colwell K A, Collins L M and Throckmorton C S 2014 Utilizing a language model to improve online dynamic data collection in P300 spellers *IEEE Trans. Neural Syst. Rehabil. Eng.* **22** 837–46
[39] Ryan D B *et al* 2010 Predictive spelling with a P300-based brain–computer interface: increasing the rate of communication *Int. J. Hum. Comput. Interact.* **27** 69–84
[40] Speier W *et al* 2011 Natural language processing with dynamic classification improves P300 speller accuracy and bit rate *J. Neural Eng.* **9** 16004
[41] Speier W *et al* 2018 Improving P300 spelling rate using language models and predictive spelling *Brain Comput. Interf.* **5** 13–22
[42] Cecotti H and Gräser A 2011 Convolutional neural networks for P300 detection with application to brain–computer interfaces *IEEE Trans. Pattern Anal. Mach. Intell.* **33** 433–45
[43] Liu M *et al* 2018 Deep learning based on batch normalization for P300 signal detection *Neurocomputing* **275** 288–97
[44] Khare S K and Bajaj V 2020 Time-frequency representation and convolutional neural network-based emotion recognition *IEEE Trans. Neural Netw. Learn. Syst.* https://doi.org/10.1109/tnnls.2020.3008938
[45] Ullo S L, Khare S K, Bajaj V and Sinha G R 2020 Hybrid computerized method for environmental sound classification *IEEE Access* **8** 124055–65

[46] Demir F, Şengür A, Bajaj V and Polat K 2019 Towards the classification of heart sounds based on convolutional deep neural network *Health Inform. Sci. Syst.* **7** 16

[47] Bajaj V, Taran S, Tanyildizi E and Sengur A 2019 Robust approach based on convolutional neural networks for identification of focal EEG signals *IEEE Sens. Lett.* **3** 1–4

[48] Chaudhary S, Taran S, Bajaj V and Sengur A 2019 Convolutional neural network-based approach towards motor imagery tasks EEG signals classification *IEEE Sensors J.* **19** 4494–500

[49] Chen S and Zhao Q 2018 Shallowing deep networks: layer-wise pruning based on feature representations *IEEE Trans. Pattern Anal. Mach. Intell.* **41** 3048–56

[50] Lawhern V J, Solon A J, Waytowich N R, Gordon S M, Hung C P and Lance B J 2018 EEGNet: a compact convolutional neural network for EEG-based brain–computer interfaces *J. Neural Eng.* **15** 056013

[51] Clevert D A, Unterthiner T and Hochreiter S 2015 Fast and accurate deep network learning by exponential linear units (ELUs), arXiv: 1511.07289

[52] Hinton G E, Srivastava N and Swersky K 2012 Lecture 6e—RMSProp: divide the gradient by a running average of its recent magnitude *COURSERA Neural Networks Mach. Learn.* p 31

[53] Chollet F 2015 Keras www.Keras.io

[54] Abadi M *et al* 2016 TensorFlow: a system for large-scale machine learning *OSDI* **16** 265–83

[55] Glorot X and Bengio Y 2010 Understanding the difficulty of training deep feedforward neural networks *Proc. 13th Int. Conf. on Artifi. Intell. Statist.* pp 249–56

[56] Srivastava N, Hinton G, Krizhevsky A, Sutskever I and Salakhutdinov R 2014 Dropout: a simple way to prevent neural networks from overfitting *J. Mach. Learn. Res.* **15** 1929–58

[57] Hart S G and Staveland L E 1988 Development of NASA-TLX (task load index): results of empirical and theoretical research *Adv. Psychol.* **52** 139–83

**IOP** Publishing

Modelling and Analysis of Active Biopotential Signals in Healthcare, Volume 2

**Varun Bajaj and G R Sinha**

# Chapter 8

# A comprehensive review of the fabrication and performance evaluation of dry electrodes for long-term ECG monitoring

**Yogita Maithani[1], Bijit Choudhuri[2], B R Mehta[1] and J P Singh[1]**

[1]*Department of Physics, Indian Institute of Technology Delhi, Hauz Khas, New Delhi-110016, India*

[2]*Department of Electronics and Communication Engineering, NIT Silchar, Silchar, Assam –788010, India*

The busy and unhealthy lifestyles of our fast-growing modern society contribute to mental stress in its inhabitants, and this chronic stress exposure makes the human body more vulnerable to disease. Moreover, unhealthy diets and a lack of exercise result in abnormalities in heart function and subsequent development of cardiovascular disease. Electrocardiography (ECG) is an effective technique for the early detection of heart function aberrations. It measures the electrical activity of the heart using a biopotential sensor called an electrode. To improve the ECG monitoring system, different types of electrodes have been designed and developed. The conventional ECG electrodes use conducting gel to eliminate the noise associated with lead artifacts, but this gel itself becomes a source of noise in the case of prolonged operation. After a long time this gel becomes dry, resulting in irritation in the skin and a subsequent reduction in the conductivity. This event leads to distortion in the ECG output, which poses a hindrance to long-term ECG monitoring. The recent development of alternative electrodes is of great interest in the emerging field of wearable and portable healthcare systems for long-term ECG monitoring. Detailed knowledge of and comparisons between different electrodes are necessary for their acceptance from research to commercial medical use. This chapter reviews the evolution of electrodes, including some recent works on the dry electrodes used in ECG monitoring devices to discuss and identify the challenges, and key and critical features in designing and developing better electrodes for long-term health monitoring devices.

## 8.1 Introduction

The frenetic pace of modern society contributes to mental stress in its inhabitants, and chronic stress exposes the body to high levels of stress hormones. Moreover, an unhealthy diet and lack of exercise result in anomalies in heart function and subsequent development of cardiovascular disease (CVD) [1]. Between 1990 and 2016, the death rate due to cardiovascular diseases rose by around 34% from 156 to 209 deaths per one-hundred thousand people in India compared to a 41% decline in the United States [2]. Moreover, it is expected that the projected deaths from CVD may increase by 2.4 million to 4 million from 2004 to 2030 in India [3]. Although the Indian healthcare sector is progressing at an optimistic pace, it is still far from achieving international standards. Most major healthcare infrastructure (60%) is located in urban areas and only 10% of the total population can benefit from these facilities. Additionally, the healthcare system places greater focus on infectious diseases, infants, accidents and maternal health compared to CVD [4]. There are a large proportion of patients with cardiovascular disease and cardiac risk factors due to coronavirus disease 2019 (COVID-19) [5]. A study carried out on 416 hospitalized COVID-19 positive patients within 20 days in Wuhan showed that 19.7% had cardiac injury and these patients had higher mortality (51.2%) than those without cardiac injury (4.5%) [6]. Electrocardiography (ECG) is a useful technique for the early detection of heart function aberrations. ECG can disclose heart diseases such as arrhythmias, which are difficult to detect in a single moment of time. Atrial fibrillation, for example, if not detected and medicated early, can increase the risk of blood clotting and stroke over a period of time [7]. Therefore, long-term continuous ECG signal monitoring of cardiac patients or affected persons with heart-associated problems is essential for accurate diagnosis. ECG is recorded by placing conductive electrodes on the skin and measuring the electrical activity of the heart with a very high time resolution and millisecond accuracy.

This chapter presents a systematic overview of electrocardiography (ECG) monitoring systems. It reviews the evolution and recent literature in the field of biopotential dry ECG electrodes to determine the critical functions for designing and developing electrodes for long-term monitoring devices. Additionally, the technical capabilities and components of conventional ECG monitoring systems, with the primary techniques of electrodes placement, instrumentation and the typical waveforms produced in the system, are briefly discussed. Section 8.2 state the purpose of this review, followed by overviews of the working principles of ECG leads and instrumentation in section 8.3. Section 8.4 categorizes the biopotential electrodes with an investigation of alternative electrodes and recent progress in this field. Section 8.5 describes the issues with dry electrodes and outlines the future scope. It also discusses the key challenges and suggestions for improving dry electrodes for long-term ECG monitoring systems and is followed by the conclusion in section 8.6.

## 8.2 Why this review?

The heart's electrical activity is measured using a biopotential sensor called an electrode. Over the years, different kinds of electrodes have been designed and modified for the improvement of ECG monitoring devices. Biopotentials are measured using conventional wet electrodes such as silver/silver chloride (Ag/AgCl) electrodes [8]. These electrodes use conducting gel to reduce the skin–electrode contact impedance between the electrode and the stratum corneum (SC) layer of the skin [9]. These electrodes have low contact impedance as well as providing excellent signal quality for short-term monitoring. However, they have many drawbacks and limitations for long-term monitoring, such as the presence of sensitizing compounds in the gel and the adhesive that can cause skin allergies and irritation [10]. Usually, skin preparation is needed before applying the electrodes on the skin, such as shaving hairy sites and cleansing with alcohol. After a time, the gel becomes dry, reducing the conductivity which distorts the ECG signal output [11]. The limitations of conventional wet electrodes and the need for long-term monitoring has motivated researchers to investigate a new substitute that can overcome these limitations and perform better. Although it can be observed that there is an ongoing trend to develop novel and efficient dry electrodes for biopotential monitoring, very few articles have been reported that present a comprehensive review of already developed electrodes. The available articles report on the preparation and performance of a developed electrode, but no article has discussed the origin of the biopotential signal and the physics associated with the lead electrode placement. This knowledge is essential to determine and eliminate any type of non-linearity in sensor attachment and the corresponding noise signal. Moreover, a review article discussing comparisons between the available growth techniques, novel material development routes, optimized morphologies of nanostructures, the selection of suitable flexible substrates and their role in efficient detection is expected to guide emerging researchers to focus on an optimized combination of these parameters. Finally, an understanding of the instrumentation associated with the ECG data conversion circuit is helpful to design an efficient electrode network while keeping commercialization constraints at acceptable limits.

This chapter will be particularly helpful to new researchers for investigating the design, development and function of the new electrodes with better performance for long-term ECG monitoring systems.

## 8.3 Working principles of the ECG lead and necessary instrumentation

### 8.3.1 Origin of the cardiac potential and ECG signal

The human heart contains four chambers, two atria at the top and two ventricles at the bottom (figure 8.1). An atrioventricular septum separates the atria and ventricles. At first, the atria are filled with blood and the muscles surrounding the heart contract, pushing the blood down to the ventricles, followed by contraction of muscles surrounding the ventricles and pushing blood out of the heart. The tricuspid valve is situated between the right atrium and the right ventricle. The unidirectional

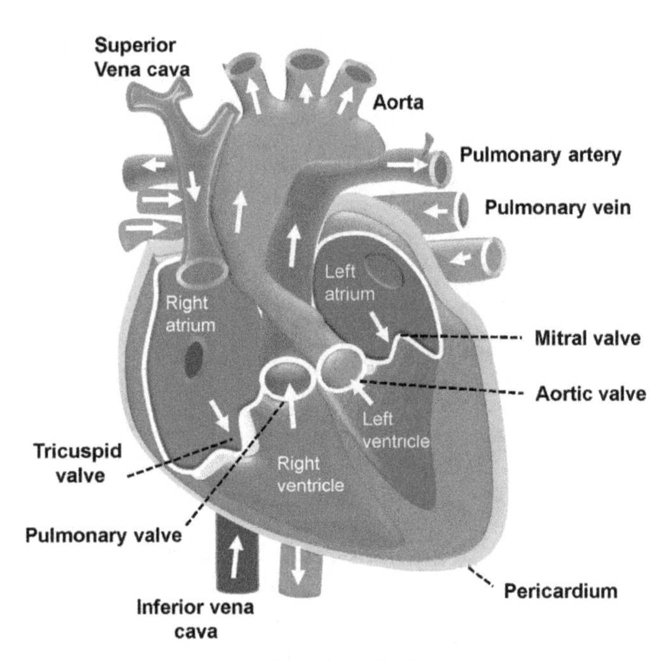


**Figure 8.1.** Schematic of the human heart.

flow of blood from the left atrium to the left ventricle is regulated by the mitral valve. The deoxygenated blood from the whole body is received by the heart through the two major veins, the superior vena cava and inferior vena cava. When the right atrium contracts, the tricuspid valve opens and blood enters into the right ventricle. During the contraction of the right ventricle, the pulmonary valve opens and blood flows through the pulmonary artery to arrive at the lungs. In the lungs, the blood is oxygenated and this oxygenated blood travels through the pulmonary vein to the left atrium chamber of the heart. When the left atrium contracts, the mitral valve opens and oxygenated blood enters into the left ventricle. With the subsequent contraction of the left ventricle, the oxygenated blood is supplied into the entire body and this blood circulation process continues. As the left ventricle needs to pump oxygenated blood through to the entire body, the cardiac muscle in the left ventricle wall is much thicker compared to the muscle of the right ventricle wall.

It can be noted that the contraction of the heart is an inevitable step in the blood circulation process. This contraction process is controlled by cation exchange induced electrical signal propagation through heart cells. At rest, the heart muscle cells have a −90 mV potential with respect to their reference. This potential difference arises due to the injection and emission of $Na^+$ and $K^+$ ionic charges due to the sodium/potassium ATPase pump. This potential level is known as the resting membrane potential [12, 13].

When a heart muscle cell is stimulated, $Na^+$ ions enter into the cell via slightly opened $Na^+$ channels and the introduction of positive charges increases the cell potential, as shown in figure 8.2. When the cell potential becomes −70 mV, all the $Na^+$ channels are opened, resulting in a rapid injection of $Na^+$ in the cell. This build-

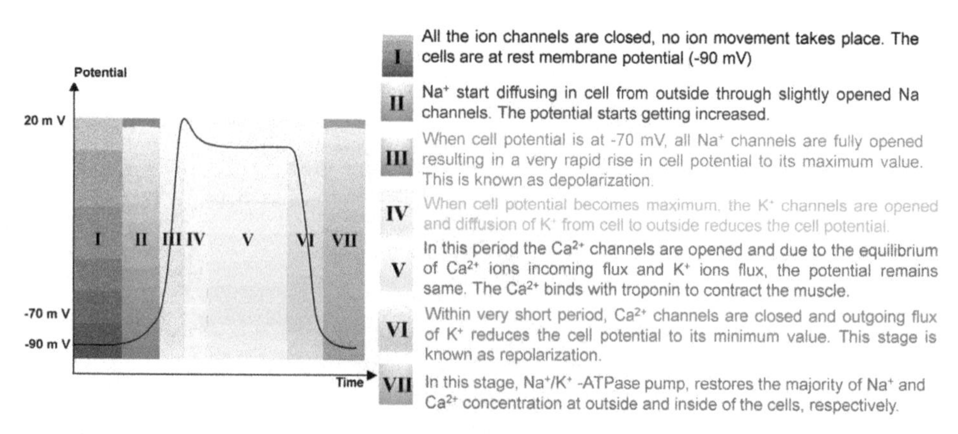


**Figure 8.2.** Potential–time characteristic for the heart cell due to the ion diffusion process [16].

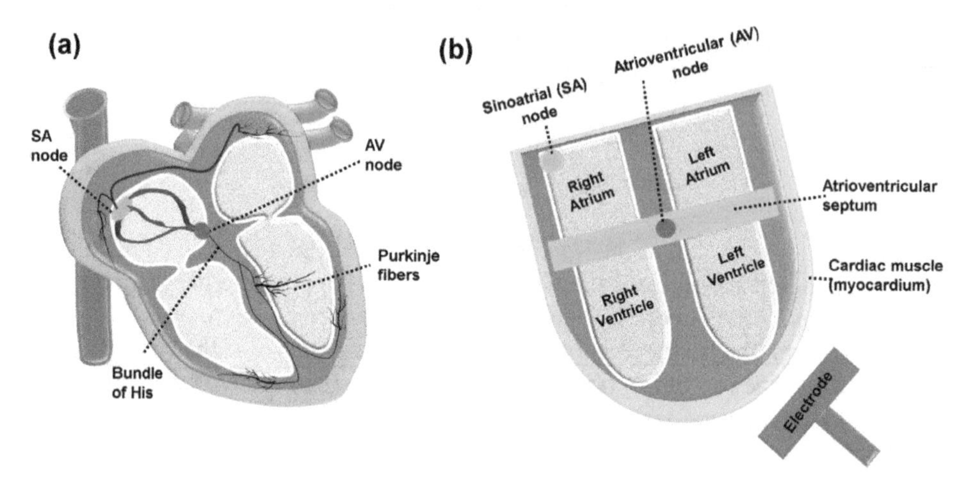


**Figure 8.3.** (a) The conduction system of the heart. (b) A schematic of the human heart with the four chambers and the electrical nodes.

up of positive potential in the muscle cell is known as depolarization. This depolarization in a cell is spread in its neighbor cells. When injection of $Na^+$ reaches its maximum value, the cell attains a maximum of about −20 mV potential, and the $Na^+$ channels start to be closed. Simultaneously, the $K^+$ channels are opened and $K^+$ ions are drained out of the heart. Subsequently, $Ca^{2+}$ channels are opened, thus allowing the flow of $Ca^{2+}$ ions into cells. The entry of $Ca^{2+}$ ions results in binding to the myosin head and muscle contraction.

As a consequence, a depolarization event is always followed by contraction of the chamber. After a very short time, the leaking of potassium ions from the cells results in the development of a negative potential in a cell again. This is known as repolarization. After a cycle of depolarization and repolarization, the sodium/potassium ATPase pump restores the ion concentration in the cell and its surroundings [14, 15].

The sinoatrial (SA) node is known as the natural pacemaker of the heart. The SA node is located in the wall of the right atrium, as shown in figure 8.3. Let us recall that

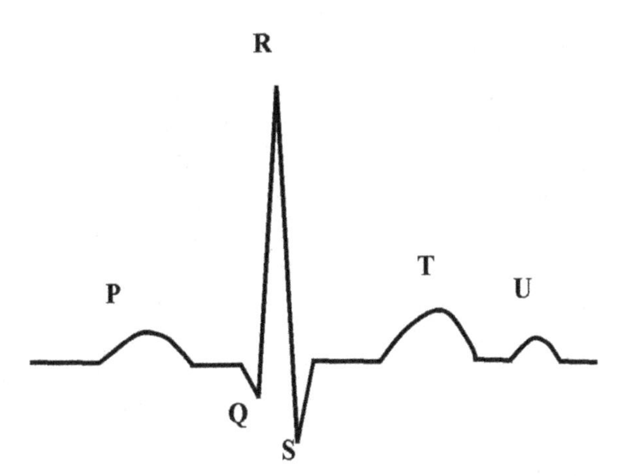


**Figure 8.4.** ECG characteristics of the human heart [17].

the ECG electrode is placed at theV5 lead (discussed in a subsequent section) location on the bottom left region of the heart. The SA node is spontaneously depolarized, leading to depolarization of the myocardium in a direction towards the electrode. The flux of positive ions leads to a positive potential peak (P peak) in the ECG characteristics, as shown in figure 8.4. When the depolarization wave propagates through the walls of the atria, it is blocked by the atrioventricular septum. As a result, the flow of positive ions through the wall is prevented and a flat profile (PR segment) is observed in the ECG characteristics. Next, with the stimulation of the atrioventricular (AV) node, the depolarization wave propagates through the 'His bundle' very rapidly, and the ECG profile exhibits a very short duration negative peak (Q peak). Subsequently, the flow of depolarization and ions through the Purkinje fibers in the ventricular walls takes place. As the myocardium in the left ventricle is thicker than in the right ventricle, there is an overall flow of depolarization towards the electrode and a sharp, high amplitude peak (R peak) is observed in the ECG profile. At the end of the depolarization cycle, the wave starts moving far from the electrode, which results in a short small negative peak (S peak) in the ECG graph. After full depolarization, the heart muscles contain positive potential and repolarization events take place to achieve the resting membrane potential in the muscles. This repolarization event moving far from the electrode leads to a small positive pulse (T peak) in the ECG profile. Further investigations have revealed the presence of a small peak (U wave) after the T peak in ECG characteristic. Although the origin of the U wave is not confirmed, it is considered to be associated with repolarization of the Purkinje fibers [17].

Thus, it can be concluded that the P peak in the ECG represents atrial depolarization. The next peak (the QRS complex) in ECG represents ventricular depolarization. The last peak, T, represents ventricular repolarization. During the ventricular depolarization, atrial repolarization also takes place. However, due to the thick wall of the ventricles, the ventricular depolarization dominates over the atrial repolarization and its corresponding peak is not observed in the ECG graph.

### 8.3.2 Placement of leads

In the previous section, it has been discussed that the electrical activity of heart results in the generation of electrical potential pulses. Now if the heart is considered as an active electric source and the tissues of the body as a passive conductor, this biopotential can be felt throughout the human body tissues. This time-varying potential difference can be detected with a pair of electrodes. These electrodes form a lead system which is based on theoretical considerations. The placement of leads on different parts of the body can allow the electrical measurement system to act as load, and thus the heart's electrical activity can be recorded. In this regard, the lead theory is a concept to establish the relationship between the biopotential generated by the heart, and the biopotential felt at the location of the electrode. Also, the lead theory presents the idea of estimating the influence of electrode locations on the human body to observe the heart's activity from various perspectives. This section describes the basic mathematical relations and components of the 12-lead system.

Einthoven and his co-workers suggested that the biopotential experienced at three extreme points in the human limbs due to the cardiac action of the heart can also be represented by a model of the heart in the centroid of a triangle and the limb extremities in the three vertices of the same triangle [18]. The model developed by Einthoven considers the human body as a homogeneous conductor and the heart as the only active dipole. However, later work by different research groups modified the model by considering the distributed nature of the heart's electric action and the inhomogeneous and irregular pattern of the human body [19–21].

Burger and van Milaan showed the relationship between dipolar sources $\vec{p}$ ($\vec{p}$ which represents the grouped cardiac electric sources of a chosen cardiac region) and the lead voltages (V). The Burger equation states that the lead voltage can be represented as the dot product of two vectors:

$$V = \vec{c} \cdot \vec{p}. \tag{8.1}$$

$\vec{c}$ is the lead vector which represents the relationship between the active electrical source and the voltage in the passive tissue at the location where the electrode is attached. The electrical excitation process is assumed to be confined within the heart region only and this results in the introduction of an electric current in the body through the conductive tissues. However, due to irregularities in the human body shape and inhomogeneities in composition, the biopotential generated in the passive extra-cardiac tissue in one part of the human body does not follow a linear relationship with the active source potential at another location. Hence, the concept of the lead vector is introduced to establish a relationship between the biopotential in passive tissue and its source of origin. It takes into account the influences of shape irregularities and inhomogeneities of the human body, the locations of the cardiac source and the electrodes.

The heart is considered to be a bioelectric source due to its time-varying cardiac actions. These types of sources are represented by the volume distribution of the impressed current density or current dipole moment per unit volume [22].

Due to the volume distributions of adjoining bio-current sources and sinks in the heart, the heart is represented by the current dipole source. $\vec{p}$ represents the current dipole moment generated due to the time-dependent electrocardiac excitation process. Hence, Einthoven described the heart using a time-dependent current dipole which contributed to the generation of biopotential throughout the conductor, the human body.

If $\vec{c} = c_x\hat{i} + \vec{c}_y\hat{j} + \vec{c}_z\hat{k}$ and $\vec{p} = p_x\hat{i} + p_y\hat{j} + p_z\hat{k}$, the voltage $V$ at the lead can also be expressed as

$$V = c_x p_x + c_y p_y + c_z p_z. \tag{8.2}$$

The right hand, left hand and the left leg are considered as the three extremities (vertices) of the human body. Let the potential at the right arm, left arm and left leg be $V_R$, $V_L$ and $V_F$, respectively. From Burger's equation, we can write

$$V_R = \vec{c}_R.\vec{p};\ V_L = \vec{c}_L.\vec{p};\ V_F = \vec{c}_F.\vec{p}, \tag{8.3}$$

where $\vec{c}_R$, $\vec{c}_L$ and $\vec{c}_F$ represent the lead vectors at the right arm, left arm and left leg, respectively [23–25]. The 12-lead system views the heart and records ECG from 12 different viewpoints. It consists of three standard bipolar limb leads (VI, VII and VIII), three unipolar augmented limb leads ($V_{aVR}$, $V_{aVl}$ and $V_{aVF}$) and six unipolar chest leads (V1, V2, V3, V4, V5 and V6). All six limb leads are related to the $V_R$, $V_L$ and $V_F$. For this system a total of ten electrodes (points of contact with the body) are used to perform an ECG.

### *8.3.2.1 Bipolar and augmented unipolar limb leads*

Two electrodes are placed on the arms and one on the legs when recording ECG. These provide the basis for the six limb leads (three standard bipolar limb leads and three augmented leads) and record six different views of electrical activity in the

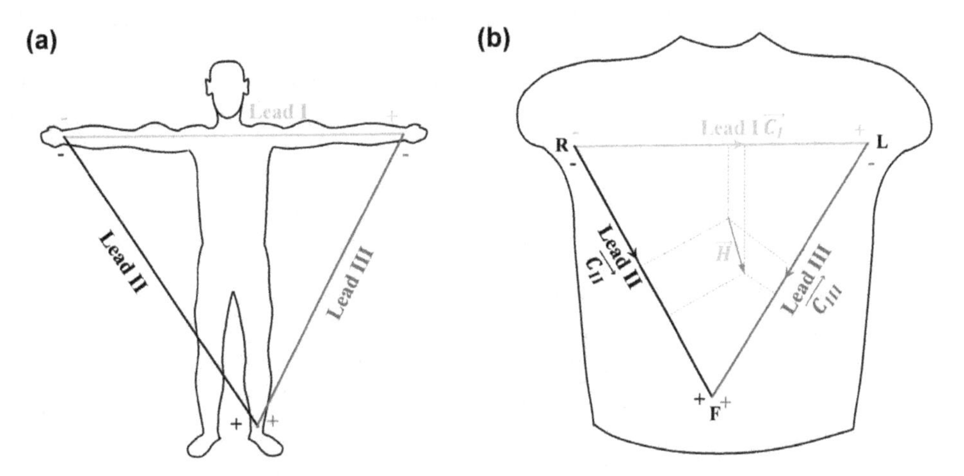


**Figure 8.5.** (a) Schematic of a bipolar lead system for measuring cardiac potential in the human body and (b) a representation of Einthoven's triangle in the human torso. The heart vector's ($\vec{H}$) projection on the limb leads can be estimated from the voltage recording at bipolar limb leads.

heart. The three bipolar limb leads are denoted as lead I, lead II and lead III (figure 8.5(a)). In bipolar leads, the potential is expressed as the potential difference between the limbs. According to Einthoven, the potential at three bipolar limb leads can be written as

$$V_{\mathrm{I}} = V_{\mathrm{L}} - V_{\mathrm{R}} \tag{8.4a}$$

$$V_{\mathrm{II}} = V_{\mathrm{F}} - V_{\mathrm{R}} \tag{8.4b}$$

$$V_{\mathrm{III}} = V_{\mathrm{F}} - V_{\mathrm{L}}. \tag{8.4c}$$

Considering all current dipole source as a single unit, we can write from Burger's equation

$$\vec{c}_{\mathrm{I}} = \vec{c}_{\mathrm{L}} - \vec{c}_{\mathrm{R}} \tag{8.5a}$$

$$\vec{c}_{\mathrm{II}} = \vec{c}_{\mathrm{F}} - \vec{c}_{\mathrm{R}} \tag{8.5b}$$

$$\vec{c}_{\mathrm{III}} = \vec{c}_{\mathrm{F}} - \vec{c}_{\mathrm{L}}. \tag{8.5c}$$

A careful observation of equation (8.4) can lead to the conclusion that

$$V_{\mathrm{I}} + V_{\mathrm{III}} = V_{\mathrm{II}}. \tag{8.6}$$

Equation (8.6) is also known as Einthoven's law. This equation can also be visualized by a lead triangle, as shown in figure 8.5(a). Selecting limb potential subtraction as the lead voltage allows us to visualize the direction of the lead vectors. For example, if the movement of electric excitation of the heart takes place from right to left, an ascending deflection will be observed in the lead I recording. As can be seen from figure 8.5(b), the heart vector's ($\vec{H}$) projection on the side lead vectors of Einthoven's triangle can be interpreted by the observation of bipolar limb lead voltages. Thus, the voltage recording at the bipolar limb leads can help to estimate the electrical activities of the heart.

Figure 8.6 shows the electrical circuit equivalent model of the human body with cardiac potential and body resistance. Using Kirchhoff's current law and node analysis at the node, the equation can be written as

$$\frac{V_1 - V_x}{R_1} + \frac{V_2 - V_x}{R_2} + \ldots + \frac{V_k - V_x}{R_k} = 0. \tag{8.7a}$$

In another way, it can be expressed as

$$V_x = \frac{\left(\frac{V_1}{R_1} + \frac{V_2}{R_2} + \ldots + \frac{V_k}{R_k}\right)}{\left(\frac{1}{R_1} + \frac{1}{R_2} + \ldots + \frac{1}{R_k}\right)}. \tag{8.7b}$$

Similarly,

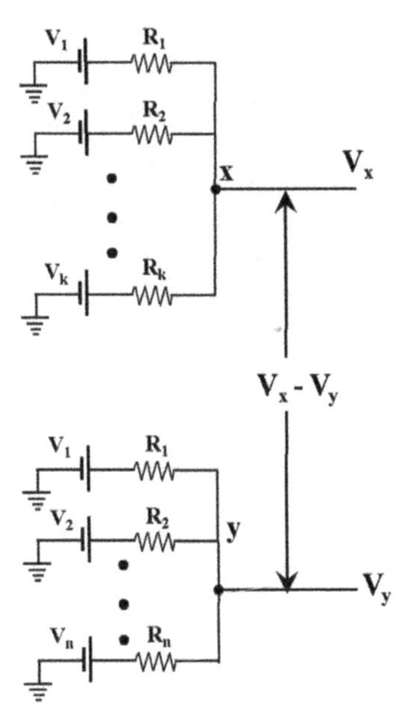


**Figure 8.6.** Electrical circuit equivalent model of the human body with cardiac potential and body resistance.

$$V_y = \frac{\left(\frac{V_1}{R_1} + \frac{V_2}{R_2} + \ldots + \frac{V_n}{R_n}\right)}{\left(\frac{1}{R_1} + \frac{1}{R_2} + \ldots + \frac{1}{R_n}\right)}. \tag{8.7c}$$

Thus, the potential difference between the nodes $x$ and $y$ can be expressed as

$$\begin{aligned} V_x - V_y &= \frac{\left(\frac{V_1}{R_1} + \frac{V_2}{R_2} + \ldots + \frac{V_k}{R_k}\right)}{\left(\frac{1}{R_1} + \frac{1}{R_2} + \ldots + \frac{1}{R_k}\right)} - \frac{\left(\frac{V_1}{R_1} + \frac{V_2}{R_2} + \ldots + \frac{V_n}{R_n}\right)}{\left(\frac{1}{R_1} + \frac{1}{R_2} + \ldots + \frac{1}{R_n}\right)} \\ &= \frac{\sum_1^k \frac{V_i}{R_i}}{\sum_1^k \frac{1}{R_i}} - \frac{\sum_1^n \frac{V_i}{R_i}}{\sum_1^n \frac{1}{R_i}}. \end{aligned} \tag{8.7d}$$

Wilson further developed the ECG system by adding a horizontal axis through the heart. He assumes a virtual reference point, located in the middle of the heart. This reference point was called the Wilson central terminal (WCT). The WCT can be assumed as a junction of three extreme terminals connected by resistances (figure 8.7). Hence, the potential at WCT ($V_T$) can be defined as the average of the potential at the right hand ($V_R$), left hand ($V_L$) and leg ($V_F$) [26],

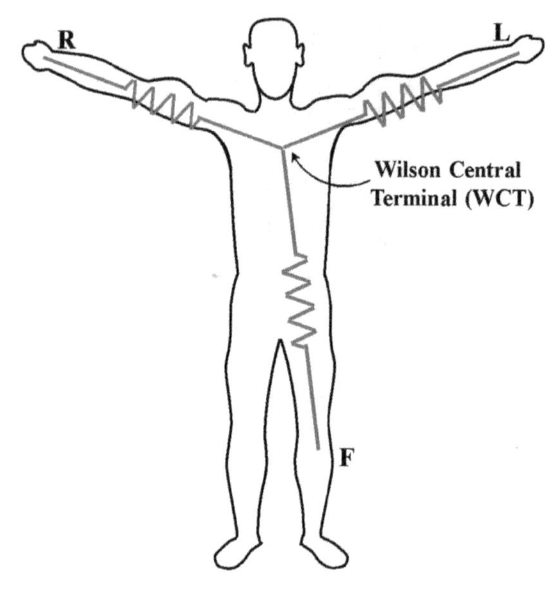


**Figure 8.7.** The location of the WCT in the human body.

$$V_T = \frac{V_R + V_L + V_F}{3},$$

and the corresponding lead vector, $C_T = \frac{\vec{C}_R + \vec{C}_L + \vec{C}_F}{3}$,

When the WCT potential is used as a reference potential, the expression becomes

$$V_R - V_T = V_R - \frac{V_R + V_L + V_F}{3} = \frac{2V_R - V_L - V_F}{3}. \tag{8.8a}$$

Similarly,

$$V_L - V_T = \frac{2V_L - V_R - V_F}{3} \tag{8.8b}$$

$$V_F - V_T = \frac{2V_F - V_R - V_L}{3}. \tag{8.8c}$$

Now, if we consider the reference terminal potential to be the average of the remaining two extreme limb terminal potentials, as shown in figure 8.8, the potential difference becomes

$$V_{aV_R} = V_R - \frac{V_L + V_F}{2} = \frac{2V_R - V_L - V_F}{2}. \tag{8.9a}$$

Similarly,

$$V_{aV_L} = \frac{2V_L - V_R - V_F}{2} \tag{8.9b}$$

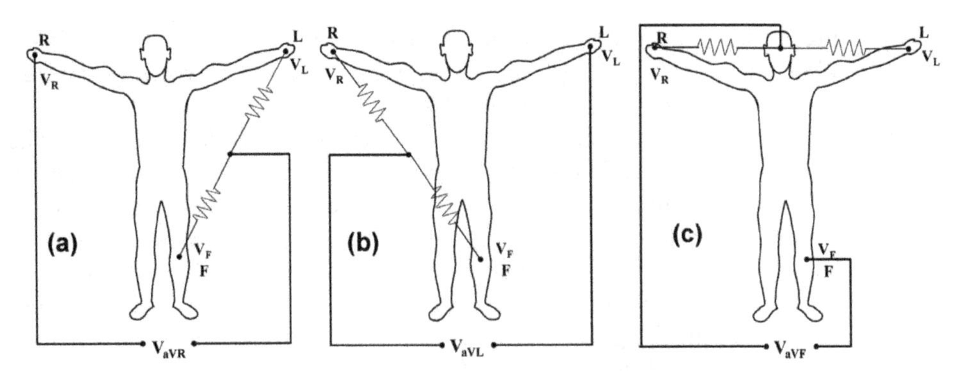


**Figure 8.8.** Schematic representation of the augmented unipolar lead system by measuring the potential difference between a limb electrode and the average of the potentials at the other two limb electrodes: (a) $aV_R$, (b) $aV_L$, (c) $aV_F$.

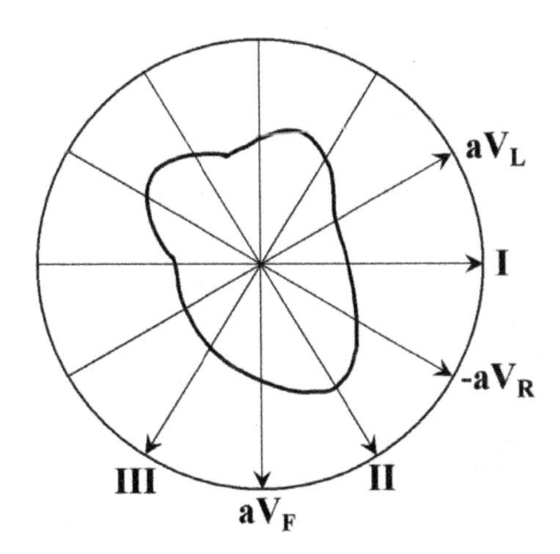


**Figure 8.9.** The frontal plane view of the heart with unipolar and bipolar limb leads.

$$V_{aV_F} = \frac{2V_F - V_R - V_L}{2}. \tag{8.9c}$$

Here, it can be observed from equations (8.8) and (8.9) that the change of reference from WCT to the average of the remaining terminals improves the potential difference by 50%. Hence these limb lead arrangements are known as augmented unipolar limb leads and an ‘a’ is used in their notation. The electrical view of the heart in its frontal plane with its unipolar and bipolar limb leads are shown in figure 8.9.

*8.3.2.2 Chest (precordial) leads*

As per the recommendation of the American Heart Association, six more leads in addition to the three limb leads are connected to the chest of a human to obtain a three-dimensional view of the electrical activities of the heart. These six leads are known as precordial (chest) leads (as shown in figure 8.10(a)). This lead offers a horizontal plane view of the heart (figure 8.10(b)) [22]. The arrangement of the lead placement is carried out as described in the table 8.1.

*Note.* There may be confusion between the terms electrode and lead. The electrode is the small electrical conductor attached to different parts of the body to allow the flow of biopotential signal from the body to the measuring instruments. In contrast, the lead system is the configuration of the electrode connections to obtain better cardiac electrical observation. In the standard ECG measurement system, ten electrodes (three on the limbs, six on the chest and one on the right leg) are connected to obtain a 12-lead ECG system (three bipolar limb leads, three augmented unipolar limb leads and six chest leads). An electrode is connected to the right leg (right leg drive) as a reference electrode to eliminate the common-mode noise, as discussed in the following section.

### 8.3.3 Integrated circuit associated with waveform generation

The amplitude of the cardiac potential generated by our heart is in millivolts. Due to tissue attenuation, the potential amplitude at the human body surface becomes of the order of a few millivolts. Thus the electrodes are connected to an amplifier circuit to boost the signal. Differential amplifiers are used for processing the biopotential signal.

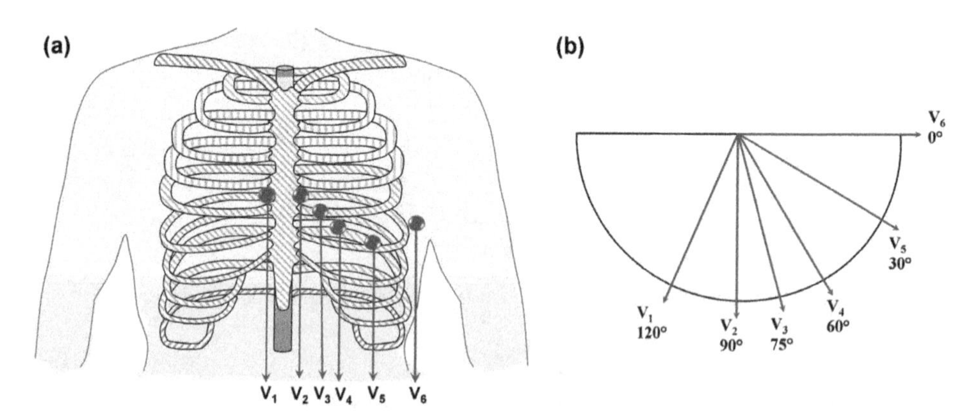


**Figure 8.10.** (a) Placement of the chest leads. (b) A horizontal plane view of the heart with chest leads.

**Table 8.1.** The details of lead placement.

| Sl. no. | Electrode | Electrode placement location |
|---|---|---|
| 1 | $V_1$ | Fourth intercostal space on the right of the sternum. |
| 2 | $V_2$ | Fourth intercostal space on the left of the sternum. |
| 3 | $V_3$ | In the middle between the locations of $V_2$ and $V_4$. |
| 4 | $V_4$ | At the intersection of the fifth intercostal space and the left midclavicular line. |
| 5 | $V_5$ | Left anterior axillary line, at the same level as $V_4$. |
| 6 | $V_6$ | Left midaxillary line, at the same level as $V_4$ and $V_5$. |

*8.3.3.1 Differential amplifiers and the common-mode rejection ratio*

When signals are fed to the inputs of a differential amplifier, they are expected to generate an amplified version of an input signal at the output terminal. However, due to the presence of a noise signal common to both input terminals, there is also an amplified noise component in the output signal. This can be seen in the circuit in figure 8.11—the output of the following differential amplifier consists of the amplified differential as well as the common-mode signal.

This is a serious issue as the differential input from the cardiac action is in the millivolt range and matching the resistance ratio in the positive and negative terminals of an operational amplifier is a challenging task. Hence, for reduction of common-mode noise, an instrumentation amplifier is utilized to implement the circuit, as shown in figure 8.12.

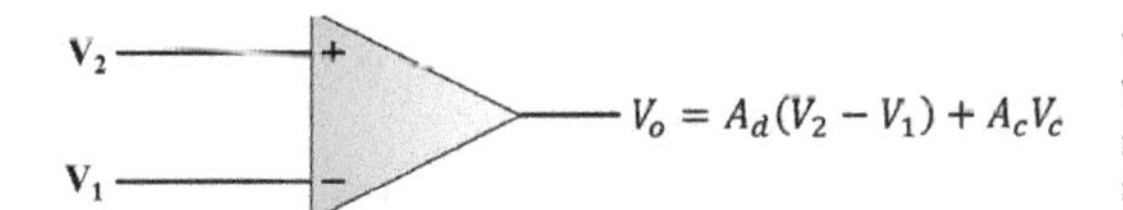


$V_2$ and $V_1$ are potentials at the input of the op-amp, $V_c$ is the common potential present in both of the inputs, $A_d$ is the differential voltage gain and the $A_c$ is the common voltage gain

**Figure 8.11.** Schematic representation of a differential amplifier with differential and common-mode voltage gain.

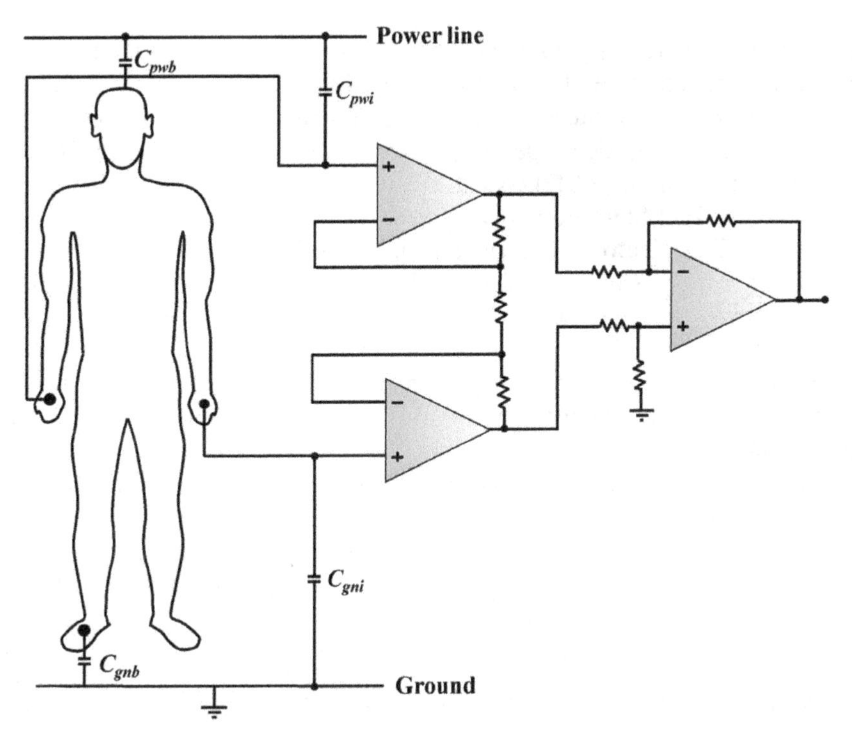


**Figure 8.12.** The application of an instrumentation amplifier for differential amplification of ECG signals and the origin of capacitances between the power/ground lines and the human body/interconnection.

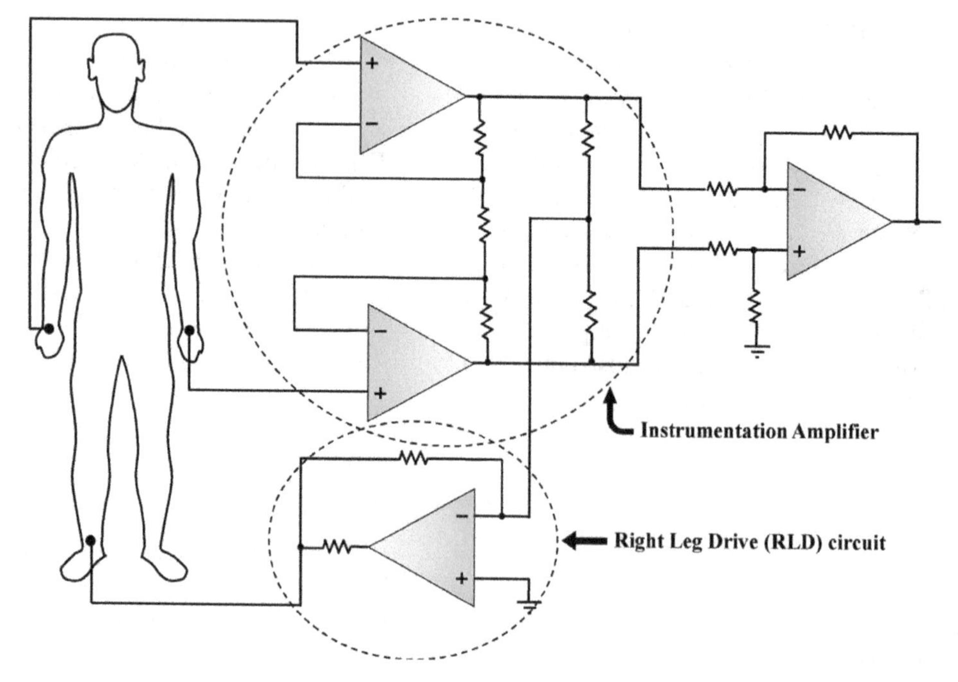


**Figure 8.13.** Implementation of the RLD circuit for noise (due to power/ground line capacitance) reduction.

However, the electrical signal flowing through the power lines in the housing gives rise to capacitances between the power lines and human body/amplifier connecting wires. This leads to noisy interference in the biopotential signal and the noisy signal is inherited by all of the electrodes [22]. To remove this noise from the differential input, the right leg drive (RLD) circuit is used, as shown in figure 8.13.

In this modified configuration, an additional electrode is connected to the right leg of the patient. This electrode collects all the noise due to the surrounding electric field and this voltage source is connected to a node common to both the input terminals via resistors. Due to the differential nature of the amplifier, the noise component is eliminated from the final output. First, this output is fed to an isolation amplifier which prevents the flow of current at a harmful level to the patient from the recording instrument. The next stage of the system is a high pass filter which eliminates low-frequency noise from the ECG signal originating in sudden movements of the patient, respiratory movement, conducting gel leakage, etc. In the next stage, a low pass filter removes the skeletal muscle generated high-frequency noise. The last stage in the filter network is a notch filter which separates the 50/60 Hz signal from the signal, and the filtered signal is provided to a digital signal processor for necessary operations.

## 8.4 Categories of biopotential ECG electrodes

Biopotential electrodes can be classified in different ways such as non-polarized and polarized electrodes according to the charge transfer mechanism between the electrode and skin interface. In a non-polarized electrode, the charge can freely pass through the skin–electrode interface, and the impedance between the skin–electrode interface can be reduced using electrolyte gel. The reversible electrochemical reduction/oxidation reactions occurs at the electrode–skin interface [27, 28]. These electrodes are also called wet electrodes and have low electrode–skin impedance, low motion artifact and low noise during monitoring [29]. The conventional Ag/AgCl electrode is an example of a wet electrode [30]. A polarized electrode operates without gel and ideally has no direct current flow between the electrode–skin interface and is called a dry electrode.

### 8.4.1 Introduction and advantages of the dry electrode

To overcome the difficulties of wet electrodes and for long-term use of biopotential ECG electrodes, researchers are investigating the opportunities for new electrodes that can be acceptable for medical use. Among them, dry electrodes which can operate without gel and skin preparation could be the best alternative. Numerous designs of dry electrodes are reported in the literature. However, they still need more quantitative evaluation to be utilized for clinical use. This section describes the evolution of dry electrodes and comparisons with conventional electrodes to find ways for its medical use. Based on the literature information and the electrode's coupling with the skin, dry electrodes can be categorized as (a) contact surface electrodes which have direct contact with only the skin surface, (b) contact penetrating electrodes which can penetrate the skin's upper layer and (c) noncontact capacity electrodes which have indirect insulating contact with the skin (figure 8.14).

The biopotential electrode works as a transducer; it converts the ionic currents of the bio-signal into electric currents within the electrodes [27, 28]. The schematics of the electrode–skin interface and its simplified electrical equivalent for wet and different types of dry electrode are shown in figure 8.15. For electrodes, the impedance at the electrode–skin interface can be simplified as a combination of two RC components, in which one is due to the electrode ($R_{wet}$, $C_{wet}$: wet electrodes; $R_{SE}$, $C_{SE}$: contact surface electrodes; $R_{ME}$, $C_{ME}$: contact penetrating electrodes; and $R_{IE}$, $C_{IE}$: nonconducting electrodes) and the second is due to the stratum corneum ($R_{sc}$, $C_{sc}$) in series with resistance due to the gel ($R_{gel}$), epidermis ($R_e$) and half-cell potential.

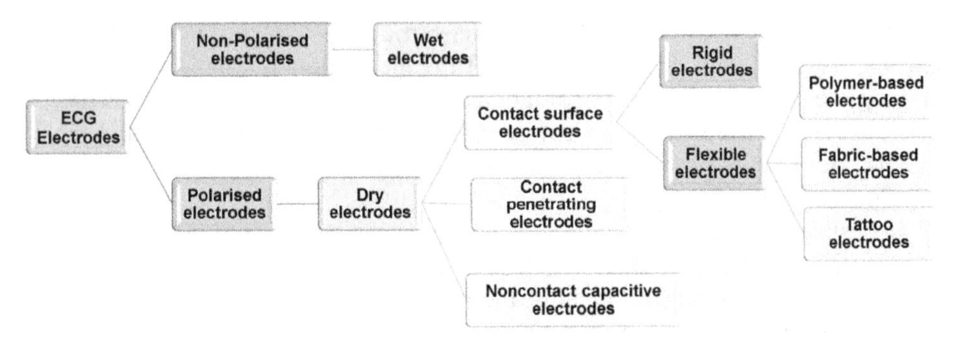


**Figure 8.14.** Types of ECG electrodes.

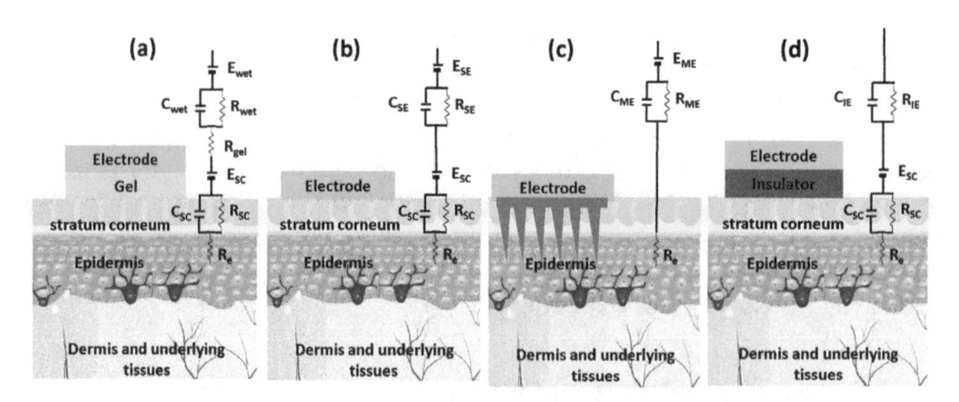


**Figure 8.15.** Schematics of electrode–skin interface models for (a) a wet (gel) electrode, and dry electrodes: (b) contact surface electrode, (c) contact penetrating electrode and (d) noncontact capacitive electrode.

The ion concentration gradient across the electrode–stratum corneum induces the potential difference ($E_{sc}$).

### 8.4.2 Fabrication and performance evaluation of contact surface electrodes

The dry contact surface electrodes operate only on the skin surface, and the electrode–skin interface model is shown in figure 8.15. Depending on the flexibility and stretchability, these electrodes can be further classified into two categories: rigid and flexible electrodes.

The rigid biopotential electrodes are generally fabricated using metals. In 1971, Bergey *et al* examined the skin–metal interface characteristics for Ag, Au, Al, anodize Al, stainless steel (SS) and brass. They reported that among these metals, SS performed the best as dry electrodes with fewer motion artifacts [31]. In 2013, Meziane *et al* reviewed the dry textile electrodes along with the stiff material based electrodes [32]. Many researchers have used stainless steel for electrode fabrication due to its good conductivity, availability and lower price [31, 33, 34]. Gondran *et al* proposed a different approach and designed a dry electrode using Na superionic conductor (NASICON) type ceramic. However, this electrode showed slightly polarized behavior and deteriorated from low-frequency noise [35]. In 2000, Searle and Kirkup compared metal (stainless steel, titanium and aluminum), insulating and gel electrodes. They reported that wet electrodes showed more motion artifacts than the dry and insulating electrodes. All the metal electrodes compared showed approximately the same average contact impedance values but higher than those of wet electrodes [36]. These rigid electrodes suffer from charging effects and motion artifacts, and can lose contact due to stiffness [34], which limits their use for real-time clinical applications [34].

The flexible dry electrodes can adjust quickly to the body shape, which can reduce motion artifacts. These flexible electrodes can be further classified into three categories depending on the material used in fabrication, such as polymer-based electrodes, fabric-based electrodes and tattoo electrodes. Some recent works on these electrodes are discussed in this section and are summarized in table 8.2.

**Table 8.2.** Flexible dry contact surface electrode comparison.

| References | Material | Size/shape | Findings | Comments |
|---|---|---|---|---|
| Gruetzmann *et al* [37] | Dry Ag foil, conductive foam (polyester, polyethylene) | Diameter: 20 mm | Fewer motion artifacts for dry foam. | Polymer-based |
| Chlaihawi *et al* [41] | CNTs, PDMS, Ag, PET | Diameter: 32 mm | Reduced motion artifacts on the measured ECG signals. | Polymer-based |
| Cheng *et al* [47] | PDMS, carbon black (CB) | Rectangular shape | Similar ECG as obtained using Ag/AgCl gel electrodes. | Polymer-based |
| Kisannagar *et al* [46] | PET, PDMS, AgNWs | Dimension: 15 × 15 mm square area | The skin–electrode impedance of the AgNW/PDMS electrodes is less than that of the wet Ag/AgCl electrodes. | Polymer-based |
| Lu *et al* [48] | Graphite nanosheet (GN), polyamide (PA) | Dimension: 30 mm × 10 mm × 10 $\mu$m | Comparable EGG signal with commercial Ag/AgCl electrodes. | Polymer-based |
| Kim *et al* [49] | AgNWs, ethylcellulose (EtC) | Rectangular shape | Transparent electrodes, comparable EGG signal with commercial Ag/AgCl electrodes. | Polymer-based |
| La *et al* [54] | Ag particles, fluoropolymer, PDMS, PU | Rectangular fabric | Use of eco-friendly and recyclable paper. | Fabric-based |
| Yapici *et al* [57] | Nylon fabric, reduced graphene oxide (rGO) | Rectangular fabric | Impedance range from 87.5 kΩ to 11.6 kΩ. | Fabric-based |
| Kabiri *et al* [60] | Graphene, PMMA, Cu tattoo paper (silhouette) | serpentine shapes | No adhesive is applied, comparable SNR with wet electrodes. | Tattoo-based |
| Wang *et al* [61] | PET, Au, Cr, tattoo paper | Thickness: 1.5 $\mu$m | Fewer motion artifacts than thicker e-tattoos and even gel electrodes. | Tattoo-based |
| Min *et al* [65] | Polyurethane, CNT, Ag f lakes, PDMS | Hexagonal mesh-patterned | Water resistant, skin-adhesive and air/water-permeable, comparable EGG signal with commercial Ag/AgCl electrodes. | Tattoo-based |

### *8.4.2.1 Polymer-based electrodes*

Numerous biopotential electrodes have been designed using polymers due to their flexible and stretchable nature. These electrodes can be fabricated using different methods, such as metal-coated polymers and polymeric composites with conductive fillers. In 2007 Gruetzmann *et al* proposed an electrically conductive foam using polymer and Ag coating on all surfaces by thermal evaporation [37]. The polymer-based electrodes with metal coating have high conductivity but most coatings are generally prepared using the high-cost photolithography and vacuum deposition techniques. These conductive coatings exhibit poor adhesion with the polymer substrate [38, 39]. The other method uses a conductive filler such as carbon nanotubes [40, 41], reduced graphene oxide [42, 43] or silver nanowires [44, 45] in a polymeric matrix but it is a challenging task to uniformly disperse these conductive fillers in the polymer matrix [46]. In 2015 Chlaihawi *et al* reported the development of a flexible dry electrode based on a multi-walled carbon nanotube polydimethylsiloxane (MWCNT/PDMS) composite [41]. Cheng *et al* reported a PDMS and carbon black conductive composite and showed stable long-term ECG signal monitoring [47]. In 2020 Kisannagar *et al* designed an AgNW/PDMS dry electrode. They showed that its signal-to-noise (SNR) value was 25.4 dB, which is comparable to a wet Ag/AgCl electrode with a value of 24.6 dB [46]. Lu *et al* reported a flexible, conductive and ultrathin electrode based on a graphite nanosheet and polyamide nanofiber composite, shown in figure 8.16(A) [48]. Kim *et al* proposed a new approach for highly transparent electrode fabrication using biodegradable ethylcellulose (EtC) and Ag nanowire (AgNW) composites for ECG monitoring, shown in figure 8.16(B) [49].

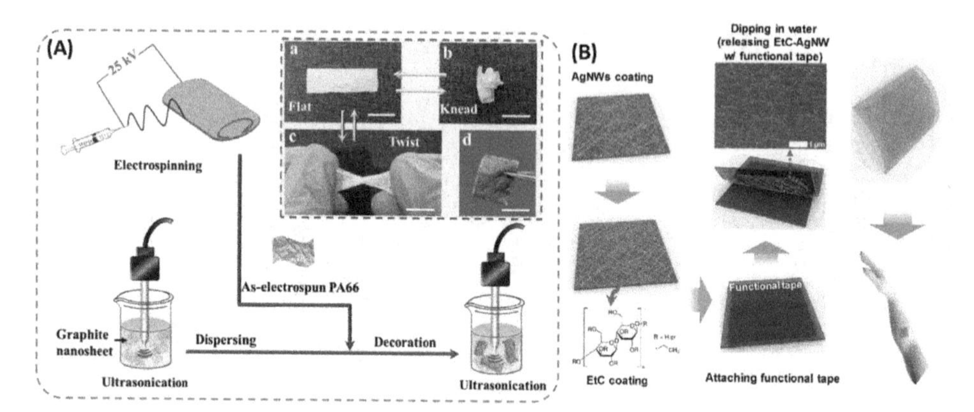


**Figure 8.16.** (A) Schematic of the graphite nanosheet/polyamide (GN/PA66) nanocomposite fabrication process. Optical images of the different conditions of the as-prepared electrospun PA66 mat: (a) flat, (b) kneaded, (c) twisted and (d) as-prepared nanocomposite after ultrasonic treatment. Reproduced with permission from [48]. Copyright 2020 Elsevier. (B) Schematic of the (EtC) and AgNW composite fabrication process. SEM images of the AgNW surface covered with EtC film. Reproduced with permission from [49]. Copyright 2020 Elsevier.

#### 8.4.2.2 Fabric-based electrodes

The fabric-based electrodes are flexible, more conforming to the body as well as being potentially washable, and these characteristics can make them a potential candidate for biopotential electrodes for long-term monitoring. Arquilla *et al* explored dry textile electrodes and classified them into two categories based on their fabrication methods: one in which conductive yarns and threads are integrated into fabrics and a second in which conductive pastes and inks are applied to fabrics using different methods, such as stenciling, sputtering and screen printing [50]. This section describes some recently published fabric-based electrode designs. In 2019 Ankhili *et al* designed dry ECG electrodes by embroidering silver-plated polyamide threads onto cotton woven fabrics and showed a good ECG signal even after 50 washes [51]. Rajanna *et al* designed a conductive knitted jersey made up of silver yarn, spandex and cotton as a substrate [52]. Recently, Li *et al* designed a silver-coated nylon yarn knitted into the fabric of underwear to record ECG signals on the body surface [53]. Many fabric electrodes based on conductive inks have been reported, such as La *et al*'s two-layered e-textile patches produced by applying nanocomposite ink made of silver powder and fluoropolymer on porous polyurethane textile substrate, as shown in figure 8.17(A) [54]. Xu *et al* applied water-based graphene ink on a cotton textile by screen printing [55]. Soroudi *et al* coated the surface of a knitted textile using a conductive elastic paste made of silver-plated copper flake and polyurethane [56]. Carbon nanotubes (CNTs) are also used as a filler in a polymer matrix for dry electrode fabrication, as shown in figure 8.17(B) [57].

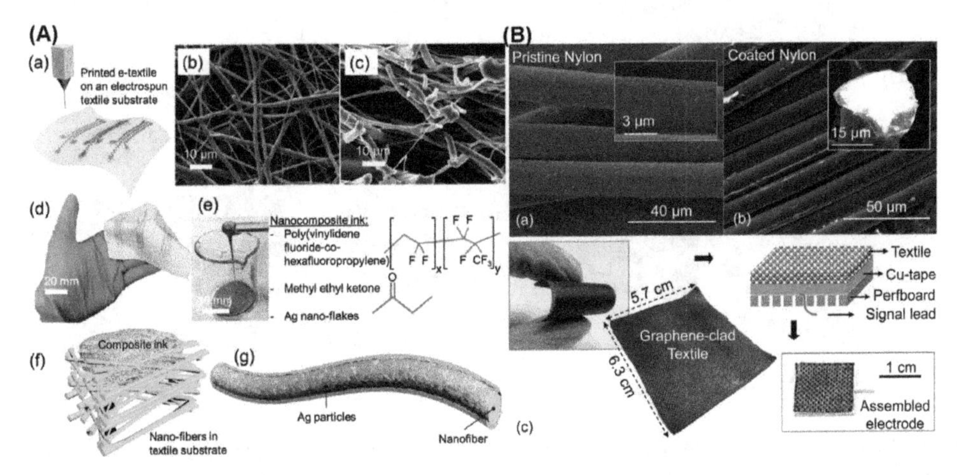


**Figure 8.17.** (A) e-textile electrodes: (a) schematic for the jet-printing of nanocomposite inks onto a textile substrate; SEM images of the compressed electrospun nanofibers of a polyurethane-based textile substrate: (b) top and (c) cross-section views; (d) e-textile sheet with printed conductive serpentine traces; (e) as-prepared nanocomposite ink consisting of silver particles, fluoroelastomer and methyl ethyl ketone; (f) schematic of the textile substrate with printed ink; and (g) schematic illustration of a single cladded nanofiber in the cladded-layer of a printed e-textile. Reproduced with permission from [54]. Copyright 2018 Wiley. (B) SEM images of (a) pristine nylon fiber, (b) graphene-coated nylon fiber and (c) optical image of a flexible nylon textile with a rGO coating and its small pieces as an electrode for ECG testing. Reproduced with permission from [57]. Copyright 2020 Elsevier.

Li *et al* designed electrodes by integrating CNT and polyurethane composites onto knitted textile substrates [58]. Similarly, Saleh *et al* reported the fabrication of flexible and highly conductive rGO-coated cotton fabric (rGOC) and showed its better performance in ECG monitoring compared to wet Ag/AgCl and metal-based rigid electrodes [59].

### *8.4.2.3 Tattoo electrodes*

A special type of epidermal electronic sensor, also known as an electronic or e-tattoo, is emerging as a bipotential electrode due to its ultra-soft, ultrathin and high flexibility properties. These e-tattoos can follow skin displacement with skin deformation, which can minimize interfacial slippage and motion artifacts and effectively lower the contact impedance [60–62]. In 2017 Kabiri *et al* designed a sub-micrometer thick transparent and stretchable electronic tattoo using graphene through a wet transfer and dry patterning method, shown in figure 8.18(A) [60]. Wang *et al* designed a tape free e-tattoo with 1.5 $\mu$m thickness using the cut-and-paste method shown in figure 8.18(B). They reported that it is capable of signal monitoring with reduced motion artifacts compared with thicker e-tattoos and even gel electrodes. Its open mesh design prevents sweat accumulating between e-tattoo and skin, which minimizes the sweat artifacts [61].

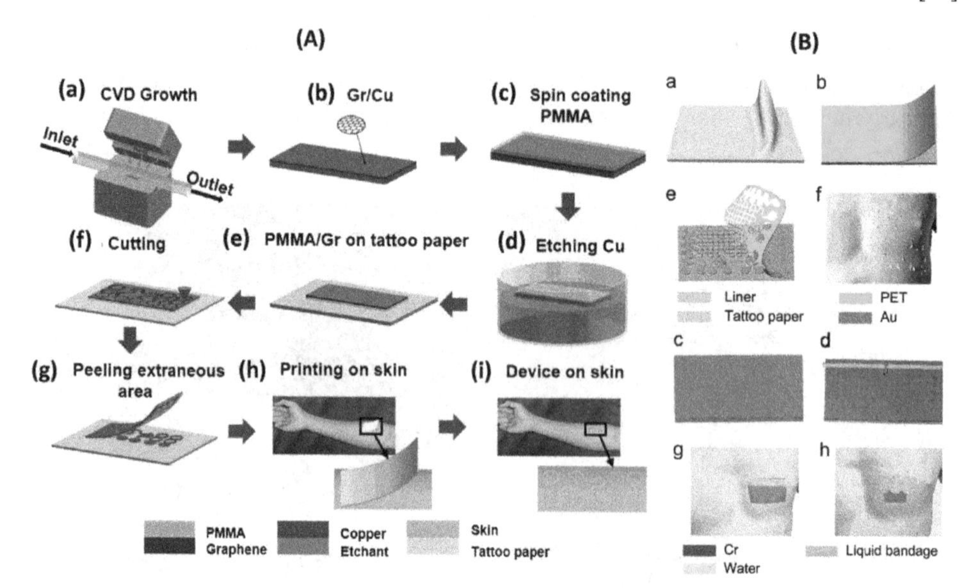


**Figure 8.18.** (A) Schematic of the graphene-based e-tattoo fabrication process: (a) and (b) Growth of graphene on Cu foil; (c) spin coating of PMMA on graphene; (d) process of etching Cu; (e) peeling and transferring process of graphene/PMMA onto tattoo paper with PMMA touching the paper and graphene facing up; (f) cutting process of graphene/PMMA; (g) Gr/PMMA peeling from the tattoo paper; (h) mounting the as-prepared e-tattoo on the skin; and (i) an image of an e-tattoo on the skin. Reproduced (adapted) with permission from [60]. Copyright 2020 American Chemical Society. (B) Schematics of the fabrication process for a tape free e-tattoo: (a) liner removal from commercial tattoo paper; (b) laminating a transparent PET sheet on the tattoo paper; (c) deposition of Cr (10 nm) and Au (100 nm) on the PET; (d) patterning of the Au/Cr/PET sheet with a commercial cutter plotter; (e) peeling off the extraneous parts from the tattoo paper; (f) a small amount applied on human skin; (g) pasting the e-tattoo on the human skin from the tattoo paper; and (h) spraying liquid bandage over the e-tattoo with the hydration sensors being covered by a stencil [61].

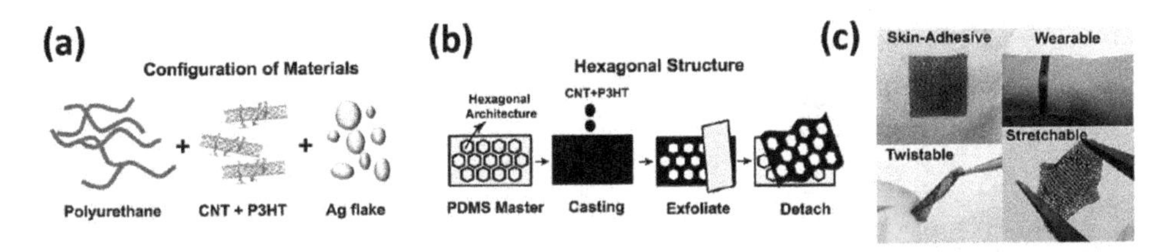


**Figure 8.19.** A carbon-based conductive polymer composite (CPC) patch with hierarchical architecture. (a) The materials used in fabrication: polyurethane, carbon nanotubes and Ag flakes. (b) Steps for fabricating CPC films. (c) As-prepared CPC film with octopus-inspired structures as a skin adhesive on rough and wet human skin, and its bendability, twistability and stretchability. Reproduced (adapted) with permission from [65]. Copyright 2020 American Chemical Society.

Huang *et al* designed a self-adsorbing electrode using a sandwich structure of Au (0.3 $\mu$m) and polyimide (PI) (1.2 $\mu$m) on the PDMS substrate. They reported that the motion artifacts, stability and sensitivity of the monitoring signal depend on the geometric consistency between the electrode and the skin [63]. In 2020 Zhou *et al* reported a skin-mountable biopotential electrode fabricated using AgNW embedded porous thermoplastic polyurethane (TPU) film [64]. Similarly, Min *et al* 2020 designed a stretchable polyurethane and conductive filler (Ag flakes and CNT) composite with water and air permeability for ECG monitoring with a hexagonal mesh pattern, as shown in figure 8.19 [65].

### 8.4.3 Fabrication and performance evaluation of contact penetrating electrodes

Contact penetrating electrodes have promising applications for biopotential ECG measurements due to their low impedance, ease of operation and suitability for use without skin preparation. Microneedles electrodes (MEs) are an example of contact penetrating electrodes which operate by penetrating the stratum corneum layer of the skin and thus minimizing its insulating effect. These MEs can be fabricated using different materials such as (a) silicon, (b) metals and (c) polymers. Some recent works on these electrodes are discussed below and summarized in table 8.3.

As silicon is a common material in microfabrication, many microneedle electrodes have been reported using silicon as a substrate material [66–70], followed by wet etch or dry etch photolithography. In 2000 Griss *et al* proposed the idea of using MEs for biopotential recording, which can penetrate the stratum corneum of the skin without affecting the dermis layer [66]. They fabricated MEs based on silicon using deep reactive ion etching and these showed better performance than standard electrodes. Similarly, in 2008 Yu *et al* designed hollow microneedles through deep reactive ion etching and showed an ECG signal with a high SNR [69]. O'Mahony *et al* proposed silicon MEs with sharp geometry fabricated through patterning and wet etching, which could accurately monitor the ECG signal [70]. The rigid nature of Si-based microneedles limits their perfect adaptability on the skin. These limits can be overcome by combining Si microneedles with a flexible substrate.

In 2016 Zhang *et al* proposed a microneedle array on a polymeric PDMS substrate and showed its suitability for ECG signal monitoring with low contact

**Table 8.3.** Contact penetrating electrode comparison.

| References | Material | Size/shape | Findings | Comments/methods |
|---|---|---|---|---|
| Griss *et al* [66] | Silicon, Ag/AgCl | Length of microneedles: 100 to 210 $\mu$m; diameter: 30 to 50 $\mu$m | Low impedance, remains about 18 k$\Omega$ at 10 Hz | DRIE, wet etching, evaporation, thermal oxidation methods |
| O'Mahony *et al* [70] | Silicon, Ag | Length of microneedle: 300 $\mu$m | Comparable ECG signals with standard wet electrodes | Anisotropic etching, thermal evaporation; risk of infection |
| Zhang *et al* [71] | Silicon, PEDOT/PSS, PDMS | Electrode diameter: 1.2 cm | Mean impedance at 10 Hz are 61.2 ± 31.3 k$\Omega$·cm$^2$ for the flexible dry electrode and 114.9 ± 36.1 k$\Omega$·cm$^2$ for the wet electrode | Dicing saw, isotopic etching, dip coating |
| Wang *et al* [72] | Silicon, parylene, Cr/Au | Octahedral pyramid-shape microneedle with a length of 190 $\mu$m | Lower electrode-interface impedance than for the wet electrode | LPCVD, wet etching, lift-off process, sputtering |
| Guo *et al* [73] | PDMS, metal | Average length of the needles: 269.985 ± 25.569 $\mu$m | The impedance of MEs at 10 Hz is 2.357 ± 0.198 M$\Omega$ | Phase transition method, 3D printing |
| Sun *et al* [77] | Ti, SU-8 negative photoresist, Au | Microneedle array length: 500 $\mu$m; base diameter: 200 $\mu$m | Ability to be inserted in the skin 100 times without breaking | Spinning coating, sputter coating, laser cutting |
| Ren *et al* [78] | Epoxy novolac resin, iron particles, Ti/Au, polyimide | Average height: 600 $\mu$m; average tip radius: 12 $\mu$m | Reduced motion artifacts | Magnetorheological drawing lithography, magnetron sputtering |
| Stauffer *et al* [79] | PMMA, Ag nanoparticles | Micropillar shaped base with a flat tip | Self-adhesive, high mean SNR, can be used under water | Inspired by grasshopper feet |

impedance [71]. Similarly, Wang *et al* fabricated silicon microneedles on a flexible parylene substrate [72]. The flexible microneedle-based electrode array is a promising candidate for biopotential recording. However, the risk of infection due silicon needles breaking off during the penetration and the complicated and expensive fabrication process, such as lithography and etching technology, limit their use. Therefore, researchers have considered metals to fabricate microneedles because of their excellent strength. In 2019 Guo *et al* reported a flexible ME based on a PDMS substrate and a Bi/In/Sn/Zn alloy to make microneedles, as shown in figure 8.20(A). Here, the low melting point alloys were used to form microneedles and their rigidity ensured effective skin piercing. They reported an impedance value comparable to the value of conventional wet electrodes [73]. Many microneedles have also been fabricated using polymers such as SU-8 photoresist [74], PLA [75] and poly(lactic-co-glycolic acid) (PLGA) [76] due to their biocompatibility and excellent toughness. In 2018 Sun *et al* fabricated a composite microneedle array electrode (CMAE) using three layers: an innermost microstructure layer of Ti, then a middle SU-8 insulating

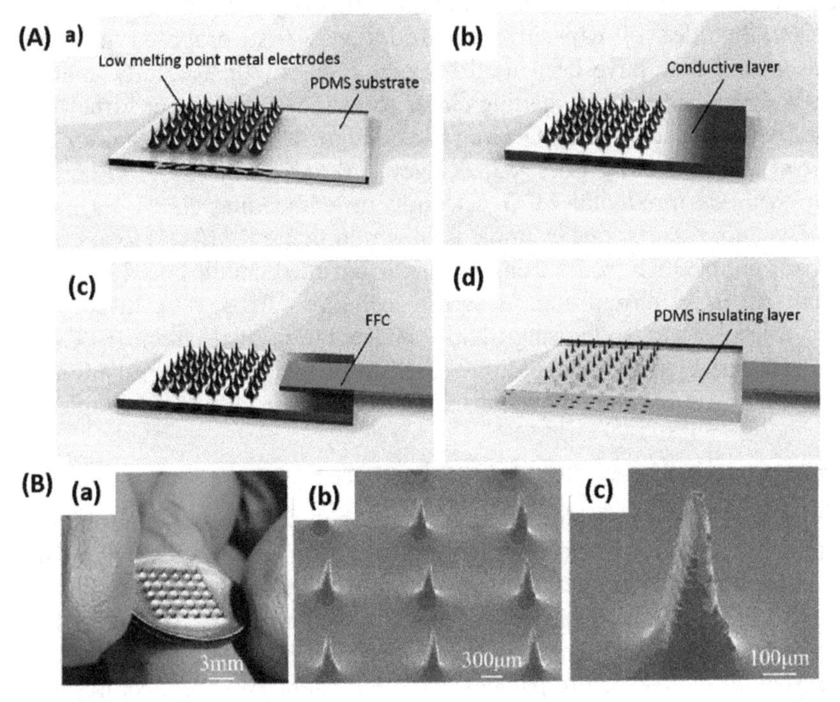


**Figure 8.20.** (A) Fabrication process of a flexible ME based on a PDMS substrate and Bi/In/Sn/Zn alloy: (a) different array sizes of droplet-shaped metal with sharp needles can be fabricated using a dispensing system; (b) spraying of liquid Ga/In alloy on the surface of the electrode array to make it conductive; (c) placing a flexible flat cable (FFC) on the surface of the conductive layer; and (d) coating of PDMS on the electrode and conductive layer to achieve isolation. Reproduced with permission from [73]. Copyright 2020 Elsevier. (B) Images of the fabricated composite microneedle array electrode (CMAE): (a) optical image; (b) SEM image of the CMAE; and (c) SEM image of a single microneedle [77].

layer and an outer layer of conductive Au (100 nm) for ECG monitoring, as shown in figure 8.20(B) [77].

Recently, different methods have been proposed for fabricating conductive patterns. In 2017 Ren *et al* used laser-directed writing (LDW) technology and magneto-rheological drawing lithography (MRDL) to fabricate microneedle electrodes on flexible PET film [78]. Stauffer *et al* used silicone rubbers with conductive silver particle composites and molded it into electrode shapes. They reported that these electrodes have a comparable skin–electrode impedance to conventional wet electrodes [79].

### 8.4.4 Fabrication and performance evaluation of noncontact capacitive electrodes

With the advancements in miniaturization and wireless technology, noncontact electrodes that can operate without direct contact with the skin surface have attracted more attention. These electrodes can be applied with air, clothing or other insulating materials between the electrode and the skin and measure biopotential signals based on the capacitive coupling between the electrode–skin interface, as shown in figure 8.15. These electrodes are more comfortable than wet and other dry contact electrodes

In 1969 the idea of capacitive electrodes was first proposed by Lopez and Richardson and they have been used for many years [80]. Recently, many designs have been developed that are getting closer to use in daily life, such the office chair combined with noncontact ECG electrodes designed by Aleksandrowicz *et al* [81], Singh *et al* [82] and Fong *et al* [83]. Kelgey [84] and Babusiak *et al* [85] proposed different methods for digital ECG detection by embedding electrodes into fabrics, sports vests and T-shirts; one example is shown in figure 8.21(A). Hou *et al* designed capacitively coupled electrodes using a standard printed circuit board (PCB) attached to a chair for ECG monitoring, as shown in figure 8.21(B) [86]. In 2019 Liu *et al* designed a flexible electrode using multilayer flexible printed circuit (FPC) materials and showed measured ECG signal reliability up to five layers of insulation materials [87]. A comparison of noncontact capacitive electrodes is provided in table 8.4.

## 8.5 Issues with dry electrodes and future scope

The discussions in the preceding section reveal that substantial research attention is being paid to the design and development of dry electrodes. As all of the novel designs offer some advantages, they also impose some hurdles for efficient monitoring. Due to the elimination of conducting gel, the skin–electrode contact has a very high impedance that degrades the reception of the biopotential signal. Numerous designs, such as nanostructure patches, nanopillars and 3D architectures, have been proposed that possess a mechanical interlock to maintain the adhesion on the skin.

The surface penetrating microneedle electrodes have the advantage of bypassing the highly resistive stratum corneum layer and show the lowest contact impedance among other dry and wet electrodes. The microneedle electrodes made using silicon show excellent and accurate results in biopotential measurement. However, these electrodes have several limitations, such as high cost, a complicated fabrication process, a risk of infection and the length of the microneedles, which should be

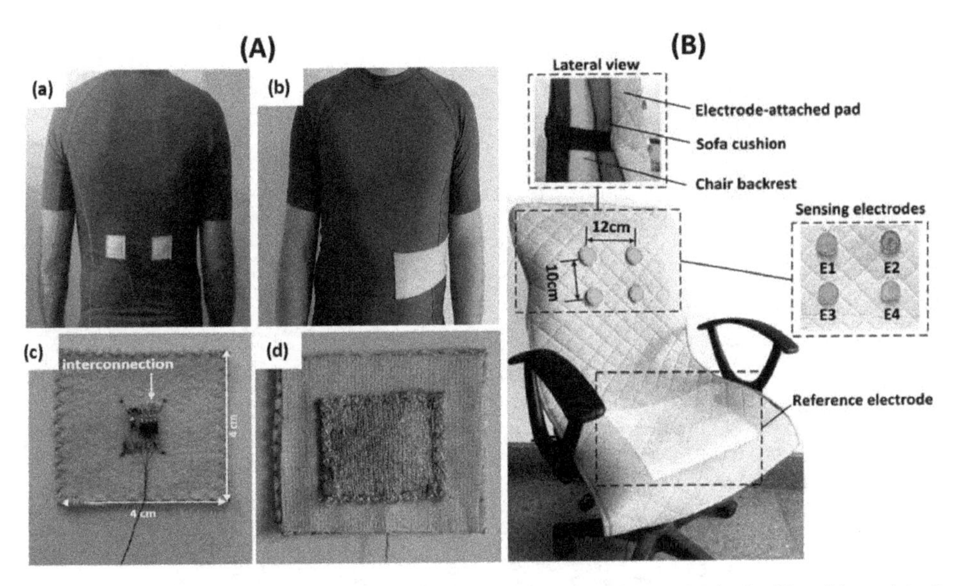


**Figure 8.21.** (A) Fabric-based electrodes: (a) a pair of ECG electrodes on the back, (b) a driven right leg (DRL) electrode on the front, (c) placement of electronics on the back of the electrode and (d) grounded shield made of ELITEX® textile on the back of the electrode. Reproduced with permission from [85]. Copyright 2020 Elsevier. (B) The electrode positions on the office chair proposed in [86].

suitable for penetration without pain or breaking. Metal-based microneedles show high conductivity and high accuracy in biopotential measurement with good microneedle strength during penetration. A metal electrode is easy to design in any shape, but its poor biocompatibility limits its use for long-term monitoring. To overcome these limitations, polymer-based electrodes have been designed with excellent biocompatibility. However, these polymers require extra methods such as sputtering, thermal evaporation and electroless plating, to make them conductive. Noncontact capacitive electrodes operate without direct conductive contact with the skin surface and can be more comfortable during measurement but the signal is easily affected by noise and motion artifacts [17]. A flexible contact surface electrode can significantly improve the contact area at the skin–electrode interface, reduce noise and motion artifacts, as well as monitoring the biopotential signal with a high SNR. These developing surface contact electrodes could present an alternative to the conventional Ag/AgCl wet electrodes for future clinical use.

In addition, the absence of conducting gel can lead to severe motion artifacts. As the electrodes are developed with the goal of long-term monitoring, it is also essential to explore the impact of stress and sweat environments on their performance. At present, the rapid advancement of material science and the discovery of efficient functional materials is leading to the achievement of superior performance. However, the biocompatibility of the majority of these materials is yet to be investigated thoroughly, and hence they cannot be used immediately for healthcare applications. Also, mechanical cracks are formed within flexible polymers (e.g. PDMS), which increases the contact impedance. For smart textiles, the packaging of

**Table 8.4.** Noncontact capacitive electrode comparison.

| References | Material | Size/shape | Findings | Comments/methods |
|---|---|---|---|---|
| Fong *et al* [83] | Fabric electrode with embedded polymer (FEEP) H-600 polymer | Area of the fabric electrode: 4 cm × 4 cm | ECG signal with high SNR and QRS amplitude. | Smart chair, long stabilization time |
| Singh *et al* [82] | Copper sheets | Dimensions: 8 cm × 3 cm × 0.1 mm | SNR is low for synthetic clothes and high for cotton clothes, increases with moisture. | Electrodes on Seat belt and chair |
| Babusiak *et al* [85] | Shieldex® and ELITEX® electro conductive fiber, polyamide, Ag | Dimensions: 4× 4 cm | ECG signal's segments and intervals are clearly notable. | Electrodes placed on a T-shirt |
| Hou *et al* [86] | Conductive layer of PCB, Al | Diameter: 42 mm | Good ECG signals for a 0.3 mm cotton layer. | The ECG signal quality declined with increasing of shirt thickness |
| Liu *et al* [87] | Copper, polyimide | Diameter: 2.5 cm | ECG signals can be measured with five layers of insulating material between the skin and the electrode, and a flexible FPC electrode achieved a higher SNR compared to traditional hard PCB electrodes. | The noise amplitude increases and the SNR decreases with increases in the insulation layer from 1 to 5 |

electronic components remains a critical challenge to prevent damage by washing. Finally, these dry electrodes require a higher setting time than gel electrodes to provide satisfactory and stable cardiac monitoring. Hence, addressing all these research problems can help dry electrodes to compete with the gel electrodes, and they can be utilized to implement commercial-grade healthcare monitoring devices [32].

## 8.6 Conclusion

This chapter provides a systematic overview of an electrocardiography (ECG) monitoring system and reviews the evolution and recent progress of biopotential dry electrodes for ECG signal monitoring. These dry electrodes are categorized into three types according to their interaction with the skin surface: surface contact electrodes, surface penetrating electrodes and capacitive noncontact electrodes. These dry electrodes show potential applications for long-term ECG monitoring, although there still needs to be more effort to enable their clinical use. It is believed that dry electrodes with better performance and more convenient operation will occupy the role of electrodes for daily and long-term monitoring of bioelectrical signals.

## Bibliography

[1] Taouk Y, Spittal M J, Lamontagne A D and Milner A J 2020 Psychosocial work stressors and risk of all-cause and coronary heart disease mortality: a systematic review and meta-analysis *Scand. J. Work. Environ. Heal.* **46** 19–31

[2] Prabhakaran D, Singh K, Roth G A, Banerjee A, Pagidipati N J and Huffman M D 2018 Cardiovascular diseases in India compared with the United States *J. Am. Coll. Cardiol.* **72** 79–95

[3] Patel V, Chatterji S, Chisholm D, Ebrahim S, Gopalakrishna G, Mathers C, Mohan V, Prabhakaran D, Ravindran R D and Reddy K S 2011 Chronic diseases and injuries in India *Lancet* **377** 413–28

[4] George C E, Ramadas D, Norman G, Mukherjee D and Rao T 2016 Barriers to cardiovascular disease risk reduction: does physicians' perspective matter? *Indian Heart J.* **68** 278–85

[5] Tan W and Aboulhosn J 2020 The cardiovascular burden of coronavirus disease 2019 (COVID-19) with a focus on congenital heart disease *Int. J. Cardiol.* **309** 70–7

[6] Shi S *et al* 2020 Association of cardiac injury with mortality in hospitalized patients with COVID-19 in Wuhan, China *JAMA Cardiol* 1–8

[7] Vieau S and Iaizzo P A 2015 Basic ECG theory, 12-lead recordings, and their interpretation *Handbook of Cardiac Anatomy, Physiology, and Devices* ed P Iaizzo (Cham: Springer) pp 321–24

[8] Preidel W 1993 Silver-chloride reference electrode *US Patent* 5230786

[9] Eggins B R 1993 Skin contact electrodes for medical applications *Analyst* **118** 439–42

[10] Tam H W and Webster J G 1977 Minimizing electrode motion artifact by skin abrasion *IEEE Trans. Biomed. Eng.* **BME-24** 134–9

[11] Marozas V, Petrenas A, Daukantas S and Lukosevicius A 2011 A comparison of conductive textile-based and silver/silver chloride gel electrodes in exercise electrocardiogram recordings *J. Electrocardiol.* **44** 189–94
[12] Santana L F, Cheng E P and Lederer W J 2010 How does the shape of the cardiac action potential control calcium signaling and contraction in the heart? *J. Mol. Cell. Cardiol.* **49** 901–3
[13] Purves D, Augustine G J, Fitzpatrick D, Katz L C, LaMantia A-S, McNamara J O and Williams S M 2001 *Neuroscience* 2nd edn (Sunderland, MA: Sinauer Associates)
[14] Grunnet M 2010 Repolarization of the cardiac action potential. Does an increase in repolarization capacity constitute a new anti-arrhythmic principle? *Acta Physiol.* **198** 1–48
[15] Joung B, Chen P S and Lin S F 2011 The role of the calcium and the voltage clocks in sinoatrial node dysfunction *Yonsei Med. J.* **52** 211–9
[16] Richard E K 2012 *Cardiovascular Physiology Concepts* 2nd edn (Philadelphia, PA: Lippincott, Williams and Wilkins)
[17] Pérez Riera A R *et al* 2008 The enigmatic sixth wave of the electrocardiogram: the U wave *Cardiol J.* **15** 408–21
[18] Einthoven W 1912 The different forms of the human electrocardiogram and their signification *Lancet* **179** 853–61
[19] McFee R and Johnston F D 1953 Electrocardiographic leads: I. Introduction *Circulation* **8** 554–68
[20] McFee R and Johnston F D 1954 Electrocardiographic leads: II. Analysis *Circulation* **9** 255–66
[21] McFee R and Johnston F D 1954 Electrocardiographic leads: III. Synthesis *Circulation* **9** 868–80
[22] Macfarlane P W, van Oosterom A, Pahlm O, Kligfield P, Janse M and Camm J 2010 *Comprehensive Electrocardiology* 2nd edn (London: Springer)
[23] Burger H C and Van Milaan J B 1946 Heart-vector and leads *Br. Heart J.* **8** 157–61
[24] Burger H C and Van Milaan J B 1947 Heart-vector leads. Part II *Br. Heart J.* **9** 154–60
[25] Burger H C and Van Milaan J B 1948 Heart-vector and leads; geometrical representation *Br. Heart J.* **10** 229–33
[26] Okamoto Y and Mashima S 1998 The zero potential and Wilson's central terminal in electrocardiography *Bioelectrochem. Bioenerg.* **47** 291–5
[27] Grimnes S 1983 Impedance measurement of individual skin surface electrodes *Med. Biol. Eng. Comput.* **21** 750–5
[28] Rosell J, Colominas J, Riu P, Pallas-Areny R and Webster J G 1988 *IEEE Trans. Biomed. Eng.* **35** 649–51
[29] Tallgren P, Vanhatalo S, Kaila K and Voipio J 2005 Evaluation of commercially available electrodes and gels for recording of slow EEG potentials *Clin. Neurophysiol.* **116** 799–806
[30] Martinsen O G and Grimnes S 2011 *Bioimpedance and Bioelectricity Basics* (Oxford: Academic)
[31] Bergey G E, Squires R D and Sipple W C 1971 Electrocardiogram recording with pasteless electrodes *IEEE Trans. Biomed. Eng.* **BME-18** 206–11
[32] Meziane N, Webster J G, Attari M and Nimunkar A J 2013 Dry electrodes for electrocardiography *Physiol. Meas.* **34** R47
[33] Godin D T, Parker P A and Scott R N 1991 Noise characteristics of stainless-steel surface electrodes *Med. Biol. Eng. Comput.* **29** 585–90
[34] De Luca C J, Le Fever R S and Stulen F B 1979 Pasteless electrode for clinical use *Med. Biol. Eng. Comput.* **17** 387–90

[35] Gondran C, Siebert E, Fabry P, Novakov E and Gumery P Y 1995 Non-polarisable dry electrode based on NASICON ceramic *Med. Biol. Eng. Comput.* **33** 452–7
[36] Searle A and Kirkup L 2000 A direct comparison of wet, dry and insulating bioelectric recording electrodes *Physiol. Meas.* **21** 271–83
[37] Gruetzmann A, Hansen S and Müller J 2007 Novel dry electrodes for ECG monitoring *Physiol. Meas.* **28** 1375–90
[38] Chen C Y, Chang C L, Chang C W, Lai S C, Chien T F, Huang H Y, Chiou J C and Luo C H 2013 A low-power bio-potential acquisition system with flexible PDMS dry electrodes for portable ubiquitous healthcare applications *Sensors* **13** 3077–91
[39] Meng Y, Li Z and Chen J 2016 A flexible dry electrode based on APTES-anchored PDMS substrate for portable ECG acquisition system *Microsyst. Technol.* **22** 2027–34
[40] Jung H, Moon J, Baek D, Lee J, Choi Y and Hong J 2012 CNT/PDMS composite flexible dry electrodes for long-term ECG monitoring *IEEE Trans. Biomed. Eng.* **59** 1472–9
[41] Chlaihawi A A, Narakathu B B, Eshkeiti A, Emamian S, Avuthu S G R and Atashbar M Z 2015 Screen printed MWCNT/PDMS based dry electrode sensor for electrocardiogram (ECG) measurements *2015 IEEE Int. Conf. on Electro/Information Technology (EIT) (Dekalb, IL)* pp 526–9
[42] Bong J *et al* 2019 Radiolucent implantable electrocardiographic monitoring device based on graphene *Carbon* **152** 946–53
[43] Zahed M A, Das P S, Maharjan P, Barman S C, Sharifuzzaman M, Yoon S H and yeong P J 2020 Flexible and robust dry electrodes based on electroconductive polymer spray-coated 3D porous graphene for long-term electrocardiogram signal monitoring system *Carbon* **165** 26–36
[44] Liu B, Luo Z, Zhang W, Tu Q and Jin X 2016 Silver nanowire-composite electrodes for long-term electrocardiogram measurements *Sens. Actuators* A **247** 459–64
[45] Qiao Y *et al* 2020 Multifunctional and high-performance electronic skin based on silver nanowires bridging graphene *Carbon* **156** 253–60
[46] Kisannagar R R, Jha P, Navalkar A, Maji S K and Gupta D 2020 Fabrication of silver nanowire/polydimethylsiloxane dry electrodes by a vacuum filtration method for electrophysiological signal monitoring *ACS Omega* **5** 10260–5
[47] Cheng X, Bao C, Wang X, Zhang F and Dong W 2019 Soft surface electrode based on PDMS-CB conductive polymer for electrocardiogram recordings *Appl. Phys.* A **125** 1–7
[48] Lu L, Yang B and Liu J 2020 Flexible multifunctional graphite nanosheet/electrospun-polyamide 66 nanocomposite sensor for ECG, strain, temperature and gas measurements *Chem. Eng. J.* **400** 125928
[49] Kim S, Lee H, Kim D, Ha H, Qaiser N, Yi H and Hwang B 2020 Ethylcellulose/Ag nanowire composites as multifunctional patchable transparent electrodes *Surf. Coat. Technol.* **394** 125898
[50] Arquilla K, Webb A K and Anderson A P 2020 Textile electrocardiogram (ECG) electrodes for wearable health monitoring *Sensors* **20** 1013
[51] Ankhili A, Zaman S U, Tao X, Cochrane C, Koncar V and Coulon D 2019 How to connect conductive flexible textile tracks to skin electrocardiography electrodes and protect them against washing *IEEE Sens. J.* **19** 11995–2002
[52] Rajanna R R, Sriraam N, Vittal P R and Arun U 2020 Performance evaluation of woven conductive dry textile electrodes for continuous ECG signals acquisition *IEEE Sens. J.* **20** 1573–81

[53] Li M, Xiong W and Li Y 2020 Wearable measurement of ECG signals based on smart clothing *Int. J. Telemed. Appl.* **2020** 1–9

[54] La T G, Qiu S, Scott D K, Bakhtiari R, Kuziek J W P, Mathewson K E, Rieger J and Chung H J 2018 Two-layered and stretchable e-textile patches for wearable healthcare electronics *Adv. Healthc. Mater* **7** 1–11

[55] Xu X, Luo M, He P, Guo X and Yang J 2019 Screen printed graphene electrodes on textile for wearable electrocardiogram monitoring *Appl. Phys.* A **125** 714

[56] Soroudi A, Hernández N, Wipenmyr J and Nierstrasz V 2019 Surface modification of textile electrodes to improve electrocardiography signals in wearable smart garment *J. Mater. Sci., Mater. Electron.* **30** 16666–75

[57] Yapici M K, Alkhidir T, Samad Y A and Liao K 2015 Graphene-clad textile electrodes for electrocardiogram monitoring *Sens. Actuators* B **221** 1469–74

[58] Li B M, Yildiz O, Mills A C, Flewwellin T J, Philip D and Jur J S 2020 Iron-on carbon nanotube (CNT) thin films for biosensing e-textile applications *Carbon* **168** 673–83

[59] Saleh S M, Jusob S M, Harun F K C, Yuliati L and Wicaksono D H B 2020 Optimization of reduced GO-based cotton electrodes for wearable electrocardiography *IEEE Sens. J.* **20** 7774–82

[60] Kabiri Ameri S, Ho R, Jang H, Tao L, Wang Y, Wang L, Schnyer D M, Akinwande D and Lu N 2017 Graphene electronic tattoo sensors *ACS Nano* **11** 7634–41

[61] Wang Y, Qiu Y, Ameri S K, Jang H, Dai Z, Huang Y and Lu N 2018 Low-cost, $\mu$m-thick, tape-free electronic tattoo sensors with minimized motion and sweat artifacts *NPJ Flex. Electron.* **2** 1–7

[62] Jeong J W, Kim M K, Cheng H, Yeo W H, Huang X, Liu Y, Zhang Y, Huang Y and Rogers J A 2014 Capacitive epidermal electronics for electrically safe, long-term electrophysiological measurements *Adv. Healthc. Mater* **3** 642–8

[63] Huang Y A, Dong W, Zhu C and Xiao L 2018 Electromechanical design of self-similar inspired surface electrodes for human–machine interaction *Complexity* **2018** 1–14

[64] Zhou W, Yao S, Wang H, Du Q, Ma Y and Zhu Y 2020 Gas-permeable, ultrathin, stretchable epidermal electronics with porous electrodes *ACS Nano* **14** 5798–805

[65] Min H, Jang S, Kim D W, Kim J, Baik S, Chun S and Pang C 2020 Highly air/water-permeable hierarchical mesh architectures for stretchable underwater electronic skin patches *ACS Appl. Mater. Interfaces* **12** 14425–32

[66] Griss P, Enoksson P, Tolvanen-Laakso H K, Meriläinen P, Ollmar S and Stemme G 2001 Micromachined electrodes for biopotential measurements *J. Microelectromech. Syst.* **10** 10–6

[67] Wang Y, Pei W H, Guo K, Gui Q, Li X Q, Da C H and Yang J H 2011 Dry electrode for the measurement of biopotential signals *Sci. China Inf. Sci.* **54** 2435–42

[68] Pei W, Zhang H, Wang Y, Guo X, Xing X, Huang Y, Xie Y, Yang X and Chen H 2017 Skin-potential variation insensitive dry electrodes for ECG recording *IEEE Trans. Biomed. Eng.* **64** 463–70

[69] Yu L M, Tay F E H, Guo D G, Xu L and Yap K L 2009 A microfabricated electrode with hollow microneedles for ECG measurement *Sens. Actuators* A **151** 17–22

[70] O'Mahony C, Pini F, Blake A, Webster C, O'Brien J and McCarthy K G 2012 Microneedle-based electrodes with integrated through-silicon via for biopotential recording *Sens. Actuators* A **186** 130–6

[71] Zhang H, Pei W, Chen Y, Guo X, Wu X, Yang X and Chen H 2016 A motion interference-insensitive flexible dry electrode *IEEE Trans. Biomed. Eng.* **63** 1136–44

[72] Wang R, Jiang X, Wang W and Li Z 2017 A microneedle electrode array on flexible substrate for long-term EEG monitoring *Sens. Actuators* B **244** 750–8
[73] Guo S, Lin R, Wang L, Lau S, Wang Q and Liu R 2019 Low melting point metal-based flexible 3D biomedical microelectrode array by phase transition method *Mater. Sci. Eng.* C **99** 735–9
[74] Kumar A, Bhartia B and Mukhopadhyay K 2015 Long term biopotential recording by body conformable photolithography fabricated low cost polymeric microneedle arrays *Sens. Actuators* A **236** 164–72
[75] Kim M, Kim T, Kim D S and Chung W K 2015 Curved microneedle array-based sEMG electrode for robust long-term measurements and high selectivity *Sensors* **15** 16265–80
[76] Ren L, Jiang Q, Chen K, Chen Z, Pan C and Jiang L 2016 Fabrication of a micro-needle array electrode by thermal drawing for bio-signals monitoring *Sensors* **16** 908
[77] Sun Y, Ren L, Jiang L, Tang Y and Liu B 2018 Fabrication of composite microneedle array electrode for temperature and bio-signal monitoring *Sensors* **18** 1–12
[78] Ren L, Jiang Q, Chen Z, Chen K, Xu S, Gao J and Jiang L 2017 Flexible microneedle array electrode using magnetorheological drawing lithography for bio-signal monitoring *Sens. Actuators* A **268** 38–45
[79] Stauffer F, Thielen M, Sauter C, Chardonnens S, Bachmann S, Tybrandt K, Peters C, Hierold C and Vörös J 2018 Skin conformal polymer electrodes for clinical ECG and EEG recordings *Adv. Healthc. Mater* **7** 1–10
[80] Lopez A and Richardson P C 1969 Capacitive electrocardiographic and bioelectric electrodes *IEEE Trans. Biomed. Eng.* **1** 99
[81] Aleksandrowicz A and Leonhardt S 2007 Wireless and non-contact ECG measurement system—the 'Aachen SmartChair ' *Acta Polytechn.* **47** 68–71
[82] Singh R K, Sarkar A and Anoop C S 2016 A health monitoring system using multiple non-contact ECG sensors for automotive drivers *IEEE Int. Instrum. Meas. Technol. Conf. Proc.* pp 1–6
[83] Fong E M and Chung W Y 2015 A hygroscopic sensor electrode for fast stabilized non-contact ECG signal acquisition *Sensors* **15** 19237–50
[84] Varadan V K, Kumar P S, Oh S, Mathur G N, Rai P and Kegley L 2011 e-bra with nanosensors, smart electronics and smartphone communication network for real time cardiac health monitoring *Proc. SPIE* **7980** 121–7
[85] Babusiak B, Borik S and Balogova L 2018 Textile electrodes in capacitive signal sensing applications *Meas. J. Int. Meas. Confed.* **114** 69–77
[86] Hou Z, Xiang J, Dong Y, Xue X, Xiong H and Yang B 2018 Capturing electrocardiogram signals from chairs by multiple capacitively coupled unipolar electrodes *Sensors* **18** 2835
[87] Liu S, Zhu M, Liu X, Samuel O W, Wang X, Huang Z, Wu W, Chen S and Li G 2019 Flexible noncontact electrodes for comfortable monitoring of physiological signals *Int. J. Adapt. Control Signal Process.* **33** 1307–18

**IOP** Publishing

Modelling and Analysis of Active Biopotential Signals in Healthcare, Volume 2

**Varun Bajaj and G R Sinha**

# Chapter 9

# Effective cardiac health diagnosis using event-driven ECG processing with subband feature extraction and machine learning techniques

**Saeed Mian Qaisar[1] and Syed Fawad Hussain[2]**

[1]*College of Engineering, Effat University, Jeddah, 22332, Saudi Arabia*

[2]*Ghulam Ishaq Khan Institute of Engineering Sciences and Technology, Topi, Khyber Pakhtunkhwa, Pakistan*

This chapter describes a test bed for the measurement, conditioning, analysis and automatic categorization of electrocardiogram (ECG) signals. The novelty is to intelligently incorporate the event-driven features in the system. The goal is to build an adequate method through introducing real-time compression and computationally efficient signal processing and data transmission. The method realizes the efficient acquisition and segmentation of continuous time ECG signals by utilizing the novel event-driven principle. It also incorporates adaptive-rate conditioning and subband decomposition. The dimensions of the extracted subband features are reduced by intelligently extracting the useful features from each selected subband. The MIT-BIH dataset is utilized to demonstrate the interesting capabilities of the suggested framework. The proposed event-driven acquisition chain is seen to add a significant real-time gain in compression relative to traditional fixed-rate equivalents. Consequently, a remarkable reduction of arithmetic complexity is attained which results in a power-efficient system realization with a reduced activity of data transmission towards the cloud. Therefore, a significant transmitted power consumption reduction is promised along with an efficient post-cloud based classification. The performance of classification is also quantified in terms of percentage accuracy, F-measure and kappa statistics.

doi:10.1088/978-0-7503-3411-2ch9

## 9.1 Introduction

All hospitals and healthcenters provide a variety of health services and devices to diagnose disease. In order to take care of and monitor your health, a variety of smart wearable devices have recently been proposed that can be used remotely by the patient.

In most developed countries and in many developing countries, the most common cause of death in adults and the elderly is cardiovascular disease [1]. Therefore, early diagnosis of the signs of a heart attack is very important to allow medical attention to be called for as quickly as possible. Also, most clinical diagnoses of heart disease rely on a patient's history and sometimes they depend on physical examination. It is important to use ECG devices to help people detect cardiovascular diseases as well as to achieve an appropriate diagnosis in a real-time manner [2, 3].

This work leads to the creation and implementation of an electrocardiogram (ECG) test bed for event-driven processing and diagnosis of cardiovascular diseases [1]. It is a contribution towards the development of a smart and portable cloud-based mobile healthcare system, which will help cardiac patients to determine their health condition in real time. In the case of an emergency, alarms will be generated and sent to nearby healthcare centers and to the patient's relatives. The system can also be used by medial practitioners for better and easy diagnosis.

Sophisticated and specialized implants are used to store the biomedical signals. This is particularly popular in the case of modern hospitals and homes for the elderly. It increases the accuracy of the diagnosis and decision-making process and speeds it up. New biomedical devices have provided a possible future opportunity for remote cardiovascular and respiratory healthcare tracking using wireless sensing tactics. This is particularly useful for the elderly and patients with severe heart problems [4]. The successful implementation and incorporation of wireless sensing developments has attracted numerous industries, such as biomedical implant-based fabric architecture and computer-based health diagnosis [5, 6].

Many advances have recently been made in the area of surgical instruments and testing technologies to meet the requirements of healthcare and to improve prognosis. The modern tools and devices in the healthcare sector provide accurate and faster health assessment. In addition, wireless sensing technology and modern networking strategies have revolutionized the realization of the non-stop real-time analysis of the elderly and patients with critical health conditions. Mobile cloud-based healthcare devices play a key role in tracking and analysing the patient condition. A real-time patient surveillance program continuously gathers the physiological signals of patients to analyse the risks associated with critical health problems in real time [5]. In recent years, mobile patient tracking based on sensors has become increasingly popular in customized healthcare systems, carried out using cloud repositories and sensing tools [6].

The concept of a cloud-based patient monitoring framework for medical treatment and care is shown in figure 9.1. In this approach biomedical implants are constantly gathering, processing and transmitting vital signs from patients and further transmitting them to a cloud server. Subsequently, a specially developed

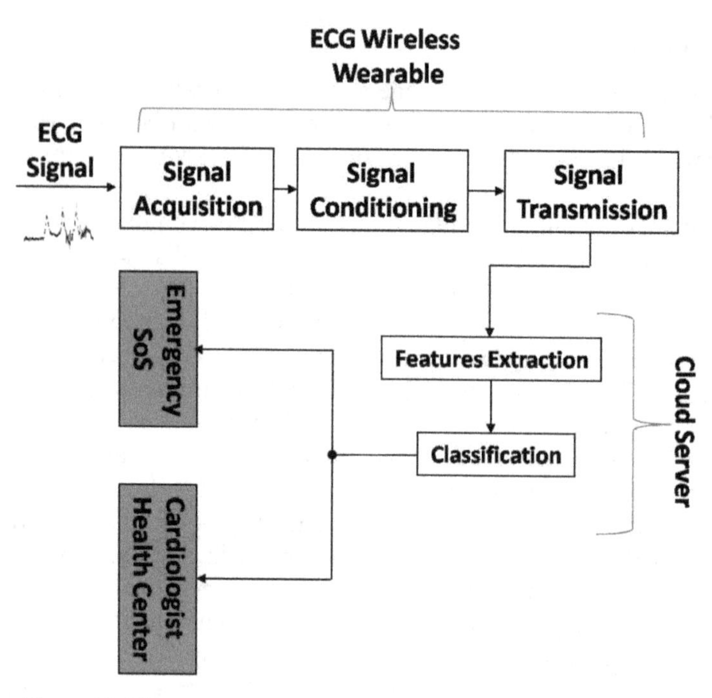


**Figure 9.1.** The concept of cloud-based cardiovascular disease monitoring.

could-based application is used to extract beneficial information and prepare a report, which is conveyed to the relevant healthcenter in real time [6].

Multiple signal processing and artificial intelligence algorithms may be utilized to evaluate physiological data for automatic diagnosis. In [7] the authors used a field programmable gate array (FPGA) based approach for ECG signal identification. In [8] an ECG acquisition sensor was used to perform ECG signal analysis based on computer technology and to deliver the research results to the cloud. Cloud-based approaches for cardiac failure diagnosis have become popular in monitoring cardiovascular disease patients [8].

Remote healthcare monitoring schemes require the use of wireless carrier services and the Global Positioning System (GPS). This enables the biomedical sensors to convey biomedical signals in real time to the cloud server. The data are then stored, decoded and classified using a cloud-based application. The connection between the implanted biomedical sensors and the cloud server can be established via a 5G network [9].

New achievements in cloud-based mobile patient tracking services have become possible thanks to the remarkable advances in smart biomedical wireless implants, remote and mobile processing modules, and wireless communications networks. A large number of precise, wireless and smart sensors are available to collect various biomedical data relating to health. Such devices can be conveniently mounted in a smartphone or smartwatch. These smart devices will make critical and important contributions to the realization of modern and effective integrated cloud-based healthcare solutions [9]. The vital signs of the patient are collected using an appropriate smart biomedical sensor. Then the programmed critical health information is preconditioned, processed and distributed to the smartphone. The phone then starts sending the collected information to the server via a wireless mobile network to continuously verify the patient's status. In emergency situations, the cloud service application sends details to the emergency room, the clinician and the family of the patient [10].

This chapter incorporates an innovative event-driven concept in an arrhythmia monitoring framework for efficient recording and analysis of ECG signals through front-end wireless wearables in the context of cloud-assisted healthcare solutions. The focus is to enhance system output and wearable power consumption while maintaining sufficient detection precision by achieving real-time compression, improved processing capacity and reduction of transmission. The approach is promising and can contribute to accurate, wearable applications of the ECG along with precise decision support for the detection of arrhythmia [2, 3, 11–13].

## 9.2 Background and literature review

Cardiac disease is one of the world's leading causes of death [14]. Ventricular atrial fibrillations are the cause of irregular heartbeats and most sudden heart attacks. The electrocardiogram (ECG) is a complicated signal and provides essential information about the functioning of cardiovascular systems [15]. Numerous pre-processing techniques for ECG signals such as quadratic filtering [16], Kalman filtering [17], main component analysis [18], empirical mode decomposition, Riegmann Liouvelle fractional integral filtering and Savitzky–Golay filtering [52, 53] have been incorporated. The preconditioned ECG signals are analysed using effective tactics of signal processing to accurately classify cardiac diseases [15, 19, 54]. ECG feature extraction has been explored for a long time and various advanced methods have been discussed in previous studies for accurate and fast ECG feature extraction. The parameters that must be reviewed while evolving an algorithm for feature extraction of an ECG signal are the simplicity of the algorithm and accuracy in providing superior results in feature extraction and classification. Several of the main approaches for extraction are presented in [15, 20]. The features produced during the investigative process are used to effectively identify heart arrhythmia. Various data mining algorithms are used for ECG signal classification [20]. Several of the main approaches for classification of ECGs are presented in [15, 20, 21, 54].

The rapid diagnosis of arrhythmia disorders will achieve the effective treatment of heart failure. Consequently, patients suffering from chronic heart disease require regular care. One of the best tools for constant monitoring of cardiac conditions is a portable ECG system [22]. For an effective emergency response, the timely and correct reporting of adverse circumstances to healthcare experts is critical. In this way, a mobile health service based on the cloud could expand healthcare outreach, decision-making, serious illness care and emergency management. Power saving is crucial in wireless selfpowered biomedical sensors. This is attainable by reducing the data dimensions and the activity of transmission towards the cloud. A variety of efforts have been reported in this regard [23–26, 32–35].

## 9.3 The electrocardiograph (ECG) in healthcare

An ECG pulse represents a cardiac cycle (also called a PQRST cycle). The inflection points of the signal are signified by the letters P, Q, R, S and T. The cycle is divided into several segments. The main segments are the P-wave, T-wave and QRS-complex. This mechanism is further depicted with the help of figure 9.2. These segments in the cardiac cycle are based on the electrical activity of the heart, i.e. the polarization and depolarization of different heart muscles and chambers.

In [27] the author proposed a variety of approaches and ideas for ECG recording and processing. He developed some useful ideas for preparing, encoding and automatically categorizing various forms of heart pulses. He also assessed the effects of the positioning of electrodes at different body regions on the efficiency of an automatic heartbeat descriptor. In this process different positions were considered on the arms, legs and head. A theoretical model of the system is described in [27]. Moreover, the author has also reported the outcome of a differential lead based ECG signal recording [27].

In medical care the ECG signal is perhaps the most essential factor. The scientific developments in electroencephalography have been significantly improved with the advent of computers, electronic components and cell networks. Further ECG

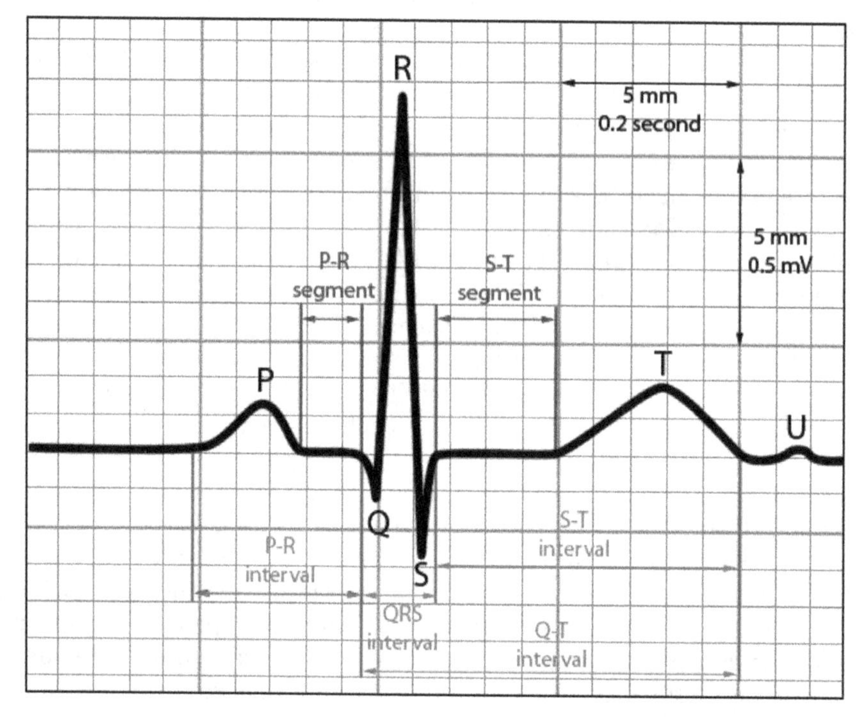


**Figure 9.2.** A PQRST cycle.

advancement became possible with the advent of cellular networking and cloud services. ECG sensors are fabricated using a variety of advanced materials. Various nanotools are available today which can receive, interpret and relay ECG signals to the cloud. These are wireless sensors and super-light devices which enable effective tracking of the physiological functioning of affected individuals. Because of its relatively small size and lighter weight, when using these implants the patient does not experience too much discomfort. Each of these implants contributes to an ECG test which is not invasive. The independent working life of these tools is increased by low-power designs and electronics technologies. They enable a greatly prolonged period of continuous and constant tracking of the cardiac activity of the patient relative to the conventional Holter recorders. The use of such sensors is not only limited to cardiac system tracking. They can also be used for other healthcare applications such as the monitoring of patients with epilepsy, Alzheimer's disease, diabetes, high blood pressure and temperature [28–30]. In addition, they can also provide the patient's geographic location. Identification of the location of a patient is particularly valuable for emergency cases [9].

The clinical biotechnology signals for the management of infections and diseases are generally considered universal standard practice. Artifact-free signals are processed by technical staff before they can be delivered to the medical practitioner. For continuous live observation, tracking, examination and evaluation of cardiac system functionality, it is essential to use wearable biomedical sensors. The multichannel ECG signals are collected by smart implantable devices to efficiently diagnose cardiofunctionality. This is a very advanced and complicated method which encompasses a number of cross-disciplinary advanced methods and devices. The results of the process should be successful and reliable or it could lead to irreversible patient harm, as it may result in a misjudgment during treatment. This can be prevented by using a cloud platform and software through developing a monitoring system between patients and doctors. Additionally, some of the currently available wearables are infrequently used by doctors. This is partly related to the validity of the research obtained so far and the challenges of interpreting big data. Such shortcomings have been acknowledged by numerous scientific institutions and research and development teams, and the creation of software for critical signal processing has begun. Other common wearables now advertised for the analysis of ECG signals are reported in [28–30]. In the research of intelligent healthcare monitoring there are many interesting experiments that have contributed to the creation of several applications, methods and structures. Smart sensors and smartphone applications offer opportunities in automated health management programs for providing healthcare services to remote patients [31].

## 9.4 The proposed approach

A block diagram of the proposed system is shown in figure 9.3. The blocks, enclosed in a red line, are suggested to be incorporated into the wireless biomedical sensor. The post-cloud server based application, enclosed in a blue line, solely performs the

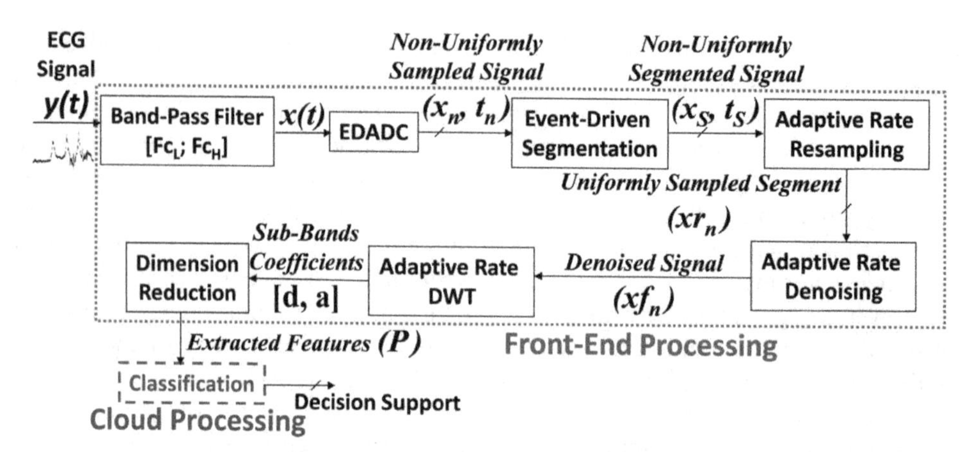


**Figure 9.3.** The proposed system block diagram.

classification task on the data received via the wireless cellular network. The final outcome can be used for decision support and emergency notification.

### 9.4.1 Dataset

In this study the MIT-BIH arrhythmia database is used to show the promising characteristics of the proposed framework [39]. It is an ECG archive with several channels. Each ECG channel is band-limited to 60 Hz and is then recorded at a sampling rate of 360 Hz using an analog-to-digital converter (ADC) of 11 bit resolution. To evaluate the efficacy of the recommended solution in terms of recognizing numerous cardiac pulse groups, five different kinds of ECG recordings are employed. The types considered are right bundle branch block (RBBB), normal (N), premature ventricular contraction (PVC), left bundle branch block (LBBB) and atrial premature contraction (APC).

### 9.4.2 The ECG signal reconstruction

In many applications analog signals are converted into discrete signals (analog-to-digital) using sampling and quantization operations (ADCs) [37]. Then the discrete signals are processed using digital signal processors. Afterward, the processed signals are reconstructed back from digital to analog by utilizing digital-to-analog converters (DACs).

Using Fourier analysis, the sampling process can be described from a frequency domain perspective, its causes can be analysed and the reconstruction phenomenon discussed [45]. Let $x_a(t)$ be an analog signal. Its spectral representation is achieveable using

$$x_a(j\Omega) \triangleq \int_{-\infty}^{\infty} x_a(t)e^{-j\Omega t}\,dt, \tag{9.1}$$

where $\Omega$ is the frequency. The inverse of equation (9.1) is given by

$$x_a(t) = \frac{1}{2\pi} \int_{-\infty}^{\infty} x_a(j\Omega) e^{j\Omega t} \, d\Omega. \tag{9.2}$$

To find $(n)$ we have to sample $x_a(t)$ at sampling interval $T_s$. The process can be described as $x(n) \triangleq x_a(nT_s)$. Let $X(e^{jw_n})$ is the frequency domain transformed version of $x(n)$. According to [45] the aliasing formula is

$$X(e^{j\omega}) = \frac{1}{T_s} \sum_{l=-\infty}^{\infty} x_a\left[j\left(\frac{\omega}{T_s} - \frac{2\pi}{T_s} l\right)\right] \tag{9.3}$$

$$\omega = \Omega T.$$

Therefore, the discrete signal could be an aliased version of the identical analog signal because there will be overlap if the higher frequencies are not properly sampled and they overlap with lower frequencies. The phenomenon is further described with the help of figure 9.4 [47].

The used ECG dataset waveforms are up-sampled with a factor of 100 to evaluate the event-driven ADC (EDADC) module. A specifically designed array of cascaded cubic-spline interpolators is used for up-sampling [42]. It produces a quasi-analog representation of the incoming signal which is used as the input for the EDADC.

### 9.4.3 The event-driven acquisition

The reconstructed signal $x(t)$ is conveyed to the EDADC. The classical arrhythmia diagnosis systems are based on uniform sampling based ADCs whereas the proposed

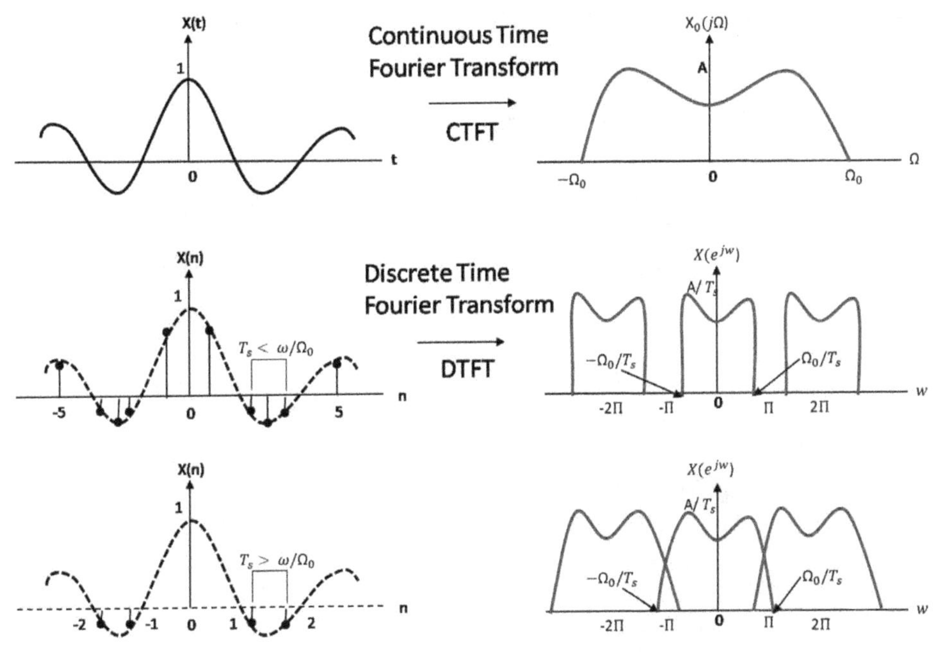


**Figure 9.4.** Sampling and its impact in the time and frequency domains [47].

arrhythmia system is based on EDADCs. Time invariant sampling is performed by a traditional ADC which takes samples from a given signal at constant time intervals, whereas in the EDADC the sampling process is time varying as a function of the incoming signal variations [36, 37].

The concept behind the event-driven sampling is to only obtain the sample if the reconstructed input signal $x(t)$ intersects with one of the thresholds. The distributed levels among the analog signals are designed to span all of the amplitude signal range $\Delta x(t)$. The spaces between the levels are equally distributed and separated by a quantum $q$. For a selected EDADC amplitude dynamics, $\Delta V$, and resolution, $M$, its quantum, $q$, can be calculated as $q = \frac{\Delta V}{2^{M-1}}$. All samples are a couple of $(x_n, t_n)$, where $n$ is the amplitude and $t_n$ is the time. $x_n$ is one of the predefined thresholds and $t_n$ is measured by using a timer circuit [39]. For clarification of the difference between the acquisition principles of ADCs and EDADCs, the following example is considered.

In the case of the ADC, samples are taken at constant intervals. Let us consider the ECG signal, $x(t)$, shown in figure 9.5. Since the samples are taken at a constant time interval there are not many samples around the peak of information and there are a lot of unneeded samples where there is no information in the signal. The black dots in the figure below represent the samples taken at regular time intervals.

In the case of the EDADC, samples are not taken at constant intervals. Let us consider the ECG signal, $x(t)$, shown in figure 9.6. An EDADC is used and the samples are taken only where there is activity in the signal. This allows for more efficient sampling such that the samples represent valuable information. This technique is better than time invariant ADC because it requires a lower number of samples and does a better job at obtaining samples that contain information. The

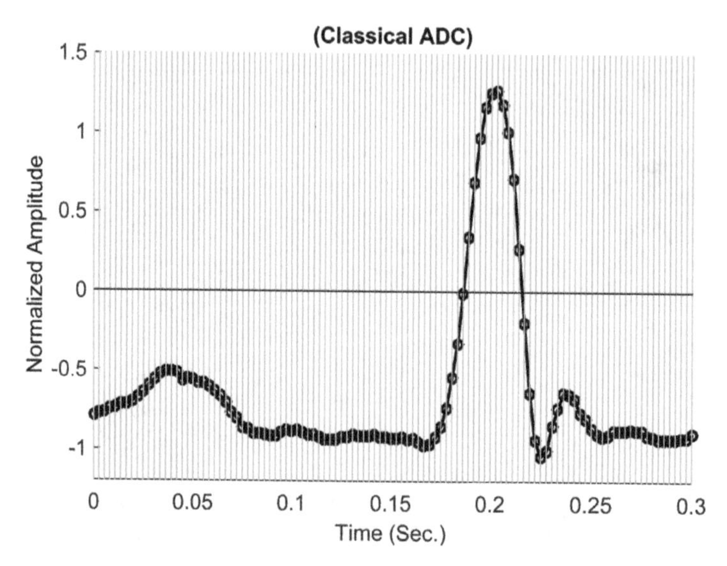


**Figure 9.5.** The ECG acquisition principle of the classical ADC.

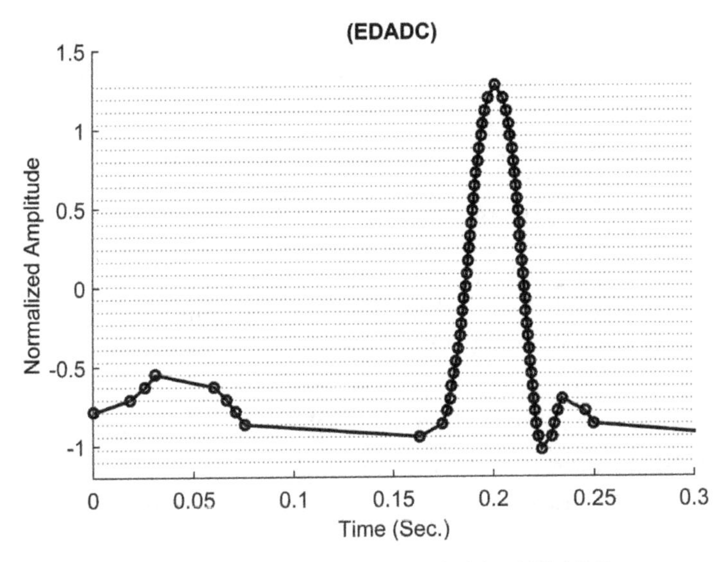


**Figure 9.6.** The ECG acquisition principle of EDADC.

black dots in the figure below represent the samples taken at the peak of activity in the signal.

Figures 9.5 and 9.6 demonstrate the difference in the number of samples extracted from the incoming data using EDADC and the ADC. For the classical ADC to be able to extract as much important information as acquired by the EDADC, the sampling frequency must be many times higher than the one shown in figure 9.5.

### 9.4.4 The event-driven segmentation

The data points obtained from EDADC do not have an equal distance between them in terms of time. Thus classical techniques cannot be used to process or analyse the data [36, 37, 41]. Windowing is an essential operation and it is required for the limited time data acquisition to fulfill the practical system implementation specifications [42]. To obtain the windowed version of a sampled signal $x_n$ the $N$ sample segment selected must be centered on $\tau$ by using the following equation:

$$xw_n = \sum_{n=\tau+\frac{L}{2}}^{\tau+\frac{L}{2}} x_n \cdot w_{n-\tau}, \tag{9.4}$$

where $xw_n$ is the windowed version of $x_n$, $L$ is the effective length in seconds and $\tau$ is the central time of the window function $w_n$. $N$ can be calculated using the following equation, where $F_s$ is the sampling frequency:

$$N = L \cdot F_s. \tag{9.5}$$

Performing digital signal processing and analysis requires a finite set of data. Windowing functions are applied to capture a limited frame of data [36] as shown in figure 9.7.

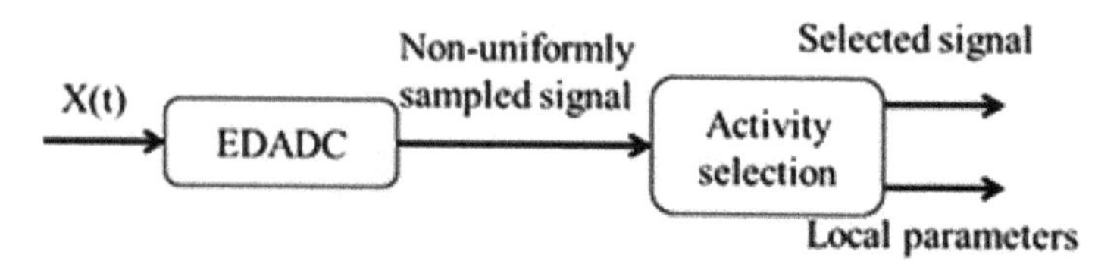


**Figure 9.7.** The sequence of activity selection based windowing.

Segmentation of the EDADC output is realized by using the activity selection algorithm (ASA) [40, 44, 55]. This is built on the sampling non-uniformity and permits the realization of adaptive-rate hybrid feature extraction.

### 9.4.5 The adaptive-rate resampling and denoising

This method denoises the segmented signal with an adaptive-rate digital band-pass filter. This approach uses the original event-driven adaptive-rate filtering and adaptive resolution short time Fourier transformation techniques [40]. These are described in the following.

Techniques for adaptive-rate filtration are being developed. They are based on event-driven sensing (EDS) [36, 37, 41]. The sampling frequency of the EDADC follows $x(t)$ temporal disparities. Equation (9.6) represents the mathematical equation of the EDADC maximum sampling frequency [37], where $f_{\max}$ is the highest component of the $x(t)$ frequency, $A_{\text{in}}$ represents the amplitude of the incoming signal and $F_{s_{\max}}$ is the EDADC maximum sampling frequency:

$$F_{s_{\max}} = 2f_{\max}(2^M - 1)\frac{A_{\text{in}}}{\Delta V}. \tag{9.6}$$

Let the $i$th selected window acquired with the ASA be $W^i$ and $F_s^i$ represent the sampling frequency of this selected window, then

$$F_s^i = N^i/L^i, \tag{9.7}$$

where $N^i$ is the number of samples in the chosen window, $W^i$, and $L^i$ is the length of this window measured in seconds. The $W^i$ is being resampled uniformly. Locally mined parameters for each segment are utilized to tune the frequency of the resampling. Compared to classical analogs, the selected segments of the signal in this manner are being resampled and processed at the same or lower rates. Contrasting with the traditional techniques, the arithmetic complexity reduction of the suggested technique is thus increased substantially. The selection of the resampling frequency of ($W^i$), $F_{rs}{}^i$, depends on the reference sampling frequency, $F_r$, and the sampling frequency of the selected window, $F_s{}^i$.

The resampling frequency must remain higher than and closer to the Nyquist frequency of sampling which is $F_{\text{Nyq}} = 2{\cdot}f_{\max}$. Then $N_r{}^i$ is known to be the number of samples existing in $W^i$ after the resampling. Interpolation is used to achieve the online resampling. For the online resampling of the segmented signal, the simplified linear interpolation (SLI) technique is used. The process is clear from figure 9.8. The SLI is a special case of linear interpolation (LI) which performs curve fitting by using

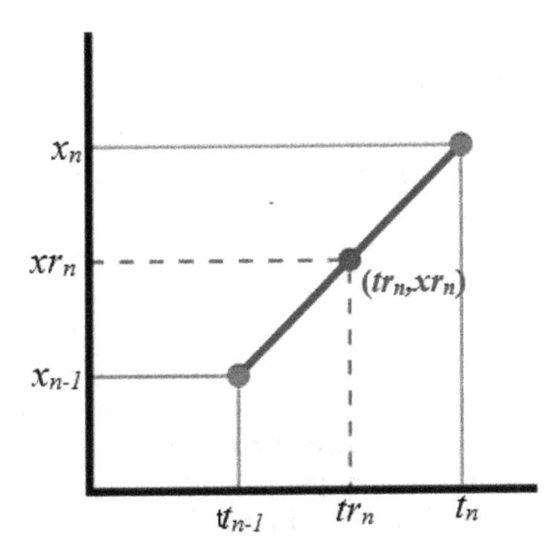


**Figure 9.8.** The principle of the simplified linear interpolator.

linear polynomials to construct new data points within the range of a discrete set of known data points. In the case of SLI, the $n$th resampling instant, $t_{r,n}$, lies exactly in the middle of the known $n$th and $(n-1)$th sampling instants, namely $t_n$ and $t_{r,n-1}$. $x_{r,n}$ is the value of the interpolated sample for SLI and it is equal to the mean of its prior and next non-uniform samples. In SLI, $\frac{q}{2}$ limits the maximum error per interpolation [43].

For the proposed method, a reference finite impulse response (FIR) filter bank is designed offline. The realization of the architecture for an appropriate parameter range is obtained when changing the reference sampling frequencies $F_{ref}$. The lower and upper cut-off frequencies of the reference band-pass filters are $Fc_{min}$ and $Fc_{max}$, respectively. Then, the upper bound of $F_{ref}$ is chosen to be $F_r$, and the lower bound is selected to be ($Fs_{min} \geqslant 2Fc_{max}$) to guarantee a suitable digital filtering process.

The mathematical equation used to compute $F_{ref}$ is represented in equation (9.8), where $Q$ is the length of the $F_{ref}$. The value of $Q$ is always selected to be a weighted binary. Equation (9.9) is then used to calculate the unique offset, Δ:

$$F_{ref} = \{Fs_{min}, Fs_{min} + \Delta, \ \ldots, \ Fs_{min} + (Q-1)\Delta = F_r\} \tag{9.8}$$

$$\Delta = \frac{F_r - Fs_{min}}{Q-1}. \tag{9.9}$$

For each $W^i$ a proper filter is selected online from the reference set. $h_{c_k}$ represents the selected filter for $W^i$ and is its sampling frequency. This choice is made depending on the $F_{ref}$ and $F_s^i$. $F_{ref_c}$ should be equal to $F_{rs}^{\ i}$ to guarantee a proper filtering process [40, 44].

### 9.4.6 Extraction of features and dimension reduction

*9.4.6.1 Adaptive-rate discrete wavelet transform*

The wavelet transform (WT) is commonly used to investigate the time varying ECG signals. It allows achieving the multi-resolution time-frequency representation of signals [4]. This can be represented in mathematical form using equation (9.10), where $s$ and $u$, respectively, represent the dilation and the translation parameters:

$$W_x^{\psi}(u, s) = \frac{1}{\sqrt{S}} \int_{-\infty}^{+\infty} x(t)\psi * \left(\frac{(t-u)}{s}\right) dt. \tag{9.10}$$

A discrete-time wavelet transform (DWT) is used for analysing the digital signals. In the studied case the Daubechies algorithm based wavelet decomposition is utilized for the subband decomposition. The approximation, $a_m^i$, and detail, $d_m^i$, coefficients are calculated by utilizing half-band high-pass and low-pass decimation filters. This can be represented in mathematical form via equations (9.11) and (9.12). In these equations, $m$ represents the level of decomposition. In this study, a fourth level of decomposition is performed, i.e. $m \in \{1, 2, 3, 4\}$. $g_{2n-k}$ and $h_{2n-k}$ are, respectively, the low-pass and high-pass FIR filters with a subsampling factor of 2:

$$a_m^i = \sum_{k=1}^{K_g^i} xf_n^i \cdot g_{2n-k} \tag{9.11}$$

$$d_m^i = \sum_{k=1}^{K_g^i} xf_n^i \cdot h_{2n-k} \tag{9.12}$$

According to [3], the subbands extracted by wavelet decomposition are functions of the incoming signal sampling rate. The $W^i$ can be resampled at a specific frequency $F_{rs}{}^i$ in the suggested solution. Consequently, $W^i$ is decomposed at a specific rate and the system could achieve subbands with a lower arithmetic cost. In fact, the system needs to treat a lower count of samples compared to the conventional counter fix-rate decomposition based approaches [3]. Additionally, tuning of $F_{rs}{}^i$ permits better focus on the signal band of interest compared to the fix-rate counterparts [3].

#### *9.4.6.2 Dimension reduction*

The discrete wavelet transformation (DWT) is used to analyse the ECG signals. First, each ECG signal is decomposed using DWT into subbands. The decomposition levels is set to 4. The four coefficients d1–d4 of the detail subbands and A4 approximation subband are used. An important aspect that determines the effectiveness of classification is the type of selected wavelet. Here, the Daubechies wavelet of order 1 (db1) is chosen as the wavelet type.

The dimension of the obtained feature vectors is too large (150) when using DWT. Therefore, we extract statistical features from the detail subbands d1–d4 and the approximate subband a4. The important question here is, which statistical features are important? Because there is generally no *a priori* way of clearly deciding which features to remove, we use some features that were previously used for related issues in the research. Remember that not all applications are similarly important or lead to a strong classification. Fortunately, methods exist in the area of supervised learning which will allow us to select the 'important features'. A feature is deemed important if it contributes to the classification accuracy of the data. We explore all the features studied here and employ feature selection to reduce the feature set.

The following sixteen statistical features are considered in this study:

*Power spectrum (PS).* A signal's power spectrum is the distribution of the signal's energy per unit time plotted against its given frequency bins. We compute the power spectrum by using the fast Fourier transform (FFT) and computing the eaverage absolute value. This is done for each subband.

*Mean absolute value (MAV).* The MAV of a signal is computed by summing the absolute of the values of the signal and normalizing it by its length. It measures the central tendency and represents a probability frequency distribution of the signal.

*Standard deviation (STD).* The SD of the signal measures the spread of the signal around its mean value. It represents the amount of changes in the frequency distribution.

*Skewness (SK).* The skewness of a signal is determined by calculating a signal's frequency distribution over its mean. That is a function of this distribution's asymmetry. A negative value implies that the left tail is longer while a positive value indicates that the right tail is longer.

*Kurtosis (K).* Kurtosis is a measure of the 'tailedness' or curved shape of the signal. Similar to skewness, it measures the shape of the signal. It is seen as a measure of the dispersion of its normalized value along its expected value.

*Ratio (R).* It is calculated as the ratio of the average value computed from the detailed signal to the average value computed from the approximate signal.

*Zero crossings (ZC).* The ZC value is the number of zero crossings in a signal, i.e. the number of times the signal value changes sign.

*Number of zeroes (NZ).* The NZ is the number of zero elements in the signal vector. It is a measure of the time the signal is at rest, i.e. neither positive nor negative.

*Peak positive value (PV1).* The PV1 is the highest positive amplitude attained by the signal. It measures the highest value in the signal.

*Peak positive index (PI1).* The PI1 is the index (location) where the peak positive value occurs. It is used in combination with the peak value.

*Second peak positive value (PV2).* The PV2 is the second maximum positive amplitude of the signal. It measures the second highest value in the signal.

*Second peak positive index (PI2).* The PI2 is the index (location) where the second peak positive value occurs. It is used along with the second peak value.

*Peak negative value (NV1).* The NV1 value is the maximum negative amplitude attained by the signal. It measures the lowest value of the signal.

*Peak negative index (NI1).* The NI1 is the index (location) where the peak negative value occurs. It is used in combination with the peak value.

*Second peak negative value (NV2).* The NV2 is the second highest negative amplitude attained by the signal. It is also used in combination with the index of this value.

*Second peak vegative index (NI2).* The NI2 is the index (location) where the second peak positive value occurs. It is used in combination with the second peak value.

### 9.4.7 Machine learning algorithms

The features are used to represent the signals and once the relevant features have been extracted, the signal data form a reduced matrix with 16 features extracted from each of the five selected subbands giving a total of 80 features. Therefore, since there are five ECG classes each having 300 instances, the final data matrix has a dimension of 1500×80. This dataset, instead of the original signals, is used for classification.

The many classification techniques used may require tuning the parameters to obtain maximum results. Therefore, we employ the standard and popular way of validation to tune the parameters of the classifier during the training phase. For the test set, the value that results in the optimal output is used. In the following we briefly describe the different machine learning classifiers used in our experiments along with the parameter settings.

#### *9.4.7.1 k-nearest neighbors (kNN)*

The $k$NN algorithm [38, 39] is a popular classifier that finds the distance of the test data to the $k$ nearest neighbors in the training data to decide its category label. Let $v_j$ represents a sample and $\langle v_j, l_j \rangle$ represents the pair of the data vectors along with its label, $l_j \in [1, C]$ where $C$ denotes the maximum classes in the dataset. Then the $k$NN process to classify a new vector $z$ is given by

$$\text{argmin dist}_j(\boldsymbol{v}_j, z)\ \forall\, j = 1, \ldots, N, \tag{9.13}$$

where dist(,) is a distance measure. The value of $k$ is usually set to a small odd number to ensure a majority consensus. Similarly, any distance metric can be used, including classical metrics such as Manhattan, Euclidean, Chebyshev, Hamming, Minkowski, etc, or advanced metrics such as the $\chi$-sim, cosine similarity, etc [49–51]. A good combination for the ECG data, determined using cross-validation, is to set $k = 3$ and use the Euclidean distance metric. This is because the data are dense and contain real values. Other measures might be more suitable for sparse data, physical distances and angles between vectors.

#### *9.4.7.2 Artificial neural network (ANN)*

Artificial neural networks (ANNs) have become a widely used classifier in the domain of machine learning. They are based on the perceived working neurons and synapses in the human brain [46]. An ANN is a network of interconnected, parallel processing units (called neurons or nodes), which work together to produce a classifier. One of the most popular types of ANNs is the multi-layer perceptron (MLP) in which the neurons are stacked in layers. The units from the initial layers are fully connected to units from the later layers and the flow of information flows in a forward fashion. A neuron may or may not fire (i.e. pass on information to the next node) depending on its activation function. The connections are in the form of a link between nodes with certain weights associated with them. A back propagation algorithm is usually employed to transfer error to other layers so they may correct the connection weights.

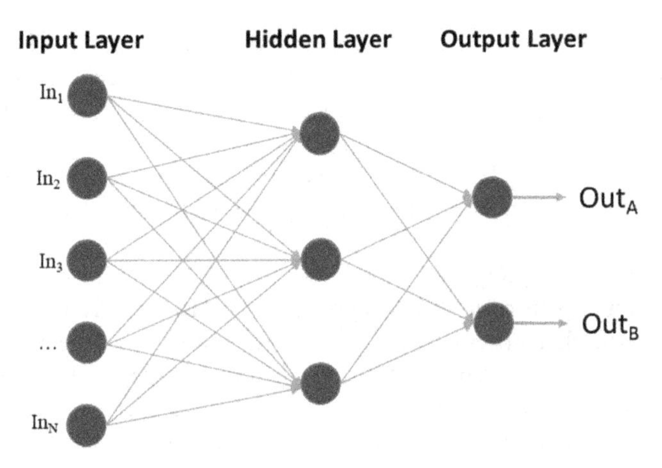


**Figure 9.9.** An ANN with $N$ inputs, one hidden layer with three neurons and two outputs.

The initial set of neurons usually form the input layer and the final set of neurons are the output layers. There may be several intermediate layers, known as hidden layers. The advantage of these hidden layers is that they can help model more complex problems. The input and output layers are typically set based on the number of features per sample and the number of groups available, respectively. Ideally, the greater the number of neurons in the hidden layer(s) and the higher the number of layers, the nicer the problem can possibly be modeled by the algorithm. A simple model of the ANN is shown in figure 9.9.

*9.4.7.3 Support vector machines (SVMs)*

The SVM was developed by Cortes and Vapnik [48]. In SVM, the first task is to find those samples that represent the boundary conditions—known as the support vectors, which are then used to search for the optimal hyperplane separating these vectors of the opposite classes. To achieve this, we maximize the margin between the closest points, i.e. the points lying closest to the boundary. This is done by optimally separating the hyperplane that divides the data of the two classes described in the following equation, where $x$ represents the data vector ($x = [x_1, x_2, \ldots, x_p]$) having $p$ attributes, where the weight vector is given by $w = [w_1, w_2, \ldots, w_p]$, and the scalar bias is $b$:

$$h(x) = \text{sign}(w \cdot x^{\text{T}} + b). \tag{9.14}$$

In an SVM, we train the weights $w$ such that a test sample $x$ when input to the classifier function $h$ gives either a positive value (belonging to the first class) or a negative value (belonging to the second class). Other parameters are also involved, such as the $c$ parameter (called the regularization parameter) which determines the penalty for misclassification during training and the slack variables.

When the two classes are not linearly separable, a Kernel function [48] is used that transforms the samples into a higher dimension were they may be separable. Moreover, for classifying data with more than two classes, multiple approaches can be used. For instance, we can keep it as a two-class problem by keeping the label of one class (positive class) and relabeling all other classes with the same label (negative class). This may be repeated for each class during classification to identify samples of that particular class. It is known as the one-versus-all approach. Another commonly used strategy is taking samples of only two classes at any given time and using a round-robin manner to classify the data. This is known as the one-versus-one approach [48, 49].

##### *9.4.7.4 Random forest (RF)*

The RF algorithm has its roots in ideas proposed by Ho Tim Kam [51]. It belongs to the family of ensemble learners which construct multiple sets of decision trees during training, and the output generated using this multitude of trees is used for a 'consensus' decision. The consensus can be either the mode or the mean of the classifier outputs, depending on the problem and classifiers used.

RF uses the bagging technique for classification. In bagging, multiple samples of the data (with replacements) are taken for learning different classifiers (in our case decision trees), thereby reducing the risk of over-fitting. As opposed to bagging, however, it is permissible in RF to employ different tree classification algorithms in each of the multitude of trees.

In our approach, the number of trees is set to 60 and we retained the out-of-bag predictions for each tree. The maximum number of splits at each branch is set to 10. For the split at the nodes, we select the interaction-curvature method that minimizes the $p$-value generated from using the chi-square tests-of-independence between each of the predictor–response pairs.

##### *9.4.7.5 Bagging (BG)*

Bagging is derived from the 'bootstrap aggregation' of classification trees. Bootstrap aggregation is a generic method used to reduce the variance for those algorithms showing a high variance. In classical bagging, given a dataset, multiple subsets are created from the dataset (including with replacement) to generate a larger dataset and an algorithm is trained on the larger dataset. Decision trees, in particular, are rather sensitive to the training data on which the algorithm is trained. Therefore, when their dataset is changed, the corresponding tree generated can be quite different [52, 53].

In bagging, multiple trees are generated using multiple larger datasets created from the given training data. Each of the trees is allowed to fit the training data (i.e. with little or no pruning) since any bias (over-fitting) can be reduced by employing a collection of trees. For this reason, bagging is considered as a powerful classifier.

## 9.5 The performance evaluation measures

The following evaluation measures are used.

### 9.5.1 Compression ratio

In traditional sensing the incoming analog signal is received at a set rate. The number of samples $N$ acquired for the considered period $L_{\mathrm{T}}$ can therefore be calculated using

$$N = F_{\text{ref}} \times L_{\text{T}}. \tag{9.15}$$

The sampling rate for the proposed method [11] is not fixed. Therefore, the number of samples can vary for a considered time period $L_{\text{T}}$ and depend on the signal characteristics [36, 37, 40, 44]. Let $N_{\text{ED}}$ be the count of collected samples. Then the gain in compression is

$$G_{\text{COMP}} = \frac{N}{N_{\text{ED}}}. \tag{9.16}$$

### 9.5.2 Computational complexity

The front-end processing module's computational complexity is analysed in detail, until the denoising level. The complexity of the extraction of features and classification is evaluated intuitively by taking into consideration the compression performance of of the suggested tactic. The adaptive-rate FIR (ARFIR) method is used to denoise the resampled signal [11, 40, 44].

The numerical complexity of a classical $K$ order FIR filter is well known, so it is possible to calculate the whole numerical complexity $C_{\text{FIR}}$ by using

$$C_{\text{FIR}} = \underbrace{(K-1).\,N}_{\text{Additions}} + \underbrace{K \cdot N}_{\text{Multiplications}}. \tag{9.17}$$

In the case of an invented solution the online adaptation requires additional operations. First, a filter $h_{\text{ck}}$ is picked up for $W^i$. Using the successive approximation algorithm, this filter selection is performed in real time. In the worst scenario it conducts $\log_2(Q)$ comparisons. Here, $Q$ is the length of set $F_{\text{ref}}$, in other words it is the count of elements in this set. It is also necessary to resample the selected segments at the real-time computed resampling rate $F_{\text{rs}}{}^i$. The live resampling is achieved by using the SLI. For SLI, the count of operations for each $W^i$ is $N_{\text{r}}{}^i$ additions and $N_{\text{r}}{}^i$ binary weighted divisions. The binary weighted divisions are realized by employing the one-bit right shift mechanism. The complexity of such an implementation is insignificant in contrast to the addition and multiplication operations [45]. In this background, the computational cost of SLI is approximated as $N_{\text{r}}{}^i$ additions for $W^i$. After uniform resampling, the $W^i$ is conditioned by utilizing a $K^i$ order digital filter. The count of operations during the conditioning process is $K^i{\cdot}N_{\text{r}}{}^i$ multiplications and $K^i{\cdot}N_{\text{r}}{}^i$ additions for $W^i$. The complexity of a comparison operation is assumed to be equal to that of an addition operation. This assumption allows us to evaluate the processing cost of the devised ARFIR with respect to that of the classical one. In this context the computational complexity of the ARFIR can be represented by

$$C_{\text{ARFIR}} = \underbrace{(K^i - 1) \cdot N_{\text{r}}{}^i + N_{\text{r}}{}^i + \log_2(Q)}_{\text{Additions}} + \underbrace{K^i \cdot N_{\text{r}}{}^i}_{\text{Multiplications}}. \tag{9.18}$$

### 9.5.3 Classification accuracy

The solution used in this work has several advantages from the point of view of the hardware complexity, the overall compression, processing and transmission. The downside to this, however, is a small decrease of accuracy. We evaluate the work in terms of accuracy to focus its comparison with existing works. To ensure the results are not biased we use cross-validation. This helps in reducing any abnormal result that may occur due to the dataset used, the particular training–testing split on the dataset, parameter or classifier bias [43].We used a ten-fold cross-validation and evaluated all the algorithms on the same dataset. In particular, for each training–testing set in each fold, we save the indices of the training and testing sets and ensure the same indices are used for all the algorithms. Moreover, to avoid metric bias, we choose to evaluate the results on multiple evaluation metrics as given below.

*Accuracy* (Acc) is used to compare the results from the algorithm to the ground truth. It is defined as the percentage of test dataset labels that have been correctly predicted. We usually define accuracy in terms of true positives (TPs), true negatives (TNs), false positives (FPs) and false negatives (FNs), respectively. Mathematically, this is given as

$$\text{Accuracy} = \frac{\text{TP} + \text{TN}}{\text{TP} + \text{TN} + \text{FP} + \text{FN}}. \tag{9.19}$$

*The F-measure* (F1) is an alternative evaluation metric frequently employed to compare the results between competing algorithms. Using the values of TP, TN, FP and FN as discussed above, we can define two new scores—precision and recall. *Precision* is the ratio of the TPs to the sum of all values classified as positive (i.e. TP and FP) and is given as $\text{precision} = \frac{\text{TP}}{(\text{TP} + \text{FP})}$. It signifies how accurately we predicted the positive values. *Recall* is the ratio of the TP to all the data points that should have been positive according to the ground truth (i.e. TP and FN). It signifies how many of the positive values have been correctly predicted amongst all the true labels. The F-measure provides a balance between the values of precision and recall. Depending of whether the class size is taken into account or not, we can divide the F1 score as a macro- or a micro-score, respectively. In our case, however, since all classes have the same size, the simple *F*-measure is used as given by

$$F = \frac{2 \times \text{precision} \times \text{recall}}{\text{precision} + \text{recall}}. \tag{9.20}$$

*The kappa index* (Kappa) is a popular statistic to compare two results by judging the agreement between their respective predicted labels. It is usually a preferred method in many cases as it is seen as a better measure. This is because a simple accuracy metric does not take into account the possibility of random or probabilistic agreement between the two labels. The commonly employed kappa index is Cohen's kappa measure, as given in equation (9.21), where $p_0$ is the agreement between the predicted labels and the ground truth expressed as a percentage, and $p_e$ is the probability that the agreement simply happened randomly. Mathematically, $p_e$

is computed as $\frac{(TP+TN)(TP+FN)+(FP+TN)(FP+FN)}{(TP+TN+FP+FN)^2}$. The value of the kappa index is given as

$$\text{kappa} = 1 - \frac{1 - p_0}{1 - p_e}. \tag{9.21}$$

*Specificity* (Sp) is a measure that computes the number of TNs that were correctly classified as a ratio of all negatives (i.e. negatives that were classified as negatives and negatives that were wrong classified as positives). This is in contrast to the precision value discussed above which computes a simple score but for positives. For this reason, specificity is sometimes also called the 'true negative' rate, and is given by

$$\text{Sp} = \frac{\text{TN}}{\text{TN} + \text{FP}}. \tag{9.22}$$

## 9.6 Experimental results and discussion

### 9.6.1 Experimental results

The efficiency of the suggested tactic is tested for five groups of arrhythmias [39]. In this context, MATLAB is utilized to incorporate and test all the device modules [46]. In total 1500 instances of the ECG pulses are classified. They belong to five groups which are called RBBB, N, PVC, LBBB and APC. Three hundred instances are considered for each class. It leads to an equal representation of each class and eliminates possible biases during the process of classification.

Conventionally, the signal is a band confined to 60 Hz and is received at a sampling rate of 360 Hz with conventional 11 bit resolution ADC. The signal is divided into 0.9 s segments and each segment is composed of 320 samples.

In the case proposed, the considered ECG signals are recovered using the up-sampling factor of 100 to test the EDADC. The reconstructed signals are acquired with an EDADC of $M$ = 5 bits. The ECG signals are band-limited up to $f_{max}$ = 60 Hz.

The EDADC output is segmented using the ASA. The ASA adapts $L^i$ and $F_{rs}{}^i$ according to the $\tilde{x}(t)$ temporal variations. It contributes to enhancing the system's computational efficiency. For all 300 instances of each type, the average reductions in the number of obtained samples compared to the classical equivalent are determined. These are given in table 9.1.

Table 9.1 shows that the attained overall average reduction in the number of collected samples for all five classes is three-fold.

**Table 9.1.** Summary of the compression gain.

| CLASS | Normal | RBBB | LBBB | APC | PVC |
|---|---|---|---|---|---|
| $G_{COMP}$ | 3.1 | 2.8 | 3.0 | 3.0 | 3.0 |

The resampled signal is denoised using the ARFIR filtering technique [11]. A band-pass filters bank is designed offline for the cut-off frequencies of [$F_{Cmin}$ = 0.7; $F_{Cmax}$ = 35] Hz. The filter bank is implemented for a set of sampling frequencies, $F_{ref}$, between $F_{s_{min}} = 80$ Hz $< 2 \cdot F_{Cmax}$ to $F_r = 360$ Hz. In this case, $\Delta = 40$ Hz is chosen. It realizes a bank of $Q = 8$ band-pass filters. A summary of the designed filter bank is shown in table 9.2.

The filtering method improves the expected signal signal-to-noise ratio (SNR) and allows for a better output in classification [2]. The adaptation of the online filter is clear from table 9.2. It enables the signal to be enhanced with a decreased numerical cost relative to the standard approaches which are invariant in nature [3, 12].

Using equations (9.18) and (9.19) the numerical advantage of the proposed ARFIR is determined over the classical one. The results are summarized in table 9.3. It illustrates that compared to its traditional counterpart, the proposed ARFIR technique achieves notable numerical complexity reduction. This is accomplished by cleverly adjusting the parameter values, such as $F_{rs}{}^{i}$, $N_r{}^{i}$ and $K^i$ for each $W^i$. The incoming signal pilots those alterations. The overall arithematic complexity reduction, attained by the suggested solution compared to the conventional counterparts, is 7.56-fold and 7.68-fold in terms of additions and multiplications, respectively.

Generally, checking the output of the chosen or removed features is not straightforward. Even a feature's output is linked to its intended use. Therefore we use classification as a performance benchmark to determine the utility of the extracted features as well as the performance of the entire system. To do this, we test all the classifiers (ANN, *k*NN, SVM, RF and BG) on the datasets and record all the evaluation metrics (accuracy, F1 score, kappa and specificity). This is shown in table 9.4.

**Table 9.2.** Summary of the reference filter bank parameters.

| $h_{ck}$ | $h_{1k}$ | $h_{2k}$ | $h_{3k}$ | $h_{4k}$ | $h_{5k}$ | $h_{6k}$ | $h_{7k}$ | $h_{8k}$ |
|---|---|---|---|---|---|---|---|---|
| $\mathbf{F_{refc}}$**(Hz)** | 80 | 120 | 160 | 200 | 240 | 280 | 320 | 360 |
| $\mathbf{K_c}$ | 25 | 39 | 52 | 65 | 78 | 91 | 104 | 117 |

**Table 9.3.** Summary of the ARFIR arithematic complexity gain.

| CLASS | Gain in additions | Gain in multiplications |
|---|---|---|
| Normal | 8.0 | 8. 1 |
| RBBB | 7. 1 | 7.2 |
| LBBB | 7.6 | 7.7 |
| APC | 7.3 | 7.5 |
| PVC | 7.8 | 7.9 |

**Table 9.4.** Accuracy, F1-measure, kappa statistics and specificity for different classifiers.

| | Acc | F1 | Kappa | Sp |
|---|---|---|---|---|
| ANN | 91 | 0.74 | 0.91 | 0.71 |
| *k*NN | 63 | 0.36 | 0.62 | 0.14 |
| SVM | 89 | 0.73 | 0.89 | 0.65 |
| RF | 91 | 0.78 | 0.92 | 0.73 |
| BG | 92 | 0.79 | 0.93 | 0.75 |

Overall, using the proposed approach leads to very good accuracy, particularly with random forest, ANN and bagging. Hence, integrating adaptive-rate signal acquisition along with classification can result in better results as compared to the other methods. This is significant since the overall complexity is less when using this strategy. Moreover, as RF uses multiple classifiers, it is usually more robust and less likely to be affected as is evident when we look at the kappa statistics and specificity. As expected, the lowest accuracy is obtained using *k*NN since it is both a less sophisticated classifer and can easily be affected by the *k* neighbhors chosen. An outlier will have a significant effect on the accuracy since it can influence the nearest labels, particularly when *k* is smaller. SVM scores higher than *k*NN but lower than ANN, RF and bagging.

Detailed classification results are presented in tables 9.5–9.9. Moreover, we also analyze the results in terms of the classes to be differentiated. It is found that the the most confused classes amongst those tested are the PVC and RBBB, particulary for the ANN and *k*NN classifiers. This confusion is the lowest in bagging, which also results in the highest accuracy value. Similarly, APC and PVC as well as LBBB and RBBB are almost always never confused. In terms of false positives and false negatives, using bagging results in the overall lowest value compared to the other algorithms.

### 9.6.2 Discussion

Cloud-based biomedical signal processing for automatic diagnosis of cardiac health condition monitoring is a novel trend. It is obvious that it will play an important role in the future solutions for mobile healthcare.

From the results for our proposed approach, the benefits of the devised tactic are obvious. It employs a smart combination of the event-driven acquisition and the methods for adaptive-rate processing. The system parameters can be adjusted as a function of incoming signal variations. This ensures that the dynamic parameters of the proposed solution are adapted according to the incoming signal disparities. It is shown that the adaptation of $L^i$ and $N^i$ automates the real-time tuning of $F_{rs}{}^i$. This real-time adjustment of $F_{rs}{}^i$ enhances the processing effectiveness and the power efficiency of the designed approach in contrast to the traditional equivalent. It is obtained by excluding needless operations while performing the real-time and online interpolation of the active signal portions. It also avoids the collection and conditioning of irrelevant information and leads towards an effective denoising

**Table 9.5.** Confusion matrix using the ANN.

| | True positives | False positives | False negatives | True negatives |
|---|---|---|---|---|
| Normal | 25 | 2 | 1 | 122 |
| APC | 27 | 4 | 4 | 115 |
| LBBB | 28 | 1 | 2 | 119 |
| PVC | 25 | 4 | 0 | 121 |
| RBBB | 31 | 3 | 7 | 109 |

**Table 9.6.** Confusion matrix using *k*NN.

| | True positives | False positives | False negatives | True negatives |
|---|---|---|---|---|
| Normal | 16 | 16 | 10 | 108 |
| APC | 16 | 8 | 15 | 111 |
| LBBB | 28 | 15 | 2 | 105 |
| PVC | 17 | 13 | 8 | 112 |
| RBBB | 17 | 4 | 21 | 108 |

**Table 9.7.** Confusion matrix using the SVM.

| | True positives | False positives | False negatives | True negatives |
|---|---|---|---|---|
| Normal | 23 | 1 | 3 | 123 |
| APC | 24 | 6 | 7 | 113 |
| LBBB | 30 | 1 | 0 | 119 |
| PVC | 25 | 3 | 0 | 122 |
| RBBB | 31 | 6 | 7 | 106 |

**Table 9.8.** Confusion matrix using RF.

| | True positives | False positives | False negatives | True negatives |
|---|---|---|---|---|
| Normal | 25 | 2 | 1 | 122 |
| APC | 26 | 5 | 5 | 114 |
| LBBB | 30 | 1 | 0 | 119 |
| PVC | 25 | 0 | 0 | 125 |
| RBBB | 31 | 5 | 7 | 107 |

**Table 9.9.** Confusion matrix using bagging.

| | True positives | False positives | False negatives | True negatives |
|---|---|---|---|---|
| Normal | 24 | 0 | 2 | 124 |
| APC | 27 | 3 | 4 | 116 |
| LBBB | 28 | 1 | 2 | 119 |
| PVC | 25 | 1 | 0 | 124 |
| RBBB | 34 | 7 | 4 | 105 |

mechanism. Additionally the real-time adoption of filter orders also adds to the effectiveness of the suggested conditioning mechanism [42–44].

Thanks to the adaptive-rate acquisition, a notable compression gain of three-fold is attained by the proposed solution. The computational complexity is also reduced at every stage of the system by using adaptive-rate processing techniques instead of time invariant techniques. The average gain for both intended subjects is also computed. The gain in additions and multiplications for ARFIR is 7.56-fold and 7.68-fold, respectively. The reduction of circuit complexity is accomplished using a 5 bit resolution EDADC as opposed to the 11 bit resolution used in previous studies. The dimension reduction resulted in representing each incoming ECG instance by 80 features. On the other hand, in the traditional system each ECG instance is represented by 320 features. Our proposed approach thus assures an overall four-fold reduction in the amount of information to be transmitted towards the cloud. It also confirms a computationally effective post-classification as the system has to deal with four times fewer samples. As a result of the computational and hardware complexity reduction, the power efficiency of the system is inherently improved.

The proposed method will derive and manipulate the most appropriate criteria for defining different arrhythmias. Studies have shown that a clever combination of the EDADC, ASA, ARFIR, ARDWT, dimension reduction and bagging contributes to the highest classification outcomes for the studied dataset. There are several reasons for bagging's superior performance compared to the other studied approaches of classification. Bagging uses multiple classifiers. It is founded on the bootstrap aggregation which reduces the variance for this algorithm. It allows pruning the unwanded nodes and renders a better classification decision. The results confirmed that bagging is more robust and less likely to be affected, as is evident when we look at the kappa statistics and specificity. It is necessary for the arrhythmia diagnosis systems to correctly classify heart pulses. The model established offers the best classification results for an intelligent combination of EDADC, ASA, ARFIR, ARDWT, dimension reduction and bagging. Across all the studied classifiers it reaches the highest precision of 92 percent.

## 9.7 Conclusion

In cloud-based mobile automatic arrhythmia diagnosis one crucial function is the consistent and effective monitoring of EEGs. In this chapter we developed an event-driven and adaptive-rate arrhythmia diagnosis mechanism which utilizes the ECG signals in this regard. The suggested approach was found to be practical, accurate and successful in the measurement and evaluation of arrhythmia diagnosis. The advances in mobile devices and sensor technologies include innovative ways to improve the quality of care for people with disabilities. The proposed approach is new and has the ability to improve cardiac health monitoring and medical systems for the next age.

With the help of the results it is shown that as a whole the average reductions in the count of operations in terms of additions and multiplications, respectively, for the ARFIR in contrast to the conventional equals are 7.56-fold and 7.68-fold. In addition, the proposed approach also attains a notable reduction in the acquired count of samples; for the studied case it is three-fold. It confirms an important enhancement in the effectiveness of the post stages, such as conditioning, transmission and classification. Moreover, a significant reduction in bandwidth usage is also assured by these findings. Furthermore, a simpler five bit resolution EDADC is employed in the proposed approach for the recording of ECG signals. In contrast, a relatively complex 11 bit resolution ADC is utilized for the ECG signal recoding in the case of traditional equals. It assures a simpler circuit level realization of the suggested tactic in contrast to the conventional equivalents.

On the side of feature extraction, a four-fold reduction in sample count is realized with a diminished complexity hybrid feature extraction approach. It assures a notable reduction in the real-time count of the operations of the feature extraction process in contrast to the traditional tactics. A similar amount of gain in terms of the processing effectiveness is also expected from the classification stage.

The highest accuracy obtained by integrating the event-driven analog-to-digital converter, activity selection algorithm, adaptive-rate FIR filtering and adaptive-rate feature extraction with bagging is 92%. It indicates that the proposed approach achieves identical classification performance compared to the conventional equivalent, while attaining evident gains in compression, transmission and hardware complexity.

According to prior studies, the main advantage of this strategy is to eliminate the redundant samples for analysis and introduce a real-time compression gain. This is achieved by carefully incorporating the adaptive-rate acquisition and segmentation tools into the system. It is also demonstrated that the proposed approach achieves fairly high precision in distinction, which is comparable to other conventional approaches based on fixed-rate digitization and processing, while retaining a remarkable numerical advantage over them. It confirms the value of the proposed solution for the design and production of low-power cloud-based arrhythmia diagnosis systems.

The performance of the proposed method is dependent on the utilized methods of resampling, conditioning, feature mining and classification. A study of device effectiveness in terms of processing cost and precision while utilizing the higher order interpolators is an upcoming task. Another opportunity is to exploit alternative tactics of feature mining such as the round cosine transform and the wavelet packet decomposition (WPD). Certain other research opportunities are to explore the miniaturization, optimization and integrated development of the proposed solution.

## Acknowledgements

This work was funded by Effat University under grant number UC#7/28Feb 2018/10.2–44g.

## References

[1] Arunkumar K and Bhaskar M 2019 Heart rate estimation from photoplethysmography signal for wearable health monitoring devices *Biomed. Signal Process. Control* **50** 1–9

[2] Qaisar S M and Subasi A 2020 Cloud-based ECG monitoring using event-driven ECG acquisition and machine learning techniques *Phys. Eng. Sci. Med.* **43** 623–34

[3] Mian Qaisar S and Fawad Hussain S 2020 Arrhythmia diagnosis by using level-crossing ECG sampling and sub-bands features extraction for mobile healthcare *Sensors* **20** 2252

[4] Baig M M, Afifi S, Gholam Hosseini H and Mirza F 2019 A systematic review of wearable sensors and IoT-based monitoring applications for older adults—a focus on ageing population and independent living *J. Med. Syst.* **43** 233

[5] Venkatesan C, Karthigaikumar P and Satheeskumaran S 2018 Mobile cloud computing for ECG telemonitoring and real-time coronary heart disease risk detection *Biomed. Signal Process. Control* **44** 138–45

[6] Catarinucci L *et al* 2015 An IoT-aware architecture for smart healthcare systems *IEEE Internet Things J.* **2** 515–26

[7] Wang X, Zhu Y, Ha Y, Qiu M and Huang T 2016 An FPGA-based cloud system for massive ECG data analysis *IEEE Trans. Circuits Syst. II Exp. Briefs* **64** 309–13

[8] Wang X, Gui Q, Liu B, Jin Z and Chen Y 2013 Enabling smart personalized healthcare: a hybrid mobile-cloud approach for ECG telemonitoring *IEEE J. Biomed. Health Inform.* **18** 739–45

[9] Guzik P and Malik M 2016 ECG by mobile technologies *J. Electrocardiol.* **49** 894–901

[10] Serhani M A, El Menshawy M and Benharref A 2016 SME2EM: smart mobile end-to-end monitoring architecture for life-long diseases *Comput. Biol. Med.* **68** 137–54

[11] Qaisar S M 2019 Efficient mobile systems based on adaptive rate signal processing *Comput. Electr. Eng.* **79** 106462

[12] Qaisar S M and Subasi A 2018 An adaptive rate ECG acquisition and analysis for efficient diagnosis of the cardiovascular diseases *2018 IEEE 3rd Int. Conf. on Signal and Image Processing (ICSIP)* pp 177–81

[13] Subasi A, Bandic L and Qaisar S M 2020 Cloud-based health monitoring framework using smart sensors and smartphone *Innovation in Health Informatics* (Amsterdam: Elsevier) pp 217–43

[14] Mozaffarian D *et al* 2016 Heart disease and stroke statistics—2016 update *Circulation* **133** e38–e360

[15] Alickovic E and Subasi A 2015 Effect of multiscale PCA de-noising in ECG beat classification for diagnosis of cardiovascular diseases *Circuits Syst. Signal Process.* **34** 513–33

[16] Phukpattaranont P 2015 QRS detection algorithm based on the quadratic filter *Expert Syst. Appl.* **42** 4867–77

[17] Hesar H D and Mohebbi M 2016 ECG denoising using marginalized particle extended Kalman filter with an automatic particle weighting strategy *IEEE J. Biomed. Health Inform.* **21** 635–44
[18] Rodríguez R, Mexicano A, Bila J, Cervantes S and Ponce R 2015 Feature extraction of electrocardiogram signals by applying adaptive threshold and principal component analysis *J. Appl. Res. Technol.* **13** 261–9
[19] Linh T H 2018 A solution for improvement of ECG arrhythmia recognition using respiration information *Vietnam J. Sci. Technol.* **56** 335
[20] da S Luz E J, Schwartz W R, Cámara-Chávez G and Menotti D 2016 ECG-based heartbeat classification for arrhythmia detection: a survey *Comput. Methods Programs Biomed.* **127** 144–64
[21] Li P *et al* 2016 High-performance personalized heartbeat classification model for long-term ECG signal *IEEE Trans. Biomed. Eng.* 64 78–86
[22] Zhang X and Lian Y 2014 A 300-mV 220-nW event-driven ADC with real-time QRS detection for wearable ECG sensors *IEEE Trans. Biomed. Circuits Syst.* **8** 834–43
[23] Rezaii T Y, Beheshti S, Shamsi M and Eftekharifar S 2018 ECG signal compression and denoising via optimum sparsity order selection in compressed sensing framework *Biomed. Signal Process. Control* **41** 161–71
[24] Shaw L, Rahman D and Routray A 2018 Highly efficient compression algorithms for multichannel EEG *IEEE Trans. Neural Syst. Rehabil. Eng.* **26** 957–68
[25] Chen S, Hua W, Li Z, Li J and Gao X 2017 Heartbeat classification using projected and dynamic features of ECG signal *Biomed. Signal Process. Control* **31** 165–73
[26] Niederhauser T, Haeberlin A, Jesacher B, Fischer A and Tanner H 2017 Model-based delineation of non-uniformly sampled ECG signals *Computing in Cardiology (CinC) Conf.* pp 1–4
[27] Einthoven W 1903 The string galvanometer and the human electrocardiogram *Proc. KNAW* 6 107–15
[28] Muse Meditation made easy *Muse: the Brain Sensing Headband* http://choosemuse.com/
[29] Emotiv https://emotiv.com/
[30] GE Healthcare https://gehealthcare.co.uk/
[31] Salvador C H *et al* 2005 Airmed-cardio: a GSM and internet services-based system for out-of-hospital follow-up of cardiac patients *IEEE Trans. Inf. Technol. Biomed.* **9** 73–85
[32] Herscovici N *et al* 2007 m-health e-emergency systems: current status and future directions *IEEE Antennas Propag. Mag.* **49** 216–31
[33] Shih D-H, Chiang H-S, Lin B and Lin S-B 2010 An embedded mobile ECG reasoning system for elderly patients *IEEE Trans. Inf. Technol. Biomed.* **14** 854–65
[34] Ren Y, Werner R, Pazzi N and Boukerche A 2010 Monitoring patients via a secure and mobile healthcare system *IEEE Wirel. Commun.* **17** 59–65
[35] Xia H, Asif I and Zhao X 2013 Cloud-ECG for real time ECG monitoring and analysis *Comput. Methods Programs Biomed.* **110** 253–9
[36] Qaisar S M *et al* 2017 Time-domain characterization of a wireless ECG system event driven A/D converter *2017 IEEE Int. Instrumentation and Measurement Technology Conf. (I2MTC)* pp 1–6
[37] Qaisar S M, Yahiaoui R and Gharbi T 2013 An efficient signal acquisition with an adaptive rate A/D conversion *2013 IEEE Int. Conf. on Circuits and Systems (ICCAS)* pp 124–9

[38] Jin S-W, Li J-J, Li Z-N and Wang A-X 2017 A hysteresis comparator for level-crossing ADC *2017 29th Chinese Control and Decision Conf. (CCDC)* pp 7753–7
[39] Moody G B and Mark R G 2001 The impact of the MIT-BIH arrhythmia database *IEEE Eng. Med. Biol. Mag.* **20** 45–50
[40] Qaisar S M, Fesquet L and Renaudin M 2014 Adaptive rate filtering a computationally efficient signal processing approach *Signal Process.* **94** 620–30
[41] Allier E, Sicard G, Fesquet L and Renaudin M 2003 A new class of asynchronous A/D converters based on time quantization *Int. Symp. on Asynchronous Circuits and Systems. Proc.* pp 196–205
[42] Welch T B, Wright C H and Morrow M G 2016 *Real-time Digital Signal Processing from MATLAB to C with the TMS320C6x DSPs* (Boca Raton, FL: CRC Press)
[43] Qaisar S M, Akbar M, Beyrouthy T, Al-Habib W and Asmatulah M 2016 An error measurement for resampled level crossing signal *Int. Conf. on Event-based Control, Communication, and Signal Processing (EBCCSP)* pp 1–4
[44] Qaisar S M, Fesquet L and Renaudin M 2006 Spectral analysis of a signal driven sampling scheme *14th European Signal Processing Conf.* pp 1–5
[45] Cavanagh J 2017 *Computer Arithmetic and Verilog HDL Fundamentals* (Boca Raton, FL: CRC Press)
[46] Paluszek M and Thomas S 2016 *MATLAB Machine Learning* (New York: Apress)
[47] Herman R L 2016 *An Introduction to Fourier Analysis* (Boca Raton, FL: CRC Press)
[48] Hussain S F 2019 A novel robust kernel for classifying high-dimensional data using support vector machines *Expert Syst. Appl.* **131** 116–31
[49] Hussain S F and Iqbal S 2018 CCGA: co-similarity based co-clustering using genetic algorithm *Appl. Soft Comput.* **72** 30–42
[50] Hussain S F 2011 Bi-clustering gene expression data using co-similarity *Int. Conf. on Advanced Data Mining and Applications (ADMA)* pp 190–200
[51] Hussain S F and Haris M 2019 A *k*-means based co-clustering (kCC) algorithm for sparse, high dimensional data *Expert Syst. Appl.* **118** 20–34
[52] Jain S, Bajaj V and Kumar A 2018 Effective de-noising of ECG by optimised adaptive thresholding on noisy modes *IET Sci. Meas. Technol.* **12** 640–44
[53] Jain S, Bajaj V and Kumar A 2017 Riemann Liouvelle fractional integral based empirical mode decomposition for ECG denoising *IEEE J. Biomed. Health Informat.* **22** 1133–9
[54] Jain S, Bajaj V and Kumar A 2016 Efficient algorithm for classification of electrocardiogram beats based on artificial bee colony-based least-squares support vector machines classifier *Electron. Lett.* **52** 1198–200

**IOP** Publishing

Modelling and Analysis of Active Biopotential Signals in Healthcare, Volume 2

**Varun Bajaj and G R Sinha**

# Chapter 10

# Analysis of heart patients using a tree based ensemble model

**Dhyan Chandra Yadav and Saurabh Pal**

*VBS Purvanchal University, Jaunpur, India*

The number of deaths due to heart disease has increased significantly. Heart disease is very common in both men and women and now occurs not only in old people but also in young people. In today's society, heart disease is a great challenge to human life. Health departments have a variety of high-tech devices for testing the human body, from which it can be inferred whether the person may have heart disease or not. Machine learning algorithms provide help in disease prediction, classification of disease type and correlation of features. Heart diseases have various symptoms and categories, such as slow heartbeat, fluttering in one's chest, racing heartbeat etc. Machine learning algorithms provide help in classifying and understanding huge datasets. These learning algorithms categorize large problems into sub-problems at the class level in a target variable. Each class in the target variable provides information about affected and non-affected patients.

In this study we use heart disease related information from the Univeristy of California repository. The dataset contains 1025 instances of 14 features of ill and healthy patients in the target variable. In this chapter we propose and analyse the correlation coefficient, mean absolute error, root mean square error, relative absolute error and root relative square error using four tree based classification algorithms: the M5P, random tree, reduced error pruning and random forest ensemble methods. All the prediction based algorithms were applied after feature selection from a heart patient dataset. In this chapter, we use two feature based algorithms, extra tree and chi-squared, to obtain better relevant features. The data table is analysed using different feature selection methods for better prediction. All the analysis is performed using two experimental set-ups. The first experiment

doi:10.1088/978-0-7503-3411-2ch10

applies the extra tree on the M5P, random tree, reduced error pruning and random forest ensemble methods. In the second experiment we use chi-square applied on the above four tree based algorithms. After carrying out the experiments we analyse and calculate the correlation coefficient, mean absolute error (MAE), root mean square error (RMSE), relative absolute error (RAE) and root relative square error (RRSE).

Based on the results, we finally conclude that the feature selection method of extra tree with the random forest ensemble method has a higher correlation coefficient (0.99) compare to the other algorithms. After the feature selection technique (extra tree), we calculated the minimum error values as MAE = 0.002, RMSE = 0.054, RAE = 0.58 and RRSE = 10.81. Finally, we find that the random forest ensemble method predicts heart disease better compared to the other methods.

## 10.1 Introduction

Van Hooser *et al* [1] introduced about heart organ in the body and how it performs the actions which are taking place in the body. In the field of medicine, the identification of any disease becomes more challenging if the data are huge or complex. When the information about a disease is not fully available, it becomes necessary to complete any incomplete or uncertain information, which helps greatly in studying the disease. If any type of problem arises in the heart it is very harmful for the body; the heart is required to transport blood around the body to allow different organs to operate. Without a healthy heart, the system of the body cannot operate successfully.

If there is a blockage in the flow of blood in any part of the heart, it can cause a heart attack, because the heart muscles also transport oxygen along with blood. In humans, a heart attack can be both small or large. Our bodies gives us some signals before a heart attack:

- Chest pain.
- A feeling of tension.
- Difficulty sleeping at night.
- Chest heaviness or pressure.
- Increase of heart rate and increase.
- Excessive sweating.
- Abdominal pain with a burning sensation.
- Nausea and vomiting.
- Swelling of the feet.
- Skin inflammation, the appearance of rashes and a dry cough.
- Cold feet and hands.

## 10.2 Related works

Amin *et al* [2] discussed cardiovascular disease prediction for making informed decisions in a clinical setting. They used different features and classification based techniques for better prediction. The authors found a hybrid technique using naïve Bayes (NB) and logistic regression methods and achieved an 87.4% accuracy score. Mohan *et al* [3] studied heart disease prediction using hybrid machine learning techniques. They used machine learning support vector machine (SVM) and feature extraction methods for better prediction. They used hybrid random forest with a linear model and found an 88.7% accuracy level. Verma *et al* [4] studied skin disease using the passive aggressive classifier, linear discriminant analysis, radius neighbors classifier, Bernoulli naïve Bayesian and extra tree classifier machine learning techniques. They classified skin diseases using bagging, AdaBoost and gradient boosting classifiers and found a better accuracy (99.68%) using a gradient boosting ensemble method. Jayaraman and Sultana [5] recognized heart disease as a serious disease and determined features using cuckoo search algorithms. They used a neural network and found a 99.85% accuracy with minimum error rate and time complexity. Gokulnath and Shantharajah [6] identified heart disease related features using various machine learning algorithms. They used different feature selection algorithms to develop a model from a genetic algorithm with an SVM and found 84.34% classification accuracy. Ali *et al* [7] recognized preprocess techniques of heart disease dataset features and reduced the preprocessed problems with heart disease dataset. They generated a hybrid $\chi^2$–deep neural network (DNN) model that measured the performance of the DNN and artificial neural network (ANN) and determined a 93.33% classification accuracy. Narayan and Sathiyamoorthy [8] recommended a system based on the fast Fourier tranform (FFT) to identify heart disease through machine learning algorithms. They used an ANN, naïve Bayes and SVM. The authors calculated 93% accuracy using the FFT technique and machine learning algorithms. Verma *et al* [9] predicted skin disease using the bagging, AdaBoost and gradient boosting classification techniques. They used other algorithms, the extra tree classifier, radius neighbors classifier and passive aggressive classifier, to develop an ensemble model and found better results for the ensemble method. Wu *et al* [10] identified coronary artery heart disease using a classification model of machine learning and monocardiographs using deep learning with filters. The results were 91.73% sensitivity, 87.91% specificity and 89.81% accuracy. Gonsalves *et al* [11] analysed the features of coronary heart disease and discovered correlations between them. They used the machine learning technology of naïve Bayes, SVM and decision tree and found that naïve Bayes produced the highest accuracy compared to the other algorithms. Alaa *et al* [12] determined the risk of cardiovascular disease using machine learning pipelines of data input, feature processing and classification algorithms. They performed studies on 473 available variables and measured a receiver operating characteristics curve of 0.77 using true positive rate and false positive rate. Cömert *et al* [13] considered fetal heart rate and uterine contraction using an ANN, $k$-nearest neighbors ($k$NN), decision tree and SVM. The SVM achieved 77.40% sensitivity and 93.80% specificity. Le *et al* [14] predicted heart disease using a feature selection (infinite latent feature selection) method with an SVM. They selected 24 features from 46 features and achieved 97.87% accuracy for 'no presence' with presence classes and achieved 93.92% accuracy for five different classes. Haq *et al* [15] developed a hybrid intelligent system for better prediction of heart disease. The authors used $k$NN, ANN, decision tree (DT) and NB with Lasso feature selection algorithms for comparison and measured an 85% value area under curve. Paul *et al* [16] developed a fuzzy rule based system to identify the risk factors in heart disease. The authors used an adaptive weighted fuzzy system with the genetic algorithm and multiple-dipole-modeling-of-space-particle swarm optimization (PSO) and produced 89.47% accuracy, 88.57% sensitivity and 100% specificity. Vijayashree and Sultana [17] classified a large volume of heart disease data through PSO meta-heuristic algorithms with important feature contributions. The authors compared PSO + SVM using various feature selection algorithms and found 84.36% accuracy using the PSO + SVM method. Manogaran *et al* [18] modified a hybrid model using multiple kernel learning with adaptive neuro fuzzy inference for heart disease. The authors modified the cuckoo search algorithm to develop a model through a machine kernel learning and adaptive neuro fuzzy inference system. They calculated 98% sensitivity and 97% specificity using this modified method. Vivekanandan and Iyengar [19] discuss the prediction of heart disease through feature selection and a differential evolution algorithm. They used a big data fuzzy analytic hierarchy process with a feed-forward neural network and calculated 83% as the highest accuracy. Mustaqeem *et al* [20] analysed a hybrid model for heart disease using a statistical based machine learning algorithm. They measured the prediction model and evaluated the confusion matrix with 97.8% classification accuracy. Khateeb and Usman [21] predicted heart disease using the *k*NN classification technique. They identified risk factors using data mining technique: *k*NN (*k*-nearest neighbor) and calculated 80% accuracy with 14 features. Shah *et al* [22] achieved heart disease diagnosis using feature extraction through a parallel probabilities principal component. They used an SVM with a radial basis function and measured 91.30% accuracy. Ramotra *et al* [23] discussed heart disease using DT, NB, SVM, *k*NN and ANNs. They used the Weka and SPSS tools and calculated 85.87% accuracy using the SVM in SPSS.

## 10.3 Methodology

In this section we describe the details of our heart disease dataset, the applied techniques and the important relevant features.

### 10.3.1 Data description

In this study, we acquired the dataset from the UCI repository of heart disease. We indexed all the data features and found 14 features were available with 1025 instances. We read all the columns using a column list and drew a model through linear regression. The initial model transformed the data features to fit the the model as selected and we describe the feature index as follows:

Index([‘age’, ‘sex’, ‘cp’, ‘trestbps’, ‘chol’, ‘fbs’, ‘restecg’, ‘thalach’, ‘exang’,
‘oldpeak’, ‘slope’, ‘ca’, ‘thal’, ‘target’], dtype=‘object’).

We describe the whole dataset and calculate the mean, standard deviation (STD), minimum (min) and maximum (max) values, and numeric percentage values of 25%, 50% and 75% in table 10.1.

We observe the distribution of the features with the target variable in the heart disease dataset and plot it in figure 10.1.

**Table 10.1.** Summary of the features with the calculated mean, STD, min and max values.

| | age | sex | cp | trestbps | chol | fbs | restecg | thalach | exang | oldpeak | slope | ca | thal | target |
|---|---|---|---|---|---|---|---|---|---|---|---|---|---|---|
| **Count** | 1025 | 1025 | 1025 | 1025 | 1025 | 1025 | 1025 | 1025 | 1025 | 1025 | 1025 | 1025 | 1025 | 1025 |
| **Mean** | 54.4 | 0.7 | 0.9 | 131.6 | 246.0 | 0.2 | 0.5 | 149.1 | 0.3 | 1.1 | 1.4 | 0.8 | 2.3 | 0.5 |
| **STD** | 9.1 | 0.5 | 1.0 | 17.5 | 51.6 | 0.4 | 0.5 | 23.0 | 0.5 | 1.2 | 0.6 | 1.0 | 0.6 | 0.5 |
| **Min** | 29.0 | 0.0 | 0.0 | 94.0 | 126.0 | 0.0 | 0.0 | 71.0 | 0.0 | 0.0 | 0.0 | 0.0 | 0.0 | 0.0 |
| **25%** | 48.0 | 0.0 | 0.0 | 120.0 | 211.0 | 0.0 | 0.0 | 132.0 | 0.0 | 0.0 | 1.0 | 0.0 | 2.0 | 0.0 |
| **50%** | 56.0 | 1.0 | 1.0 | 130.0 | 240.0 | 0.0 | 1.0 | 152.0 | 0.0 | 0.8 | 1.0 | 0.0 | 2.0 | 1.0 |
| **75%** | 61.0 | 1.0 | 2.0 | 140.0 | 275.0 | 0.0 | 1.0 | 166.0 | 1.0 | 1.8 | 2.0 | 1.0 | 3.0 | 1.0 |
| **Max** | 77.0 | 1.0 | 3.0 | 200.0 | 564.0 | 1.0 | 2.0 | 202.0 | 1.0 | 6.2 | 2.0 | 4.0 | 3.0 | 1.0 |

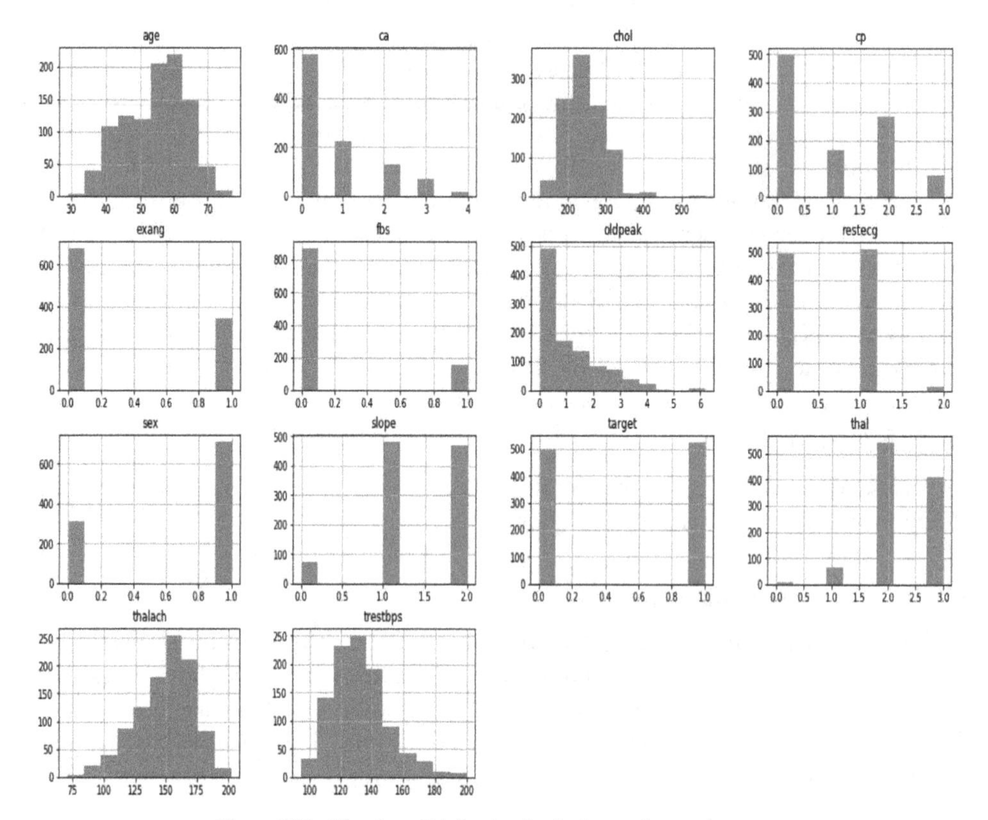


**Figure 10.1.** The data distribution in the heart disease dataset.

### 10.3.2 Algorithms

Chaurasia *et al* [24, 25] discuss machine learning techniques and their predictions. Machine learning provides various techniques for ensemble learners, predicts better results and avoids weak learners. Nandhini *et al* [26] presented various decision tree

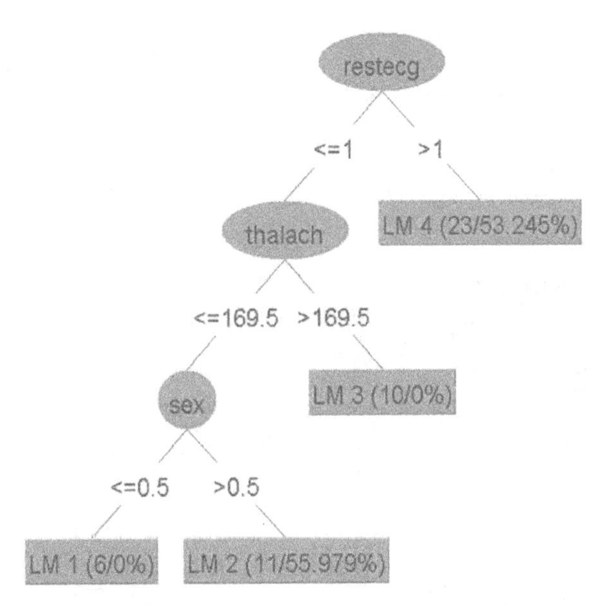


**Figure 10.2.** Representation of the M5P algorithm tree model for 100 instances of heart disease.

models of machine learning and their predictions. We used the successor model of ID3 and C4.5, the M5P tree model, random tree and reduced pruning.

*10.3.2.1 The M5P algorithm*

In machine learning, the M5P model tree is analysed as a decision tree and predicts a variable numeric response. The M5P tree model was developed by J R Quinlan in 1992. The M5P tree model has high efficiency and can handle very high dimensionality of features with expected error reduction. The M5P tree is pruned at each node of the tree during processing and improves the performance of a subtree by comparing the estimated error at each internal node. The M5P tree model estimates the error and accuracy for hidden cases of the model and computes the absolute differences between the response and predicted values (figure 10.2).

Achmad *et al* [27] introduced the standard deviation reduction (SDR) model for missing values. The tree model observes the error of the subtree through the left and right subtree. The M5P tree model evaluates the missing values as

$$\mathrm{SDR} = \frac{\mathrm{STD}(I) - \sum \mathrm{STD}(I_n) | I_n |}{| I |}, \tag{10.1}$$

where $I$ is the training set that reaches the node, $I = 1, 2, \ldots, n$ are the sets that result from splitting the node and STD is the standard deviation.

In this study, we used a heart disease dataset and created 48 clusters from LM1 to LM48. Using the Weka tool, we viewed the M5P tree graphical model and produced a large number of leaves. In this algorithmic study, we found the numeric prediction of each feature and generated rules for each level of the tree. All the predictions depend on the target variables for the 48 rules of the tree (table 10.2).

**Table 10.2.** Representation of the M5P model tree for heart disease.

<table>
<tr>
<td>cp <= 0.5 :<br>| ca <= 0.5 :<br>| | thal <= 2.5 :<br>| | | trestbps <= 117.5 : LM1 (28/0%)<br>| | | trestbps > 117.5 :<br>| | | | slope <= 1.5 :<br>| | | | | restecg <= 0.5 : LM2 (15/0%)<br>| | | | | restecg > 0.5 :<br>| | | | | | exang <= 0.5 :<br>| | | | | | | age <= 62.5 : LM3 (7/0%)<br>| | | | | | | age > 62.5 : LM4 (4/0%)<br>| | | | | | exang > 0.5 : LM5 (25/0%)<br>| | | | slope > 1.5 :<br>| | | | | chol <= 298 : LM6 (36/0%)<br>| | | | | chol > 298 : LM7 (10/0%)<br>| | thal > 2.5 :<br>| | | oldpeak <= 0.65 :<br>| | | | age <= 42 : LM8 (10/0%)<br>| | | | age > 42 :<br>| | | | | chol <= 237.5 : LM9 (15/0%)<br>| | | | | chol > 237.5 : LM10 (6/0%)<br>| | | oldpeak > 0.65 : LM11 (63/0%)<br>....... ........... ............<br>....... .......... ............<br>| | | | | | thalach <= 148 : LM37 (22/0%)<br>| | | | | | thalach > 148 :<br>| | | | | | | fbs <= 0.5 :<br>| | | | | | | | ca <= 0.5 :<br>| | | | | | | | | age <= 65.5 : LM38 (13/0%)<br>| | | | | | | | | age > 65.5 : LM39 (4/0%)<br>| | | | | | | | ca > 0.5 : LM40 (12/0%)<br>| | | | | | | fbs > 0.5 : LM41 (10/0%)<br>| | | | | oldpeak > 2 :<br>| | | | | | trestbps <= 137 : LM42 (15/0%)<br>| | | | | | trestbps > 137 : LM43 (6/0%)<br>| | | | chol > 245.5 :<br>| | | | | chol <= 279 :<br>| | | | | | chol <= 269.5 : LM44 (13/0%)<br>| | | | | | chol > 269.5 :<br>| | | | | | | trestbps <= 139 : LM45 (3/0%)<br>| | | | | | | trestbps > 139 :<br>| | | | | | | | chol <= 271.5 : LM46 (3/0%)<br>| | | | | | | | chol > 271.5 : LM47 (6/0%)<br>| | | | | chol > 279 : LM48 (24/0%)</td>
<td>LM num: 1<br>target =<br>-0.0009 * age<br>- 0.0194 * sex<br>+ 0.0033 * cp<br>- 0.0006 * trestbps<br>- 0.0003 * chol<br>- 0.0647 * restecg<br>+ 0.0001 * thalach<br>- 0.1173 * exang<br>- 0.0033 * oldpeak<br>+ 0.0995 * slope<br>- 0.0108 * ca<br>- 0.0302 * thal<br>+ 1.058<br><br>.......... ............... ...............<br>......... ............... ................<br><br>LM num: 48<br>target =<br>0.0032 * age<br>- 0.0325 * sex<br>+ 0.0031 * cp<br>- 0.0005 * trestbps<br>- 0.0014 * chol<br>+ 0.0561 * fbs<br>+ 0.0013 * restecg<br>+ 0.001 * thalach<br>- 0.0039 * exang<br>- 0.0322 * oldpeak<br>+ 0.0021 * slope<br>- 0.0312 * ca<br>- 0.0032 * thal<br>+ 0.3526<br><br>Number of Rules : 48</td>
</tr>
</table>

In the algorithms the subtree prunes the left and right children to check for error and find improvements in decision making.

The M5P tree model requires three steps:

- The model generates a regression tree from the heart disease training data.
- The tree model checks for error and deletes nodes whose demot increases error.
- The tree model reduces the tree size without loss in accuracy.

Now we can explain the M5P algorithm decision tree for heart disease features as follows:

```
1. M5' (training)
2. {
3. SD = sd (training)// standard deviation
4. for each k-valued feature convert into k – 1 binary features
5. rt = new_nd// nd represented for node
6. rt.training = training
7. split(rt)// rt represented for root
8. prune(rt)
9. }
10. split(nd)
11. {
12. If size of (nd.training) < 4 or sd (nd . training) < 0.05 + SD
13. nd . type = LEAF
14. else
15. nd . type = INTERIOR
16. for each continuous and binary feature
17. for all possible split positions
18. calculate the features' SDR
19. nd . features = feature with max SDR
20. split (nd . left)
21. split (nd . right)
22. }
23. prune (nd)
24. {
25. if nd = INTERIOR then
26. prune (nd . left_child)
27. prune (nd . right_child)
28. nd . model = linear_regression (nd)
29. if subtree_error(nd) > error (nd) then
30. nd . type = LEAF
31. }
```

*10.3.2.2 The random tree algorithm*

Yadav and Pal [28] discussed the random tree and a decision model. The random tree generated a decision node for each class with variables. In the results, we observed the random selection of heart disease in each decision node. After the examination of this algorithm, we found the tree distribution on the class level variable as a target variable (figure 10.3).

In this study we have used a heart disease dataset and created 65 levels of the tree with the help of the Weka tool, we viewed the random tree graphical model and produced a large number of leaves. For these algorithms we found a numeric prediction of each feature on each class level (table 10.3).

In this algorithm we explain the random selection of each node for the decision tree and generate a tree structure. The random algorithm performs on alphabetical class level instances and the random selection avoids weak learners and selects the strong learners for better decision tree prediction.

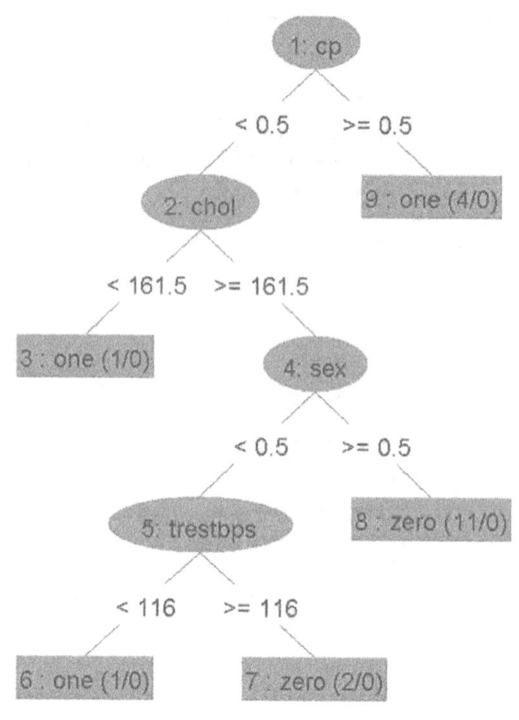


**Figure 10.3.** Representation of the random tree algorithm for 100 instances of heart disease.

1. Train $(T, Y, N)$
2. Training Set$(T)$ = $\{(y_1,t_1)\ldots\ldots.(y_n,t_n)\}$
3. Features$(Y)$ = $\{A_1,\ldots\ldots\ldots.A_k\}$.// $A$ is defined as features.
4. Result: $N$ random tree $\{x_1,\ldots\ldots\ldots\ldots\ldots.x_n\}$// $N$ is used in random selection for decision tree.
5. Begin
6. For $i$ ε$\{1,N\}$ do
7. Generate Tree Structure $(X_i, Y)$;
8. End
9. For $(y,x)$ ε $T$ do
10. For $i$ ε $\{1\ldots\ldots N\}$ do
11. Update Statistics $(X_i, (y,x))$;
12. End
13. Return $\{x_1,x_2,\ldots\ldots\ldots\ldots\ldots\ldots\ldots\ldots\ldots.x_n\}$
14. End.

*10.3.2.3 Reduced error pruning*

Otok *et al* [29] discussed reduced error pruning in a tree model. Reduced error pruning was developed by Qinlan for decision making in a tree model. This method

**Table 10.3.** Representation of the size of the random tree for heart disease.

```
RandomTree
==========

oldpeak < 1.95
|   exang < 0.5
|   |   thal < 2.5
|   |   |   ca < 0.5
|   |   |   |   chol < 325.5 : one (57/0)
|   |   |   |   chol >= 325.5
|   |   |   |   |   cp < 0.5 : zero (1/0)
|   |   |   |   |   cp >= 0.5 : one (3/0)
|   |   |   ca >= 0.5
|   |   |   |   cp < 1.5
|   |   |   |   |   slope < 1.5 : zero (1/0)
|   |   |   |   |   slope >= 1.5
|   |   |   |   |   |   trestbps < 135
|   |   |   |   |   |   |   restecg < 0.5
|   |   |   |   |   |   |   |   sex < 0.5 : one (1/0)
|   |   |   |   |   |   |   |   sex >= 0.5 : zero (3/0)
|   |   |   |   |   |   |   restecg >= 0.5 : one (3/0)
|   |   |   |   |   |   trestbps >= 135 : zero (2/0)
|   |   |   |   cp >= 1.5
|   |   |   |   |   chol < 162 : zero (1/0)
|   |   |   |   |   chol >= 162 : one (11/0)
|   |   thal >= 2.5
|   |   |   cp < 0.5 : zero (13/0)
|   |   |   cp >= 0.5
|   |   |   |   chol < 293
|   |   |   |   |   trestbps < 151
|   |   |   |   |   |   chol < 225 : one (3/0)
|   |   |   |   |   |   chol >= 225
|   |   |   |   |   |   |   cp < 1.5 : zero (1/0)
|   |   |   |   |   |   |   cp >= 1.5
|   |   |   |   |   |   |   |   age < 52 : zero (1/0)
|   |   |   |   |   |   |   |   age >= 52
|   |   |   |   |   |   |   |   |   chol < 243.5 : zero (1/0)
|   |   |   |   |   |   |   |   |   chol >= 243.5 : one (3/0)
|   |   |   |   |   trestbps >= 151 : zero (3/0)
|   |   |   |   chol >= 293 : one (7/0)
|   exang >= 0.5
|   |   slope < 1.5
|   |   |   cp < 2.5
|   |   |   |   age < 51.5
|   |   |   |   |   ca < 0.5
|   |   |   |   |   |   thalach < 134.5 : zero (2/0)
|   |   |   |   |   |   thalach >= 134.5 : one (2/0)
|   |   |   |   |   ca >= 0.5 : zero (1/0)
|   |   |   |   age >= 51.5
|   |   |   |   |   oldpeak < 0.4
|   |   |   |   |   |   age < 63.5 : zero (4/0)
|   |   |   |   |   |   age >= 63.5 : one (1/0)
|   |   |   |   |   oldpeak >= 0.4 : zero (17/0)
|   |   |   cp >= 2.5 : one (1/0)
|   |   slope >= 1.5
|   |   |   cp < 1
|   |   |   |   ca < 0.5
|   |   |   |   |   age < 52 : zero (1/0)
|   |   |   |   |   age >= 52 : one (3/0)
|   |   |   |   ca >= 0.5 : zero (8/0)
|   |   |   cp >= 1 : one (3/0)
oldpeak >= 1.95
|   chol < 212.5
|   |   ca < 0.5
|   |   |   chol < 205.5 : zero (9/0)
|   |   |   chol >= 205.5 : one (1/0)
|   |   ca >= 0.5 : zero (7/0)
|   chol >= 212.5 : zero (25/0)

Size of the tree : 65
```

analyses each decision node for pruning and removes subtrees and leaf nodes (figure 10.4):

$$\text{Cost complexity pruning (CCP)} = \frac{\text{err}(\text{Prune}(R,\, r)(S)) - \text{Prune}(R,\, D)}{|\text{ leaves}(r)| - |\text{ leaves}(\text{Prune}(R,\, r))|}, \quad (10.2)$$

where err($R$, $D$) is the error rate of tree $R$ over dataset $D$ and Prune($R$, $r$) is the tree obtained by pruning the subtrees $r$ from the tree $R$.

In this experiment, we examine each level of the node in the trees and find the error at each node. Each feature is compared for value in prediction for size and depth. We measured the size of the tree as 17 due to the class level target variable (table 10.4).

The useable trading data is divided into three parts: training, validation and testing, pruning the tree and providing an estimate of accuracy. If the new generated subtree has less error than the previous then the subtree contains no subtree with same property and the subtree is replace by a leaf node and explains the reduced error pruning as:

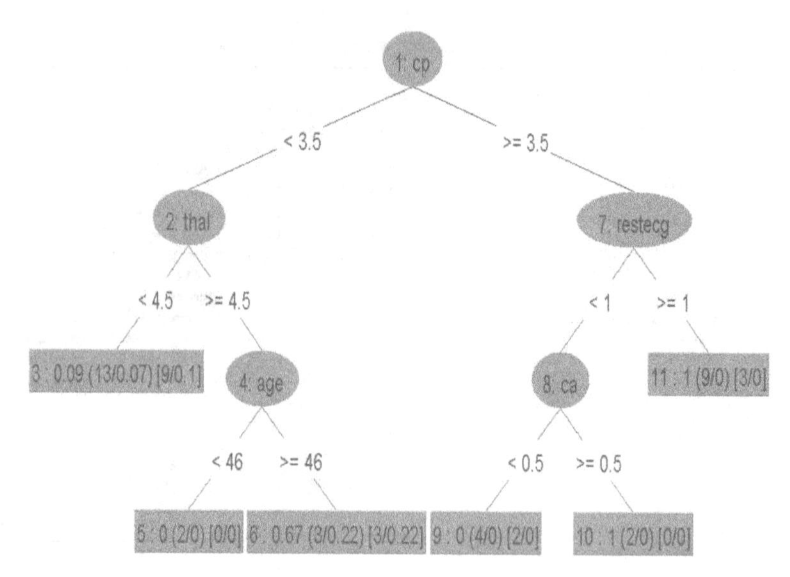


**Figure 10.4.** Representation of the reduced error pruning algorithm for 100 instances of heart disease.

**Table 10.4.** Representation of the reduced error pruned model tree for heart disease.

```
REPTree
============

thal < 4.5
|   ca < 0.5 : 0.12 (77/0.07) [42/0.17]
|   ca >= 0.5
|   |   cp < 3.5 : 0.24 (21/0.15) [8/0.27]
|   |   cp >= 3.5
|   |   |   sex < 0.5 : 0.5 (5/0.24) [1/0.36]
|   |   |   sex >= 0.5 : 1 (12/0) [2/0]
thal >= 4.5
|   oldpeak < 0.85
|   |   chol < 240.5
|   |   |   age < 41.5 : 0.67 (3/0.22) [0/0]
|   |   |   age >= 41.5
|   |   |   |   ca < 0.5 : 0.07 (9/0) [6/0.17]
|   |   |   |   ca >= 0.5 : 0.63 (4/0.25) [4/0.25]
|   |   chol >= 240.5 : 0.79 (17/0.15) [7/0.22]
|   oldpeak >= 0.85 : 0.87 (54/0.1) [31/0.14]

Size of the tree : 17
```

1. $[t_1,t_2]$ = pruning_test($k,T$)// $t_1$ as an subtree
2. Begin
3. If Leaf($L$)// find INPUT subtree without pruning
4. If Errors($L$) $\leqslant k$
5. $t_1 = 1$; // find OUTPUT with by pruning tree
6. Else $s = \infty$;
7. $t_2 = L$; // $t_2$ = the OUTPUT of pruned tree
8. return ($t_1$, $t_2$);
9. End If
10. If Errors(root($L$)) $\leqslant k$ // $k$ represents error INPUT
11. $t_1 = 1$
12. $t_2$ = root($L$)// $L$ represents Leaf node
13. return ($t_1$, $t_2$)
14. End If
15. End.

## 10.4 Formula representation

In this chapter we have used some computational formula for better prediction of heart disease. Negi *et al* discussed the correlation coefficient (CC) with error prediction [30, 31]:

$$CC_{ab} = \frac{\sum(a_i - \bar{a})(b_i - \bar{b})}{\sum(a_i - \bar{a})^2 \sum(b_i - \bar{b})^2}, \tag{10.3}$$

where $a_i$, $b_i$, $\bar{a}$ and $\bar{b}$ are variables in the CC: $\bar{a}$ is the mean of variable $a$, $\bar{b}$ is the mean of variable $b$ and $i$ is calculated from the (1, …, $n$) sample correlation in the formula.

$$\text{MAE} = 1/n \sum_{i=1}^{n} |a_i - a|, \tag{10.4}$$

where $n$ is the number of errors and $| a_i - a |$ represents the errors.

$$\text{RAE} = \frac{\left[\sum_{i=1}^{n} (V_i - A_i)^2\right]^{1/2}}{\left[\sum_{i=1}^{n} A_i^2\right]^{1/2}}, \tag{10.5}$$

where $V_i$ is the calculated predicted value and $A_i$ is the calculated actual value for ($i$ = 1, …, $n$) samples.

$$\text{RMSE} = \sqrt{1/n \sum_{i=1}^{n} (f_i - o_i)^2}, \tag{10.6}$$

where $n, f$ and $o$ represents the sample, forecast and observed values, respectively, for ($i$ = 1, …, $n$).

$$\text{RRSE} = \frac{\sum_{j=1}^{n}(V_{ij} - T_{ij})^2}{\sum_{j=1}^{n}(T_j - \bar{T})^2}, \tag{10.7}$$

where $V_{ij}$ is the value predicted by the individual program $i$ for a sample case $j$ (out of $n$ sample cases), $T_j$ is the target value for sample case $j$ and $\bar{T}$ represents a perfect fit in the sample cases:

$$\bar{T} = \sum_{j=1}^{n}(T_j - T). \tag{10.8}$$

The computational formula generated for the correlation coefficient and various errors is evaluated using statistical analysis.

## 10.5 Proposed ensemble method

The main objective of random forest is the random selection of decision nodes from trees for decision making (figure 10.5).

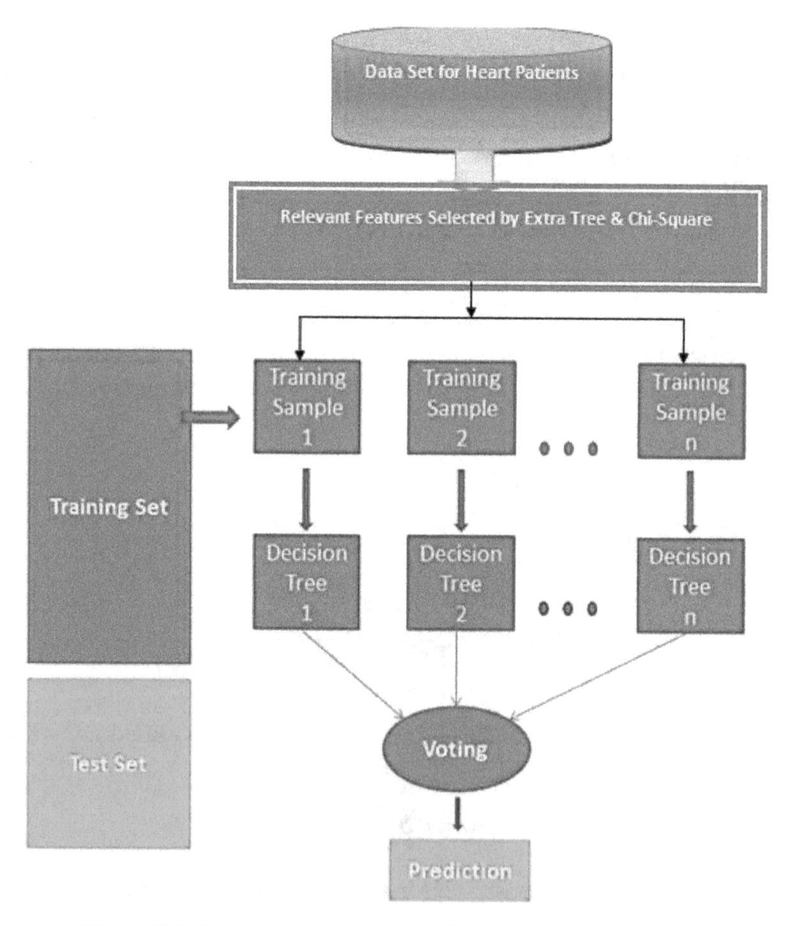


**Figure 10.5.** Proposed model of random forest as a ensemble method.

Rustam *et al* [32, 33] discussed random forest and their variable classification model. Random forest is an ensemble learning method to classify a dataset on the basis of a class variable. It provides the final decision on the data sample and finds the best prediction in the data sample. In this chapter we develop a new method as an ensemble model (random forest) using the M5P tree, random tree and error reduced pruning machine learning classifier tree algorithms. For better prediction, we selected two features techniques: extra tree and $\chi$-squared. These two feature selection based algorithms calculate the score value for relevant features and, after feature reduction, machine learning algorithms are applied for training (75%) and testing (25%) the heart disease dataset with the random forest tree based ensemble model. The voting algorithm provides help in making the final decision from the predicted results.

## 10.6 Results

In this study we have calculated the heart disease features through the extra tree feature importing and $\chi$-squared features techniques. We have found various results for different features, as shown in figure 10.6.

We created a model using the extra tree classifier and achieved all the important features in the column with the numeric importance values calculated as

[0.069 769 12 0.057 035 77 0.117 752 02 0.057 954 54
0.051 301 98 0.024 002 69 0.034 118 92 0.092 136
02 0.128 102 73 0.073 951 05 0.075 342 26 0.117 999 04 0.100 533 86]

For the above results, it is clear that two features (0.024 002 69, 0.034 118 92) have lower scores (>0.05) and the other eleven features calculated high relevant value as compared with others. The target variable is last features as a dependable variable. We find that the target variable already has a high score from the previous observation.

The $\chi$-squared feature selection technique produced results as shown in table 10.5.

Only two features (restecg: 9.739 343, fbs: 1.477 550) have a lower score (>15), while all the other features measured high relevant features value (figure 10.7).

In the $\chi$-squared feature selection technique, we find more relevant features through the best features and select the $k$ best scores and fit the best feature data frame by plotting the features. $\chi$-squared values present all the important features in a column with the important values presented in figure 10.7 and tables 10.6–10.8.

These two features selection techniques are applied on the machine learning algorithms and the results are as follows.

The random forest ensemble model calculated a high correlation coefficient (0.99) compare to the other algorithms. After the feature selection technique (extra tree), we calculated the minimum error values as MAE = 0.002, RMSE = 0.054, RAE = 0.58 and RRSE = 10.81. The algorithm M5P obtained the following results: correlation coefficient = 0.93, MAE = 0.14, RMSE = 0.23, RAE = 29.54 and RRSE = 36.24. The random tree produced the following results: correlation coefficient = 0.95, MAE = 0.03, RMSE = 0.05, RAE = 07.11 and

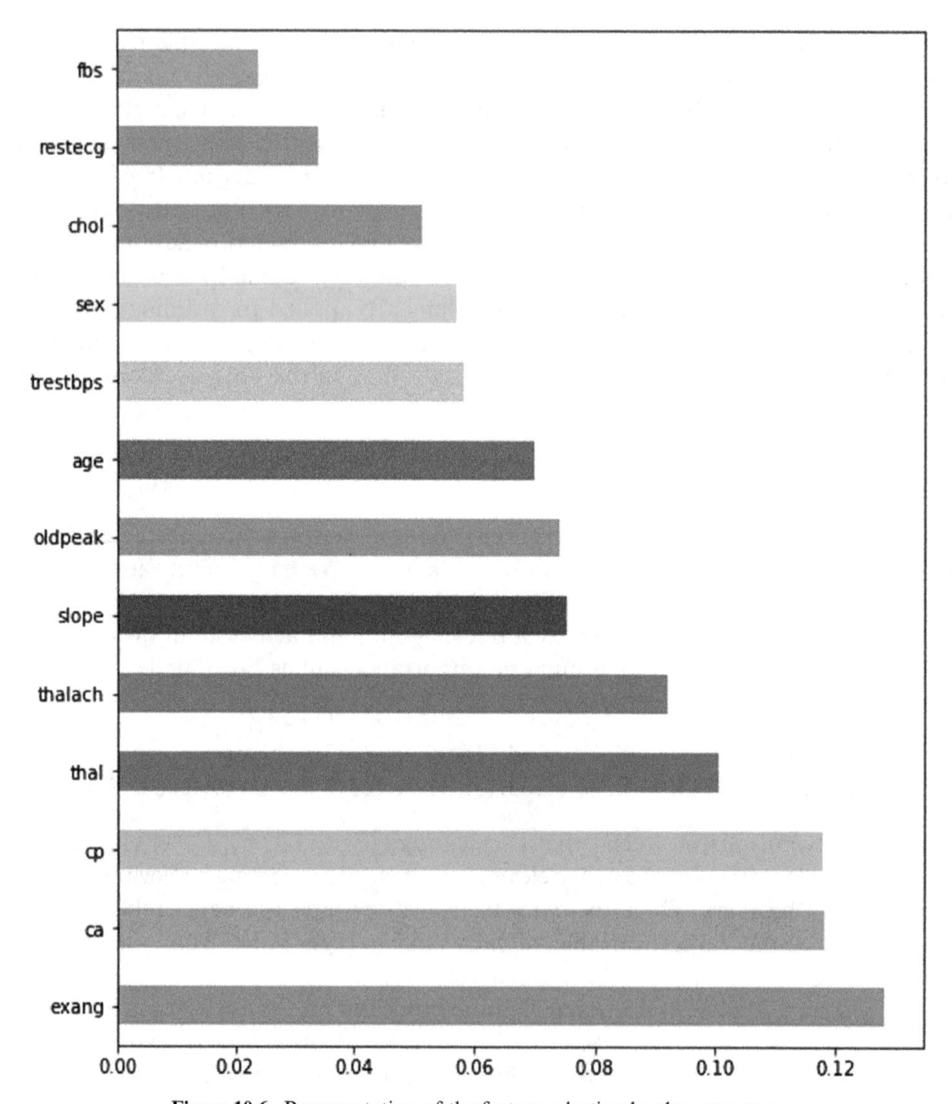


**Figure 10.6.** Representation of the feature selection by the extra tree.

RRSE = 11.43. The reduced error pruning tree calculated results as follows: correlation coefficient = 0.96, MAE = 0.09, RMSE = 0.24, RAE = 19.70 and RRSE = 38.86.

## 10.7 Discussion

Liu *et al* [34] discussed the relationship between variables using a correlation coefficient. In this chapter, we used the correlation coefficient for the numeric measurement of the statistical relation between variables. We also measured the multivariate random variables as a variable distribution. We can easily test our approach on a heart disease data sample. The numeric values of a correlation

**Table 10.5.** Results of the $\chi$-squared feature selection calculation.

| | Specs | Score |
|---|---|---|
| 7 | thalach | 650.008 493 |
| 13 | target | 499.000 000 |
| 9 | oldpeak | 253.653 461 |
| 2 | cp | 217.823 922 |
| 11 | ca | 210.625 919 |
| 8 | exang | 130.470 927 |
| 4 | chol | 110.723 364 |
| 0 | age | 81.425 368 |
| 3 | trestbps | 45.974 069 |
| 10 | slope | 33.673 948 |
| 1 | sex | 24.373 650 |
| 12 | thal | 19.373 465 |
| 6 | **restecg** | **9.739 343** |
| 5 | **fbs** | **1.477 550** |

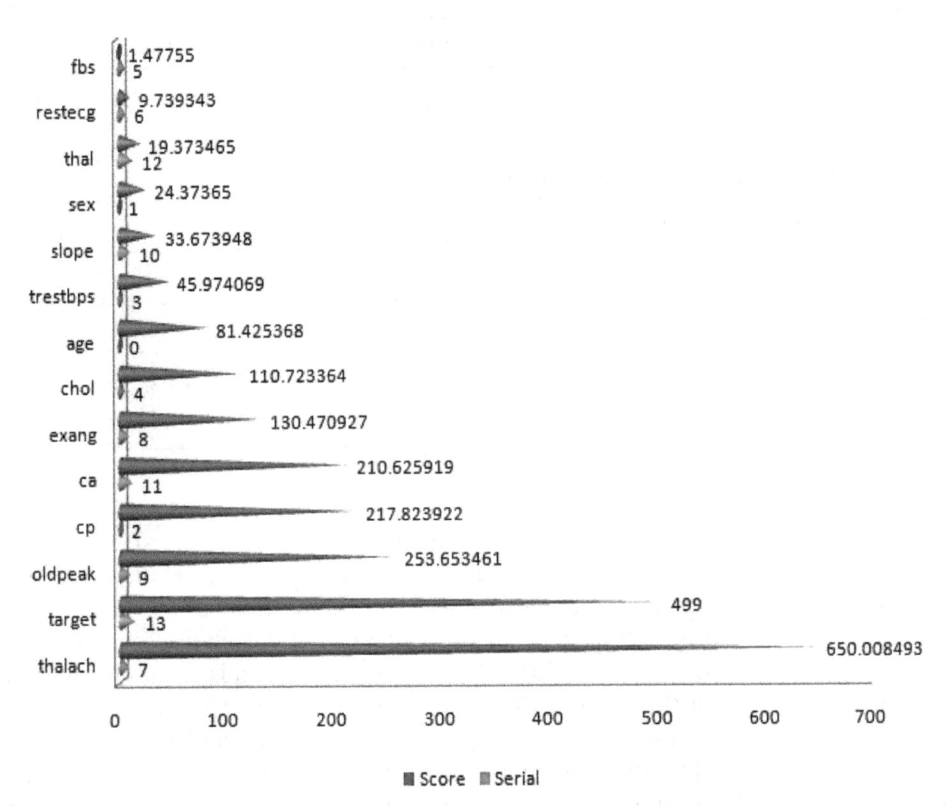


**Figure 10.7.** The $\chi$-squared graph plot of score calculation.

**Table 10.6.** Results for the extra tree feature selection computational model.

| Extra tree feature selection method | | | | | |
|---|---|---|---|---|---|
| Algorithms | Correlation coefficient | MAE | RMSE | RAE | RRSE |
| M5P | 0.93 | 0.14 | 0.23 | 29.54 | 36.24 |
| RT | 0.95 | 0.03 | 0.05 | 7.11 | 11.43 |
| REPT | 0.96 | 0.09 | 0.24 | 19.70 | 38.86 |
| RFT | **0.99** | **0.002** | **0.054** | **0.58** | **10.81** |

**Table 10.7.** Results for the $\chi$-squared feature selection computational model.

| $\chi$-squared feature selection method | | | | | |
|---|---|---|---|---|---|
| Algorithms | Correlation coefficient | MAE | RMSE | RAE | RRSE |
| M5P | 0.80 | 0.26 | 0.35 | 31.71 | 37.78 |
| RT | 0.86 | 0.15 | 0.17 | 11.34 | 17.57 |
| RFPT | 0.85 | 0.21 | 0.36 | 21.91 | 33.97 |
| RFT | **0.98** | **0.014** | **0.06** | **3.63** | **16.91** |

**Table 10.8.** Results for the computational model without feature selection.

| Without feature selection method | | | | | |
|---|---|---|---|---|---|
| Algorithms | Correlation coefficient | MAE | RMSE | RAE | RRSE |
| M5P | 0.89 | 0.37 | 0.46 | 34.17 | 35.47 |
| RT | 0.91 | 0.26 | 0.27 | 12.37 | 17.81 |
| REPT | 0.91 | 0.29 | 0.47 | 24.29 | 26.97 |
| RFT | 0.93 | 0.027 | 0.071 | 7.14 | 15.94 |

coefficient fall between −1 to +1. If the numeric value move toward +1 then it reflects a strong relationship between features, otherwise (closer to −1) there will be a weak relationship between the features and 0 shows no strength of relationship. Within our results, we find that the numeric value of the correlation coefficient is always high (it tends to 1). The random forest ensemble method is calculated as 0.99 and all the other algorithms, M5P, RT and REP, have lower correlation values compared to the random forest ensemble method (figures 10.8 and 10.9).

Kuthirummal *et al* [35] discussed accurate scale measurements. In this analysis, we found the mean absolute error measured between the same phenomena. The MAE measured how much error there was between the predictions and actual values. It analysed the difference between accurate scale measurements as predicted

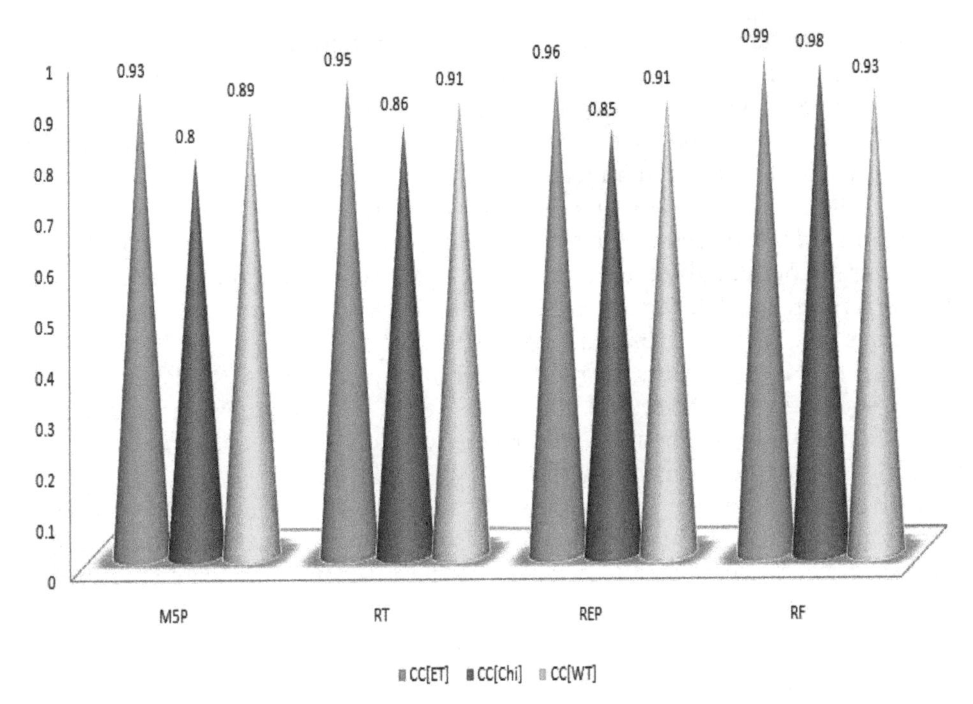


**Figure 10.8.** The performance of the correlation coefficient using M5P, RT, REP and RF.

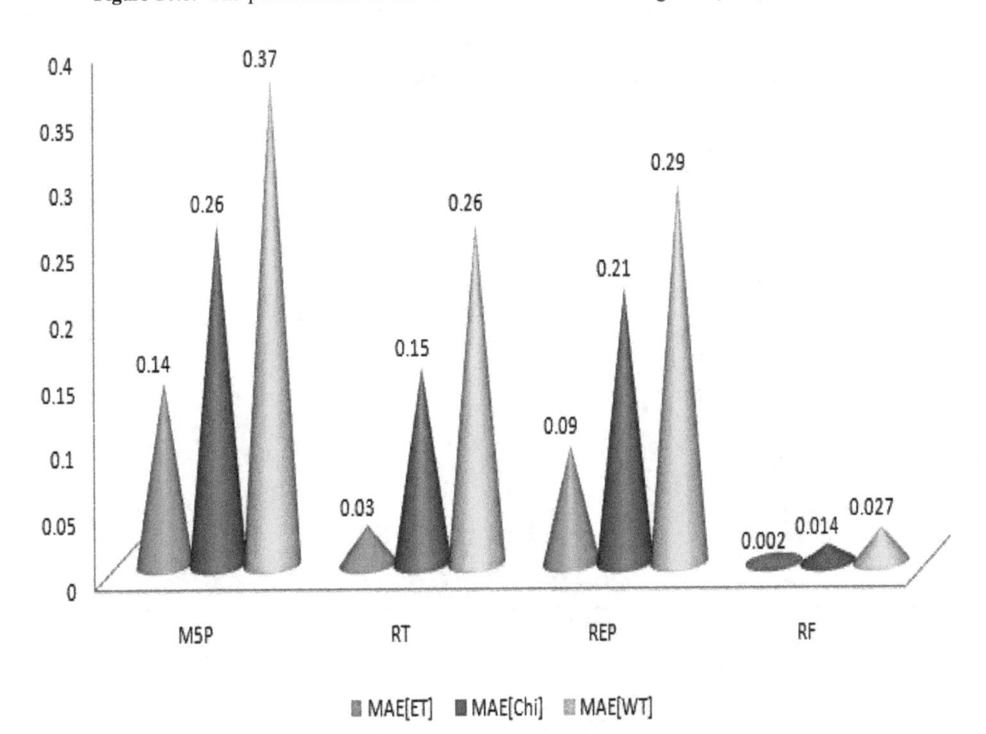


**Figure 10.9.** The performance of MAE using M5P, RT, REP and RF.

values–actual values. In this chapter, we find the lower error using the random forest ensemble method in each iteration compared to all the other methods (figure 10.10).

In this analysis, we find the error measurement as the square root of two sample moments between the predicted numeric values and the observed values. In this chapter we find that the random forest ensemble model with the extra tree produced less error (0.05) compared to the other algorithms (figure 10.11).

Kumar *et al* [36] discussed the ratio of total absolute error. The relative absolute error is a ratio of total absolute error. In this study we found that the numeric values of the error were much lower using the random forest ensemble method but the other algorithms had high relative error (figure 10.12).

Ashrafian *et al* [37] discuss the root mean prior squared error (RMPSE). The root relative square error is the ratio RMSE/RMPSE. In this chapter we calculated 10.81 as the numeric value of the random forest ensemble with the extra tree and all the other algorithms had a high error rate.

On the basis of our previous analysis [38], we calculated a high accuracy using an ensemble method and majority voting. In this chapter we calculated a high correlation coefficient and low error using the random forest ensemble method with majority voting.

## 10.8 Conclusion

In this chapter we have examined two important techniques: extra tree and $\chi$-squared. These techniques selected highly relevant important features from a heart disease dataset. This dataset was taken from the UCI repository. There were

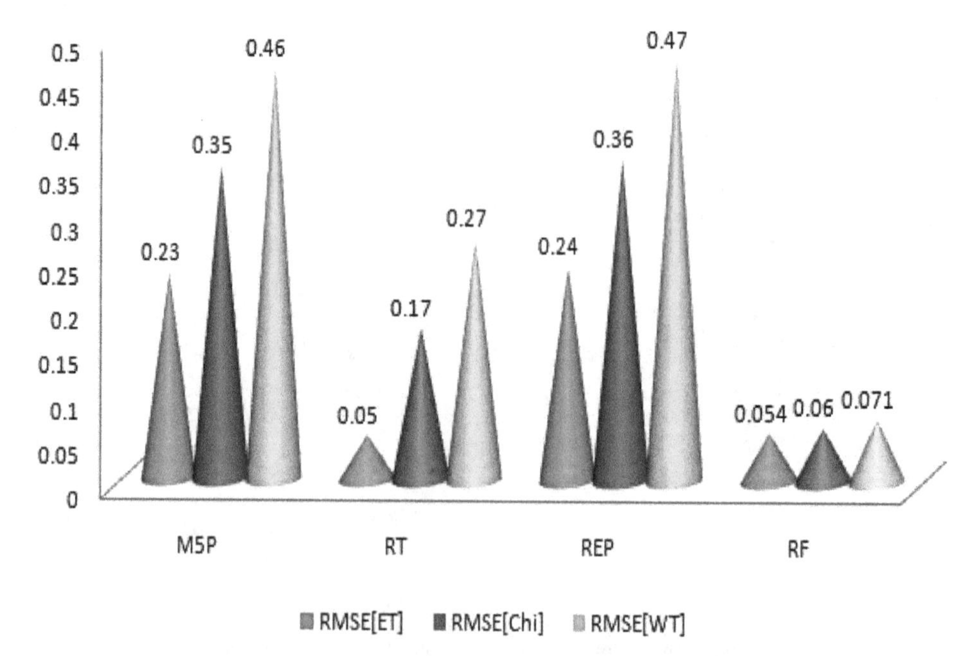


**Figure 10.10.** The performance of RMSE using M5P, RT, REP snd RF.

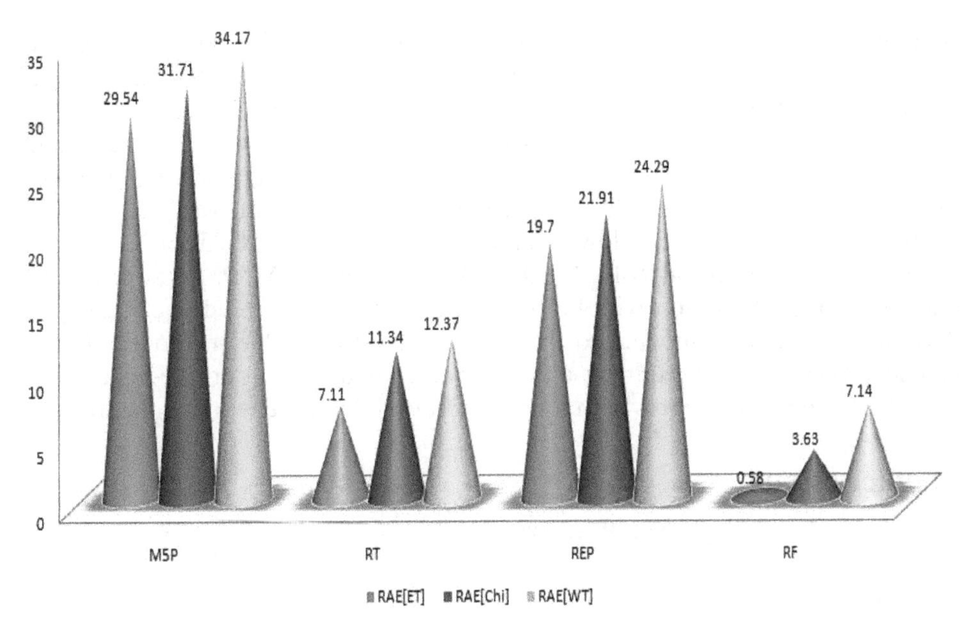


**Figure 10.11.** The performance of RAE using M5P, RT, REP and RF.

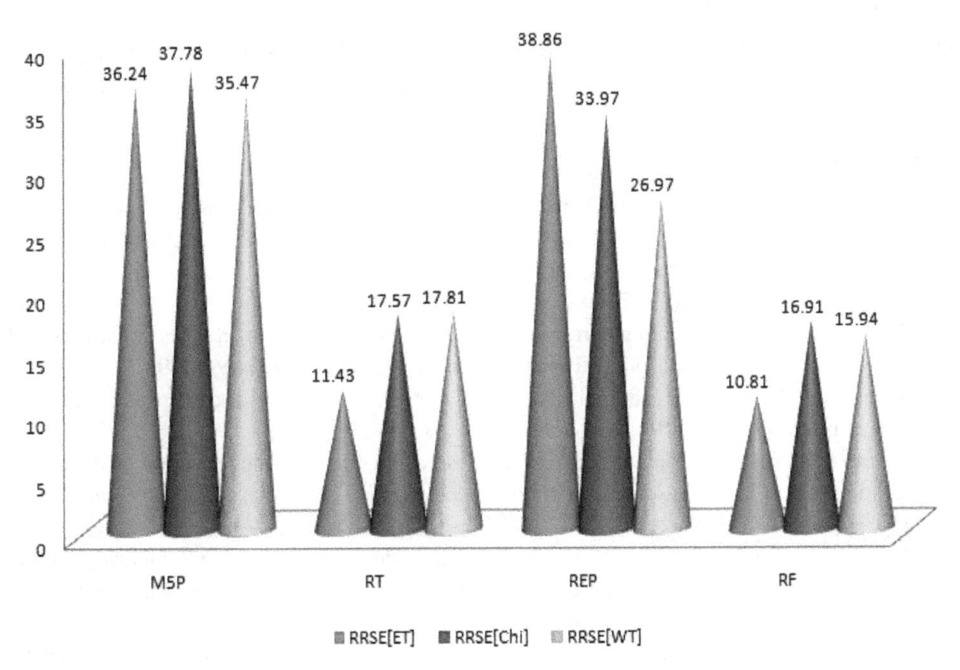


**Figure 10.12.** The performance RRSE using M5P, RT, REP and RF.

14 total calculated available features with 1025 instances. We applied the tree based algorithms M5P, random tree, reduced error pruning and random forest ensemble for better decision making on heart disease. In this analysis, we provide help in determining the severity of heart disease and the relevant features. The random forest ensemble method proved more effective in identifying disease with high calculated numeric values (0.99) of the correlation coefficient with the extra tree important features and found lower values of 0.002 for MAE, 0.05 for RMSE, 0.58 for RAE and 10.81 for RRSE. All the other algorithms, M5P, random tree and reduced error pruning tree, with both feature techniques provided good results but these were not high compared to the random forest ensemble methods. In future planned work, we will examine a huge dataset and make our prediction using machine learning algorithms with a neural network hybrid ensemble method.

## Conflict of Interest

The authors have no conflict of interest.

## Funding

This study was not funded.

## Acknowledgements

The author is grateful to Veer Bahadur Singh Purvanchal University Jaunpur, Uttar Pradesh, for providing financial support to work as a postdoctoral research fellow.

## References

[1] Van Hooser J C, Rouse K L, Meyer M L, Siegler A M, Fruehauf B M, Ballance E H, Solberg S M, Dibble M J and Lutfiyya M N 2020 Knowledge of heart attack and stroke symptoms among US Native American adults: a cross-sectional population-based study analyzing a multi-year BRFSS database *BMC Public Health* **20** 40

[2] Amin M S, Chiam Y K and Varathan K D 2019 Identification of significant features and data mining techniques in predicting heart disease *Telemat. Inform.* **36** 82–93

[3] Mohan S, Thirumalai C and Srivastava G 2019 Effective heart disease prediction using hybrid machine learning techniques *IEEE Access* **7** 81542–54

[4] Verma A K, Pal S and Kumar S 2020 Prediction of skin disease using ensemble data mining techniques and feature selection method—a comparative study *Appl. Biochem. Biotechnol.* **190** 341–59

[5] Jayaraman V and Sultana H P 2019 Artificial gravitational cuckoo search algorithm along with particle bee optimized associative memory neural network for feature selection in heart disease classification *J. Ambient Intell. Human. Comput.*

[6] Gokulnath C B and Shantharajah S P 2019 An optimized feature selection based on genetic approach and support vector machine for heart disease *Cluster Comput.* **22** 14777–87

[7] Ali L, Rahman A, Khan A, Zhou M, Javeed A and Khan J A 2019 An automated diagnostic system for heart disease prediction based on $\chi^2$ statistical model and optimally configured deep neural network *IEEE Access* **7** 34938–45

[8] Narayan S and Sathiyamoorthy E 2019 A novel recommender system based on FFT with machine learning for predicting and identifying heart diseases *Neural Comput. Appl.* **31** 93–102

[9] Verma A K, Pal S and Kumar S 2019 Comparison of skin disease prediction by feature selection using ensemble data mining techniques *Inform. Med. Unlocked* **16** 100202

[10] Wu J M T, Tsai M H, Huang Y Z, Islam S H, Hassan M M, Alelaiwi A and Fortino G 2019 Applying an ensemble convolutional neural network with Savitzky–Golay filter to construct a phonocardiogram prediction model *Appl. Soft Comput.* **78** 29–40

[11] Gonsalves A H, Thabtah F, Mohammad R M A and Singh G 2019 Prediction of coronary heart disease using machine learning: an experimental analysis *Proc. of the 2019 3rd Int. Conf. on Deep Learning Technologies* pp 51–6

[12] Alaa A M, Bolton T, Di Angelantonio E, Rudd J H and van Der Schaar M 2019 Cardiovascular disease risk prediction using automated machine learning: a prospective study of 423,604 UK Biobank participants *PLoS One* **14** e0213653

[13] Cömert Z, Şengür A, Budak Ü and Kocamaz A F 2019 Prediction of intrapartum fetal hypoxia considering feature selection algorithms and machine learning models *Health Inform. Sci. Syst.* 7 17

[14] Le H M, Tran T D and Van Tran L A N G 2018 Automatic heart disease prediction using feature selection and data mining technique *J. Comput. Sci. Cybernet.* **34** 33–48

[15] Haq A U, Li J P, Memon M H, Nazir S and Sun R 2018 A hybrid intelligent system framework for the prediction of heart disease using machine learning algorithms *Mobile Inform. Syst.* **2018** 3860146

[16] Paul A K, Shill P C, Rabin M R I and Murase K 2018 Adaptive weighted fuzzy rule-based system for the risk level assessment of heart disease *Appl. Intell.* **48** 1739–56

[17] Vijayashree J and Sultana H P 2018 A machine learning framework for feature selection in heart disease classification using improved particle swarm optimization with support vector machine classifier *Prog. Comput. Software* **44** 388–97

[18] Manogaran G, Varatharajan R and Priyan M K 2018 Hybrid recommendation system for heart disease diagnosis based on multiple kernel learning with adaptive neuro-fuzzy inference system *Multimedia Tools Appl.* **77** 4379–99

[19] Vivekanandan T and Iyengar N C S N 2017 Optimal feature selection using a modified differential evolution algorithm and its effectiveness for prediction of heart disease *Comput. Biol. Med.* **90** 125–36

[20] Mustaqeem A, Anwar S M, Khan A R and Majid M 2017 A statistical analysis based recommender model for heart disease patients *Int. J. Med. Informat.* **108** 134–45

[21] Khateeb N and Usman M 2017 Efficient heart disease prediction system using *k*-nearest neighbor classification technique *Proc. of the Int. Conf. on Big Data and Internet of Things* pp 21–6

[22] Shah S M S, Batool S, Khan I, Ashraf M U, Abbas S H and Hussain S A 2017 Feature extraction through parallel probabilistic principal component analysis for heart disease diagnosis *Physica* A **482** 796–807

[23] Ramotra A K, Mahajan A, Kumar R and Mansotra V 2020 Comparative analysis of data mining classification techniques for prediction of heart disease using the Weka and SPSS modeler tools *Smart Trends in Computing and Communications* (Singapore: Springer) pp 89–97

[24] Chaurasia V and Pal S 2020 Covid-19 pandemic: application of machine learning time series analysis for prediction of human future *Res. Biomed. Eng.* https://doi.org/10.1007/s42600-020-00105-4

[25] Yadav D C and Pal S 2020 Prediction of thyroid disease using decision tree ensemble method *Human-Intell. Syst. Integr.* https://doi.org/10.1007/s42454-020-00006-y

[26] Nandhini C U and Tamilselvi P R 2020 Hybrid framework of ID3 with multivariate attribute selection for heart disease analysis *Materials Today: Proc.* https://doi.org/10.1016/j.matpr.2020.06.254

[27] Achmad H, Tanumihardja M, Sartini R S, Singgih M F, Ramadhany Y F and Mutmainnah N 2020 Chewable lozenges using white shrimp waste (*Litopenaeus vannamei*) in reduce colonization of bacteria *Streptococcus mutans* in the case of early childhood caries *System. Rev. Pharm.* **11** 293–9

[28] Yadav D C and Pal S 2020 Discovery of hidden pattern in thyroid disease by machine learning algorithms *Indian J. Public Health Res. Develop.* **11** 61

[29] Otok B W, Musa M and Yasmirullah S D P 2020 Propensity score stratification using bootstrap aggregating classification trees analysis *Heliyon* **6** e04288

[30] Negi H S and Kanda N 2020 An appraisal of spatio-temporal characteristics of temperature and precipitation using gridded datasets over NW-Himalaya *Climate Change and the White World* (Cham: Springer) pp 219–38

[31] Yadav D C and Pal S 2019 To generate an ensemble model for women thyroid prediction using data mining techniques *Asian Pacific J. Cancer Prevent. APJCP* **20** 1275

[32] Rustam Z and Saragih G S 2020 Prediction schizophrenia using random forest *Telkomnika* **18** 1433–8

[33] Yadav D C and Pal S 2020 Prediction of heart disease using feature selection and random forest ensemble method *Int. J. Pharm. Res.* **12** 56–66

[34] Liu Y, Mu Y, Chen K, Li Y and Guo J 2020 Daily activity feature selection in smart homes based on Pearson correlation coefficient *Neural Process. Lett.* **51** 177–87

[35] Kuthirummal N, Vanathi M, Mukhija R, Gupta N, Meel R, Saxena R and Tandon R 2020 Evaluation of Barrett universal II formula for intraocular lens power calculation in Asian Indian population *Indian J. Ophthalmol.* **68** 59

[36] Kumar V, Mishra B K, Mazzara M, Thanh D N and Verma A 2020 Prediction of malignant and benign breast cancer: a data mining approach in healthcare applications *Advances in Data Science and Management* (Singapore: Springer) pp 435–42

[37] Ashrafian A, Gandomi A H, Rezaie-Balf M and Emadi M 2020 An evolutionary approach to formulate the compressive strength of roller compacted concrete pavement *Measurement* **152** 107309

[38] Yadav D C and Pal S 2019 Thyroid prediction using ensemble data mining techniques *Int. J. Inf. Technol.* https://doi.org/10.1007/s41870-019-00395-7

**IOP** Publishing

Modelling and Analysis of Active Biopotential Signals in Healthcare, Volume 2

**Varun Bajaj and G R Sinha**

# Chapter 11

# Heartbeat classification using parametric and time–frequency methods

**Abdulhamit Subasi and Saeed Mian Qaisar**

*Effat University, College of Engineering, Jeddah, 21478, Saudi Arabia*

The heart is a muscle which beats in a rhythmic way, pumping blood through the whole body. This contraction begins at the sinoatrial node, which functions as a natural pacemaker, and propagates through the rest of the muscle. The propagation of the electrical signal follows a pattern. As a consequence of this operation electrical currents are produced on the body surface, causing variations in the skin surface's electrical potential. With the use of electrodes and suitable equipment these signals can be detected and measured. Electrocardiogram (ECG) signals describe the human heart's electrical mobility. Computer-aided diagnosis helps cardiologists identify, classify and diagnose conditions using ECG. This chapter presents an easy and accurate approach for classifying heartbeats from ECG signals. For automated diagnosis, valuable features and well-designed classification systems are essential. Consequently, several frequency and time–frequency based feature extraction methods are used to extract features and identify heartbeats to enhance ECG beat classification performance. The main problems faced in automated heartbeat classification are the large variations between different patients and the high cost of labeling clinical ECG data. Heartbeat classification aims to correctly identify the ECG signal category by evaluating and extracting features and carrying out dimension reduction for efficiency, as well as predicting possible pathological or functional events using machine learning techniques. The purpose of this chapter is to present the steps of heartbeat classification using ECG signals.

doi:10.1088/978-0-7503-3411-2ch11  

## 11.1 Introduction

One of the leading cause of mortality is cardiovascular disease (CVD). It was responsible for 31% of global deaths in 2016, 85% of which were caused by heart attack. The estimated cost of CVD to the European and American economies is calculated as €210 billion and $555 billion, respectively [1]. The conventional approach to diagnosing CVD is based on the medical history and clinical evaluation of individual patients. Such findings are defined according to a set of objective scientific criteria for classifying patients on the basis of medical disease taxonomy. For certain cases, due to working with large volumes of heterogeneous data, the current rule-based diagnostic approach is ineffective and requires extensive analysis and medical knowledge to achieve adequate diagnostic accuracy. In places where there is a shortage of medical expertise and clinical equipment, particularly in developing countries, the problem is more pronounced. This motivates the need for a reliable, automated, and low-cost monitoring and diagnostic system. Healthcare providers are increasingly demanding this provision, so that accurate medical tests can be related to the use of computer-aided diagnosis (CAD). CAD consists of automated monitoring of the health conditions on the basis of analysing physiological signals to track and assess the functioning of the corresponding organ. CAD offers compact and simple tools to educate individuals on their illnesses [2].

CVD has high morbidity, particularly for people over the age of fifty. Arrhythmia is a large category of CVDs. Arrhythmias are anomalous electrical behaviors in the heartbeat cycle, e.g. irregular electric signal source and abnormal electrical signal conduction. While most arrhythmias are non-lifethreatening, serious cardiovascular complications and sudden death are associated with some arrhythmias. The early diagnosis of arrhythmia thus helps to avoid premature death and to treat cardiovascular disease. Experts usually treat arrhythmias based on electrocardiography (ECG). The ECG is obtained by electrodes mounted on the chest and limb surfaces. During the heartbeat cycle, the ECG will monitor changes in electrical potential. Diagnosis based on ECG is, however, a difficult and tedious task for doctors. Such arrhythmias are also unpredictable and they may not occur within a short window of time. Thus a CAD system is needed [3].

Arrhythmia is a disruption in the heartrate in terms of regularity, place of origin or conduction of electrical impulses. ECG diagnosis of arrhythmia has attracted considerable interest from cardiologists. Many algorithms have been studied extensively to increase the performance of arrhythmia diagnosis [4]. Cardiac diseases predominate among the various diseases that affect human beings. Heart disease may affect a person regardless of age. Arrhythmia is an irregular heartbeat, and there are several forms of cardiac arrhythmia, which range from normal to fatal. Early detection of cardiac arrhythmia can save a life. ECG is utilized to determine cardiac conditions and to predict arrhythmias. The interpretation of the ECG remains the greatest obstacle today, because there is no ideal solution for accurate prediction. Today researchers are designing many algorithms and models to assist in this, but research is still ongoing to discover an appropriate solution for clinical settings in real time. A comprehensive analysis was performed on the existing methods for minimizing the effects of noise [5–7], feature extraction and classification of ECG signals. The introduction of novel analysis techniques and also the changes made by researchers to current methods are identified by comparing the output indicators obtained. The advances in technology inspired researchers to apply them in the study of arrhythmia to reduce the load on doctors and to improve the life expectancy of cardiac patients [8].

Cardiac arrhythmias present severe threats to patients. Many types of arrhythmias endanger life and can lead to cardiac arrest and sudden death. Consequently, the effective automated identification and recognition of ECG patterns is crucial for the diagnosis and treatment of patients with lifethreatening cardiac arrhythmias. An appropriate and non-invasive technique for the detection of these disorders is to analyse ECGs which provide valuable electrophysiological details about the functionality of cardiovascular systems. In recent decades, the computerized identification of ECGs has become a mature procedure which helps cardiologists to categorize long-term ECG records [9].

Arrhythmias are usually not dangerous, but may also lead to congestive heart failure or death. Detecting and diagnosing arrhythmia by using ECG signal examination is very time-consuming and tedious. Hence, automated arrhythmia diagnosis is critical for detection systems. The ECG signal is typically contaminated with many types of interference which changes its characteristics, making significant features difficult to extract and affecting the performance of classification. The preprocessing phase eliminates the various types of noise related to the signal. After the preprocessing phase, heartbeat segmentation is an important stage in generating a series of different heartbeats. R-peaks are then observed and segmentation is performed. ECG heartbeats include various types of time-domain, frequency-domain and cardiac rhythm morphology characteristics. Feature extraction methods include statistical approaches, frequency-domain analysis, time-domain analysis, time–frequency domain analysis and hybrid techniques [10].

An irregularity in the rates or rhythms of the heartbeat is called arrhythmia, where the heart can beat too slowly, too quickly or at an irregular pace. Many arrhythmias are not harmful, but others may be lifethreatening or worse. The heart cannot pump enough blood into the body in an arrhythmia. Having a loss of blood flow can affect the heart, brain and other organs. In certain cases, arrhythmias only occur suddenly in a patient's regular activities. Usually, a Holter device is used to capture a long-term ECG signal to identify these unusual occurrences. Consequently, the automated identification of abnormal heartbeats from a large volume of ECG data is crucial. Furthermore, beat-by-beat manual examination is time-consuming and tedious in reviewing long-term ECG data, particularly in the case of wearable healthcare monitoring or bedside monitoring, in which real-time assessment is a challenging activity for medical professionals. Hence, physicians usually use computer-aided software to determine and analyse the ECG signals. Additionally, technologies are required to manage and evaluate these vast and complicated datasets in a reasonable time and suitable storage space. Big data analytics, a concept applied to large and complicated data collections, plays a vital role in handling the vast volume of healthcare data and enhancing the level of healthcare facilities available to patients. One of the difficulties in this framework is the classification of ECG signals, which is focused on efficiently distributed computing systems and scientific methods such as artificial intelligence and machine learning [11]. This chapter concentrates on the automated identification of ECG heartbeats using various techniques of frequency, time–frequency and machine learning.

## 11.2 Background and literature review

An effective method of treating and avoiding cardiac failure and arrhythmia is the electrocardiogram (ECG). Several of the approaches proposed in the literature include conditioning, segmentation of heartbeats, extraction of features, reduction of dimensions and classification. Romdhane *et al* [12] proposed a form of deep learning based on a convolutionary neural network (CNN) model. For the classification perspective, CNN models can perform the extraction of features automatically and in tandem. The suggested solution is based on a segmentation algorithm for the heartbeat that is distinct from other existing approaches. The model is trained and validated using the INCART and MIT-BIH datasets to identify five categories of arrhythmias. The assessment findings revealed the classification success for the majority of groups. The proposed method achieved a total accuracy of 98.41%, a total F1-score of 98.38%, an overall accuracy of 98.37% and an overall recall of 98.41%.

ECG is a beneficial tool in the diagnosis of heart disease. Automated ECG diagnosis helps in the monitoring of the heart using small devices, in particular wearables. To automatically detect arrhythmias, Shi *et al* [13] proposed an objective method of classifying heartbeats using ECG. A hierarchical classification approach is utilized, based on weighted extreme gradient boosting (XGBoost). The preprocessed heartbeats derive a large number of features from six groups. Then, the recursive deletion of features is used to select features. A hierarchical classifier is created in the classification phase. The achieved sensitivities for normal (N), ventricular (V), supraventricular (S) and unknown beats (Q) were 92.1%, 95.1%, 91.7% and 61.6%, respectively.

Arrhythmia can be categorized into several classes, including lifethreatening and non-lifethreatening. Precise identification of the categories of arrhythmias can help avoid heart failure and reduce mortality. Li *et al* [14] proposed an algorithm consisting of four residual stages, each consisting of three layers of 1D convolution, three layers of batch normalization (BP), three layers of rectified linear unit (ReLU) and a structure of 'identity shortcut links'. With single-lead ECG heartbeats, they obtained an average precision, sensitivity and positive predictivity of 99.06%, 93.21% and 96.76%, respectively. The results revealed that the deep ResNet model has high classification efficiency in the two-lead datasets, achieving 99.38% accuracy, 94.54% sensitivity and 98.14% specificity.

An automated method for classifying heartbeats using ECG is critical to assist doctors and experts in diagnosing heart disease. Lee *et al* [15] implemented an algorithm that utilizes a local transform pattern (LTP) with a self-organizing map and hybrid neural-fuzzy (NF) logic method. A multidimensional histogram structure is derived using three methods of LTP-based feature extraction as a 1D local gradient pattern (1DLGP), 1D local binary pattern and local neighbor descriptor model. Then, the self-organizing map was employed on a fuzzy-logic program to improve labeling consistency and the time latency. The system performance is demonstrated with the help of experimental results. The efficiency of the proposed system using 1DLGP + NF with 196 features measurements showed robust performance at 87% (sensitivity) and 98.84% (accuracy).

ECG is an effective procedure for the treatment of arrhythmias. Computer-aided treatment is quickly increasing to overcome the shortcomings of manual inspection. To categorize various arrythmias, Shi *et al* [16] have suggested an inter-patient automated diagnosis tactic. The proposed technique is based on region feature extraction and ensemble classification. There are four steps to the proposed method. During the preprocessing phase, the ECG signal is conditioned and segmented. Heartbeats are then separated into three areas, and regional characteristics are removed. Eventually the dimensions of the features are diminished and fused. Finally, an ensemble classifier is used for the automated identification. The sensitivities obtained for normal, LBBB, RBBB, APV and VPC were 95.0%, 27.9%, 79.6%, 81.8% and 88.1%, respectively.

Sudden cardiac failure is a result of an irregular heartbeat. Accurate diagnosis of such an abnormal condition is therefore important in detecting heart problems. Asgharzadeh-Bonab *et al* [17] developed a CAD approach based on time–frequency analysis of ECG signals and deep neural networks for the detection of arrhythmia. Time–frequency techniques are capable of supplying spectral information at various intervals that is helpful for evaluating the non-stationary signals. Entropy is a good measure of the ECG signals and can differentiate between different forms of heartbeats. Time–frequency spectral entropy is recommended to suppress the effective characteristics of the ECG signals. As all calculated entropies cannot have separability between different types, an application of two-dimensional principle component analysis can be utilized to decrease the extracted feature dimensions. A convolution neural network is utilized for diagnosing the ECG beat signals and predicting arrhythmias.

Mondéjar-Guerra *et al* [18] developed a system for automated ECG classification utilizing a combination of multiple support vector machines (SVMs). For the characterization of the ECG, the time intervals between subsequent beats and their morphology is utilized. Various wavelet dependent descriptors, higher order statistics (HOS), local binary patterns (LBP) and multiple amplitude values were used. Rather than combining all of these features to feed a single SVM model, they suggested training different SVM models for each feature type. To obtain the final prediction, the decisions of the various models are combined using the rules of majority, sum and product.

Rajesh and Dhuli [19] suggested a system for the classification of five heartbeat classes. Given the complexity of the ECG signal, they used a non-stationary and nonlinear decomposition method. Then sample entropy measurements and higher order statistics are used with the intrinsic mode functions (IMFs) created from

ICEEMD on every ECG segment. In addition, three preprocessing techniques at the code level are implemented on the derived feature set to match the distribution of the heartbeat class. Such features are eventually fed into the AdaBoost ensemble classifier to distinguish heartbeats. Jha and Kolekar [10] reported a novel and successful ECG beat testing procedure for normal heartbeat and seven forms of arrhythmia. The technique suggested allows the use of tunable Q-wavelet based features of ECG beats which are derived from various ECG signals. ECG beat features are used to classify via an SVM. The suggested classifier's average accuracy, sensitivity and specificity for eight different classes of ECG beats is 99.27%, 96.22% and 99.58%, respectively.

Alickovic and Subasi [20] suggested the random forest (RF) classifier for automated categorization of ECG pulse patterns in the treatment of cardiac arrhythmias. Discrete wavelet transformation (DWT) is employed to decompose ECG signals into consecutive bands of several frequencies. The frequency bands obtained with time–frequency analysis are used. Their statistical features are extracted and employed for the precise identification of the ECG signals. For the MIT-BIH database, the RF classifier achieved an overall accuracy of 99.33%. Atal and Singh [21] proposed a system for automatic detection of arrhythmia. They used the deep convolutional neural network (DCNN) for automatic classification. The bat-rider optimization algorithm (BaROA) that uses the multi-objective bat algorithm and the rider optimization algorithm is developed. Gabor features are derived from the ECG signals in such a way that these features match the real ECG features. Eventually, the signals are supplied to the BaROA-centered DCNN classifier which determines the condition of the person as arrhythmia or no-arrhythmia by using the ECG signals. An automatic framework for the treatment of normal sinus rhythm, LBBB, RBBB, APB and PVC on ECG signals using a mixture of a convolutionary neural network (CNN) and long-short-term memory (LSTM) was introduced by Oh *et al* [22]. The suggested approach showed high classification efficiency when managing data of variable length, attaining 98.10% accuracy, 97.50% sensitivity and 98.70% precision using ten-fold cross validation.

Shi *et al* [3] suggested a novel numerical classification system to maximize the ECG heartbeat rating. A deep structure based on the CNN and LSTM network, with several input layers, is suggested. Four input layers are assembled based on the properties of various regions of a heartbeat and RR interval. The first three inputs are processed by employing several steps. The three outputs of the CNN are combined, going through an LSTM network. Finally, the expected tag is produced from the last completely connected layer. The suggested solution reaches an average precision of 99.26%.

## 11.3 Materials and methods

### 11.3.1 Dataset

The functionality of the proposed framework is verified using the MIT-BIH Arrhythmia database [23]. A collection of 24 hour outpatient ECG records is utilized during this analysis. It contains arrhythmias which are clinically significant.

The signal is band-limited to 60Hz and is captured at a sampling frequency of 360 Hz with a standard 11-bit resolution ADC. Every considered ECG record is divided into short time length windows. The window lengths are selected equal to 0.9 s. In order to study the suggested method's performance in terms of identifying various categories of the heart pulses, the ECG records of five different categories are employed. The types of categories considered are right bundle branch block (RBBB), normal (N), premature ventricular contraction (PVC), left bundle branch block (LBBB) and atrial premature contraction (APC). A total of 1500 ECG instances are considered and they belong to the five different categories considered. For each class 300 instances are considered. This equal representation of each class avoids any possible bias during the classification process.

### 11.3.2 Wavelet transform

The wavelet transform (WT) is a widely used method for the study of time–frequency which facilitates the achievement of a multi-resolution time–frequency study that the traditional short-term Fourier transform cannot accomplish. The wavelet transform is a promising choice for processing signals with a changing spectrum as opposed to the short time Fourier transform. It tells us not only which frequencies are present in a signal but also when these frequencies took place. Working with different scales allows this to be achieved. The WT has a superior time resolution and inferior frequency resolution for high frequencies and vice versa [24]. For an efficient evaluation, it generates a compact representation of a signal in the time and frequency domains.

The basic functions are called wavelets, which are repartitioned in time and frequency. They can be signified mathematically using the following equation, where, $s$ and $u$ are the dilation and translation factors, respectively [24], and the dilation and translation factor values are continuous, meaning there can be an infinite number of wavelets:

$$\Psi(t) = \frac{1}{\sqrt{S}}\psi((t-u)/s). \tag{11.1}$$

The process of analysing a signal $x(t)$ with the help of wavelets is mathematically described as

$$W \times (u, s) = \frac{1}{\sqrt{S}} \int_{-\infty}^{+\infty} x(t)\psi^*((t-u)/s)\mathrm{d}t. \tag{11.2}$$

Any signal can be decomposed in various time–frequency bands by the application of wavelets. Since the wavelet is located in time, we can multiply our signal time-wise with the wavelet at various locations. It is the convolution procedure. Once the procedure is carried out with the mother wavelet, it can be scaled and the procedure can be repeated again at different scales.

### 11.3.3 Denoising with multiscale principal component analysis

The multiscale principal component analysis (MSPCA) is an enhanced version of principal component analysis (PCA). It is a combination with benefits from both the PCA and the wavelet analysis. It is often used for denoising by feature reduction in a variety of applications, such as biomedical signal processing and image processing. It eliminates the correlation among coefficients, obtained using the wavelet analysis. The MSPCA carries out the principal component analysis of the wavelet coefficients at every scale. Its multiscale nature makes it suitable for the denoising and feature reduction of a variety of time varying signals such as ACG, electromyogram and electroencephalogram. It focuses on the pertinent scales while avoiding others. This results in these changes being effectively discovered in the intended data. In addition, it enables the effective and adaptive retrieval of the residuals and ratings. The MSPCA not only enhances the identification of determinist shifts but also extracts the characteristics that indicate anomalies [25].

### 11.3.4 Feature extraction

The following robust, parametric and time–frequency analysis based features extraction techniques are considered.

#### *11.3.4.1 The autoregressive Burg approach*

The autoregressive (AR) model based Burg approach determines the reflection coefficients, $r_k$, while minimizing the backward ($e_b$) and forward ($e_f$) prediction errors [26, 27, 28]. The procedure for the $k$th-order model is defined by the following equations, where $a_k$ are the coefficients of the AR model:

$$e_{f,k}(n) = x(n) + \sum_{i=1}^{k} a_{k,i} x(n-i) \ldots\ldots n = k+1, \ldots\ldots, N \tag{11.3}$$

$$e_{b,k}(n) = x(n-k) + \sum_{i=1}^{k} a_{k,i}^{*} x(n-k+i) \ldots\ldots n = k+1, \ldots\ldots, N. \tag{11.4}$$

The relationship between the $a_k$ and $r_k$ parameters of the AR model can be mathematically described by

$$\begin{cases} a_{k,i} = a_{k-1,i} + r_k a_{k-1,k-i}^{*}, \; i = 1, \ldots, k-1 \\ r_k, \; i = k \end{cases}. \tag{11.5}$$

It takes into consideration the estimation of $r_k$ by recursive-in-order approaches such that the coefficients of the $(k-1)$th order model of AR are determined. It justifies the errors of approximation by utilizing the mathematical expressions given by

$$e_{f,k}(n) = e_{f,k-1}(n) + r_k e_{b,k-1}(n-1) \tag{11.6}$$

$$e_{b,k}(n) = e_{b,k-1}(n-1) + r_k^{*} e_{b,k-1}(n). \tag{11.7}$$

The prediction errors satisfy the above mathematical expressions. Equations (11.6) and (11.7) are utilized for determining the coefficients of the model of AR. Then, these determined coefficients are utilized for estimating the power spectral density (PSD). The process can be mathematically depicted with the help of the following equation, where $e_k = e_{f,k} + e_{b,k}$ and it presents the accumulative least-squares error [26, 27, 28]:

$$P_{\mathrm{BURG}}(f) = \frac{e_k}{\left|1 + \sum_{m=1}^{k} a(m)\mathrm{e}^{-j2\pi fm}\right|^2}. \tag{11.8}$$

*11.3.4.2 The autoregressive Yulear*

The Yulear approach based on the AR model determines its parameters by adopting and reducing the power of estimation error. Equation (11.5) is differentiated in accordance with the imaginary and real portions of $a(m)$ [26]. It is realized by employing a complex gradient. The process can be mathematically depicted as

$$\frac{1}{N}\sum_{n=-\infty}^{\infty}\left(x(n) + \sum_{m=1}^{k} a(m)x(n-m)\right)x^{*}(n-l) = 0,\ l = 1, 2, \ldots, k. \tag{11.9}$$

In the form of a matrix the process can be expressed as

$$\begin{bmatrix} z(1) \\ \vdots \\ z(k) \end{bmatrix} + \begin{bmatrix} z(0)\ldots\ldots\ldots\ldots z(-k+1) \\ \vdots \\ z(k-1)\ldots\ldots\ldots\ldots z(0) \end{bmatrix}\begin{bmatrix} a(1) \\ \vdots \\ a(k) \end{bmatrix} = \begin{bmatrix} 0 \\ \vdots \\ 0 \end{bmatrix}. \tag{11.10}$$

Equation (11.10) can be symbolized as

$$z_k + Z_k a = 0. \tag{11.11}$$

In equation (11.10), $z(m)$ is given by

$$z(m) = \begin{cases} \frac{1}{N}\sum_{n=0}^{N-1-m} x^{*}(n)x(n+m),\ m = 0, 1, \ldots., k \\ z^{*}(-m),\ m = (-k+1), (-k+2), \ldots.., -1 \end{cases}. \tag{11.12}$$

Parameters of the AR model can be determined from equation (11.11) by using the equation

$$a = -z_k Z_k^{-1}. \tag{11.13}$$

By utilizing the estimated AR model parameters, the power spectral density (PSD) could be calculated using the following equation, where $\sigma^2$ represents the variance of white noise [26]:

$$P_{\mathrm{YW}}(f) = \frac{\sigma^2}{\left|1 + \sum_{m=1}^{k} a(m)\mathrm{e}^{-j2\pi fm}\right|^2}. \tag{11.14}$$

*11.3.4.3 Covariance*

The covariance approach is based on a similar principle of the correlation dependent Yule–Walker algorithm. It is also determined by minimizing the approximation of the determination error of power [29]. The process is depicted mathematically as

$$\rho = \frac{1}{N-k}\sum_{n=k}^{N-1}\left|\left(x(n) + \sum_{m=1}^{k} a(m)x(n-m)\right)\right|^2. \tag{11.15}$$

The major difference among the autocorrelation and the covariance approaches is the limit of the coefficients employed for predicting the estimated error power. For the covariance technique, all data points are required to be computed from the measured $\rho^2$. Therefore, the minimization of equation (11.15) could be performed by determining the complex gradient to drive the AR model parameters. The process is depicted with the help of the following [29]:

$$\begin{bmatrix} c(1, 0) \\ \vdots \\ c(k, 0) \end{bmatrix} + \begin{bmatrix} c(1, 1)\ldots\ldots\ldots\ldots c(1, k) \\ \vdots \\ c(k, 1)\ldots\ldots\ldots\ldots c(k, k) \end{bmatrix}\begin{bmatrix} a(1) \\ \vdots \\ a(k) \end{bmatrix} = \begin{bmatrix} 0 \\ \vdots \\ 0 \end{bmatrix}. \tag{11.16}$$

Equation (11.16) can be symbolized as

$$c_k + C_k a = 0. \tag{11.17}$$

In equation (11.16), $c(j, m)$ is given by

$$c(j, m) = \frac{1}{N-k}\sum_{n=k}^{N-1} x^*(n-j)x(n-m). \tag{11.18}$$

Parameters of the AR model can be determined from the following equation using equation (11.17):

$$a = -c_k C_k^{-1}. \tag{11.19}$$

By utilizing the estimated AR model parameters, the power spectral density (PSD) could be calculated using the following equation, where $\sigma^2$ represents the variance of white noise [29]:

$$P_{\text{COV}}(f) = \frac{\sigma^2}{\left|1 + \sum_{m=1}^{k} a(m)e^{-j2\pi fm}\right|^2}. \tag{11.20}$$

*11.3.4.4 Modified covariance*

The modified covarience approach determines the parameters of the AR model by reducing the mean of the approximated error powers of the forward and the backward predictors [29, 30]. The forward and backward predictors are, respectively, given by

$$\tilde{x}(n) = -\sum_{m=1}^{k} a(m)x(n-m) \tag{11.21}$$

$$\tilde{x}(n) = -\sum_{m=1}^{k} a^*(m)x(n+m). \tag{11.22}$$

The process of approximating the AR model parameters based on the procedure of reducing the mean approximation error of forward and backward predictors is clear from equation (11.23), where $\rho_f$ and $\rho_b$ can be expressed, respectively, via equations (11.24) and (11.25). The minimization of equation (11.23) can be realized by performing the complex gradient [29, 30]:

$$\rho = \frac{1}{2}(\rho_f + \rho_b) \tag{11.23}$$

$$\rho_f = \frac{1}{N-k}\sum_{n=k}^{N-1}\left|\left(x(n) + \sum_{m=1}^{k} a(m)x(n-m)\right)\right|^2 \tag{11.24}$$

$$\rho_b = \frac{1}{N-k}\sum_{n=0}^{N-1-k}\left|\left(x(n) + \sum_{m=1}^{k} a^*(m)x(n+m)\right)\right|^2. \tag{11.25}$$

The process is depicted, in matrix form, with the help of the following equation:

$$\begin{bmatrix} c(1,0) \\ \vdots \\ c(k,0) \end{bmatrix} + \begin{bmatrix} c(1,1) \ldots\ldots\ldots\ldots c(1,k) \\ \vdots \\ c(k,1) \ldots\ldots\ldots\ldots c(k,k) \end{bmatrix} \begin{bmatrix} a(1) \\ \vdots \\ a(k) \end{bmatrix} = \begin{bmatrix} 0 \\ \vdots \\ 0 \end{bmatrix}. \tag{11.26}$$

Equation (11.26) can be symbolized as

$$c_k + C_k a = 0. \tag{11.27}$$

In equation (11.26), $c(j, m)$ is given by

$$c(j,m) = \frac{1}{2(N-k)}\left(\sum_{n=k}^{N-1} x^*(n-j)x(n-m) + \sum_{n=0}^{N-1-k} x(n+j)x^*(n+m)\right). \tag{11.28}$$

The parameters of the AR model can be determined from the following equation by using equation (11.27):

$$a = -c_k C_k^{-1}. \tag{11.29}$$

By utilizing the estimated AR model parameters, the power spectral density (PSD) could be calculated using the following equation, where $\sigma^2$ represents the variance of white noise [29, 30]:

$$P_{\mathrm{MCOV}}(f) = \frac{\sigma^2}{\left|1 + \sum_{m=1}^{k} a(m)\mathrm{e}^{-j2\pi fm}\right|^2}. \tag{11.30}$$

*11.3.4.5 Multiple signal classification*

Multiple signal classification (MUSIC) is one of the frequently used subspace based approaches for spectral analysis. It is an effective eigenstructure based estimator of the noise subspace frequency [31]. This can be achieved either by decomposing using the eigenvalues of the approximate array correlation matrix or through decomposing the data matrix in singular values, with its columns becoming the array signal vectors. For numeric reasons the second approach is favored. When the subspace of noise is determined, a search for angle pairs within the range is performed by searching for eigenvectors that are as orthogonal as possible to the noise subspace. Normally this is realized by looking for peaks in the MUSIC spectrum.

It diminishes the impact of inauthentic coefficients by employing the mean of the eigenvector spectra of all the subspaces of the noise. By utilizing the MUSIC approach the power spectral density (PSD) could be calculated using the following equation [31], where $M$ is the dimension of the noise subspace and $A_i(f)$ is the polynomial to all the eigenvectors of the noise subspace:

$$P_{\mathrm{MUSIC}}(f) = \frac{1}{\frac{1}{M}\sum_{i=0}^{M-1}\left|A_i(f)\right|^2}. \tag{11.31}$$

*11.3.4.6 Eigenvector method*

It is a well known and frequently used subspace based approach for the spectral analysis. To discern spurious zeros from real zeros, the eigenvector based procedure induces spurious zeros within the unit circle and employs either the noise subspace or the signal subspace for determining a wanted noise subspace vector $a$ [32].

By utilizing the eigenvector approach the PSD could be calculated using the following equation [32], where $M$ is dimensioning the noise subspace and $A_i(f)$ is the polynomial to all the eigenvectors of the noise subspace:

$$P_{\mathrm{ev}}(f) = \frac{1}{\sum_{i=0}^{M-1}\left|A_i(f)\right|^2/\lambda_i}. \tag{11.32}$$

*11.3.4.7 Discrete wavelet transform*

The discrete wavelet transform (DWT) is the sampled version of the WT and isolates a discrete time signal into subbands and these isolated bands are represented in terms of wavelet coefficients [33, 34, 35]. Sampling of the continuous time wavelet parameters would then produce the DWT. The problem of synthesizing the signal from its transformed version obviously depends on the resolution of the sampling grid. A fine resolution of the sampling grid would allow convenient reconstruction but with obvious redundancy, that is, oversampling.

#### *11.3.4.8 Wavelet packet decomposition*

The wavelet packet decomposition (WPD) is frequently employed for extracting the classifiable features of the ECG, such as intermittent signals. It could decompose a signal with a superior time–frequency resolution compared to Fourier's short time transformation [36, 37]. Any signal is split into its subbands. A single function $\psi$ is transformed into 'wavelets' with the use of dilations and translations [38]. The WPD is a DWT extension. The DWT only decomposes the components of a signal at low frequencies. WPD divides the components of low frequencies or approximations and the components of high frequencies or details into sublevels [33, 36, 39].

#### *11.3.4.9 Stationary wavelet transform*

The stationary wavelet transform (SWT) is an enhancement of the DWT and possesses one feature that has advantages over the DWT—it is almost shift invariant. Any time series can be decomposed effectively by utilizing the DWT [40]. The process consists of the cascaded stages of band-pass filtering and decimation. It is a quick operation in computational terms and can be put into place by subsequent filter banks. Regrettably, as applied to discrete time series, the DWT is not shift invariant. If the incoming time series is shifted in time then the DWT output coefficients could change significantly. This shortfall is treated to a certain extent by utilizing the SWT. It is achieved by replacing the down-sampling step of the DWT with the up-sampling process. $l_1$ and $h_1$ are, respectively, the low and high pass half-band filters used at the first level of decomposition. $l_2$ and $h_2$ are, respectively, the low and high pass half-band filters used at the second level of decomposition and $l_3$ and $h_3$ are, respectively, the low and high pass half-band filters used at the third level of decomposition [40].

#### *11.3.4.10 Tunable Q-factor wavelet transform*

The tunable Q-wavelet transform (TQWT) is an important way of analysing oscillating signals [41, 42]. As a feature of guided deployment it can be conveniently modified. $Q$, $r$ and $j$ are its major adjustable parameters. $Q$ represents the Q-factor, $r$ represents the amount of oversampling and the degree of decomposition is $j$. $Q$ performs the count of wavelet oscillations. Unwanted over-oscillation is governed by $r$. The TQWT is designed using digital filter banks, developed in particular for the targeted application. Such functions are non-rational functions of transformation and are known for their practical implementation [41, 43].

The oscillatory activity of the signal to be analysed is used to pick an effective Q-factor. The wavelet transformations most commonly display a mild tendency to change the Q-factor, except for continuous wavelet transformation. Due to this restriction their use is limited to a particular system. The discrete time TQWT handles finite length sampled signals. The principle is similar to that of the short time

Fourier transform and could be effectively realized with the FFT. In [41] Selesnick has shown that by using moderate oversampling ratios the TQWT can be efficiently realized [41, 42].

*11.3.4.11 Dual tree complex wavelet transform*

The dual tree complex wavelet transform (DT-CWT) is an enhanced version of the DWT and possesses two features that have advantages over the DWT. It is almost shift invariant and is directionally selective in two and higher dimensions. It is carried out with a redundancy factor of just 2 and is significantly smaller than the redundancy factors utilized in the stationary DWT [44]. It shows a fine movement invariant property and improves the directional estimation in contrast to the DWT directional estimation. The DT-CWT provides fantastic restoration by making use of two simultaneously devastated channel bank trees with coefficients with adding value created at each tree [45]. The DT-CWT calculates the complex signal transformation by means of two separate DWT decompositions, namely tree 1 and tree 2. The filters used in tree 1 could be designed differently compared to the filters designed in tree 2. In this way the real coefficients can be produced by one DWT and the imaginary coefficients can be produced by the second DWT. It shows the DT-CWT with a second level of decomposition [46]. $l_1$ and $h_1$ are, respectively, the low and high pass half-band filters for the tree 1. $l_2$ and $h_2$ are, respectively, the low and high pass half-band filters for tree 2. $W_{110}(k)$ and $W_{111}(k)$ are subband coefficients after the first level of decomposition for tree 1. $W_{120}(k)$, $W_{121}(k)$, $W_{122}(k)$ and $W_{123}(k)$ are subband coefficients after the second level of decomposition for tree 1. $W_{210}(k)$ and $W_{211}(k)$ are subband coefficients after the first level of decomposition for tree 2. $W_{220}(k)$, $W_{221}(k)$, $W_{222}(k)$ and $W_{223}(k)$ are subband coefficients after the second level of decomposition for tree 2 [44].

### 11.3.5 Machine learning methods

*11.3.5.1 Artificial neural network*

Artificial neural networks (ANNs) are a group of connected input and output units and every connection has a specific weight. An ANN is a computational model that depends on the structure and functions of biological neural networks and can include long training times so it is more appropriate for applications. Neural networks have several advantages. First they have a high tolerance of noisy data. Second, they even have the ability to classify patterns that are not trained. Third, even a little knowledge of the relationships between attributes and classes can be used. Fourth, it is considered different to most decision tree algorithms because it is more appropriate for continuous-valued inputs and outputs [47].

*11.3.5.2* k*-nearest neighbors*

*k*-nearest neighbors (*k*NN) is a set of *k* objects and it exists in the training group that is considered the closest to the test objects. It has the ability to manage well when given huge training datasets. These classifiers are based on learning by relationship, that is, by comparing a given test tuple with training tuples that are like it. It has three key elements: a set of labeled instances, e.g. a set of stored records, a similarity metric or distance to calculate the distance between instances and the value of $k$, the number of nearest neighbors [48].

#### *11.3.5.3 Support vector machine*

The support vector machine (SVM) is the most accurate method among the other machine learning methods. It is an algorithm for the classification of linear and also nonlinear data. In a two-class learning task, the purpose or aim of the used method is to find the best classification technique to discriminate between members of two classes in the training data [48].

#### *11.3.5.4 Classification and regression tree*

The classification and regression tree (CART) has the ability to evolve artificial intelligence, machine learning, data mining and non-parametric statistics. Also, it is significant because of the technical innovations it introduces, the extensiveness of its study of decision trees, its authoritative treatment of large sample theory for trees and its complex discussion of tree-structured data analysis. The CART decision tree is a binary recursive partitioning procedure and it is able to process continuous and nominal attributes both as targets and predictors. Its purpose is to produce a sequence of nested pruned trees. In CART the missing value handling process is completely automated and locally adaptive at each node. At every node in the tree, the selected splitter convinces a binary partition of the data [48].

#### *11.3.5.5 REP tree*

This type of method utilizes information gain/variance reduction and prunes it employing reduced-error pruning to build a decision or regression tree. Also, it has some tools such as optimization for the speed used to sort values for numeric attributes once and it can be split into instances or pieces this make it able to deal with missing values [49].

#### *11.3.5.6 LAD tree*

The LAD tree learning algorithm is applied to find the error criterion and acquire the regression trees. It is a parallel binary classifier and henceforth can separate between positive and negative examples. In each iteration, a single feature is picked at the splitter node for testing purposes. The effective response and weights are stored in each training instance. The working principle is to calculate the mean value of the instances by restraining the least-squares value in a particular subset. The best drop in this calculation is relevant by selecting the maximum gain carefully in the tests [50].

#### *11.3.5.7 Random tree classifier*

The random tree is a supervised classifier. It is an ensemble learning technique which produces many individual learners. Moreover, it is a collection (ensemble) of tree predictors, which is termed a forest. It can produce a random set of data for building a decision tree by using a bagging idea. The standard tree utilizes the best split among all instances to split each node, but the random tree utilizes the best among the subset of predictors randomly chosen at that node to split each node. The algorithm has the ability to deal with both classification and regression problems [51, 52].

*11.3.5.8 Random forest*

In the random forest (RF), each of the classifiers in the collection is a choice tree classifier with the goal that the accumulation of classifiers is a 'forest'. The individual choice trees are created utilizing an irregular determination of attributes at every node to decide the split. All the more formally, each tree relies on the estimations of an irregular vector tested freely and with a similar assumption for all trees in the forest [53, 54].

*11.3.5.9 Rotation forest*

Rotation forest (RoF) is a method for generating a classifier group depend on feature extraction. For a base classifier, we can randomly split the feature set into $K$ subsets ($K$ is a parameter of the calculation) and principal component analysis (PCA) is applied to each subset to create the training data. Therefore, to maintain the variability information in the data, all the main parts are retained [55].

### 11.3.6 The performance evaluation measures

The performance of a sample dataset is measured when the chosen output parameters of a given dataset are determined by taking into account the true class marks. The output of the data collection characterizes the degree of similarity with the goal definition of the dataset. By testing the model on the testing dataset, the training efficiency is calculated. This success analysis is useful for a deeper understanding of process. The predicted performance of the model essentially reflects the true output of the entire framework. The true success of the model means it is capable of correctly classifying random new instances from the context that is being considered. Reasonable assessment techniques are needed for evaluating the true results, which often include previously unseen instances [56].

The $k$-fold cross validation is a highly sophisticated procedure for model evaluation. This technique exploits the correlation between biases and variances. Its mode of operation entails splitting of the dataset into equal portions known as $k$ subsets, where the value $k$ refers to the total number of the fold's iterations. A single iteration of $k$-fold cross validation corresponds to the hold-out procedure whereby $(k - 1)/k$ data are selected for the training and $1/k$ of the data are selected for the evaluation. The training set size does not reduce to a point where it would adversely impact the model quality, particularly for sufficiently large values of $k$ since the validation set is small. The $k$-fold cross validation procedure effectively implements the training or testing sets [56]. In this study, ten-fold cross validation is utilized to calculate the performance of the classifier. To evaluate the performance of the models, the accuracy, precision, recall, F-measure, receiver operating characteristics (ROCs) and kappa statistic are utilized:

$$\text{Accuracy} = \frac{\text{TP} + \text{TN}}{\text{TP} + \text{TN} + \text{FP} + \text{FN}} \tag{11.33}$$

$$\text{Precision} = \frac{\text{TP}}{\text{TP} + \text{FP}} \tag{11.34}$$

$$\text{Recall} = \frac{\text{TP}}{\text{TP} + \text{FN}} \tag{11.35}$$

$$\text{F-measure} = \frac{\text{Precision} \times \text{Recall}}{\text{Precision} + \text{Recall}}. \tag{11.36}$$

An overview of the ROCs is one of the correct tools that enhances the classifier performance assessment/evaluation at multiple points of operation, identification and selection. The graph displays the full spectrum of various operating points, which in a single plot have consequently varying rates of the TPR and FPR tradeoff. The output of the scoring classifier which relies solely on its scoring function aspect can be graphically indicated using the ROC curve. The concept of scoring incorporates the information surrounding the relationship between qualities and class values [56].

The kappa statistic takes account of the predicted outcome by subtracting them from the results of predictors. For a complete prediction model, it describes the outcomes as part of the sum. This assessment is more rigorous than ordinary percentage agreement estimation as it assumes the probability of an arrangement happening by chance. The degree of agreement between the description of the expected and observed dataset, as well as the adjustment of the difference that occurs by accident, is performed using the kappa statistic [52] which is presented as

$$k = \frac{p_o - p_e}{1 - p_e} = 1 - \frac{1 - p_o}{1 - p_e}. \tag{11.37}$$

In the above model equation, $p_o$ is the agreement observed relatively among raters (the same as accuracy) and $p_e$ is the theoretical probability of chance agreement.

## 11.4 Experimental results

A summary of the classification performance obtained for the AR Yulear approach is presented in table 11.1. It shows that on average, for all five arrhythmia classes, the random forest and rotation forest classifiers achieve the highest classification performance of 94.7%. Random forest and rotation forest reach the highest F-measure values of 0.948 and 0.947, respectively. Both the random forest and rotation forest classifiers obtain the maximum ROC area value of 0.995. Random forest and rotation forest obtain the maximum kappa value of 0.934.

It is evident from these results that random forest and rotation forest outperform in terms of the classification performance in contrast to the other considered classifiers such as *k*NN, ANN, SVM, CART, REP tree, random tree and LDA tree. While considering all the evaluation measures, such as accuracy, F-measure,

**Table 11.1.** Classification results for the AR Yulear method.

| | Normal | APC | PVC | RBBB | LBBB | Average | F-measure | ROC area | Kappa |
|---|---|---|---|---|---|---|---|---|---|
| SVM | 1 | 0.847 | 0.87 | 0.83 | 0.907 | 0.891 | 0.892 | 0.956 | 0.8633 |
| *k*NN | 1 | 0.917 | 0.877 | 0.807 | 0.89 | 0.898 | 0.898 | 0.973 | 0.8725 |
| ANN | 1 | 0.85 | 0.837 | 0.777 | 0.877 | 0.868 | 0.868 | 0.971 | 0.835 |
| RF | 1 | 0.937 | 0.92 | 0.93 | 0.95 | 0.947 | 0.948 | 0.995 | 0.9342 |
| CART | 1 | 0.857 | 0.85 | 0.863 | 0.88 | 0.89 | 0.89 | 0.952 | 0.8625 |
| RoF | 1 | 0.937 | 0.93 | 0.927 | 0.943 | 0.947 | 0.947 | 0.995 | 0.9342 |
| REP tree | 1 | 0.883 | 0.787 | 0.817 | 0.9 | 0.877 | 0.877 | 0.961 | 0.8467 |
| Random tree | 1 | 0.877 | 0.863 | 0.783 | 0.86 | 0.877 | 0.876 | 0.923 | 0.8458 |
| LAD tree | 1 | 0.873 | 0.82 | 0.773 | 0.857 | 0.865 | 0.866 | 0.973 | 0.8308 |

**Table 11.2.** Classification results for the AR Burg method.

| | Normal | APC | PVC | RBBB | LBBB | Average | F-measure | ROC area | Kappa |
|---|---|---|---|---|---|---|---|---|---|
| SVM | 1 | 0.837 | 0.863 | 0.827 | 0.877 | 0.881 | 0.882 | 0.952 | 0.8508 |
| *k*NN | 1 | 0.937 | 0.893 | 0.76 | 0.91 | 0.9 | 0.899 | 0.975 | 0.875 |
| ANN | 1 | 0.88 | 0.843 | 0.7 | 0.917 | 0.868 | 0.866 | 0.97 | 0.835 |
| RF | 1 | 0.943 | 0.947 | 0.947 | 0.97 | 0.961 | 0.961 | 0.997 | 0.9517 |
| CART | 1 | 0.88 | 0.877 | 0.847 | 0.903 | 0.901 | 0.901 | 0.964 | 0.8767 |
| RoF | 1 | 0.94 | 0.957 | 0.933 | 0.957 | 0.957 | 0.957 | 0.996 | 0.9467 |
| REP tree | 1 | 0.9 | 0.873 | 0.86 | 0.917 | 0.91 | 0.91 | 0.965 | 0.8875 |
| Random tree | 1 | 0.887 | 0.863 | 0.833 | 0.873 | 0.891 | 0.891 | 0.932 | 0.8642 |
| LAD tree | 1 | 0.833 | 0.83 | 0.857 | 0.913 | 0.887 | 0.887 | 0.978 | 0.8583 |

ROC area and kappa statistics, the random forest achieves a superior performance even compared to the rotation forest, which lags behind the random forest in terms of the F-measure value.

A summary of the classification performance obtained for the AR Burg approach is presented in table 11.2. It shows that on average for all five arrhythmia classes, the best classification accuracy of 96.1% is realized with the random forest and the second highest average classification accuracy of 95.7% is achieved by the rotation forest. The highest F-measure value of 0.961 is achieved by the random forest and the rotation forest achieves the second highest F-measure value of 0.957. The highest ROC area value of 0.997 is realized by the random forest and the second highest ROC area value of 0.996 is realized with the rotation forest classifier. The highest kappa value of 0.952 is also realized by the random forest and the second highest kappa value of 0.947 is achieved by the rotation forest classifier.

It is evident from these results that random forest and rotation forest outperform in terms of the classification performance in contrast to the other considered classifiers such as *k*NN, ANN, SVM, CART, REP tree, random tree and LDA tree. When considering all the evaluation measures, such as accuracy, F-measure, ROC area and kappa statistics, the random forest achieved a superior performance even compared to the rotation forest, which lags behind the random forest in all classification evaluation measures.

A summary of the classification performance obtained for the covariance approach is presented in table 11.3. It shows that on average for all five arrhythmia classes, the best classification accuracy of 100% is achieved with the random tree and the second highest average classification accuracy of 99.9% is achieved by the rotation forest. The highest F-measure value of 1.0 is achieved by the random tree and the second highest F-measure value of 0.999 is realized by the rotation forest. The highest ROC area value of 1.0 is realized by the random tree and the rotation forest classifiers. The highest kappa value of 1 is also realized by the random tree and the second highest kappa value of 0.998 is achieved by the rotation forest.

It is evident from these results that the random tree and rotation forest outperform in terms of the classification performance in contrast to the other considered classifiers, such as *k*NN, ANN, SVM, CART, REP tree, random forest and LDA tree. When considering all the evaluation measures, such as accuracy, F-measure, ROC area and kappa statistics, the random tree achieves the superior performance even compared to the rotation forest, which lags behind the random tree in terms of accuracy, F-measure and the kappa statistic.

A summary of classification performance obtained for the modified covariance approach is presented in table 11.4. It shows that on average for all five arrhythmia classes, the best classification accuracy of 96.3% is realized with the rotation forest and the second highest average classification accuracy of 95.6% is achieved by the random forest. The highest F-measure value of 0.963 is achieved by the rotation forest and the second highest F-measure value of 0.956 is realized by the random forest. The highest ROC area value of 0.997 is realized by both the random forest and the rotation forest classifiers. The highest kappa value of 0.954 is realized using the rotation forest and the second highest kappa value of 0.945 is achieved by the random forest.

**Table 11.3.** Classification results for the covariance method.

| | Normal | APC | PVC | RBBB | LBBB | Average | F-measure | ROC area | Kappa |
|---|---|---|---|---|---|---|---|---|---|
| SVM | 1 | 0.91 | 0.883 | 0.857 | 0.947 | 0.919 | 0.919 | 0.969 | 0.8992 |
| *k*NN | 1 | 0.933 | 0.88 | 0.777 | 0.907 | 0.899 | 0.899 | 0.972 | 0.8742 |
| ANN | 1 | 0.87 | 0.843 | 0.693 | 0.913 | 0.864 | 0.863 | 0.974 | 0.83 |
| RF | 1 | 0.953 | 0.92 | 0.93 | 0.957 | 0.952 | 0.952 | 0.996 | 0.94 |
| CART | 1 | 0.88 | 0.87 | 0.887 | 0.943 | 0.916 | 0.916 | 0.974 | 0.89 |
| RoF | 1 | 0.997 | 0.997 | 1 | 1 | 0.999 | 0.999 | 1 | 0.9983 |
| REP tree | 1 | 0.917 | 0.95 | 0.887 | 0.963 | 0.943 | 0.943 | 0.994 | 0.9292 |
| Random tree | 1 | 1 | 1 | 1 | 1 | 1 | 1 | 1 | 1 |
| LAD tree | 1 | 0.897 | 0.877 | 0.77 | 0.92 | 0.893 | 0.892 | 0.983 | 0.8658 |

**Table 11.4.** Classification results for modified covariance method.

| | Normal | APC | PVC | RBBB | LBBB | Average | F-measure | ROC area | Kappa |
|---|---|---|---|---|---|---|---|---|---|
| SVM | 1 | 0.897 | 0.88 | 0.87 | 0.957 | 0.921 | 0.921 | 0.971 | 0.9008 |
| *k*NN | 1 | 0.927 | 0.873 | 0.763 | 0.907 | 0.894 | 0.893 | 0.972 | 0.8675 |
| ANN | 1 | 0.89 | 0.803 | 0.737 | 0.893 | 0.865 | 0.864 | 0.971 | 0.8308 |
| RF | 1 | 0.943 | 0.923 | 0.95 | 0.963 | 0.956 | 0.956 | 0.997 | 0.945 |
| CART | 1 | 0.883 | 0.85 | 0.823 | 0.92 | 0.895 | 0.895 | 0.956 | 0.8692 |
| RoF | 1 | 0.953 | 0.947 | 0.947 | 0.97 | 0.963 | 0.963 | 0.997 | 0.9542 |
| REP tree | 1 | 0.86 | 0.86 | 0.803 | 0.917 | 0.888 | 0.888 | 0.967 | 0.86 |
| Random tree | 1 | 0.917 | 0.867 | 0.84 | 0.863 | 0.897 | 0.897 | 0.936 | 0.8717 |
| LAD tree | 1 | 0.837 | 0.837 | 0.83 | 0.893 | 0.879 | 0.88 | 0.977 | 0.8492 |

**Table 11.5.** Classification results for the eigenvector method.

| | Normal | APC | PVC | RBBB | LBBB | Average | F-measure | ROC area | Kappa |
|---|---|---|---|---|---|---|---|---|---|
| SVM | 0.993 | 0.79 | 0.82 | 0.903 | 0.777 | 0.857 | 0.859 | 0.937 | 0.8208 |
| *k*NN | 0.987 | 0.897 | 0.867 | 0.827 | 0.867 | 0.889 | 0.889 | 0.972 | 0.8608 |
| ANN | 0.97 | 0.833 | 0.777 | 0.763 | 0.797 | 0.828 | 0.829 | 0.965 | 0.785 |
| RF | 1 | 0.95 | 0.98 | 0.977 | 0.98 | 0.977 | 0.977 | 0.999 | 0.9717 |
| CART | 1 | 0.893 | 0.923 | 0.877 | 0.937 | 0.926 | 0.926 | 0.966 | 0.9075 |
| RoF | 1 | 0.943 | 0.977 | 0.97 | 0.993 | 0.977 | 0.977 | 0.998 | 0.9708 |
| REP tree | 1 | 0.87 | 0.907 | 0.88 | 0.917 | 0.915 | 0.915 | 0.974 | 0.8933 |
| Random tree | 1 | 0.893 | 0.917 | 0.887 | 0.927 | 0.925 | 0.925 | 0.953 | 0.9058 |
| LAD tree | 1 | 0.827 | 0.893 | 0.9 | 0.87 | 0.898 | 0.898 | 0.979 | 0.8725 |

It is evident from these results that the random forest and rotation forest outperform in terms of the classification performance in contrast to other benchmark classifiers, such as *k*NN, ANN, SVM, CART, REP tree, random tree and LDA tree. When considering all the evaluation measures such as accuracy, F-measure, ROC area and kappa statistic, the rotation forest achieves a superior performance even compared to the random forest, which lags behind the rotation forest in terms of accuracy, F-measure and kappa statistic.

A summary of classification performance obtained for the eigenvector approach is presented in table 11.5. It shows that on average for all five arrhythmia classes, the best classification accuracy of 97.7% is realized with the rotation forest and the random forest. The highest F-measure value of 0.977 is achieved by the rotation forest and the random forest. The highest ROC area value of 0.999 is realized by the random forest and the second highest ROC area value of 0.998 is obtained with the rotation forest classifiers. The highest kappa value of 0.972 is realized by the random forest and the second highest kappa value of 0.971 is achieved by the rotation forest.

It is evident from these results that the random forest and rotation forest outperform in terms of the classification performance in contrast to other considered classifiers such as *k*NN, ANN, SVM, CART, REP tree, random tree and LDA tree. When considering all the evaluation measures, such as accuracy, F-measure, ROC area and kappa statistic, the random forest achieves superior performance even compared to the rotation forest, which lags behind the random forest in terms of the ROC area and kappa statistic.

A summary of classification performance obtained for the MUSIC approach is presented in table 11.6. It shows that on average for all five arrhythmia classes, the best classification accuracy of 96.9% is realized with the random forest. The second highest average classification accuracy of 96.5% is realized by the SVM and the rotation forest closely follows the SVM by achieving an average classification accuracy of 96.3%. The highest F-measure value of 0.969 is achieved by the random forest and the second highest F-measure value of 0.965 is achieved by the SVM. The rotation forest closely follows the SVM with the F-measure value of 0.963. The highest ROC area value of 0.998 is realized by the random forest and the rotation forest and the second highest ROC area value of 0.993 is obtained with the ANN classifier. The highest kappa value of 0.961 is realized by the random forest and the second highest kappa value of 0.957 is achieved by the SVM. The rotation forest closely follows the SVM with the kappa value of 0.953.

It is evident from these results that the SVM, random forest and rotation forest outperform in terms of the classification performance in contrast to other considered classifiers, such as *k*NN, ANN, CART, REP tree, random tree and LDA tree. When considering all the evaluation measures, such as accuracy, F-measure, ROC area and kappa statistic, the random forest achieves superior performance even compared to the SVM and the rotation forest.

A summary of classification performance obtained for the DWT approach is presented in table 11.7. It shows that on average for all five arrhythmia classes, the best classification accuracy of 98.1% is realized with the ANN. The second highest average classification accuracy of 97.8% is realized by the SVM. The random forest

**Table 11.6.** Classification results for the MUSIC method.

| | Normal | APC | PVC | RBBB | LBBB | Average | F-measure | ROC area | Kappa |
|---|---|---|---|---|---|---|---|---|---|
| SVM | 1 | 0.95 | 0.957 | 0.95 | 0.97 | 0.965 | 0.965 | 0.989 | 0.9567 |
| *k*NN | 0.993 | 0.937 | 0.943 | 0.933 | 0.93 | 0.947 | 0.948 | 0.967 | 0.9342 |
| ANN | 0.99 | 0.913 | 0.96 | 0.937 | 0.947 | 0.949 | 0.95 | 0.993 | 0.9367 |
| RF | 1 | 0.963 | 0.957 | 0.957 | 0.967 | 0.969 | 0.969 | 0.998 | 0.9608 |
| CART | 1 | 0.887 | 0.883 | 0.863 | 0.9 | 0.907 | 0.907 | 0.962 | 0.8833 |
| RoF | 1 | 0.95 | 0.957 | 0.937 | 0.97 | 0.963 | 0.963 | 0.998 | 0.9533 |
| REP tree | 1 | 0.86 | 0.873 | 0.863 | 0.89 | 0.897 | 0.897 | 0.973 | 0.8717 |
| Random tree | 0.997 | 0.867 | 0.883 | 0.847 | 0.89 | 0.897 | 0.897 | 0.935 | 0.8708 |
| LAD tree | 1 | 0.863 | 0.897 | 0.9 | 0.9 | 0.913 | 0.913 | 0.985 | 0.8917 |

**Table 11.7.** Classification results for the DWT method.

| | Normal | APC | PVC | RBBB | LBBB | Average | F-measure | ROC area | Kappa |
|---|---|---|---|---|---|---|---|---|---|
| SVM | 0.997 | 0.973 | 0.963 | 0.97 | 0.987 | 0.978 | 0.978 | 0.991 | 0.9725 |
| *k*NN | 0.993 | 0.943 | 0.91 | 0.943 | 0.99 | 0.956 | 0.956 | 0.993 | 0.945 |
| ANN | 1 | 0.963 | 0.963 | 0.983 | 0.997 | 0.981 | 0.981 | 0.997 | 0.9767 |
| RF | 1 | 0.953 | 0.963 | 0.98 | 0.987 | 0.977 | 0.977 | 0.999 | 0.9708 |
| CART | 0.947 | 0.85 | 0.863 | 0.867 | 0.92 | 0.889 | 0.951 | 0.973 | 0.8617 |
| RoF | 1 | 0.94 | 0.967 | 0.987 | 0.983 | 0.975 | 0.975 | 0.999 | 0.9692 |
| REP tree | 0.933 | 0.833 | 0.847 | 0.87 | 0.92 | 0.881 | 0.881 | 0.953 | 0.8508 |
| Random tree | 0.94 | 0.883 | 0.877 | 0.81 | 0.91 | 0.884 | 0.884 | 0.928 | 0.855 |
| LAD tree | 0.947 | 0.827 | 0.83 | 0.87 | 0.91 | 0.877 | 0.877 | 0.981 | 0.8458 |

and the rotation forest closely follow the SVM by achieving respective average classification accuracies of 97.7% and 97.5%. The highest F-measure value of 0.981 is achieved by the ANN. The second highest F-measure value of 0.978 is achieved by the SVM. The random forest and the rotation forest closely follow the SVM with the respective F-measure values of 0.977 and 0.975. The highest ROC area value of 0.999 is realized by the random forest and the rotation forest, and the second highest ROC area value of 0.997 is obtained with the ANN classifier. The highest kappa value of 0.977 is realized by the ANN. The second highest kappa value of 0.973 is achieved by the SVM. The random forest and the rotation forest closely follow the SVM with the respective kappa values of 0.971 and 0.969.

It is evident from these results that ANN, SVM, random forest and rotation forest outperform in terms of the classification performance in contrast to the other classifiers considered, such as *k*NN, CART, REP tree, random tree and LDA tree. When considering the majority of the evaluation measures, such as accuracy, F-measure and kappa statistic, the ANN achieves superior performance even compared to the SVM, random forest and rotation forest.

A summary of the classification performance obtained for the WPD approach is presented in table 11.8. It shows that on average for all five arrhythmia classes, the best classification accuracy of 98.5% is realized with the ANN. The second highest average classification accuracy of 98.3% is realized by the SVM. The random forest and the rotation forest closely follow the SVM by achieving average classification accuracies of 97.3% and 97.1%, respectively. The highest F-measure value of 0.985 is achieved by the ANN. The second highest F-measure value of 0.983 is achieved by the SVM. The random forest and the rotation forest closely follow the SVM with the F-measure values of 0.973 and 0.971, respectively. The highest ROC area value of 0.999 is achieved by the ANN. The random forest achieved the second highest ROC area value of 0.998. The rotation forest closely follows the random forest with an ROC area value of 0.997. The highest kappa value of 0.982 is achieved by the ANN. The second highest kappa value of 0.978 is achieved by the SVM. The random forest closely follows the SVM with the kappa value of 0.966.

**Table 11.8.** Classification results for the WPD method.

| | Normal | APC | PVC | RBBB | LBBB | Average | F-measure | ROC area | Kappa |
|---|---|---|---|---|---|---|---|---|---|
| SVM | 1 | 0.973 | 0.993 | 0.98 | 0.967 | 0.983 | 0.983 | 0.995 | 0.9783 |
| *k*NN | 0.987 | 0.91 | 0.943 | 0.95 | 0.993 | 0.957 | 0.957 | 0.995 | 0.9458 |
| ANN | 0.997 | 0.97 | 0.98 | 0.987 | 0.993 | 0.985 | 0.985 | 0.999 | 0.9817 |
| RF | 1 | 0.951 | 0.943 | 0.981 | 0.989 | 0.973 | 0.973 | 0.998 | 0.9657 |
| CART | 0.856 | 0.748 | 0.821 | 0.827 | 0.849 | 0.82 | 0.82 | 0.916 | 0.9657 |
| RoF | 1 | 0.942 | 0.962 | 0.962 | 0.989 | 0.971 | 0.971 | 0.997 | 0.9632 |
| REP tree | 0.87 | 0.827 | 0.753 | 0.843 | 0.877 | 0.834 | 0.835 | 0.942 | 0.7925 |
| Random tree | 0.873 | 0.753 | 0.75 | 0.817 | 0.843 | 0.807 | 0.807 | 0.88 | 0.9558 |
| LAD tree | 0.885 | 0.816 | 0.755 | 0.827 | 0.849 | 0.849 | 0.826 | 0.965 | 0.7817 |

**Table 11.9.** Classification results for the SWT method.

| | Normal | APC | PVC | RBBB | LBBB | AVERAGE | F-measure | ROC area | Kappa |
|---|---|---|---|---|---|---|---|---|---|
| SVM | 0.983 | 0.973 | 0.953 | 0.983 | 1 | 0.979 | 0.979 | 0.992 | 0.9733 |
| *k*NN | 0.983 | 0.943 | 0.953 | 0.993 | 0.997 | 0.974 | 0.974 | 0.984 | 0.9675 |
| ANN | 0.98 | 0.963 | 0.96 | 0.99 | 0.997 | 0.978 | 0.978 | 0.997 | 0.9725 |
| RF | 0.96 | 0.96 | 0.95 | 0.983 | 1 | 0.971 | 0.971 | 0.997 | 0.9633 |
| CART | 0.897 | 0.9 | 0.877 | 0.93 | 0.987 | 0.918 | 0.918 | 0.96 | 0.8975 |
| RoF | 0.96 | 0.953 | 0.95 | 0.997 | 1 | 0.972 | 0.972 | 0.998 | 0.965 |
| REP tree | 0.88 | 0.857 | 0.853 | 0.923 | 0.993 | 0.901 | 0.902 | 0.964 | 0.8767 |
| Random tree | 0.867 | 0.863 | 0.823 | 0.86 | 0.98 | 0.879 | 0.879 | 0.924 | 0.8483 |
| LAD tree | 0.87 | 0.843 | 0.81 | 0.947 | 0.98 | 0.89 | 0.89 | 0.977 | 0.8625 |

It is evident from these results that the ANN, SVM, random forest and rotation forest outperform in terms of the classification performance in contrast to the other considered classifiers, such as *k*NN, CART, REP tree, random tree and LDA tree. When considering all of the evaluation measures, such as accuracy, F-measure, ROC area and kappa statistic, the ANN achieved superior performance even compared to the SVM, random forest and rotation forest.

A summary of the classification performance obtained for the SWT approach is presented in table 11.9. It shows that on average for all five arrhythmia classes, the best classification accuracy of 97.9% is achieved with the SVM. The second highest average classification accuracy of 97.8% is achieved by the ANN. The *k*NN and the rotation forest closely follow the ANN by achieving the respective average classification accuracies of 97.4% and 97.2%. The highest F-measure value of 0.979 is achieved by the SVM. The second highest F-measure value of 0.978 is achieved by the ANN. The *k*NN and the rotation forest closely follow the ANN with the respective F-measure values of 0.974 and 0.972. The highest ROC area value of 0.998 is achieved by the rotation forest. The ANN and the random forest achieved the second highest ROC area value of 0.997. The SVM closely follows with an ROC area value of 0.992. The highest kappa value of 0.9733 is achieved by the SVM. The second highest kappa value of 0.9725 is achieved by the ANN. The *k*NN and the rotation forest closely follow the ANN with respective kappa values of 0.968 and 0.965.

It is evident from these results that the SVM, ANN, *k*NN and rotation forest outperform in terms of classification performance in contrast to the other classifiers, such as random forest, CART, REP tree, random tree and LDA tree. When considering the majority of the evaluation measures, such as accuracy, F-measure and kappa statistic, the SVM achieves superior performance even compared to the ANN, *k*NN and rotation forest.

A summary of the classification performance obtained for the TQWT approach is presented in table 11.10. It shows that on average for all five arrhythmia classes, the best classification accuracy of 99.5% is achieved with the ANN. The second highest average classification accuracy of 98.9% is achieved by the SVM. The rotation forest closely follows the SVM by achieving an average classification accuracy of 98.2%. The highest F-measure value of 0.995 is achieved by the ANN. The second highest F-measure value of 0.989 is achieved by the SVM. The rotation forest closely follows the SVM with an F-measure value of 0.982. The highest ROC area value of 1.0 is achieved by the ANN. The rotation forest and the random forest achieve the second highest ROC area value of 0.999. The SVM closely follows with an ROC area value of 0.997. The highest kappa value of 0.993 is realized by the ANN. The second highest kappa value of 0.987 is achieved by the SVM. The rotation forest closely follows the SVM with a kappa value of 0.978.

It is evident from these results that the ANN, SVM and rotation forest outperform in terms of the classification performance in contrast to the other classifiers, such as random forest, *k*NN, CART, REP tree, random tree and LDA tree. When considering all of the evaluation measures, such as accuracy, F-measure, ROC area and kappa statistic, the ANN achieves superior performance even compared to the SVM and the rotation forest.

**Table 11.10.** Classification results for the TQWT method.

| | Normal | APC | PVC | RBBB | LBBB | AVERAGE | F-measure | ROC area | Kappa |
|---|---|---|---|---|---|---|---|---|---|
| SVM | 1 | 0.967 | 0.983 | 1 | 0.997 | 0.989 | 0.989 | 0.997 | 0.9867 |
| *k*NN | 0.997 | 0.94 | 0.96 | 0.95 | 0.987 | 0.967 | 0.967 | 0.995 | 0.9583 |
| ANN | 1 | 0.987 | 0.993 | 0.997 | 0.997 | 0.995 | 0.995 | 1 | 0.9933 |
| RF | 0.997 | 0.96 | 0.943 | 0.977 | 0.98 | 0.971 | 0.971 | 0.999 | 0.9642 |
| CART | 0.98 | 0.913 | 0.91 | 0.847 | 0.923 | 0.915 | 0.915 | 0.958 | 0.8933 |
| RoF | 0.997 | 0.963 | 0.97 | 0.993 | 0.987 | 0.982 | 0.982 | 0.999 | 0.9775 |
| REP tree | 0.977 | 0.87 | 0.867 | 0.867 | 0.9 | 0.896 | 0.897 | 0.963 | 0.87 |
| Random tree | 0.967 | 0.897 | 0.873 | 0.86 | 0.903 | 0.9 | 0.9 | 0.938 | 0.875 |
| LAD tree | 0.98 | 0.927 | 0.883 | 0.843 | 0.947 | 0.916 | 0.916 | 0.988 | 0.895 |

**Table 11.11.** Classification results for the DT-CWT method.

| | Normal | APC | PVC | RBBB | LBBB | Average | F-measure | ROC area | Kappa |
|---|---|---|---|---|---|---|---|---|---|
| SVM | 1 | 0.983 | 0.983 | 0.99 | 0.997 | 0.991 | 0.991 | 0.997 | 0.9883 |
| *k*NN | 0.983 | 0.937 | 0.903 | 0.93 | 0.98 | 0.947 | 0.947 | 0.967 | 0.9333 |
| ANN | 0.983 | 0.98 | 0.97 | 0.973 | 0.997 | 0.981 | 0.981 | 0.998 | 0.9758 |
| RF | 0.983 | 0.997 | 0.977 | 0.993 | 0.99 | 0.988 | 0.988 | 1 | 0.985 |
| CART | 0.927 | 0.923 | 0.903 | 0.89 | 0.94 | 0.917 | 0.917 | 0.954 | 0.8958 |
| RoF | 0.987 | 1 | 0.96 | 0.993 | 0.987 | 0.985 | 0.985 | 0.999 | 0.9817 |
| REP tree | 0.917 | 0.917 | 0.86 | 0.897 | 0.937 | 0.905 | 0.905 | 0.959 | 0.8817 |
| Random tree | 0.913 | 0.883 | 0.86 | 0.813 | 0.88 | 0.87 | 0.87 | 0.919 | 0.8375 |
| LAD tree | 0.943 | 0.927 | 0.877 | 0.913 | 0.923 | 0.917 | 0.917 | 0.989 | 0.8958 |

A summary of the classification performance obtained for the DT-DWT approach is presented in table 11.11. It shows that for the average for all five arrhythmia classes, the best classification accuracy of 99.1% is achieved with the SVM. The second highest average classification accuracy of 98.8% is achieved by the random forest. The rotation forest follows the random forest closely by attaining an average classification accuracy of 98.5%. The highest F-measure value of 0.991 is achieved by the SVM. The second highest F-measure value of 0.988 is attained by the random forest. The rotation forest closely follows the random forest with the F-measure value of 0.985. The highest ROC area value of 1.0 is achieved by the random forest. The rotation forest achieves the second highest ROC area value of 0.999. The ANN and the SVM follow closely with respective ROC area values of 0.998 and 0.997. The highest kappa value of 0.988 is realized by the SVM. The second highest kappa value of 0.985 is achieved by the random forest. The rotation forest closely follows the random forest with a kappa value of 0.982.

It is evident from these results that the SVM, random forest and rotation forest outperform in terms of the classification performance in contrast to the other classifiers, such as random forest, *k*NN, ANN, CART, REP tree, random tree and LDA tree. When considering the majority of the evaluation measures, such as accuracy, F-measure and kappa statistic, the SVM achieves superior performance even compared to the random forest and the rotation forest.

## 11.5 Discussion

The classification results reported in the above section show that for the five class arrhythmia dataset studied the overall best classification performance is achieved within the designed framework using MSPCA based ECG signal denoising, covariance based feature extraction and random tree based classification. It resulted in 100% average classification accuracy, and 1.0 F-measure, 1.0 ROC area and 1.0 kappa statistic values for the five class arrhythmia dataset considered.

The framework with MSPCA based ECG signal denoising, covariance based feature extraction and rotation forest based classification achieves the second best classification performance. It resulted in 99.9% average classification accuracy, 0.999 F-measure, 1.0 ROC area and 0.998 kappa statistic values.

The framework with MSPCA based ECG signal denoising, TQWT based feature extraction and ANN based classification achieves the third best classification performance. It resulted in 99.5% average classification accuracy, 0.995 F-measure, 1.0 ROC area and 0.993 kappa statistic values.

The framework with MSPCA based ECG signal denoising, WPD based feature extraction and ANN based classification achieves the fourth best classification performance. It resulted in 98.5% average classification accuracy, 0.985 F-measure, 0.999 ROC area and 0.982 kappa statistic values.

From the above discussion it can be revealed that for the proposed framework of automatic ECG based arrhythmia classification, among the studied parametric and time–frequency features extraction methods, the covariance method leads to the best classification performance with the random tree classifier. It resulted in 100% average classification accuracy, 1.0 F-measure, 1.0 ROC area and 1.0 kappa statistic values for the five class arrhythmia dataset considered.

Moreover, it can be summarized that among the time–frequency analysis based approaches the best performance was achieved by the framework with MSPCA based ECG signal denoising, the TQWT based features extraction technique and ANN based classification. It shows that the TQWT based feature extraction technique significantly enhances the ECG based arrhythmia classification accuracy, since TQWT is characteristically robust against input signal artifacts and noise.

When considering an overall performance comparison between all of the feature extraction techniques considered, it is evident that for the designed framework of automatic ECG based arrhythmia classification, the parametric concept based techniques secure the best classification performance. The best performance is attained by the framework with MSPCA based ECG signal denoising, covariance based features extraction and random tree based classification.

## 11.6 Conclusion

In this chapter a rich comparison is presented between the variety of robust parametric and time–frequency based feature extraction methods in the context of ECG based automatic arrhythmia classification. The results demonstrate that among the time–frequency analysis based methods the proposed framework with MSPCA based ECG signal denoising, TQWT based features extraction and ANN based classification achieves the best classification performance. It resulted in 99.5% average classification accuracy, 0.995 F-measure, 1.0 ROC area and 0.993 kappa statistic values. Among all the feature extraction approaches considered, the covariance method leads to the best classification performance when carefully employed within the proposed framework with MSPCA based ECG signal denoising and the random tree based classifier. It resulted in 100% average classification accuracy, 1.0 F-measure, 1.0 ROC area and 1.0 kappa statistic values for the five class arrhythmia dataset considered. It confirms the benefit of integrating the proposed covariance based approach in concurrent ECG analysis based automatic arrhythmia diagnosis mechanisms in modern artificial intelligence based healthcare frameworks.

## Bibliography


[1] Benjamin E J *et al* 2018 Heart disease and stroke statistics—2018 update: a report from the American Heart Association *Circulation* **137** e67–e492

[2] Ebrahimi Z, Loni M, Daneshtalab M and Gharehbaghi A 2020 A review on deep learning methods for ECG arrhythmia classification *Expert Syst. Appl.* X **7** 100033

[3] Shi H, Qin C, Xiao D, Zhao L and Liu C 2020 Automated heartbeat classification based on deep neural network with multiple input layers *Knowl.-Based Syst.* **188** 105036

[4] Wang G, Zhang C, Liu Y, Yang H, Fu D, Wang H and Zhang P 2019 A global and updatable ECG beat classification system based on recurrent neural networks and active learning *Inf. Sci.* **501** 523–42

[5] Jain S, Bajaj V and Kumar A 2018 Effective de-noising of ECG by optimised adaptive thresholding on noisy modes *IET Sci. Meas. Technol.* **12** 640–4

[6] Jain S, Bajaj V and Kumar A 2017 Riemann Liouvelle fractional integral based empirical mode decomposition for ECG denoising *IEEE J. Biomed. Health Inform.* **22** 1133–9

[7] Jain S, Bajaj V and Kumar A 2016 Efficient algorithm for classification of electrocardiogram beats based on artificial bee colony-based least-squares support vector machines classifier *Electron. Lett.* **52** 1198–200

[8] Sangaiah A K, Arumugam M and Bian G-B 2020 An intelligent learning approach for improving ECG signal classification and arrhythmia analysis *Artif. Intell. Med.* **103** 101788

[9] Mathews S M, Kambhamettu C and Barner K E 2018 A novel application of deep learning for single-lead ECG classification *Comput. Biol. Med.* **99** 53–62

[10] Jha C K and Kolekar M H 2020 Cardiac arrhythmia classification using tunable Q-wavelet transform based features and support vector machine classifier *Biomed. Signal Process. Control* **59** 101875

[11] Sannino G and De Pietro G 2018 A deep learning approach for ECG-based heartbeat classification for arrhythmia detection *Future Gener. Comput. Syst.* **86** 446–55

[12] Romdhane T F, Alhichri H, Ouni R and Atri M 2020 Electrocardiogram heartbeat classification based on a deep convolutional neural network and focal loss *Comput. Biol. Med.* 103866

[13] Shi H, Wang H, Huang Y, Zhao L, Qin C and Liu C 2019 A hierarchical method based on weighted extreme gradient boosting in ECG heartbeat classification *Comput. Methods Programs Biomed.* **171** 1–10

[14] Li Z, Zhou D, Wan L, Li J and Mou W 2020 Heartbeat classification using deep residual convolutional neural network from 2-lead electrocardiogram *J. Electrocardiol.* **58** 105–12

[15] Lee M, Song T-G and Lee J-H 2020 Heartbeat classification using local transform pattern feature and hybrid neural fuzzy-logic system based on self-organizing map *Biomed. Signal Process. Control* **57** 101690

[16] Shi H, Wang H, Zhang F, Huang Y, Zhao L and Liu C 2019 Inter-patient heartbeat classification based on region feature extraction and ensemble classifier *Biomed. Signal Process. Control* **51** 97–105

[17] Asgharzadeh-Bonab A, Amirani M C and Mehri A 2020 Spectral entropy and deep convolutional neural network for ECG beat classification *Biocybern. Biomed. Eng.* **40** 691–700

[18] Mondéjar-Guerra V, Novo J, Rouco J, Penedo M G and Ortega M 2019 Heartbeat classification fusing temporal and morphological information of ECGs via ensemble of classifiers *Biomed. Signal Process. Control* **47** 41–8

[19] Rajesh K N V P S and Dhuli R 2018 Classification of imbalanced ECG beats using re-sampling techniques and AdaBoost ensemble classifier *Biomed. Signal Process. Control* **41** 242–54

[20] Alickovic E and Subasi A 2016 Medical decision support system for diagnosis of heart arrhythmia using DWT and random forests classifier *J. Med. Syst.* **40** 1–12

[21] Atal D K and Singh M 2020 Arrhythmia classification with ECG signals based on the optimization-enabled deep convolutional neural network *Comput. Methods Programs Biomed.* **196** 105607

[22] Oh S L, Ng E Y, San Tan R and Acharya U R 2018 Automated diagnosis of arrhythmia using combination of CNN and LSTM techniques with variable length heart beats *Comput. Biol. Med.* **102** 278–87

[23] Moody G B and Mark R G 2001 The impact of the MIT-BIH arrhythmia database *IEEE Eng. Med. Biol. Mag.* **20** 45–50

[24] Addison P S 2017 *The Illustrated Wavelet Transform Handbook: Introductory Theory and Applications in Science, Engineering, Medicine and Finance* (Boca Raton, FL: CRC Press)

[25] Bakshi B R 1998 Multiscale PCA with application to multivariate statistical process monitoring *AIChE J.* **44** 1596–610

[26] Bore J C, Ayedh W M A, Li P, Yao D and Xu P 2019 Sparse autoregressive modeling via the least absolute LP-norm penalized solution *IEEE Access* **7** 40959–68

[27] Mian Qaisar S and Subasi A 2020 Effective epileptic seizure detection based on the event-driven processing and machine learning for mobile healthcare *J. Ambient Intell. Humaniz. Comput.*

[28] Mian Qaisar S and Subasi A 2020 Cloud-based ECG monitoring using event-driven ECG acquisition and machine learning techniques *Phys. Eng. Sci. Med.* **43** 623–34

[29] Bonat W H and Jørgensen B 2016 Multivariate covariance generalized linear models *J. R. Stat. Soc.* C **65** 649–75

[30] Seth D, Chakraborty D, Ghosal P and Sanyal S K 2017 Brain computer interfacing: a spectrum estimation based neurophysiological signal interpretation *2017 4th Int. Conf. on Signal Processing and Integrated Networks (SPIN) (IEEE)* pp 534–9

[31] Wei Z, Xiaolong L, Jin Z, Xueyun W and Hongxing L 2018 Foetal heart rate estimation by empirical mode decomposition and MUSIC spectrum *Biomed. Signal Process. Control* **42** 287–96

[32] Prabhakar S K, Rajaguru H and Lee S-W 2020 Eigen vector method with swarm and non swarm intelligence techniques for epileptic seizure classification *2020 8th Int. Winter Conf. on Brain-Computer Interface (BCI) (IEEE)* pp 1–6

[33] Kevric J and Subasi A 2017 Comparison of signal decomposition methods in classification of EEG signals for motor-imagery BCI system *Biomed. Signal Process. Control* **31** 398–406

[34] Chen D, Wan S, Xiang J and Bao F S 2017 A high-performance seizure detection algorithm based on discrete wavelet transform (DWT) and EEG *PLoS One* **12** e0173138

[35] Alickovic E, Kevric J and Subasi A 2018 Performance evaluation of empirical mode decomposition, discrete wavelet transform, and wavelet packed decomposition for automated epileptic seizure detection and prediction *Biomed. Signal Process. Control* **39** 94–102

[36] Thakor N V, Gramatikov B, Sherman D and Bronzino J 2000 Wavelet (time-scale) analysis in biomedical signal processing *Biomed. Eng. Handb.* **56** 1–56

[37] Ingle V K and Proakis J G 2016 *Digital Signal Processing Using Matlab: a Problem Solving Companion* (Boston, MA: Cengage Learning)

[38] Khokhar S, Zin A A B M, Mokhtar A S B and Pesaran M 2015 A comprehensive overview on signal processing and artificial intelligence techniques applications in classification of power quality disturbances *Renew. Sustain. Energy Rev.* **51** 1650–63

[39] Subasi A 2013 Classification of EMG signals using PSO optimized SVM for diagnosis of neuromuscular disorders *Comput. Biol. Med.* **43** 576–86

[40] Wang S, Du S, Atangana A, Liu A and Lu Z 2018 Application of stationary wavelet entropy in pathological brain detection *Multimed. Tools Appl.* **77** 3701–14

[41] Selesnick I W 2011 Wavelet transform with tunable Q-factor *IEEE Trans. Signal Process.* **59** 3560–75

[42] Subasi A and Qaisar S M 2020 Surface EMG signal classification using TQWT, bagging and boosting for hand movement recognition *J. Ambient Intell. Humaniz. Comput.* https://doi.org/10.1007/s12652-020-01980-6

[43] Patidar S and Pachori R B 2014 Classification of cardiac sound signals using constrained tunable-Q wavelet transform *Expert Syst. Appl.* **41** 7161–70

[44] Wang S, Lu S, Dong Z, Yang J, Yang M and Zhang Y 2016 Dual-tree complex wavelet transform and twin support vector machine for pathological brain detection *Appl. Sci.* **6** 169

[45] Celik T and Ma K-K 2010 Unsupervised change detection for satellite images using dual-tree complex wavelet transform *IEEE Trans. Geosci. Remote Sens.* **48** 1199–210

[46] Selesnick I W, Baraniuk R G and Kingsbury N C 2005 The dual-tree complex wavelet transform *IEEE Signal Process Mag.* **22** 123–51

[47] Han J, Pei J and Kamber M 2011 *Data Mining: Concepts and Techniques* (Amsterdam: Elsevier)

[48] Wu X, Kumar V, Quinlan J R, Ghosh J, Yang Q, Motoda H, McLachlan G J, Ng A, Liu B and Philip S Y 2008 Top 10 algorithms in data mining *Knowl. Inf. Syst.* **14** 1–37

[49] Witten I H, Frank E, Hall M A and Pal C J 2016 *Data Mining: Practical Machine Learning Tools and Techniques* (San Mateo, CA: Morgan Kaufmann)

[50] Holmes G, Pfahringer B, Kirkby R, Frank E and Hall M 2002 *Multi Class Alternating Decision Trees Machine Learning: ECML 2002* (Berlin: Springer) pp 161–72

[51] Kalmegh S 2015 Analysis of WEKA data mining algorithm REPTree, simple CART and random tree for classification of Indian news *Int. J. Innov. Sci. Eng. Technol.* **2** 438–46

[52] Ngo T 2011 Data mining: practical machine learning tools and technique third edition by I H Witten, E Frank and M A Hall *ACM SIGSOFT Softw. Eng. Notes* **36** 51–2

[53] Breiman L 2001 Random forests *Mach. Learn.* **45** 5–32

[54] Han J, Pei J and Kamber M 2011 *Data Mining: Concepts and Techniques* (Amsterdam: Elsevier)

[55] Rodriguez J J, Kuncheva L I and Alonso C J 2006 Rotation forest: a new classifier ensemble method *IEEE Trans. Pattern Anal. Mach. Intell.* **28** 1619–30

[56] Cichosz P 2014 *Data Mining Algorithms: Explained Using R* (New York: Wiley)

**IOP** Publishing

Modelling and Analysis of Active Biopotential Signals in Healthcare, Volume 2

**Varun Bajaj and G R Sinha**

# Chapter 12

# Segmentation of ECG waves using LSTM networks

**Aboli N Londhe and Mithilesh Atulkar**

*National Institute of Technology Raipur, Chhattisgarh, India, 492010*

The electrocardiogram (ECG) signal consists of key waves, i.e. the P wave, QRS wave and T wave, and carry critical cardiac information. Further derivation of standard intervals and segments from these waves alerts cardiologists to any abberations and guides them in diagnosis and treatment. The detection of these waves is very challenging due to the complexity of the cardiac complexes. Despite the advanced machine learning and deep learning methods used in the literature, current methods are less than reliable and robust.

A semantic segmentation approach has been implemented here for reliable and robust detection of critical ECG waves in a continuous signal. For the sample-wise classification of ECG signals in this study, long short-term memory (LSTM) networks are attempted as they are well suited to classifying time series data. The QT database (QTDB) from PhysioNet is used in this study, which is designed for delineation of ECG signal. It includes roughly 15 min long excerpts of two-channelled ECG recordings sampled at 250 Hz from a total of 105 patients. The QTDB is an excellent database that includes cases of related abnormalities properly illustrated with annotations of the important wave boundaries using an automated intelligent system. Exhaustive experiments are conducted using LSTM, bidirectional LSTM (BiLSTM) and stacked BiLSTM networks which are trained with 70% of the data (73 patients) and tested with the remaining 30% (32 patients).

The average and weighted averages of performance parameters, i.e. accuracy, recall, precision, specificity and F1-score, are measured. The stacked BiLSTM network outperforms the other networks with an accuracy of 89.77% compared to accuracies of 83.26% and 87.46% for the LSTM and BiLSTM networks, respectively. The results of semantic segmentation are also compared to conventional segmentation which fails to segment continuous ECG signals in real time.

## 12.1 Introduction

The recording of rhythmic activity due to the polarization and depolarization of the ventricles and atrium of the heart is known as an electrocardiogram (ECG). ECG is the recording of electrical currents generated primarily due to the propagation of stimuli through the pairs of atria and ventricles, called activation, followed by their reversal to the resting state, called recovery. This rhythmic activity acts as an important biomarker for the diagnosis of various cardiac diseases [1]. With expert clinical experience, some critical cardiac diseases can be detected using accurate ECG diagnoses while many others can be estimated with an acceptable probability [2]. This rhythmic activity follows a pattern of waveforms termed the P wave, QRS complex and T wave. The durations of these waves, which define the cardiac subprocesses (intervals) and periods of silence (segments), provide clinical information about the functioning of the heart.

### 12.1.1 ECG waves, intervals and segments

ECG carries clinical information with standard intervals of time, i.e. PR segment, PR interval, QRS interval, QT interval and ST segment [1–5], as shown in figure 12.1.

The first wave of ECG is the P wave and it represents the atrial depolarization that involves the successive activation of the right and left atrium of the heart. The height of the P wave is no greater than 2.5 mm (0.25 mV) while its width does not exceed more than 0.12 s. An abnormal P wave morphology in the ECG signal indicates issues in atrial depolarization. The enlargement of atria due to aberrant

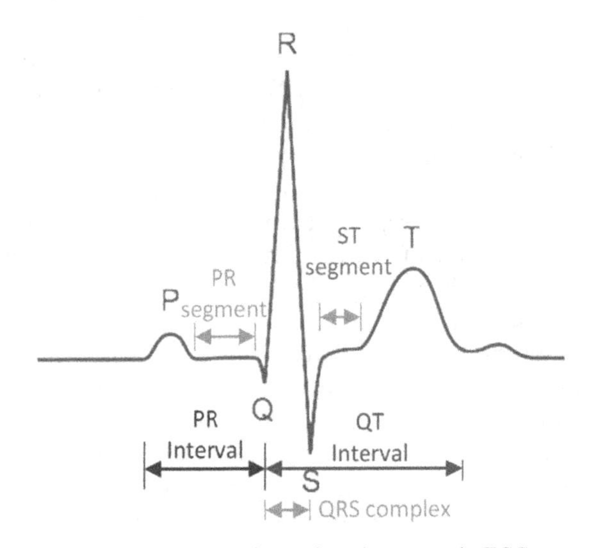


**Figure 12.1.** Waves, intervals and segments in ECG.

valves is a condition of reduced ability to empty blood into the ventricles. Pulmonary valve stenosis or enhanced pulmonary artery pressure enlarge the right atrium, called hypertrophy [1]. This condition leads to increased electrical currents and consequently increases the amplitude of the P wave, termed P pulmonale. Similarly, the enlargement of the left atrium due to mitral valve stenosis amplifies its contribution to the P wave, termed P mitrale [1, 3]. In another abnormality, the morphology of the P wave changes when cells far away from the sinoatrial node (the ectopic focus) depolarize the atria. Thereafter, the electrical stimulus propagates in the opposite direction to normal, thus making a normal P wave negative.

The QRS wave in ECG reflects the depolarization of the ventricles, i.e. the spread of stimulus through the ventricles. Its duration is measured from the onset to the end of the QRS wave. The QRS amplitude varies among patients and measured ECG leads. It depends on the size of the ventricular chambers and the closer placement of ECG electrodes. Its duration should not exceed 0.1 s. Abnormal changes in the QRS of ECG are produced by both right and left ventricular hypertrophy due to persistent pressure leading to increased wall thickness. Typically, there is simultaneous depolarization of the left and right ventricles, but due to the larger mass, the left ventricle shares the major electrical current. A broad QRS wave also indicates slow ventricular depolarization which reflects the dysfunction of the conduction system [3].

The third normal wave component of ECG is the T wave, which represents the repolarization of ventricles. The duration of the T wave ranges from 0.10 to 0.25 s or sometimes even longer. Its amplitude typically does not exceed more than 5 mm while its shape is sharp-rounded or asymmetrical. The T wave also suffers from abnormalities of low amplitude (flat), high amplitude and inversion, which may be the result of cardiac and non-cardiac conditions, i.e. occlusion in coronary arteries, hyperkalaemia and ischemia, respectively [1, 4].

The span of time that includes one or more waves and their connected silent periods is called an interval. There are three significant intervals PR, QT and RR that are assessed by cardiologists [1]. The PR interval is generally measured from the start of the P wave to the start of the QRS wave, ranging from 0.12 to 0.22 s. It reflects the time required for the stimulus to spread through the atria and further pass through the atrioventricular junction.

The QT interval ranges from the beginning of the QRS wave and continues to the end of the T wave. It primarily represents the total duration of de- and repolarization of the ventricles. Extended QT intervals suggest potentially deadly ventricular arrythmias and hence must be assessed frequently [4]. The normal value of the QT interval depends inversely on the heart rate.

The RR interval is the interval between two consecutive QRS complexes and represents the heart rate. It reflects cardiac conditions such as atrial arrythmias, tachycardia and prolonged QT interval, as mentioned above [1].

The isoelectric line between the offset point of the P wave and the onset of the QRS complex is called the PR segment. It indicates the slow spread of stimulus through the atrioventricular node. PR segment deviations may indicate acute pericarditis [1, 3].

Another isoelectric line between the offset of the QRS complex and the T wave is called the ST segment. This segment is altered in many cardiac conditions and hence requires careful assessment. Its deviations indicate acute myocardial ischemia. Deviations are either of depression type, i.e. a shift below the reference PR segment, or the elevation type [1, 3]. ST depressions represent cardiac conditions such as digoxin, sympathetic tone and hypokalemia, heart failure, tachycardia, etc, while ischemia typically causes ST elevations [1].

### 12.1.2 ECG wave segmentation

Each of these ECG wave components have their standard amplitude and duration which indicate the various cardiac biomarkers. Analysis of these biomarkers guides medical experts in the diagnosis and treatment of health conditions [6]. However, the subjective analysis of ECG signals requires expertise and is time consuming. Therefore, researchers are focusing on objective solutions which overcome the limitations of subjective analysis. Moreover, advancements in signal processing and computing techniques enables the automatic interpretation of ECG which further helps in the accurate and fast diagnosis of cardiovascular disease [6–11].

In the literature several methods have been proposed for automatic ECG segmentation and characterization. They can be categorized as the detection of (i) peaks, (ii) fiducial points or (iii) waves in the ECG signal [12–26]. These methods were targeted either to measure amplitude and duration or to understand wave morphologies. However, for accurate measurement of the amplitude and duration for determining cardiac abnormalities from ECG, the determination of morphology can provide a complete assessment of intervals or/and segments [1–5]. Therefore, the focus of research has shifted towards the segmentation of individual ECG waves [27–59].

Conventional signal processing techniques, such as low-pass differentiators [27, 28], derivative based methods [29–37], wavelet-based methods [38–54], the Hilbert transform (HT) [37, 53–57], piecewise linear approximations [58, 59], multiple higher order moments [51], the multiple-order derivative wavelet (MDWM) [48], the multi-scale morphological derivative [60], the phasor transform (PT) [61], bandwidth features with a support vector machine (SVM) [68] and amplitude-based methods [33] have been attempted to extract the ECG waves and related intervals/segments. Even though these methods achieved state-of-the-art results, these approaches were still mainly based on the detection of the QRS complex. Neglecting the P and T waves restricts them to finding limited information. Hence, in order to extract other waves, such as thee P wave and T wave, various feature-based machine learning approaches such as neural networks (NNs) [24], adaptive thresholding [44, 61–65], naïve Bayes [66], hidden Markov model (HMM) [42–46, 48, 50, 71], logistic regression [12] and ruled-based [17] methods have been explored to detect and further classify ECG waves. Although these state-of-the-art machine learning techniques could achieve accuracies ranging from 91%–98%, the primary limitation of the conventional machine learning based classification of ECG intervals is their feature dependency. Machine learning based approaches are highly feature driven and require suitable features to provide satisfactory results [6–8].

In recent times, to overcome the limitation of the feature dependency of conventional machine learning methods, deep learning (DL)-based techniques are being adopted for various applications of pattern recognition and classification. Good work based on DL for ECG [6–8, 14, 69, 70] and EEG signals [71, 72] has been published in recent years. Among the various DL methods, the convolution neural network [6, 7] and long short-term memory (LSTM) [8] have been found to be efficient in ECG segmentation.

In [6] a four-layered basic CNN model was used to localize the P, QRS and T waves in the ECG signal. This model achieved an average accuracy of 96.2%. However, the work contains very superficial experimentation and insufficient evaluation to justify the proposed approach. In [7] a deep CNN was applied to segment the ECG waves. The experimental evaluation was performed by calculating the true positive rate (TPR) and positive predictive value (PPV). It has been found that the segmentation of the QRS complex achieved a very high TPR and PPV, but the performance was found to decrease for the P and T waves. In [8] bidirectional LSTM (BiLSTM) was used to classify ECG waves, however it used manually based derivatives. This semi-automated technique could achieve an average accuracy of 92.00%.

In the above literature, irrespective of the advanced machine learning or deep learning technique, the existing methods mainly suffer from four major problems. First, and most importantly, these works were less reliable as they were mainly focused on detecting the fiducial points, which are highly challenging due to several variables such as shape, amplitude, frequency and time in the ECG signals. Second, none of the existing methods has presented the experimental evaluation of different ECG leads, as there is high variability from lead to lead [73, 74]. Third, there is a lack of robustness as the existing experiments were performed on only single clean datasets. Hence, there remains the problem of generalization. Lastly, the conventional segmentation approaches were designed for classification or detection of individual intervals and cannot be adopted directly on continuous raw ECG signals due to the biphasic nature of the P wave and T wave.

Therefore, the current authors are motivated to provide a deep learning method which can segment all the waves in a continuous signal. For that purpose, the concept of a pixel-wise classification approach for images, called semantic segmentation, has been adopted here [75]. Basically, the concept of the pixel-wise classification of images has been modified for the sample-wise classification of signals.

In this work, possible variants of LSTM networks have been designed and investigated to address the semantic segmentation of the ECG intervals in continuous ECG signals. The study involves the design and implementation of three variants, i.e. LSTM, BiLSTM and stacked BiLSTM. The proposed approach includes LSTM to extract temporal dependencies as well as short- and long-time dependencies in the forward and backward time stamps for semantic segmentation. The main purpose of using BiLSTM [76–81] is to compute information in the forward and backward time stamps, which helps to classify each prior and future data point which further helps to handle lead variability [74]. The experiments have been performed using the QT database [82] which is a standard publicly available ECG signal dataset with annotation of the P wave, QRS complex and T wave. These LSTM networks are applied on the raw ECG signals without any pre-processing and feature engineering.

This chapter is further organized as follows. Section 12.2 describes the methodology for the semantic segmentation of ECG intervals using LSTM networks. Hence, the network architectures and the method of semantic segmentation are provided in this section. A detailed description of the dataset and the experimental set-up are provided in section 12.3. The performance measurement of the LSTM networks with semantic segmentation and the conventional segmentation is presented in section 12.4. The comparative analysis, the difference between conventional segmentation and semantic segmentation, and the reliability index are discussed in section 12.5. Finally, this chapter ends with the conclusion and consideration of future research directions.

## 12.2 Methodology

This section provides an overview of the LSTM networks proposed for ECG waves, i.e. P wave, QRS complex, T wave and neural segmentation. A recurrent neural network (RNN) is a more precise architecture of the human neural networks in which information persists in loops. RNNs refer to the past information but make inferences from information with gaps, i.e. long-term dependencies are not possible. LSTM is a special type of RNN which has the capability of learning these long-term dependencies. Their default behaviour is to remember information for a long period. The structure of an LSTM cell is depicted in figure 12.2. There are three gates in the memory cell, namely the input $i_t$, output $o_t$ and forget $f_t$ gates. These gates are capable of adding or removing information by regulating the flow of information in and out of the cell. As a result, the cell can remember arbitrary time interval values.

From figure 12.2, at time $t$, the unit components of LSTM have been updated as follows:

$$i_t = \sigma(W_{Xi}X_t + W_{Gi}G_{t-1} + W_{ci}c_{t-1} + b_i) \tag{12.1}$$

$$f_t = \sigma(W_{Xf}X_t + W_{Gf}G_{t-1} + W_{cf}c_{t-1} + b_f) \tag{12.2}$$

$$o_t = \sigma(W_{Xo}X_t + W_{Go}G_{t-1} + W_{co}c_{t-1} + b_o) \tag{12.3}$$

$$c_t = f_t c_{t-1} + i_t \tanh(W_{Xc}X_t + W_{Gc}G_{t-1} + b_c) \tag{12.4}$$

$$G_t = o_t \tanh(c_t), \tag{12.5}$$

where $\sigma(\cdot)$ is the sigmoid activation function. The input gate, forget gate, output gate, cell input activation and cell state vectors are represented by $i, f, o, a$ and $c$, respectively. All these elements have the same size as the hidden vector $G$. The weight matrices $W_{ci}$, $W_{cf}$, $W_{co}$ are for peephole connections.

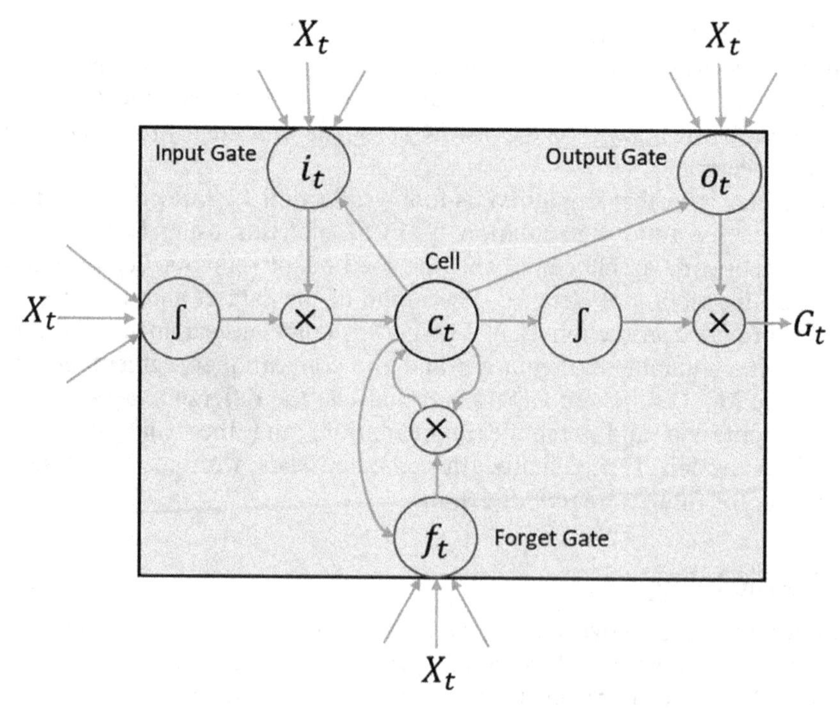


**Figure 12.2.** Architecture of a typical LSTM cell.

**Table 12.1.** Architectural details of the LSTM model.

| Layer no. | Layer | Hyperparameter | Output shape |
|---|---|---|---|
| L0 | Input layer | | (5000 × 1) |
| L1 | LSTM | 250 units | (5000 × 500) |
| L2 | Dense + softmax | 4 units | (5000 × 4) |

### 12.2.1 Architectural details of the LSTM model

The proposed LSTM based model for ECG wave segmentation consists of an input layer with a size of 5000 × 1. The hidden layer consists of single LSTM layer. For classification, a fully connected layer followed by a softmax activation function has been utilized. The implemented LSTM layer consists of 250 hidden cells. The number of units in the dense layer is set to 4. The details of each layer are given in table 12.1.

### 12.2.2 Bidirectional LSTM (BiLSTM) network

A BiLSTM network is a bidirectional RNN that utilizes LSTM cells in the forward and backward directions of the data. A BiLSTM layer learns long-term dependencies in time series data with gaps in both the forward and backward directions.

The BiLSTM utilizes both past and future data as an information source to compute backward and forward hidden sequences. The backward and forward hidden sequences are combined using the element-wise sum technique. The BiLSTM layer can classify a time series data point based on past and future data points. For example, in ECG wave segmentation the P segment can be predicted based on previous data points (earlier T waves) and future data points (the next QRS segment). Similar long-term dependencies can be learned for QRS, T, and neutral data points.

#### *12.2.2.1 Architectural details of the BiLSTM model*

In this study, to learn the long-term dependencies a BiLSTM based hidden layer has been utilized, as shown in figure 12.3. The input ECG sequence is fed to the BiLSTM layer. The layerwise details are presented in table 12.2. The BiLSTM layer includes 250 LSTM cells. Hence, there are a total of 500 cells in the L1 layer of the proposed model. The features generated by the BiLSTM layer are then passed through the batch normalization [6] layer followed by a fully connected layer with the softmax activation function. The number of units in the L2 layer is set to 4 which represents the number of classes, i.e. ECG waves.

In the last layer, a fully connected layer followed by the softmax activation function are utilized to segment the ECG time series into four classes, namely P wave, QRS complex, T wave and neutral. The ECG time series has a dimension of size of 5000. As a result, the size of the segmented ECG wave is also 5000. Suppose we have an ECG time series signal $x$ of length $T$. After training the proposed model the network produces an output sequence $y$ of length $T$. There are $|C|$ possible

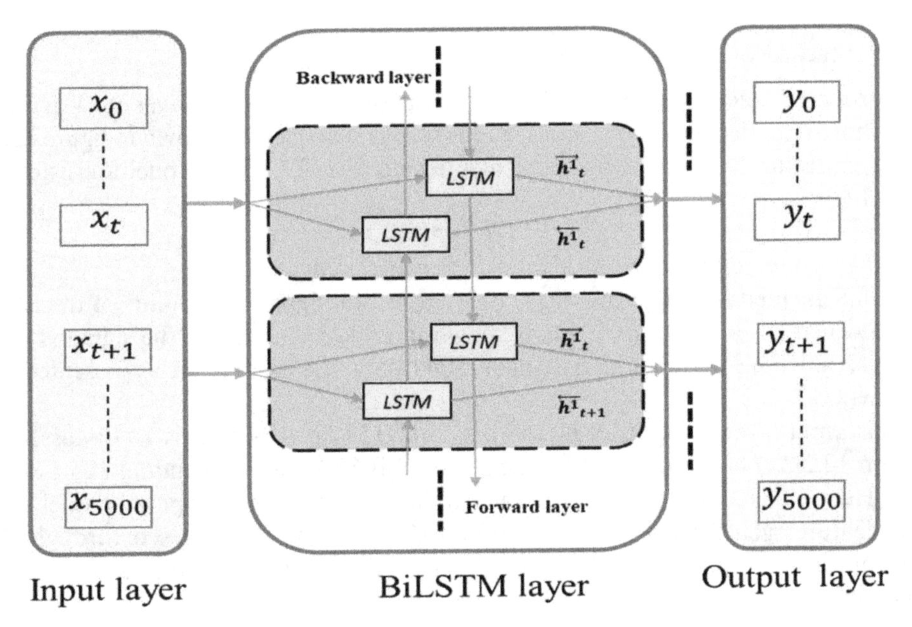


**Figure 12.3.** The architecture of the BiLSTM model.

**Table 12.2.** Architectural details of the single BiLSTM model.

| Layer no. | Layer | Hyperparameter | Output shape |
|---|---|---|---|
| L0 | Input Layer | | (5000 × 1) |
| L1 | BiLSTM | 250 units | (5000 × 500) |
| L2 | Dense + softmax | 4 units | (5000 × 4) |

classes where $C = (1, 2, 3, 4)$ and $c \in C$ for each probability distribution $y_t$. For a given $x$ state $k$ at time $t$, $y_t^c$ is the model estimation for the probability. The model has been trained to minimize the negative log probability with the length $T$ and target sequence $v$, as shown in equation (12.6). Moreover, the error derivative at the output is calculated from equation (12.7):

$$-\log P\left(\frac{v}{x}\right) = \sum_{t=1}^{T} \log y_t^{vt} \tag{12.6}$$

$$-\frac{\partial \log P\left(\frac{-}{x}\right)}{\partial \bar{y}_t^k} = y_t^k - \delta_{k,v_t}, \tag{12.7}$$

where $\bar{y}^k$ is the activation before the softmax layer. To obtain the weight gradient the derivatives are then fed to the model using backpropagation through time. Once the model has been trained the final output of the segmented ECG sequence can be acquired by calculating the maximum probability.

### 12.2.3 Stacked BiLSTM network

The proposed stacked BiLSTM model consists of two BiLSTM layers to learn the long short-term dependencies from the input ECG sequence, as shown in figure 12.4. The architectural details of the implemented stacked BiLSTM model are listed in table 12.3.

#### *12.2.3.1 Architectural details of the stacked BiLSTM model*

The implemented model's input layer consists of raw signal data points of the ECG signal which has a length of 5000, as shown in table 12.3. Hence, the signal is one-dimensional time series data that have 5000 data points and are represented by $X = (x_1, x_2, x_t, x_{t+1}, \ldots, x_{5000})$.

This input layer is followed by the first BiLSTM layer, similarly as discussed in section 12.2.2. This is followed by the second BiLSTM layer including 125 LSTM cells. Hence, there are a total of 250 units in the L2 layer of the proposed model. The joint features generated by the two BiLSTM layers are then passed through the batch normalization followed by a fully connected layer with the softmax activation function.

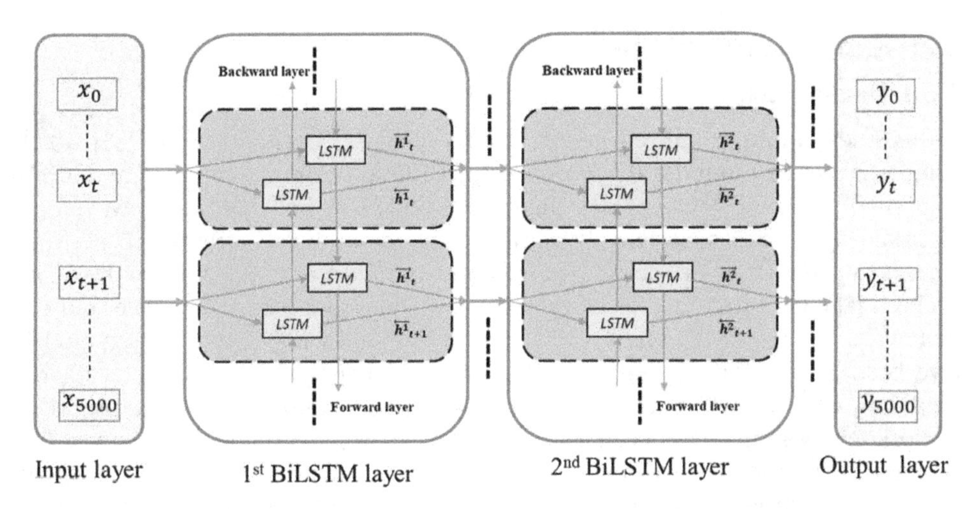


**Figure 12.4.** The architecture of the proposed stacked BiLSTM model.

**Table 12.3.** Architectural details of the stacked BiLSTM model.

| Layer no. | Layer | Hyperparameters | Output shape |
|---|---|---|---|
| L0 | Input layer | | (5000 × 1) |
| L1 | BiLSTM 1 | 250 units | (5000 × 500) |
| L2 | BiLSTM 2 + BN | 125 units | (5000 × 250) |
| L3 | Dense + softmax | 4 units | (5000 × 4) |

### 12.2.4 Semantic segmentation

Generally, segmentation is very popular in image processing for object detection applications, wherein pixel-wise classification of the image takes place to separate the foreground and background. Therefore, segmentation is also treated as a binary classification problem where each pixel is classified into foreground and background. On the other hand, in the existing segmentation of the ECG waves, various recent studies [7–12, 27–67] have implemented ECG segmentation using the conventional segmentation approach, i.e. binary classification of the sample into P/QRS/T or n/a. Although these approaches have remarkable results of segmentation of ECG intervals (see table 12.7 below), these approaches are not applicable for continuous real-time signals as they only classify the samples into either P/QRS/T or n/a. In contrast, semantic segmentation performs multiclass classification to extract or locate multiple regions/objects from singular input data (image/signal). Therefore, to overcome the limitation of the existing segmentation approaches for ECG interval detection, we have proposed a model to perform sample-wise classification to detect P, QRS, T or n/a in continuous raw ECG signal.

## 12.3 Experimental set-up

### 12.3.1 Dataset details

In this study, we have utilized the QT database (QTDB) of PhysioNet [81] which includes around 15 min duration two-channel ECG recordings sampled at 250 Hz from 105 patients. The QTDB dataset provides a well-annotated waveform generated by an expert system. By utilizing this dataset, researchers can perform experiments on ECG wave detection and segmentation. Each sample from the signal has been labelled with four classes namely, P, QRS, T and n/a. The samples outside the P wave, QRS complex and T wave are represented by n/a. The proposed models have been trained with 70% of the dataset (i.e. 73 patients) and tested with the remaining 30% (i.e. 30 patients). To provide a fixed sample size to the proposed network, every sample has been divided into a sequence of 5000 data points. The dataset distributions utilized in this study are presented in table 12.4. For training and testing, respectively, 6502 and 2844 samples have been used. The raw ECG signal with 1000 data points is depicted in figure 12.5. In this study, 5000 raw ECG data points have been used as the input and corresponding labels are used as the target for the proposed models.

**Table 12.4.** The number of data points sampled in the training and testing phases.

| Class | Training dataset (73 out of 105 patients ≅ 70%) | Testing dataset (32 out of 105 patients ≅ 30%) |
|---|---|---|
| P wave | 3,507,487 | 1,281,446 |
| QRS complex | 4,578,579 | 1,987,930 |
| T wave | 7,970,413 | 3,527,709 |
| n/a | 16,453,521 | 7,422,915 |
| Total | 32,510,000 | 14,220,000 |

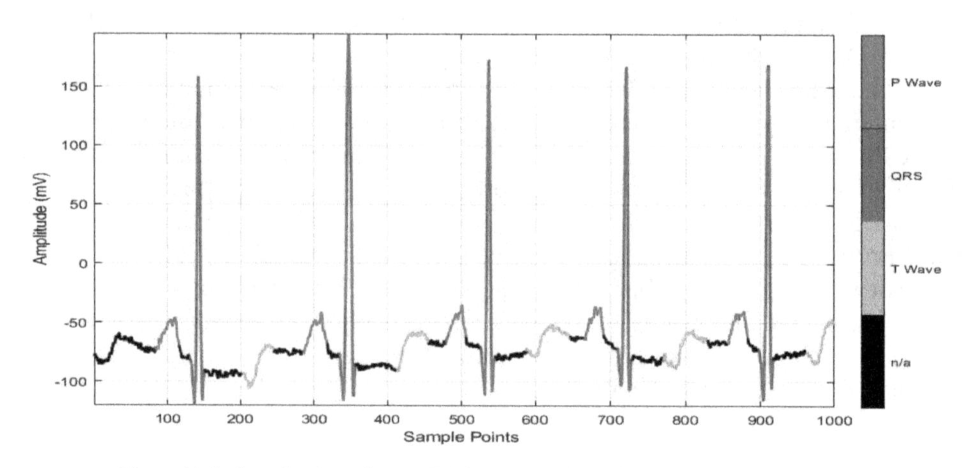


**Figure 12.5.** Sample of continuous ECG wave segmentation from raw ECG signal.

### 12.3.2 System implementation details

In this study, all the experiments were implemented using the Python Keras deep learning library [8]. The experiments were carried out on a workstation with an Intel Xeon CPU with 64 GB RAM and NVIDIA Quadro GPU with 8 GB memory. During network training, first the weights were initialized using Xavier's method [9] and then the hyperparameters have been fine-tuned to improve the performance of the proposed models. The proposed models have been trained using the Adam optimizer [10] with an initial learning rate of 0.001. The categorical cross entropy-based loss function has been utilized. All the experiments have been performed with the same dataset and hyperparameters without any pre-processing step.

## 12.4 Results

This section provides the experimental results of the proposed networks for semantic segmentation of the ECG signal. In this study we have implemented three experiments, namely semantic segmentation of ECG waves using (i) an LSTM based model, (ii) a single BiLSTM model and (iii) a stacked BiLSTM model. All the experiments have been evaluated using five metrics, namely precision, recall, F1-score, specificity and accuracy. The mathematical definitions of these evaluation metrics are tabulated in table 12.5.

### 12.4.1 Results using a single LSTM based model

The sample-wise evaluation metrics are listed in table 12.6. Moreover, we have also calculated weighted average metrics for the ECG wave segmentation. From table 12.6 it can be observed that the model has achieved accuracies of 92.54%, 92.94% and 87.47% for the P wave, QRS wave and T wave, respectively. Moreover, the average and weighted average accuracies are 87.47 and 83.26, respectively. For the neural signal, we have achieved 77.06%. The proposed model achieved F1-scores of 51.92, 72.96 and 74.29 for the P, QRS and T waves, respectively. The low F1-score indicates that the proposed model might not be able to learn the context of the time series trend properly due to noise. It can be observed from table 12.6 that the proposed method can still achieve sub-optimal segmentation results without using any pre-processing techniques.

**Table 12.5.** Mathematical formulae of the selected performance parameters.

| Parameter | Formulae |
|---|---|
| Precision (PPV) | TP/(TP + FP) |
| Recall (TPR) | TP/(TP + FN) |
| F1-score | {2 × (TPR × PPV)}/(TPR + PPV) |
| Specificity (TNR) | TN/(TN + FP) |
| Accuracy | (TP + TN)/(TP + TN + FP + FN) |

**Table 12.6.** Results of the semantic segmentation of ECG waves using LSTM.

| Class | Precision | Recall | F1-score | Specificity | Accuracy |
|---|---|---|---|---|---|
| P wave | 0.6200 | 0.4467 | 0.5192 | 0.9728 | 0.9254 |
| QRS complex | 0.7860 | 0.6807 | 0.7296 | 0.9698 | 0.9294 |
| T wave | 0.7565 | 0.7298 | 0.7429 | 0.9225 | 0.8747 |
| n/a | 0.7545 | 0.8307 | 0.7908 | 0.7049 | 0.7706 |
| Avg | 0.7292 | 0.6719 | 0.6956 | 0.8925 | 0.8750 |
| Wt. Avg. | 0.7474 | 0.7501 | 0.7460 | 0.8201 | 0.8326 |

Avg = average, Wt. Avg. = weighted average.

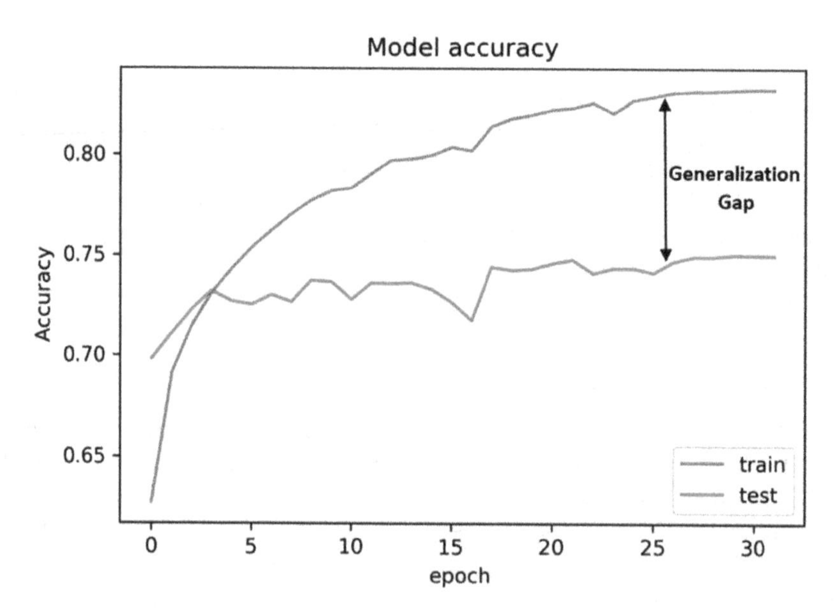


**Figure 12.6.** The learning curve for a single LSTM model.

Further, we have also analysed the trained testing curve (i.e. learning curve) for the proposed LSTM model. The curve shown in figure 12.6 indicates that there is a high generalization gap between the training and testing set. To analyse the percentage of true positives and false positives for each class, a confusion matrix has been investigated as shown in figure 12.7.

### 12.4.2 Results using a BiLSTM based model

The LSTM based ECG wave segmentation has a high generalization gap (see figure 12.6). To reduce the generalization error we have implemented another DL-based model that utilizes the BiLSTM layer. This section provides the validation results of the second model, namely the BiLSTM based model for ECG wave segmentation. The sample-wise evaluation metrics are listed in table 12.7. As for the LSTM model, we have also calculated class-wise and weighted average metrics for

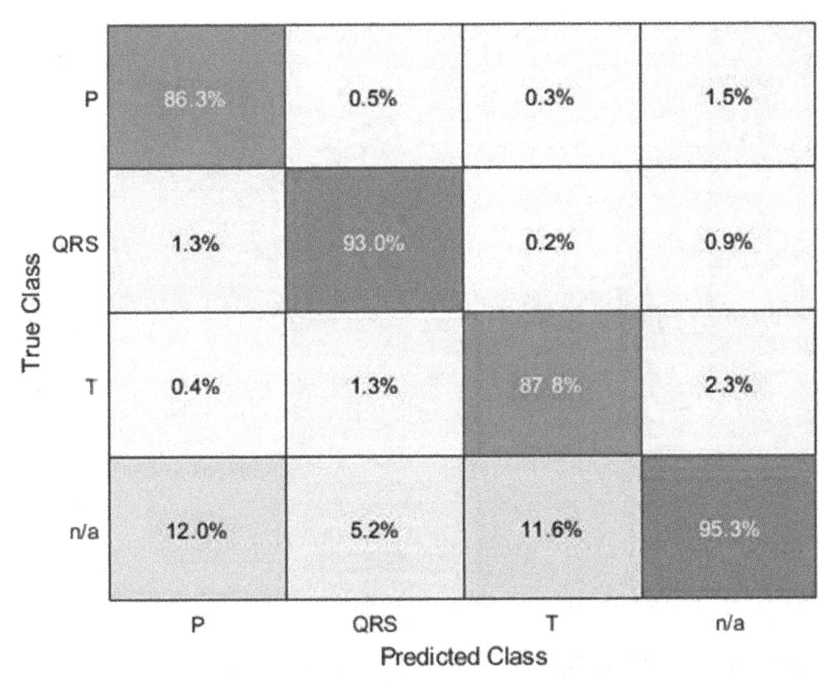


**Figure 12.7.** Normalized confusion matrix for the LSTM based model.

**Table 12.7.** Results of the semantic segmentation of ECG waves using BiLSTM.

| Class | Precision | Recall | F1-score | Specificity | Accuracy |
|---|---|---|---|---|---|
| P wave | 0.8020 | 0.6456 | 0.7142 | 0.9842 | 0.9537 |
| QRS complex | 0.8897 | 0.7772 | 0.8296 | 0.9843 | 0.9553 |
| T wave | 0.8449 | 0.700 | 0.7661 | 0.9575 | 0.8938 |
| n/a | 0.7936 | 0.9116 | 0.8485 | 0.7411 | 0.8301 |
| Avg | 0.832 25 | 0.7586 | 0.7896 | 0.9167 | 0.9082 |
| Wt. Avg. | 0.8206 | 0.8166 | 0.8135 | 0.8508 | 0.8746 |

Avg = average, Wt. Avg. = weighted average.

ECG wave segmentation. From table 12.7 it has been observed that the single BiLSTM based model has achieved accuracies of 95.37%, 95.53% and 89.38% for the P wave, QRS wave and T wave, respectively. For the neural signal, we achieved 83.0.1% accuracy. Moreover, the model achieved an average accuracy of 90.82% and a weighted average accuracy of 87.46%, which is significantly higher for the LSTM based model. The class-wise F1-scores are 71.42%, 82.96% and 76.61% for the P, QRS and T waves, respectively. The average F1-score has also been improved from 69.56% to 78.96%, which is significantly higher compared to the LSTM based model.

For the proposed BiLSTM based model we have also analysed the learning curve for the training and testing datasets. The learning curve shown in figure 12.8 indicates that there is a low generalization gap between training and testing curves

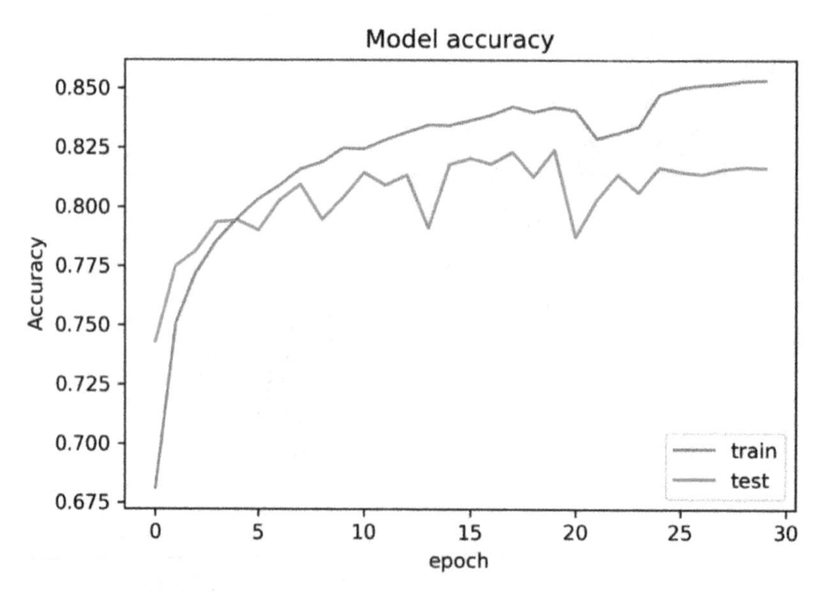


**Figure 12.8.** The learning curve for the single BiLSTM model.

compared to the LSTM based model. Further, to visualize the percentage of true positives and false positives for each class a confusion matrix has been investigated, as shown in figure 12.9.

### 12.4.3 Results using a stacked BiLSTM based model

This section provides the experimental results of our final model, which is a BiLSTM based model for ECG wave segmentation. The sample-wise evaluation metrics are listed in table 12.8. Further, the average and weighted average metrics have also been calculated for ECG wave segmentation. It can be observed from table 12.8 that the proposed stacked BiLSTM based model achieved accuracies of 96.12%, 95.76% and 91.20% for the P wave, QRS wave and T wave, respectively. For the neural signal, we achieved an accuracy of 86.38%. From the evaluation metrics it can be observed clearly that the BiLSTM based model has outperformed both the LSTM and single BiLSTM based models. The proposed method can provide optimal segmented results without using any pre- or post-processing techniques. The visual representation of the final semantic segmented ECG wave using the proposed BiLSTM model is shown in figure 12.10. It can be observed from figure 12.10 that the model has segmented the P wave and QRS wave more precisely. However, there are some data points of T which have been erroneously classified as P. There is a somewhat similar trend at the starting point of T (i.e. T-on) and the end point of the T wave (i.e. T-off). This might be the cause of the erroneous classification of some data points in the T wave.

Further, to analyse the generalization error of the proposed stacked BiLSTM based model we have explored the learning curve, as shown in figure 12.11. The learning curve indicates that there is a significantly lower generalization gap between

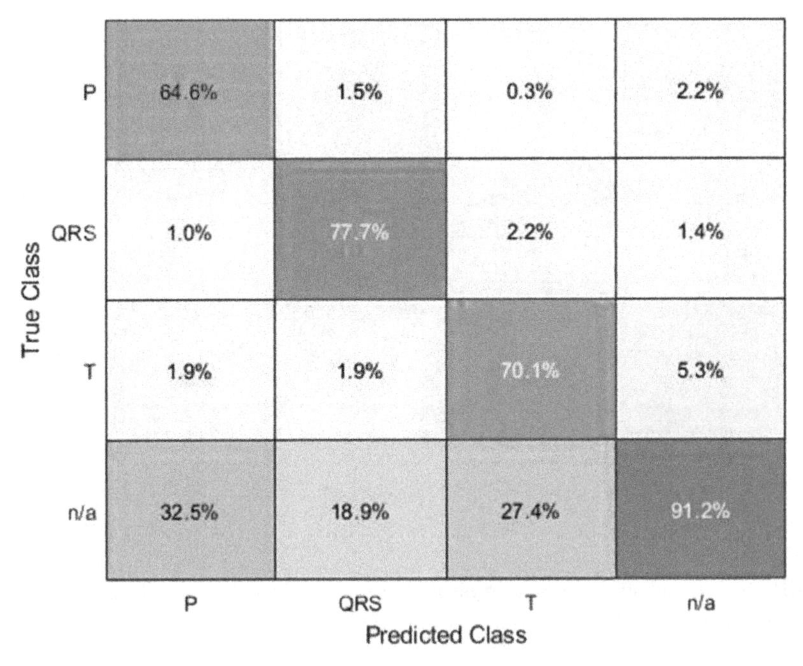


**Figure 12.9.** Normalized confusion matrix for the single BiLSTM based model.

**Table 12.8.** Results of the semantic segmentation of ECG waves.

| Class | Precision | Recall | F1-score | Specificity | Accuracy |
|---|---|---|---|---|---|
| P wave | 0.7932 | 0.770 | 0.7817 | 0.980 | 0.9612 |
| QRS complex | 0.8549 | 0.8395 | 0.8471 | 0.9768 | 0.9576 |
| T wave | 0.8552 | 0.7769 | 0.8142 | 0.9566 | 0.9120 |
| n/a | 0.8508 | 0.8962 | 0.8729 | 0.8284 | 0.8638 |
| Avg | 0.8385 | 0.8206 | 0.8289 | 0.9354 | 0.9236 |
| Wt. Avg. | 0.8473 | 0.8474 | 0.8977 | 0.8946 | 0.8977 |

the training and testing curves compared to the LSTM and single BiLSTM based models. This indicates the generalization capability of the proposed BiLSTM based model. Further, to visualize the percentage of true positives and false positives for each class a confusion matrix has been investigated, as shown in figure 12.12.

## 12.5 Discussion

This section provides a critical discussion of the proposed stacked BiLSTM based model for ECG wave segmentation. To validate the effectiveness of our proposed method an extensive comparative analysis has been carried out with other DL-based methods. Furthermore, we have also provided a comparative analysis of the proposed method with existing methods for ECG wave segmentation.

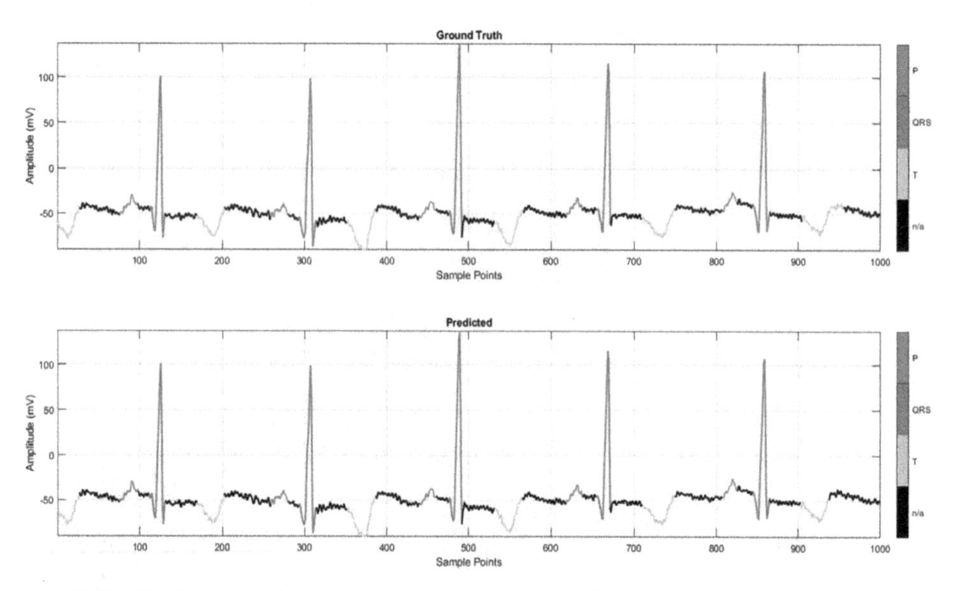


**Figure 12.10.** Visual result of the semantic segmentation of ECG signal: (a) ground truth and (b) predicted output of the proposed model.

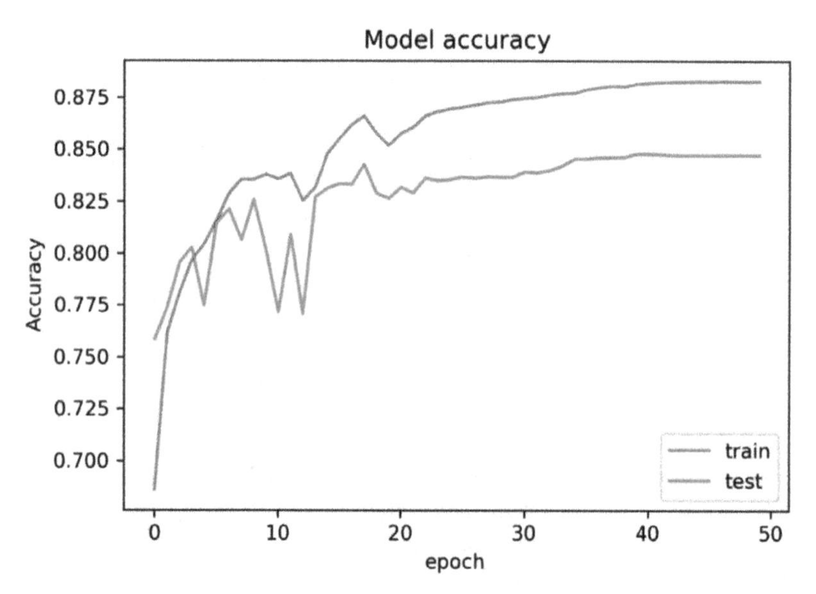


**Figure 12.11.** The learning curve for the stacked BiLSTM model.

### 12.5.1 Comparative analysis: proposed LSTM networks

The ECG wave segmentation by the proposed stacked BiLSTM method has been compared to existing methods to validate the proposed model. Since the existing methods for ECG segmentation do not provide sample-wise segmentation, in this study we have implemented different DL-based methods and the results are

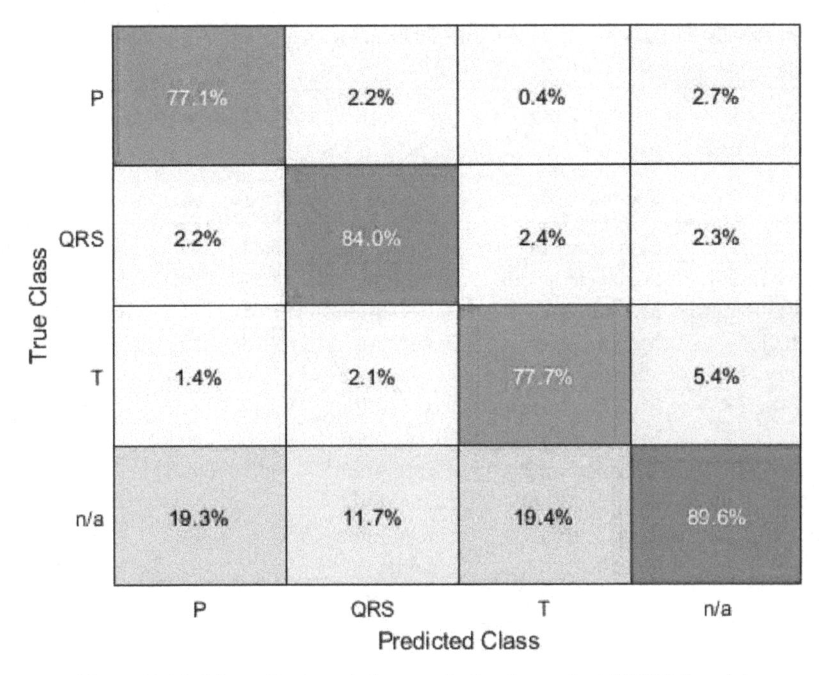


**Figure 12.12.** Normalized confusion matrix for the stacked BiLSTM model.

**Table 12.9.** Results of the semantic segmentation of the ECG interval with different DL models.

| Model | Precision | Recall | F1-score | Specificity | Accuracy |
|---|---|---|---|---|---|
| LSTM | 0.7474 | 0.7501 | 0.7460 | 0.8201 | 0.8326 |
| BiLSTM | 0.8206 | 0.8166 | 0.8135 | 0.8508 | 0.8746 |
| Stacked BiLSTM | 0.8473 | 0.8474 | 0.8465 | 0.8946 | 0.8977 |

presented in table 12.9. The weighted average accuracy for the LSTM and BiLSTM based models are 83.26% and 87.46%, respectively. For the final BiLSTM model the accuracy is 89.77%. It can be observed that there are 6.59% and 2.31% improvements in the weighted average accuracies using the proposed stacked BiLSTM based model. Moreover, the proposed model has performed significantly better in terms of the weighted sensitivity (i.e. recall), specificity, precision and F1-score. This indicates the generalization capability of the proposed stacked BiLSTM based model compared to other DL-based models.

### 12.5.2 Comparative analysis: existing state-of-the-art versus the proposed method

This section provides the comparative analysis of the proposed stacked BiLSTM based method to other conventional signal processing and machine learning methods for ECG wave segmentation. Table 12.10 presents the comparative results

**Table 12.10.** Results of ECG segmentation with different signal processing, machine learning and deep learning approaches using the QTBD.

| | | | ECG waves (accuracy in %) | | | |
|---|---|---|---|---|---|---|
| Author | Technique | Segmentation/ classification | P wave | QRS complex | T wave | Overall |
| Laguna *et al* [28] | Signal processing | Segmentation | 96.44 | 99.82 | 95.06 | 97.10 |
| Martinez *et al* [26] | Signal processing | Segmentation | 98.81 | 100 | 95.77 | 98.19 |
| Singh and Gupta [62] | Machine learning | Classification | 97.58 | | 86.26 | 91.92 |
| Sun *et al* [60] | Machine learning | Classification | 99.18 | 100 | 94.78 | 97.98 |
| Martinez *et al* [40] | Machine learning | Classification | 97.66 | 100 | 97.00 | 98.22 |
| Vazquez *et al* [63] | Machine learning | Classification | 99.47 | 100 | 95.37 | 98.28 |
| Hughes *et al* [67] | Machine learning | Classification | 96.50 | 99.74 | | 98.12 |
| Sodmann *et al* [7] | Deep learning | Segmentation | 90.0[a] | 99.9[a] | 97.7[a] | 95.86[a] |
| Abrishami *et al* [6] | Deep learning | Segmentation | | | | 96.20 |
| Abrishami *et al* [8] | Deep learning | Segmentation | 92.0 | 94.0 | 90.0 | 92.00 |
| This work | Deep learning | Semantic segmentation | 95.37 | 95.53 | 89.38 | 93.42 |

[a]Note: The values are PPV and not accuracy.

of different signal processing and machine learning based techniques for the segmentation of the ECG wave.

From table 12.10 it can be observed that some of the existing approaches have achieved higher performance in terms of accuracies. However, these methods consider the whole interval as one class. Moreover, the results have high variation (~7%–8%) in terms of average accuracy for ECG wave detection and classification using conventional methods. Hence, the semi-automated methods are found to be less preferable for real-time ECG wave segmentation. Moreover, RNN based architecture suffers from a change in signal morphology. Thus it becomes a challenging task to segment the ECG signal in a noisy database. The stacked BiLSTM based model learns the long-term dependencies in both the forward and backward paths between the time steps of time series sequences. This approach not only overcomes the problem of vanishing gradients, which occurs in the RNN model by capturing the dynamic correlation which is hidden in sequential data. Therefore the proposed stacked BiLSTM based model has achieved the highest performance compared to other conventional machine learning and DL-based approaches.

## 12.6 Conclusion

In this work an efficient deep learning architecture has been implemented based on stacked BiLSTM for semantic segmentation of the ECG signal. Previous approaches merely focused on the localization of the fiducial points of the ECG wave and might not lead to the successful measurement of ECG segments to study the cardiac conditions. The proposed model consists of BiLSTM layers capable of learning long-term dependencies in both the forward and backward directions and classifying continuous sample-wise the P wave, QRS complex and T wave from raw ECG signals. To the best of our knowledge, this is the first study that utilizes a stacked BiLSTM based model for semantic segmentation of the ECG wave. To validate our proposed method, we have performed a two-fold comparative analysis. To validate the performance, first the proposed method has been compared to other possible variants of the LSTM model and then to existing methods from the literature. The experimental results show that the proposed BiLSTM based model has shown comparable performance to other conventional machine learning and deep learning architectures. This performance has been achieved for continuous semantic segmentation of the ECG signal which other methods cannot do. In future work, the proposed model can be used for the detection of abnormal ECG waves. Further, the same architecture can be utilized for real-time disease diagnosis.

## References

[1] Goldberger A L, Goldberger Z D and Shvilkin A 2017 *Clinical Electrocardiography: A Simplified Approach* 9th edn. (St Louis, MO: Elsevier) pp 1–388

[2] Thorén A, Rawshani A, Herlitz J, Engdahl J, Kahan T, Gustafsson L and Djärv T 2020 ECG-monitoring of in-hospital cardiac arrest and factors associated with survival *Resuscitation* **150** 130–8

[3] Macfarlane P W *et al* 2012 *Electrocardiology: Comprehensive Clinical ECG* (London: Springer) pp 1–618

[4] Begg G, Willan K, Tyndall K, Pepper C and Tayebjee M 2016 Electrocardiogram interpretation and arrhythmia management: a primary and secondary care survey *British J. General Pract.* **66** e291–6

[5] Becker D E 2006 Fundamentals of electrocardiography interpretation *Anesth. Prog.* **53** 53–64

[6] Abrishami H, Campbell M, Han C, Czosek R and Zhou X 2018 P–QRS–T localization in ECG using deep learning *Proc. IEEE EMBS Int. Conf. on Biomedical Health Informatics (BHI)* pp 210–13

[7] Sodmann P, Vollmer M, Nath N and Kaderali L 2018 A convolutional neural network for ECG annotation as the basis for classification of cardiac rhythms *Physiol. Meas.* **39** 104005

[8] Abrishami H, Han C, Zhou X, Campbell M and Czosek R 2018 Supervised ECG interval segmentation using LSTM neural network *Proc. Int. Conf. on Bioinformatics & Computational Biology (BIOCOMP)* pp 71–7

[9] Campbell M J, Zhou X, Han C, Abrishami H, Webster G, Miyake Y, Sower C T, Anderson J B, Knilans T K and Czosek R J 2017 Pilot study analyzing automated ECG screening of hypertrophic cardiomyopathy *Heart Rhythm* **14** 848–52

[10] Beraza I and Romero I 2017 Comparative study of algorithms for ECG segmentation *Biomed. Signal Process. Control* **34** 166–73

[11] Karimipour A and Homaeinezhad M R 2014 Real-time electrocardiogram P–QRS–T detection–delineation algorithm based on quality-supported analysis of characteristic templates *Comput. Biol. Med.* **52** 153–65

[12] Warner R A, Ariel Y, Dalla Gasperina M and Okin P M 2002 Improved electrocardiographic detection of left ventricular hypertrophy *J. Electrocardiol.* **35** 111

[13] Kaiser W, Faber T S and Findeis M 1996 Automatic learning of rules: a practical example of using artificial intelligence to improve computer-based detection of myocardial infarction and left ventricular hypertrophy in the 12-lead ECG *J. Electrocardiol.* **29** 17–20

[14] Kiranyaz S, Ince T, Hamila R and Gabbouj M 2015 Convolutional neural networks for patient-specific ECG classification *Proc. 37th Annual Int. Conf. of the IEEE Engineering in Medicine and Biology Society (EMBC)* pp 2608–11

[15] Koski A 1996 Modelling ECG signals with hidden Markov models *Artif. Intell. Med.* **8** 453–71

[16] Crouse M S, Nowak R D and Baraniuk R G 1998 Wavelet-based statistical signal processing using hidden Markov models *IEEE Trans. Signal Process* **46** 886–902

[17] Stamkopoulos T, Maglaveras N, Bamidis P D and Pappas C 2000 Wave segmentation using nonstationary properties of ECG *Proc. Comput. Cardiol. 2000* **27** 529–32

[18] Graja S and Boucher J M 2003 Multiscale hidden Markov model applied to ECG segmentation *Proc. IEEE Int. Symp. on Intelligent Signal Processing* pp 105–9

[19] Hughes N P, Tarassenko L and Roberts S J 2004 Markov models for automated ECG interval analysis *Adv. Neural Inform. Process. Syst.* pp 611–18

[20] Andreão R V and Boudy J 2006 Combining wavelet transform and hidden Markov models for ECG segmentation *EURASIP J. Adv. Signal Process.* pp 1–8

[21] Thoraval L, Carrault G and Mora F 1992 Continuously variable duration hidden Markov models for ECG segmentation *Proc. 14th Annual Int. Conf. of the IEEE Engineering in Medicine and Biology Society* vol 2 pp 529–30

[22] Coast D A, Stern R M, Cano G G and Briller S A 1990 An approach to cardiac arrhythmia analysis using hidden Markov models *IEEE Trans. Biomed. Eng.* **37** 826–36

[23] Murthy I S N and Niranjan U C 1992 Component wave delineation of ECG by filtering in the Fourier domain *Med. Biol. Eng. Comput.* **30** 169–76

[24] Ouyang N, Yamauchi K and Ikeda M 1998 Training a NN with ECG to diagnose the hypertrophic portions of HCM *Proc. 1998 IEEE Int. Joint Conf. on Neural Networks Proc.* **1** 306–9

[25] Rahman Q A, Tereshchenko L G, Kongkatong M, Abraham T, Abraham M R and Shatkay H 2014 Identifying hypertrophic cardiomyopathy patients by classifying individual heart-beats from 12-lead ECG signals *Proc. 2014 IEEE Int. Conf. on Bioinformatics and Biomedicine (BIBM)* pp 224–9

[26] Martínez J P, Almeida R, Olmos S, Rocha A P and Laguna P 2004 A wavelet-based ECG delineator: evaluation on standard databases *IEEE Trans. Biomed. Eng.* **51** 570–81

[27] Laguna P, Thakor N V, Caminal P, Jane R, Yoon H R, de Luna A B and Guindo J 1990 New algorithm for QT interval analysis in 24-hour Holter ECG: performance and applications *Med. Biol. Eng. Comput.* **28** 67–73

[28] Laguna P, Jané R and Caminal P 1994 Automatic detection of wave boundaries in multilead ECG signals: validation with the CSE database *Comput. Biomed. Res.* **27** 45–60

[29] De Chazal P and Celler B G 1996 Automatic measurement of the QRS onset and offset in individual ECG leads *Proc. 18th Annual Int. Conf. of the IEEE Engineering in Medicine and Biology Society* vol 4 pp 1399–400

[30] Jane R, Blasi A, García J and Laguna P 1997 Evaluation of an automatic threshold based detector of waveform limits in Holter ECG with the QT database *Proc. Computers in Cardiology 1997* pp 295–8

[31] Vila J A, Gang Y, Presedo J M R, Fernández-Delgado M, Barro S and Malik M 2000 A new approach for TU complex characterization *IEEE Trans. Biomed. Eng.* **47** 764–72

[32] Schreier G, Hayn D and Lobodzinski S 2003 Development of a new QT algorithm with heterogenous ECG databases *J. Electrocardiol.* **36** 145–50

[33] Altuve M, Casanova O, Wong S, Passariello G, Hernandez A and Carrault G 2007 Evaluación de dos Métodos para la Segmentación del Ancho de la Onda T en el ECG *Proc. IV Latin American Congress on Biomedical Engineering 2007, Bioengineering Solutions for Latin America Health* pp 1254–58

[34] Frénay B, de Lannoy G and Verleysen M 2009 Emission modelling for supervised ECG segmentation using finite differences *Proc. 4th European Conf. of the Int. Federation for Medical and Biological Engineering* (Berlin: Springer) pp 1212–6

[35] Illanes-Manriquez A 2010 An automatic multi-lead electrocardiogram segmentation algorithm based on abrupt change detection *Proc. 2010 Annual Int. Conf. of the IEEE Engineering in Medicine and Biology* pp 2334–7

[36] Gupta R, Mitra M, Mondal K and Bhowmick S 2011 A derivative-based approach for QT-segment feature extraction in digitized ECG record *Proc. 2011s Int. Conf. on Emerging Applications of Information Technology* pp 63–6

[37] Mukhopadhyay S K, Mitra M and Mitra S 2011 Time plane ECG feature extraction using Hilbert transform, variable threshold and slope reversal approach *Proc. 2011 Int. Conf. on Communication and Industrial Application* pp 1–4

[38] Li C, Zheng C and Tai C 1995 Detection of ECG characteristic points using wavelet transforms *IEEE Trans. Biomed. Eng.* **42** 21–8

[39] Martinez J P, Olmos S and Laguna 2000 Evaluation of a wavelet-based ECG waveform detector on the QT database *Proc. Computers in Cardiology 2000* vol 27 pp 81–4

[40] Martínez A, Alcaraz R and Rieta J J 2010 Application of the phasor transform for automatic delineation of single-lead ECG fiducial points *Physiol. Meas.* **31** 1467

[41] Dumont J, Hernandez A I and Carrault G 2005 Parameter optimization of awavelet-based electrocardiogram delineator with an evolutionary algorithm *Proc. Computers in Cardiology 2005* pp 707–10

[42] Andreao R V, Dorizzi B and Boudy J 2006 ECG signal analysis through hidden Markov models *IEEE Trans. Biomed. Eng.* **53** 1541–9

[43] Thomas J, Rose C and Charpillet F 2006 A multi-HMM approach to ECG segmentation *Proc. 2006 18th IEEE Int. Conf. on Tools with Artificial Intelligence (ICTAI'06)* pp 609–16

[44] Madeiro J P, Cortez P C, Oliveira F I and Siqueira R S 2007 A new approach to QRS segmentation based on wavelet bases and adaptive threshold technique *Med. Eng. Phys.* **29** 26–37

[45] Thomas J, Rose C and Charpillet F 2007 A support system for ECG segmentation based on hidden Markov models *Proc. 2007 29th Annual Int. Conf. of the IEEE Engineering in Medicine and Biology Society* pp 3228–31

[46] Krimi S, Ouni K and Ellouze N 2008 An approach combining wavelet transform and hidden Markov models for ECG segmentation *Proc. 2008 3rd Int. Conf. on Information and Communication Technologies: From Theory to Applications* pp 1–6

[47] Almeida R, MartÍnez J P, Rocha A P and Laguna P 2009 Multilead ECG delineation using spatially projected leads from wavelet transform loops *IEEE Trans. Biomed. Eng.* **56** 1996–2005

[48] de Lannoy G, Frénay B, Verleysen M and Delbeke J 2009 Supervised ECG delineation using the wavelet transform and hidden Markov models *Proc. 4th European Conf. of the Int. Federation for Medical and Biological Engineering* (Berlin: Springer) pp 22–5
[49] Vítek M, Hrubeš J and Kozumplík J 2010 A wavelet-based ECG delineation with improved P wave offset detection accuracy *Anal. Biomed. Signals Images* **20** 160–5
[50] Shi W and Kheidorov I 2010 Hybrid hidden Markov models for ECG segmentation *Proc. 2010 Sixth Int. Conf. on Natural Computation* vol 6 pp 3323–28
[51] Ghaffari A, Homaeinezhad M R, Khazraee M and Daevaeiha M M 2010 Segmentation of holter ECG waves via analysis of a discrete wavelet-derived multiple skewness–kurtosis based metric *Ann. Biomed. Eng.* **38** 1497–510
[52] Dumont J, Hernandez A I and Carrault G 2008 Improving ECG beats delineation with an evolutionary optimization process *IEEE Trans. Biomed. Eng.* **57** 607–15
[53] Homaeinezhad M R, Ghaffari A, Najjaran T H, Tahmasebi M and Daevaeiha M M 2011 A unified framework for delineation of ambulatory Holter ECG events via analysis of a multiple-order derivative wavelet-based measure *Iranian J. Elect. Electro. Eng.* **7** 1–18
[54] Madeiro J P, Cortez P C, Marques J A, Seisdedos C R and Sobrinho C R 2012 An innovative approach of QRS segmentation based on first-derivative, Hilbert and wavelet transforms *Med. Eng. Phys.* **34** 1236–46
[55] Last T, Nugent C D and Owens F J 2004 Multi-component based cross correlation beat detection in electrocardiogram analysis *Biomed. Eng. Online* **3** 26
[56] Shuo Y and Desong B 2008 Automatic detection of t-wave end in ECG signals *Proc. 2008 IEEE Second Int. Symp. on Intelligent Information Technology Application* **3** 283–7
[57] Illanes-Manriquez A and Zhang Q 2008 An algorithm for robust detection of QRS onset and offset in ECG signals *Proc. 2008 Computers in Cardiology* pp 857–60
[58] Vullings H J L M, Verhaegen M H and Verbruggen H B 1998 Automated ECG segmentation with dynamic time warping *Proc. 20th Annual Int. Conf. of the IEEE Engineering in Medicine and Biology Society* **20** 163–6
[59] Zifan A, Saberi S, Moradi M H and Towhidkhah F 2006 Automated ECG segmentation using piecewise derivative dynamic time warping *Int. J. Biol. Med. Sci.* **1** 181–5
[60] Sun Y, Chan K L and Krishnan S M 2005 Characteristic wave detection in ECG signal using morphological transform *BMC Cardiovas. Disorder* **5** 1–7
[61] Martínez A, Alcaraz R and Rieta J J 2010 Application of the phasor transform for automatic delineation of single-lead ECG fiducial points *Physiol. Meas.* **31** 1467
[62] Singh Y N and Gupta P 2008 ECG to individual identification *Proc. 2008 IEEE Second Int. Conf. on Biometrics: Theory, Applications and Systems* pp 1–8
[63] Vázquez-Seisdedos C R, Neto J E, Reyes E J M, Klautau A and de Oliveira R C L 2011 New approach for T-wave end detection on electrocardiogram: performance in noisy conditions *Biomed. Eng. Online* **10** 1–11
[64] Kang W, Byun K and Kang H G 2015 Detection of fiducial points in ECG waves using iteration based adaptive thresholds *Proc. 2015 37th Annual Int. Conf. of the IEEE Engineering in Medicine and Biology Society (EMBC)* pp 2721–4
[65] Bayasi N, Tekeste T, Saleh H, Khandoker A, Mohammad B and Ismail M 2014 Adaptive technique for P and T wave delineation in electrocardiogram signals *Proc. 2014 36th Annual Int. Conf. of the IEEE Engineering in Medicine and Biology Society* pp 90–3
[66] Sayadi O and Shamsollahi M B 2009 A model-based Bayesian framework for ECG beat segmentation *Physiol. Meas.* **30** 335

[67] Hughes N P, Tarassenko L and Roberts S J 2004 Markov models for automated ECG interval analysis *Proc. Advances in Neural Information Processing Systems* pp 611–8
[68] Jain S, Bajaj V and Kumar A 2016 An efficient algorithm for classification of ECG beats based on ABC-LSSVM classifier *IET Electron. Lett.* **52** 1198–200
[69] Rajpurkar P, Hannun A Y, Haghpanahi M, Bourn C and Ng A Y 2017 Cardiologist-level arrhythmia detection with convolutional neural networks *arXiv*:1707.01836
[70] Murugesan B, Ravichandran V, Ram K, Preejith S P, Joseph J, Shankaranarayana S M and Sivaprakasam M 2018 ECGNet: deep network for arrhythmia classification *Proc. 2018 IEEE Int. Symp. on Medical Measurements and Applications (MeMeA)* pp 1–6
[71] Kshirsagar G B and Londhe N D 2019 Weighted ensemble of deep convolution neural networks for a single trial character detection in Devanagari script based P300 speller *IEEE Trans. Cognit. Develop. Syst.* **12** 551–60
[72] Kshirsagar G B and Londhe N D 2018 Improving performance of Devanagari script input-based P300 speller using deep learning *IEEE Trans. Biomed. Eng.* **66** 2992–3005
[73] Jeyhani V, Mäntysalo M, Noponen K, Seppänen T and Vehkaoja A 2019 Effect of different ECG leads on estimated R–R intervals and heart rate variability parameters *Proc. 2019 41st Annual Int. Conf. of the IEEE Engineering in Medicine and Biology Society (EMBC)* pp 3786–90
[74] Maglaveras N, Stamkopoulos T, Pappas C and Strintzis M 1998 ECG processing techniques based on neural networks and bidirectional associative memories *J. Med. Eng. Technol.* **22** 106–11
[75] Liu X, Deng Z and Yang Y 2019 Recent progress in semantic image segmentation *Artif. Intell. Rev.* **52** 1089–106
[76] Graves A, Jaitly N and Mohamed A R 2013 Hybrid speech recognition with deep bidirectional LSTM *Proc. 2013 IEEE Workshop on Automatic Speech Recognition and Understanding* pp 273–8
[77] Dohare A K, Kumar V and Kumar R 2014 An efficient new method for the detection of QRS in electrocardiogram *Comput. Electr. Eng.* **40** 1717–30
[78] Bengio Y, Simard P and Frasconi P 1994 Learning long-term dependencies with gradient descent is difficult *IEEE Trans. Neural Networks* **5** 157–66
[79] Hochreiter S and Schmidhuber J 1997 Long short-term memory *Neural Comput.* **9** 1735–80
[80] Schuster M and Paliwal K K 1997 Bidirectional recurrent neural networks *IEEE Trans. Signal Process.* **45** 2673–81
[81] Graves A and Schmidhuber J 2005 Framewise phoneme classification with bidirectional LSTM networks *Proc. 2005 IEEE Int. Joint Conf. on Neural Networks* vol 4 pp 2047–52
[82] Laguna P, Mark R G, Goldberg A and Moody G B 1997 A database for evaluation of algorithms for measurement of QT and other waveform intervals in the ECG *Proc. Computers in Cardiology* pp 673–6

**IOP** Publishing

Modelling and Analysis of Active Biopotential Signals in Healthcare, Volume 2

**Varun Bajaj and G R Sinha**

# Chapter 13

# Deep convolutional neural network based diagnosis of COVID-19 using x-ray images

**Vimal K Shrivastava[1] and Monoj K Pradhan[2]**

[1]*School of Electronics Engineering, Kalinga Institute of Industrial Technology (KIIT), Bhubaneswar, India*

[2]*Department of Agricultural Statistics and Social Sciences (L), Indira Gandhi Agricultural University, Raipur, India*

The outbreak of novel coronavirus 2019 (COVID-19) has become a pandemic, with the highly transmissible disease spreading worldwide. It is having a devastating effect on the health of the public globally and is also having a great impact on the global economy. The early diagnosis of COVID-19 cases may help in controlling its spread and treatment. In addition, the lack of sufficient COVID-19 test kits demands a rapid substitute diagnosis system. It has been observed that the chest radiological images of COVID-19 patients show abnormalities. Therefore, a deep convolutional neural network (CNN) approach based diagnosis of COVID-19 using x-ray images is presented in this chapter. The effective training of a deep CNN model requires a large training dataset, however only a limited quantity of x-ray images of COVID-19 cases are available. Therefore, our model uses a fine-tuning approach where a pre-trained model is utilized and the trained model is applied to the problem at hand. In this study, the performance of several state-of-the-art deep CNN models with fine-tuning approaches is presented and compared. The experiments are performed on a dataset with images of 125 COVID-19 cases, 500 pneumonia cases and 500 normal cases. The experiments are divided into two parts: binary classification (COVID-19 versus normal) and multiclass classification (COVID-19 versus pneumonia versus normal). Classification accuracies of 98.56% and 88.44% are achieved for binary and multiclass classification, respectively. The primarily results demonstrate that the presented model can be used in hospitals or through the Cloud to assist radiologists after its validation using additional datasets.

doi:10.1088/978-0-7503-3411-2ch13

## 13.1 Introduction

Coronavirus constitutes a large family of viruses that causes illness in animals, including humans. COVID-19 transmits from one person to another primarily through tiny droplets, either through the nose or mouth, that are expelled when a person with COVID-19 starts coughing, sneezing or speaking. The bigger droplets zoom through the air after a sneeze, but smaller droplets float. The coronavirus is thus transmitted through the air and quickly infects people when inhaled, causing serious illness. This virus causes three important symptoms: fever, cough and shortness of breath. The shortness of breath is due to Severe Acute Respiratory Syndrome Coronavirus 2 (SARS-CoV-2) which leads to serious illness and death in some cases [1]. It is also contagious. Therefore, the early detection of SARS-CoV-2 helps in effective treatment and other immunity strengthening measures to defeat the virus. Across the globe, COVID-19 positive patients with SARS-CoV-2 have lost their lives due to the severity of the disease and many other reasons such as age, comorbidities and a lack of immunity. Clinical radiology has witnessed the growing significance of SARS-CoV-2 and this field has been helping towards its diagnosis [1–3]. In the early period of the disease, most of the research gave importance to the imaging technology known as computed tomography (CT) [4]. COVID-19 positive patients are advised to have a CT scan. However, sufficient CT scan facilities are not available when the magnitude of spread of the virus across the globe is considered. As CT scanners are costly, it is expensive for service providers and patients. Thus CT scans are not viable in the large scale diagnosis and treatment of SARS-CoV-2 or COVID-19 in general. Consequently, many healthcare units have removed CT scans from their treatment planning as the primary diagnosis tool of COVID-19.

Keeping the aforementioned information in mind, a number of researchers and healthcare professionals have shown interest in the use of the faster and more cost-effective x-ray imaging of the chest, known as the chest x-ray (CXR). This kind of imaging has been identified to be viable and reliable. Since CXR is portable, it has high value in the early diagnosis of COVID-19. In particular, it is very useful in cases where the patients show symptoms of COVID-19. CXR imaging facilities are widely used and available across the globe in most healthcare units. They are faster and relatively cost-effective and have shorter decontamination times compared to CT. Due to its portability, CXR devices can be transported to COVID-19 isolation rooms. This will add to the flexibility and convenience of using of CXR images for the diagnosis of COVID-19. There is little risk of transmission in this regard. As per the standard operating procedures, CXR imaging is used frequently for patients with COVID-19 symptoms. This helps in assessing the progress in the given treatment. Thus, CXR imaging has played a significant role in treating COVID-19 affected patients. COVID-19 symptoms have been examined in many studies [5, 6] and it has been found that CXR imaging technology helps in understanding different features of the syndrome. These include interstitial abnormalities, ground-glass opacity and bilateral abnormalities. The availability of CXR imaging and its portability have led to the useful observation of disease progression and treatment outcomes. Based on the severity of the patient's condition, decisions can be made for further treatment.

Even in home-based quarantine, CXR imaging can help to determine the severity of COVID-19.

In the literature, many radiological images have been used for the detection of COVID-19. Ozturk *et al* [1] proposed DarkCovidNet using CNN for the diagnosis of COVID-19 using x-rays in term of both binary (COVID-19 versus normal) and multiclass (COVID-19 versus pneumonia versus normal). Classification accuracies of 98.08% and 87.02% have been reported in the case of binary and multiclass classification, respectively. Hemdan *et al* [2] explored seven deep CNN models to detect healthy status against COVID-19 using chest x-rays and reported that both VGG16 and DenseNet201 achieved 90% accuracy and performed better compared to the other methods. Narin *et al* [6] explored three deep learning models for detecting COVID-19 using chest x-rays and showed that ResNet50 achieved a detection accuracy of 98%. Apostolopoulos *et al* [3] explored six deep learning models with transfer learning approaches for binary and multiclass classification using x-ray images and achieved accuracies of 96.78% and 94.72%, respectively. Sethy and Behera [7] used the extracted features from different CNN models and fed these features into a support vector machine (SVM) to classify them. They achieved 95.38% accuracy using the ResNet50 model with an SVM classifier. Along the same lines, Wang and Wong [8] proposed COVID-Net which obtained 92.4% in the classification of x-rays for three classes, i.e. normal, non-COVID pneumonia and COVID-19 images.

In this chapter we explore 16 state-of-the-art models which are: VGG16, VGG19, ResNet50, ResNet101, ResNet152, ResNet50V2, ResNet101V2, ResNet152V2, InceptionV3, InceptionResNetV2, Xception, MobileNet, MobileNetV2, DenseNet121, DenseNet169 and DenseNet201. The effective training of a deep CNN model generally requires a large training dataset, however, only a limited quantity of x-ray images of COVID-19 cases are available. Therefore, our model uses a fine-tuning approach where a pre-trained model is utilized and the trained model is applied to the problem at hand. The experiments are performed in two ways: binary classification (COVID-19 versus normal) and multiclass classification (COVID-19 versus pneumonia versus normal). The contributions of this chapter are as follows: (i) the diagnosis of COVID-19 using chest x-ray images; (ii) an end-to-end architecture for the classification of COVID-19 and thus avoiding manual feature extraction; (iii) extensive comparison of the performance of 16 deep CNN models in the diagnosis of COVID-19; (iv) an encouraging performance with a small dataset by utilizing a fine-tuning approach; and (v) high accuracies of 98.56% and 88.44% for binary and multiclass classification, respectively.

The chapter is organized as follows. Section 13.2 provides an introduction to the deep CNN model and the fine-tuning approach and briefly explains various state-of-the-art deep CNN models. Section 13.3 provides the results and discussion. Finally, section 13.4 concludes our study.

## 13.2 Methodology

In this study, various deep learning models have been explored for the detection of COVID-19. Significant development of deep learning has occurred in the machine learning field over the last few decades. In these methods, end-to-end architecture has been adopted instead of defining feature extraction methods. Among the many deep learning techniques, the convolutional neural network (CNN) has achieved noteworthy performance in various image recognition tasks, namely in healthcare [9–11], agriculture [12], remote sensing [13], classification and recognition problems [14–16], etc. In this section, we first provide a brief description of CNN architecture followed by the fine-tuning approach. Finally, various state-of-the-art deep CNN models are described.

### 13.2.1 Convolutional neural network

A CNN is specifically developed for the purpose of computer vision or image recognition tasks. A CNN accelerates the learning process using convolution, pooling and fully connected layers. In the deep CNN architecture, there is alternate stacking of convolution and pooling layers and there are many layers for better performance in learning and prediction. The convolution layers are used for filtering while the pooling layers are used for subsampling. The filtering process extracts the spatial features and pooling reduces the dimensionality and thus solves the problem of the curse of dimensionality. The pooling layer also provides the translational invariance property. Each layer in the CNN helps in learning by updating the weights of associated features using back-propagation algorithms. Finally, the extracted features are passed through the softmax layer to predict the class labels. It extracts simple and general patterns from the dataset in the first few layers and then abstract features are extracted as the layers increase. In the process of CNN, each iteration is called an epoch. Figure 13.1 depicts the general architecture of a deep CNN model.

### 13.2.2 Convolution layers

The convolution layers in a CNN are the core block and have a number of filters that perform convolution operations. The filters are systematically applied to input images to create feature maps that summarize the features present in the input. The initial convolution layers learn low level features such as edges, whereas the deeper

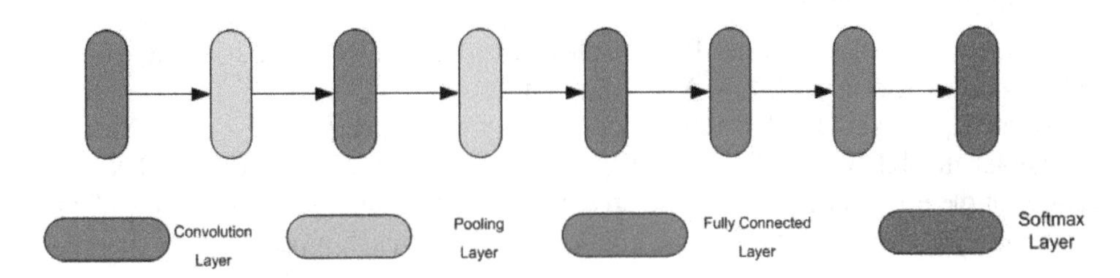


**Figure 13.1.** The general architecture of a deep CNN model.

layers learn complicated and abstract features such as shape, objects, etc. As multiple filters are used in the learning process, stacking all of them along with depth results in a complete output form pertaining to the convolution layer. After convolution operation, a rectified linear unit (ReLU) is applied to provide nonlinearity.

### 13.2.3 Pooling layers

The pooling layers summarize the presence of features in regions of feature maps by down-sampling the feature maps. Maximum pooling and average pooling are the most common methods of pooling. Max pooling activates the maximum feature. It performs the calculation of the highest or largest value in each region of a feature map. Average pooling averages the presence of features. It performs the averaging of each region of the feature map. In addition, it provides a translational invariance property.

### 13.2.4 Fully connected layer

This is the final layer of the CNN. Each neuron of the fully connected layer receives a certain weight and in turn gives a probability value as the output and the highest probability value gives the class label of the input image. The main objective of the fully connected layer is to interpret the feature vector extracted by previous layers into different classes of the image.

### 13.2.5 Fine-tuning

The limitation of the deep CNN model is that it requires a large dataset to train effectively. To mitigate this issue, a fine-tuning approach has been proposed in the literature. In the fine-tuning approach, pre-trained models are used which have already been trained on a large dataset and then the pre-trained model is reused in another task. This method is useful when a large enough training dataset is not available to train the model from scratch. Moreover, this method is particularly used in deep learning models where there are millions of parameters to train. Fine-tuning initiates the training on a new dataset by considering the trained weights as initial weights which are further updated as iteration progresses, i.e. the weights are fine-tuned during the training process.

### 13.2.6 Deep CNN models

In this study, we have explored 16 state-of-the-art deep CNN models for the detection and classification of COVID-19: VGG16, VGG19, ResNet50, ResNet101, ResNet152, ResNet50V2, ResNet101V2, ResNet152V2, InceptionV3, InceptionResNetV2, Xception, MobileNet, MobileNetV2, DenseNet121, DenseNet169 and DenseNet201. The architectures of these 16 models are depicted in figure 13.2. In addition, the parameters of all 16 models are provided in table 13.1. A brief description of each model is given below.

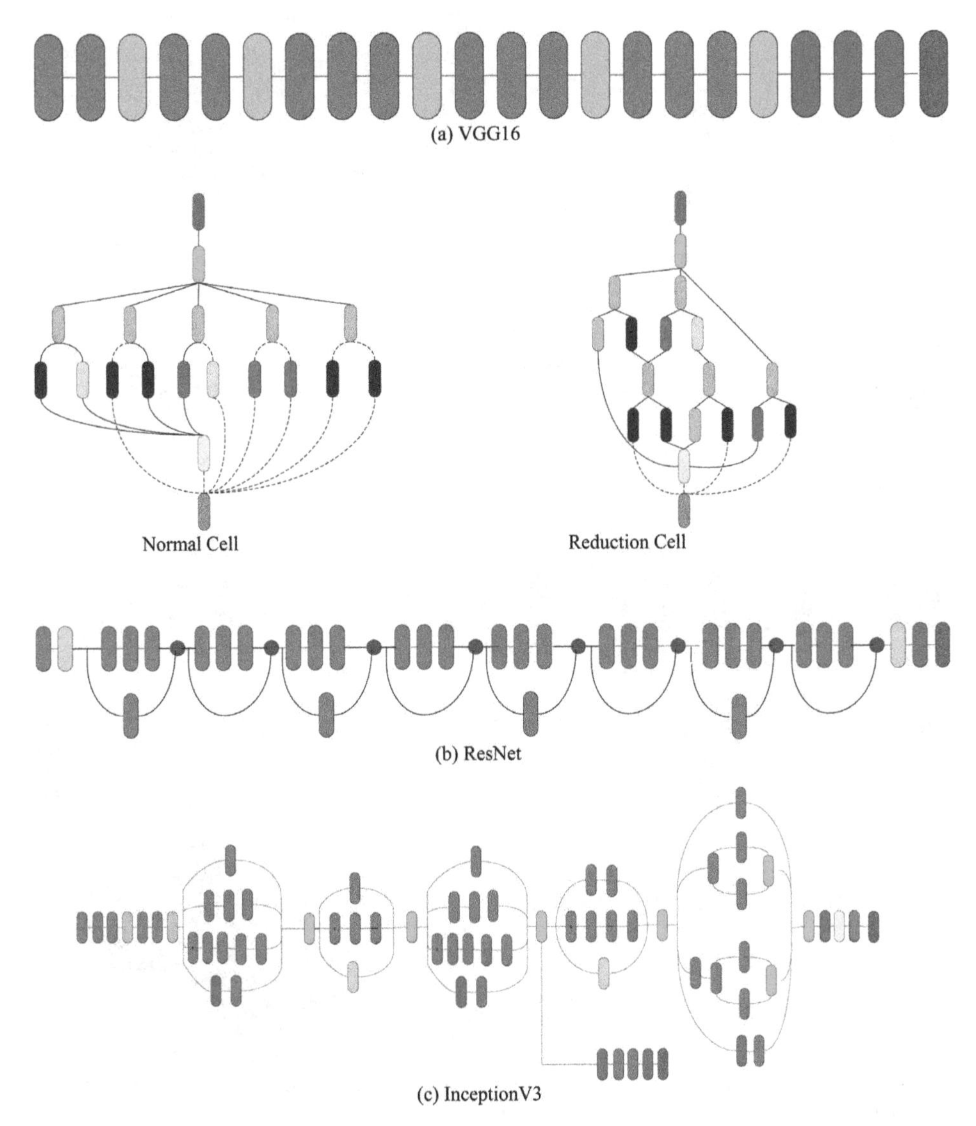


**Figure 13.2.** Architecture of state-of-the-art deep CNN models. (a) VGG16 normal cell reduction cell, (b) ResNet, (c) InceptionV3, (d) InceptionResNetV2, (e) Xception, (f) MobileNet and (g) DenseNet169.

VGG16 was proposed in [17]. It is a sequential convolutional neural network using 3×3 filters with an increasing depth of 16 layers. Max pooling was performed on a 2×2 pixel window with stride of 2. After each max pool layer, the number of convolution filters is doubled. It has three fully connected layers. The first two fully connected layers have 4096 neurons and the third one has 1000 neurons. Figure 13.2(a) shows the architecture of VGG16. Similar to VGG16, the VGG19 model was developed with addition layers having a depth of 19 layers.

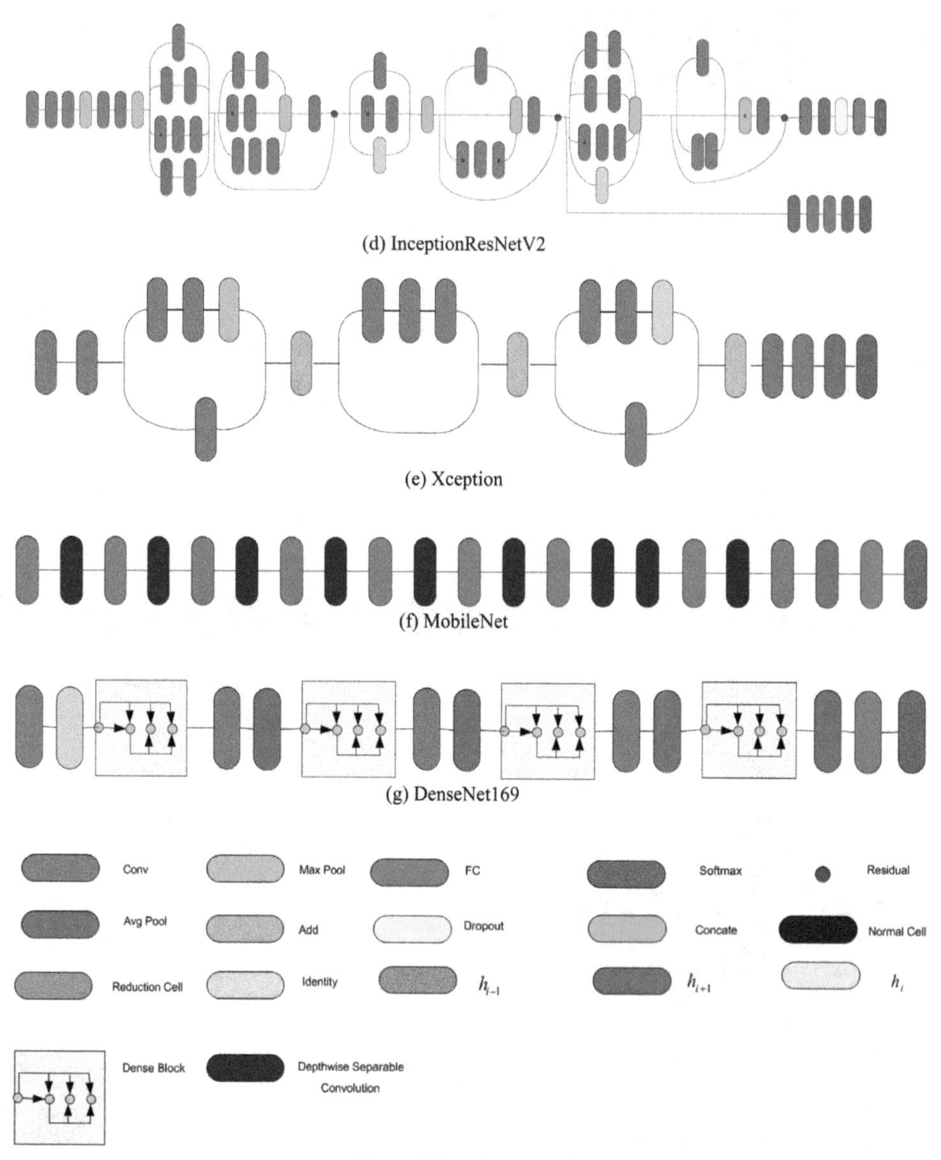


(d) InceptionResNetV2

(e) Xception

(f) MobileNet

(g) DenseNet169

**Figure 13.2.** (Continued.)

The residual neural network (ResNet) was introduced in [18]. ResNet models have been proposed to ease the training of deeper networks and they have been widely used in the recent era for the purpose of image recognition problems. It employs a residual learning framework which has skip connections from an earlier layer along with direct connection from the immediate previous layer. Each residual unit is composed of convolution and pooling layers. There are several variants of ResNet architecture, such as ResNet50, ResNet101, ResNet152, ResNet50V2, ResNet101V2 and ResNet152V2. ResNet50 is 50 layers deep, ResNet101 has 101

**Table 13.1.** Parameters of deep CNN models.

| Deep CNN model | Input shape | # Convolution layers | # Pooling layers | # Fully connected layers | Trainable parameter |
|---|---|---|---|---|---|
| Vgg16 | 224 × 224 × 3 | 13 | 5 | 3 | 134 301 514 |
| VGG19 | 224 × 224 × 3 | 13 | 5 | 3 | 139 611 210 |
| ResNet50 | 224 × 224 × 3 | 151 | 4 | 1 | 23 538 690 |
| ResNet101 | 224 × 224 × 3 | 151 | 4 | 1 | 42 556 930 |
| ResNet152 | 224 × 224 × 3 | 151 | 4 | 1 | 58 223 618 |
| ResNet50V2 | 224 × 224 × 3 | 151 | 4 | 1 | 23 539 850 |
| ResNet101V2 | 224 × 224 × 3 | 151 | 4 | 1 | 42 549 386 |
| ResNet152V2 | 224 × 224 × 3 | 151 | 4 | 1 | 58 208 394 |
| InceptionV3 | 299 × 299 × 3 | 83 | 12 | 1 | 21 788 842 |
| InceptionResNetV2 | 299 × 299 × 3 | 240 | 6 | 1 | 54 291 562 |
| Xception | 299 × 299 × 3 | 37 | 4 | 1 | 20 827 442 |
| MobileNet | 224 × 224 × 3 | 14 | 1 | 1 | 4 241 986 |
| MobileNetV2 | 224 × 224 × 3 | 14 | 1 | 1 | 2 236 682 |
| DenseNet121 | 224 × 224 × 3 | 152 | 5 | 1 | 6 964 106 |
| DenseNet169 | 224 × 224 × 3 | 152 | 5 | 1 | 12 501 130 |
| DenseNet201 | 224 × 224 × 3 | 152 | 5 | 1 | 18 112 138 |

layers and so on. Each residual block of these models has three layers with convolutions 1 × 1, 3 × 3 and 1 × 1. The 1 × 1 layer is responsible for reducing and then increasing the dimension and leaving 3 × 3 a bottleneck with smaller input and output dimensions. These models have a greater depth than VGG16/19 but have lower complexity. Figure 13.2(b) presents the basic architecture of the ResNet model.

The inception architecture was introduced in [19]. The first version is known as GoogleNet or InceptionV1. InceptionV1 was refined in various ways and was improved by adding a batch normalization layer and is now called InceptionV2 [20]. Further, this version was improved by adding the factorization idea and is known as InceptionV3 [21]. The architecture of the inception model is highly tunable. The number of filters may change in various layers in such a way that it does not affect the overall quality of the trained network. Figure 13.2(c) presents the architecture of the InceptionV3 model.

Various combinations of inception architectures and residual connections have been tried and InceptionResNetV1 and InceptionResNetV2 [22] are popular ones. It has been shown that InceptionResNetV1 matches the performance of InceptionV3 whereas InceptionResNet matches the performance of InceptionV4. Figure 13.2(d) presents the architecture of the InceptionResNetV2 model.

The architecture of Xception is based on depth-wise separable convolution layers. Here, it is assumed that mapping of the cross-channel correlation and spatial correlation in the feature maps would be completely decoupled. This model has 36 convolutional layers that perform the feature extraction from the inputs. All these 36 layers are grouped into 14 modules. Among these 14 modules, 12 modules have linear residual connections around them, except for the first and last module. This makes it easy to define and modify the structure. Moreover, its hypothesis is stronger than the hypothesis underlying the inception architecture, and thus this model is called Xception which stands for extreme Inception. The last layer is a logistic regression layer. Figure 13.2(e) presents the architecture of the Xception model.

MobileNet [23] is built from depth-wise separable convolutions. The standard convolutions factorize into depth-wise convolution and a pointwise convolution with 1×1 convolution. A standard convolution filter combines the output simultaneously in one step. However, the depth-wise separable convolution splits into two layers, one layer for filtering and a separate layer for combining inputs. This factorizing of standard convolution has a greater impact on reducing the computation time drastically as well as the model size. Figure 13.2(f) presents the architecture of MobileNet which was proposed by [24]. It consists of an initial fully convolution layer with 32 filters and these are followed by 19 residual bottleneck layers. ReLU has been used as a nonlinearity activation function. A kernel size of 3×3, dropout and batch normalization have been used in this model.

The dense convolutional network (DenseNet) was proposed by [25]. It is a densely connected CNN architecture where each layer is interconnected to each other in a feed-forward manner. Let $i$ be the $i$th layer and $x_i$ be its output. The $i$th layer receives all the preceding layer's outputs such as $x_0$, $x_1$ and $x_{i-1}$. The output of the $i$th layer can be stated as

$$x_i = H_i([x_0, x_1, x_{i-1}]), \tag{13.1}$$

where $[x_0, x_1, x_{i-1}]$ are feature maps extracted by the respective layer and $H_i(\cdot)$ is the concatenation of all these feature maps. $H_i(\cdot)$ is a composition function of three operations: batch normalization (BN) [20], followed by ReLU [26] and then a 3×3 convolution (Conv). Here, the most important part is down-sampling the feature maps and to facilitate down-sampling the architecture is divided into multiple connected blocks and each block is known as a dense block. In this chapter, three DenseNet models, i.e. DenseNet121, DenseNet169 and DenseNet201, have been explored. The architecture of the DenseNet model is depicted in figure 13.2(g).

## 13.3 Results and discussion

### 13.3.1 Dataset description

In this chapter, CXR images from two different sources have been used for the development of an automatic COVID-19 diagnosis system. One source is a dataset that has been specially created for the collection of COVID-19 data by Cohen *et al* [27]. This is being updated as-and-when by various researchers from different regions. At present, this database consists of 125 COVID-19 images. Another source of the overall dataset was collected and provided by Wang *et al* [28]. In this chapter, the dataset was collected from the two aforementioned sources and consists of 125 images of COVID-19, 500 images of pneumonia and 500 normal images. The experiments were divided into two parts: binary classification (COVID-19 versus normal) and multiclass classification (COVID-19 versus pneumonia versus normal). Figure 13.3 depicts a sample image from each class.

#### *13.3.1.1 Experimental set-up*

In this section we describe the experimental set-up to train 16 state-of-the-art deep CNN models for the diagnosis of COVID-19. These 16 models are VGG16, VGG19, ResNet50, ResNet101, ResNet152, ResNet50V2, ResNet101V2, ResNet152V2, InceptionV3, InceptionResNetV2, Xception, MobileNet, MobileNetV2, DenseNet121, DenseNet169 and DenseNet201. It can be observed from table 13.1 that the input shape varies for each model. Therefore we have reshaped each image of our dataset to the desired shape as per the requirement of each model. For example, the images have been reshaped to 224×224×3 for the VGG16 model and 299×299×3 for the InceptionV3 model and so on. We have used these models with a fine-tuning approach, where the model is loaded with pre-trained weights and the weights of all the layers are updated while training. All 16 models have been pre-trained on a large dataset known as ImageNet and the trained weights are loaded as initial weights while training. The ImageNet dataset has 1000 classes and hence the last layer of each model consists of 1000 neurons. However, the dataset used in this study has two classes in the case of binary classification (COVID-19 and normal) and three classes in the case of multiclass classification (COVID-19, pneumonia and normal). Therefore, the last fully connected layer of each model, which consists of 1000 neurons, has been replaced by a fully connected layer with two neurons for binary classification and three neurons for multiclass classification.

The dataset has been partitioned into 70% training set, 10% validation set and 20% testing set. Each model has been trained for 200 epochs with a mini-batch size of 32 and a learning rate of 0.01. The experiment was performed using the Adam (adaptive moment estimation) optimizer. Moreover, we have adopted an early stopping strategy to avoid overfitting. In the early stopping strategy, the training is stopped if a pre-defined criterion is met. Otherwise, the model is trained until the full

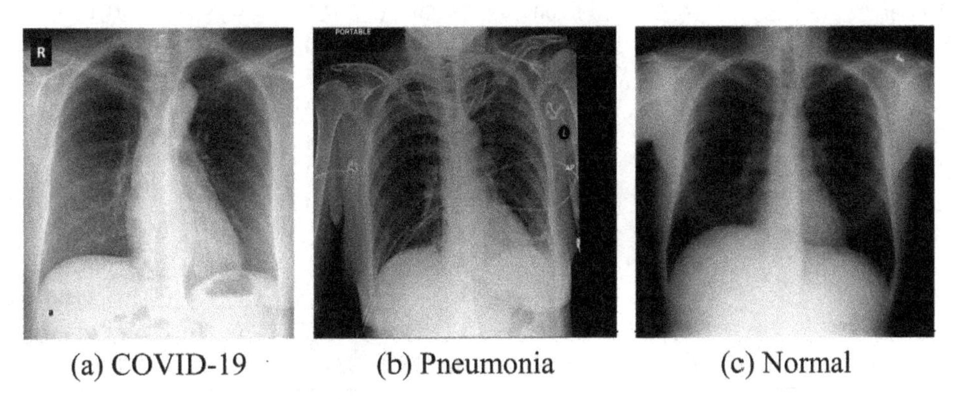


**Figure 13.3.** The sample image of each class: (a) COVID-19, (b) pneumonia and (c) normal.

epoch count. We have set the criterion of early stopping as follows: if the validation loss does not decrease to 0.001 by 50 epochs, the training should be stopped. Furthermore, we have run our model for five trials ($T$) to reduce the variability obtained in classification accuracy due to random partitioning of the training, validation and testing datasets. Finally, the overall accuracy (OA) has been calculated by averaging the accuracy of five trials. We have also computed the standard deviation (STD) of accuracy in five trials, which demonstrates the robustness of the model. All experiments have been performed in Python 3.6 with the Keras framework and a Tensorflow backend. The simulations were carried out in Google Colab, which provides an Intel(R) Xeon(R) CPU @ 2.30 GHz, 13 GB RAM and an NVIDIA Tesla K80 GPU.

#### *13.3.1.2 Experimental results*

The classification accuracy obtained using the 16 models on the testing dataset are shown in table 13.2 for binary classification and table 13.3 for multiclass classification. We have shown the classification accuracy for each trial ($T$) along with the OA and STD of five trials. It can be observed that the highest OA of 98.56% with an STD of 1.18% was obtained using the InceptionResNetV2 model for binary classification (table 13.2) and the highest OA of 88.44% with an STD of 1.16% was obtained using the Xception model for multiclass classification (table 13.3). Further, figure 13.4 depicts a graphical view of the performance comparison of these 16 models for binary and multiclass classification. For a more detailed analysis, we have shown the following parameters: class-wise accuracy, precision, sensitivity, specificity and F1-score using the

**Table 13.2.** Classification results obtained using 16 pre-trained deep CNN models in the case of binary classification (COVID-19 and normal).

| Deep CNN model | T1 | T2 | T3 | T4 | T5 | OA ± STD |
|---|---|---|---|---|---|---|
| VGG16 | 95.20 | 92.80 | 76.80 | 95.20 | 92.80 | 90.56 ± 6.96 |
| VGG19 | 81.60 | 96.80 | 77.60 | 80.80 | 96.00 | 86.56 ± 8.14 |
| ResNet50 | 93.60 | 93.60 | 93.60 | 93.60 | 96.00 | 94.08 ± 0.96 |
| ResNet101 | 87.20 | 96.80 | 94.40 | 95.20 | 88.80 | 92.48 ± 3.77 |
| ResNet152 | 96.80 | 76.00 | 96.80 | 99.20 | 96.80 | 93.12 ± 8.61 |
| ResNet50V2 | 91.20 | 96.00 | 92.00 | 93.60 | 97.60 | 94.08 ± 2.41 |
| ResNet101V2 | 96.00 | 97.60 | 95.20 | 96.79 | 98.40 | 96.80 ± 1.13 |
| ResNet152V2 | 92.79 | 99.19 | 92.00 | 97.60 | 96.00 | 95.52 ± 2.75 |
| InceptionV3 | 95.20 | 98.40 | 97.60 | 98.40 | 97.60 | 97.44 ± 1.18 |
| InceptionResNetV2 | 100 | 99.19 | 99.19 | 97.60 | 96.79 | **98.56 ± 1.18** |
| Xception | 97.60 | 99.19 | 96.00 | 98.40 | 96.79 | 97.60 ± 1.13 |
| MobileNet | 96.79 | 99.19 | 97.60 | 96.00 | 100 | 97.92 ± 1.48 |
| MobileNetV2 | 87.99 | 95.20 | 96.00 | 96.79 | 91.20 | 93.44 ± 3.33 |
| DenseNet121 | 97.60 | 98.40 | 98.40 | 97.60 | 96.00 | 97.60 ± 0.88 |
| DenseNet169 | 96.00 | 86.40 | 91.20 | 97.60 | 96.80 | 93.60 ± 4.23 |
| DenseNet201 | 92.80 | 96.00 | 96.80 | 93.60 | 92.80 | 94.40 ± 1.68 |

**Table 13.3.** Classification results obtained using 16 pre-trained deep CNN models in case of multiclass classification (COVID-19, pneumonia and normal).

| Deep CNN model | T1 | T2 | T3 | T4 | T5 | OA ± STD |
|---|---|---|---|---|---|---|
| VGG16 | 64.89 | 78.22 | 76.89 | 76.00 | 70.67 | 73.33 ± 2.67 |
| VGG19 | 72.89 | 41.78 | 56.00 | 44.44 | 72.00 | 57.42 ± 13.17 |
| ResNet50 | 80.44 | 81.78 | 76.89 | 64.00 | 83.56 | 77.33 ± 7.02 |
| ResNet101 | 85.33 | 83.56 | 76.67 | 84.00 | 80.44 | 82.40 ± 2.46 |
| ResNet152 | 76.89 | 41.78 | 64.89 | 78.67 | 84.00 | 69.25 ± 15.09 |
| ResNet50V2 | 77.78 | 85.78 | 85.33 | 85.33 | 81.33 | 83.11 ± 3.12 |
| ResNet101V2 | 83.56 | 81.33 | 80.00 | 72.00 | 87.11 | 80.80 ± 5.02 |
| ResNet152V2 | 76.89 | 87.11 | 84.00 | 82.22 | 76.67 | 81.78 ± 3.66 |
| InceptionV3 | 84.00 | 87.11 | 86.67 | 84.89 | 88.44 | 86.22 ± 1.59 |
| InceptionResNetV2 | 84.00 | 90.67 | 85.78 | 83.11 | 82.22 | 85.16 ± 3.00 |
| Xception | 89.33 | 89.77 | 88.88 | 86.66 | 87.55 | **88.44 ± 1.16** |
| MobileNet | 82.67 | 86.67 | 91.11 | 84.44 | 84.89 | 85.96 ± 2.87 |
| MobileNetV2 | 68.44 | 65.78 | 63.11 | 74.22 | 65.78 | 67.47 ± 3.78 |
| DenseNet121 | 82.67 | 84.00 | 88.00 | 79.11 | 80.44 | 82.84 ± 3.09 |
| DenseNet169 | 82.22 | 82.67 | 84.00 | 87.56 | 84.44 | 84.18 ± 1.88 |
| DenseNet201 | 85.33 | 84.89 | 82.67 | 82.22 | 86.23 | 84.27 ± 1.55 |

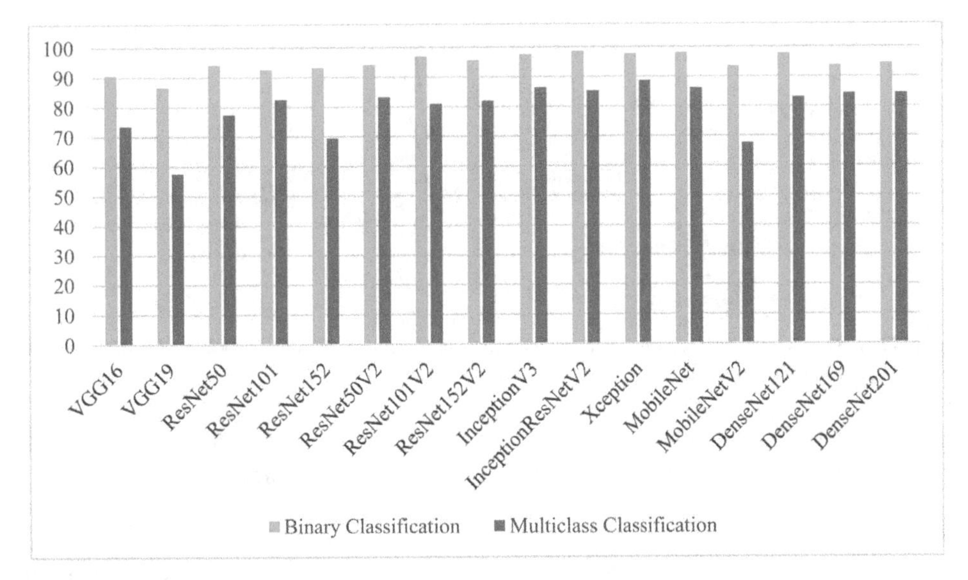


**Figure 13.4.** Performance comparison of 16 models for binary and multiclass classification.

best performing model, i.e. the InceptionResNetV2 model for binary classification (table 13.4) and the Xception model for multiclass classification (table 13.5). It can be observed that the model is able to diagnose COVID-19 cases with an accuracy of 95.65% for binary classification and 76.19% for multiclass classification. In addition, the

**Table 13.4.** Other performance parameters obtained using the InceptionResNetV2 model for binary classification.

| Classes | Class accuracy (%) | Precision (%) | Sensitivity (%) | Specificity (%) | F1-score |
|---|---|---|---|---|---|
| COVID-19 | 95.65 | 94.00 | 93.83 | 98.53 | 0.94 |
| Normal | 97.06 | 98.52 | 98.53 | 93.83 | 0.98 |

**Table 13.5.** Other performance parameters obtained using the InceptionResNetV2 model for multiclass classification.

| Classes | Class accuracy (%) | Precision (%) | Sensitivity (%) | Specificity (%) | F1-score |
|---|---|---|---|---|---|
| COVID-19 | 76.19 | 88.88 | 76.19 | 91.67 | 0.82 |
| Pneumonia | 90.00 | 78.95 | 90.00 | 80.80 | 0.84 |
| Normal | 100 | 95.24 | 100 | 99.51 | 0.98 |

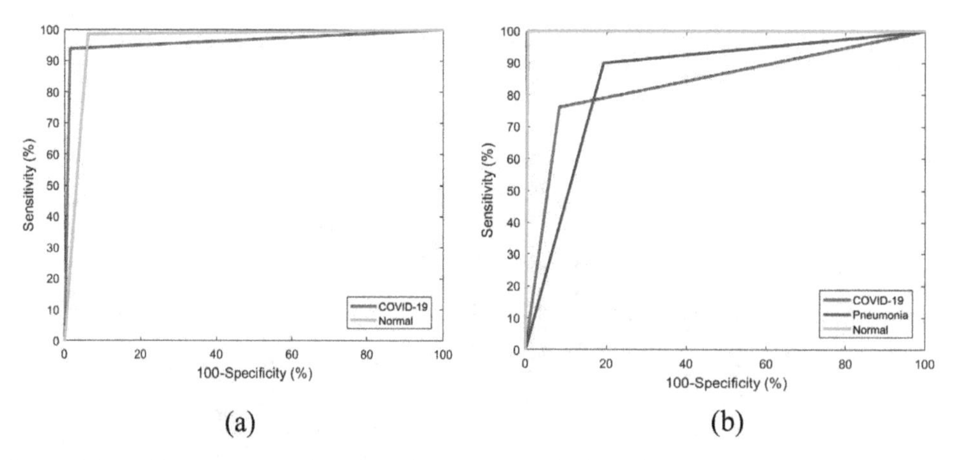


**Figure 13.5.** ROC curve for: (a) binary classification using the InceptionResNetV2 model (b) multiclass classification using the Xception model.

receiving operating characteristic (ROC) curve is shown in figure 13.5. As the curves are closer to the top-left corner, it demonstrates encouraging performance of the model.

*13.3.1.3 Discussion*

To highlight the performance of our approach for the diagnosis of COVID-19, we have presented a comparative performance analysis in table 13.6. Narin *et al* [6] explored the CXR image dataset using transfer learning with ResNet50 and achieved 98% accuracy. Apostolopoulos *et al* [3] presented a transfer learning approach using MobileNetV2 for the classification of the CXR image dataset and achieved 96.78% accuracy for binary classification and 94.72% for multiclass

**Table 13.6.** Benchmarking of our approach against the literature on the diagnosis of COVID-19.

| Authors (year) | Modality | # Classes with data size | Methodology used | Classification accuracy (%) |
|---|---|---|---|---|
| Narin *et al* (2020) [6] | CXR | 50 COVID-19, 50 normal | Transfer learning using ResNet50 | 98.00 |
| Apostolopoulos *et al* (2020) [3] | CXR | 224 COVID-19, 714 bacterial and viral pneumonia, 504 normal | Transfer learning using MobileNetV2 | 96.78 (binary), 94.72 (multiclass) |
| Sethy *et al* (2020) [7] | CXR | 133 COVID-19, 133 normal | Transfer learning using ResNet50 + SVM | 95.38 |
| Hemdan *et al* (2020) [2] | CXR | 25 COVID-19, 25 normal | VGG19 and DenseNet201 | 90.00 |
| Elasnaoui *et al* (2020) [29] | CXR | 2780 pneumonia, 1493 coronavirus, 231 COVID-19, 1583 normal | InceptionResNetV2 | 92.40 |
| Ozturk *et al* (2020) [1] | CXR | 125 COVID-19, 500 pneumonia, 500 normal | DarkCovidNet | 98.08 (binary), 87.02 (multiclass) |
| Our approach | CXR | 125 COVID-19, 500 pneumonia, 500 normal | Fine-tuning using InceptionResNetV2 Xception | 98.56 (binary), 88.44 (multiclass) |

classification. Sethy *et al* [7] used transfer learning with ResNet50 as a feature extractor and an SVM as a classifier and obtained 95.38% accuracy. Hemdan *et al* [2] examined various pre-trained models and obtained the highest classification accuracy of 90% for both the VGG16 and DenseNet201 models. In [29] various pre-trained models were utilized and the highest classification accuracy of 92.40% was reported with the InceptionResNetV2. Ozturk *et al* [1] proposed a DarkCovidNet model that achieved a classification accuracy of 98.08% for binary and 87.02% for multiclass classification. In our proposed approach, the same dataset (used by Ozturk *et al* [1]) was used and various deep CNN models with fine-tuning approaches have been explored. A better OA of 98.56% is achieved using InceptionResNetV2 for binary classification and 88.44% using Xception for multi-class classification.

## 13.4 Conclusion

In this chapter, the deep CNN based diagnosis of COVID-19 cases has been presented using CXR images. CXR images are the preferred imaging modality as the apparatus required to capture CXRs is cheaper compared to other processes such as RT-PCR and x-ray machines are also readily available in hospitals. However, the detection and diagnosis of COVID-19 patients through CXR images is not an easy task and requires experts. Therefore, automatic analysis of CXR images to diagnose COVID-19 has been presented in this study, which in turn helps doctors to diagnose and treat COVID-19 patients. An extensive comparison of the performance of 16 state-of-art deep CNN models with the fine-tuning approach has been carried out. Further, the performance of the models was evaluated based on various performance metrics such as accuracy, precision, sensitivity, specificity, F1-score and ROC curve. This comparative analysis demonstrates that InceptionResNetV2 achieved the highest classification accuracy of 98.56% with a standard deviation of 1.18% for binary classification and the Xception model achieved the highest classification accuracy of 88.44% with a standard deviation of 1.16% for multiclass classification. These promising results show that the presented models can be used in hospitals or through the Cloud to assist radiologists after further validation with more datasets.

## Bibliography

[1] Ozturk T, Talo M, Yildirim E A, Baloglu U B, Yildirim O and Acharya U R 2020 Automated detection of COVID-19 cases using deep neural networks with x-ray images *Comput. Biol. Med.* **121** 103792

[2] Hemdan E E D, Shouman M A and Karar M E 2020 COVIDX-Net: a framework of deep learning classifiers to diagnose COVID-19 in x-ray images arXiv: 2003.11055

[3] Apostolopoulos I D and Mpesiana T A 2020 Covid-19: automatic detection from x-ray images utilizing transfer learning with convolutional neural networks *Phys. Eng. Sci. Med.* **43** 635–40

[4] Ai T *et al* 2019 Correlation of chest CT and RT-PCR testing in coronavirus disease 2019 (COVID-19) in China: a report of 1014 cases *Radiology* **296** 200642

[5] Das N N, Kumar N, Kaur M, Kumar V and Singh D 2020 Automated deep transfer learning-based approach for detection of COVID-19 infection in chest x-rays *Innov. Res. Biomed. Eng.* (https://doi.org/10.1016/j.irbm.2020.07.001)

[6] Narin A, Kaya C and Pamuk Z 2020 Automatic detection of coronavirus disease (COVID-19) using x-ray images and deep convolutional neural networks *arXiv:* 200310849

[7] Sethy P K and Behera S K 2020 Detection of coronavirus disease (COVID-19) based on deep features *Preprints* 2020030300

[8] Wang L and Wong A 2020 COVID-Net: a tailored deep convolutional neural network design for detection of COVID-19 cases from chest x-ray images *arXiv:* 200309871

[9] Jadhav A R, Ghontale A G and Shrivastava V K 2019 Segmentation and border detection of melanoma lesions using convolutional neural network and SVM *Int. Conf. Computational Intelligence: Theories, Applications and Future Directions* **1** 97–108

[10] Bajaj V, Taran S, Tanyildizi E and Sengur A 2019 Robust approach based on convolutional neural networks for identification of focal EEG signals *IEEE Sens. Lett.* **3** 1–4

[11] Demir F, Şengür A, Bajaj V and Polat K 2019 Towards the classification of heart sounds based on convolutional deep neural network *Health Inf. Sci. Syst.* **7** 16

[12] Shrivastava V K, Pradhan M K, Minz S and Thakur M P 2019 Rice plant disease classification using transfer learning of deep convolution neural network *Int. Arch. Photogram. Remote Sens. Spat. Inf. Sci.* **XLII-3/W6** 631–5

[13] Bera S and Shrivastava V K 2020 Analysis of various optimizers on deep convolutional neural network model in the application of hyperspectral remote sensing image classification *Int. J. Remote Sens.* **41** 2664–83

[14] Ullo S L, Khare S K, Bajaj V and Sinha G R 2020 Hybrid computerized method for environmental sound classification *IEEE Access.* **8** 124055–65

[15] Chaudhary S, Taran S, Bajaj V and Sengur A 2019 Convolutional neural network based approach towards motor imagery tasks EEG signals classification *IEEE Sens. J.* **19** 4494–500

[16] Khare S K and Bajaj V 2020 Time–frequency representation and convolutional neural network-based emotion recognition *IEEE Trans. on Neural Networks and Learning Systems* doi:10.1109/TNNLS.2020.3008938

[17] Simonyan K and Zisserman A 2014 Very deep convolutional networks for large-scale image recognition *arXiv:* 1409.1556

[18] He K, Zhang X, Ren S and Sun J 2016 Deep residual learning for image recognition *Proc. of the IEEE Conf. on Computer Vision and Pattern Recognition* pp 770–8

[19] Szegedy C *et al* 2015 Going deeper with convolutions *Proc. of the IEEE Conf. on Computer Vision and Pattern Recognition* pp 1–9

[20] Ioffe S and Szegedy C 2015 Batch normalization: accelerating deep network training by reducing internal covariate shift *arXiv:* 1502.03167

[21] Szegedy C, Vanhoucke V, Ioffe S, Shlens J and Wojna Z 2016 Rethinking the inception architecture for computer vision *Proc. of the IEEE Conf. on Computer Vision and Pattern Recognition* pp 2818–26

[22] Szegedy C, Ioffe S, Vanhoucke V and Alemi A A 2017 Inception-V4, inception-ResNet and the impact of residual connections on learning *Thirty-First AAAI Conf. on Artificial Intelligence*

[23] Howard A G *et al* 2017 Mobilenets: efficient convolutional neural networks for mobile vision applications *arXiv:* 1704.04861

[24] Sandler M, Howard A, Zhu M, Zhmoginov A and Chen L-C 2018 MobileNetV2: inverted residuals and linear bottlenecks *Proc. of the IEEE Conf. on Computer Vision and Pattern Recognition* pp 4510–20
[25] Huang G, Liu Z, Van Der Maaten L and Weinberger K Q 2017 Densely connected convolutional networks *Proc. of the IEEE Conf. on Computer Vision and Pattern Recognition* pp 4700–08
[26] Glorot X, Bordes A and Bengio Y 2011 Deep sparse rectifier neural networks *Proc. of the Fourteenth Int. Conf. on Artificial Intelligence and Statistics* pp 315–23
[27] Cohen J P, Morrison P, Dao L, Roth K, Duong T Q and Ghassemi M 2020 COVID-19 image data collection: prospective predictions are the future *arXiv:* 2006.11988
[28] Wang X, Peng Y, Lu L, Lu Z, Bagheri M and Summers R M 2017 ChestX-ray8: hospital-scale chest x-ray database and benchmarks on weakly-supervised classification and localization of common thorax diseases *Proc. of the IEEE Conf. on Computer Vision and Pattern Recognition* pp 2097–106
[29] Elasnaoui K and Chawki Y 2020 Using x-ray images and deep learning for automated detection of coronavirus disease *J. Biomol. Struct. Dyn.* pp 1–22

**IOP** Publishing

Modelling and Analysis of Active Biopotential Signals in Healthcare, Volume 2

**Varun Bajaj and G R Sinha**

# Chapter 14

# Otitis media diagnosis model for tympanic membrane images processed in two-stage processing blocks

**Erdal Başaran[1], Zafer Cömert[2], Yüksel Çelik[3], Ümit Budak[4] and Abdulkasdir Şengür[5]**

[1]*Agri İbrahim Cecen University, Education Faculty, Computer and Instructional Technologies Education Deparment, Agri, Turkey*

[2]*Samsun University, Engineering Faculty, Software Engineering Department, Samsun, Turkey*

[3]*Karabuk University, Engineering Faculty, Computer Engineering Department, Karabuk, Turkey*

[4]*Bitlis Eren University, Engineering Faculty, Electrical and Electronics Engineering Department, Bitlis, Turkey*

[5]*Firat University, Technology Faculty, Electrical and Electronics Engineering Department, Elazig, Turkey*

Otitis media is an auditory problem caused by fluid accumulation in the middle ear. This disease is particularly common in children and is observed worldwide in 90% of children up to the age of ten. This disease can cause serious complications unless it is diagnosed and treated early. Individuals with otitis media disorder should be examined in detail by an experienced physician using an otoscope device in order to make a diagnosis if attending a clinic. Today, decision support systems are widely used for the processing of biomedical images for the diagnosis of various diseases, to obtain the properties of images, and to diagnose diseases with the help of machine learning algorithms or deep learning, where the features are automatically extracted. In this study an original dataset was created by collecting tympanic membrane images from volunteer patients using an otoscope device. The AlexNet, VGG16, GoogLeNet and ResNet50 deep transfer learning models, which have proven successful in experimental studies, were used to classify normal tympanic membrane and abnormal tympanic membrane images in this dataset. In addition, the feature vector of the images was extracted using these models. In order to obtain the features that are effective in the classification task, new feature vectors are obtained with feature selection algorithms and fed into the input of the various machine learning algorithms such as ANN, SVM, *k*NN, NB and DT. Using the proposed model, more successful classification results were obtained with fewer features. The highest classification result obtained was 83.56% using the SVM classifier.

## 14.1 Introduction

Otitis media (OM) is understood as an inflammation of the middle ear [1]. OM is a disease that causes a significant health burden [2]. As one of the most important causes of hearing loss, OM causes a decrease in the quality of life for the individual and also seriously affects a country's economy [3, 4]. Different types of OM have morphologically different properties. OM is a type of disease that includes varieties such as acute otitis media (AOM), otitis media with effusion (EOM) and chronic suppurative otitis media (CSOM) [5]. AOM usually occurs after respiratory infection in children, resulting in frequent visits to medical institutions and antibiotic use [6]. Eighty per cent of children under three years old experience this disease at least once [7]. Diagnosing AOM is difficult because the otoscopic examination of the tympanic membrane (TM) and evaluation of its movement with a pneumatic otoscope are methods that require training and expertise, so only 60% of suspected AOM patients are diagnosed correctly in children [8]. When the middle ear is examined via an otoscope, it is observed that there is redness and bulging on the membrane for AOM [9]. EOM, on the other hand, is identified by the appearance of fluid behind the membrane without encountering the symptoms and signs of AOM [10]. Although expressions such as catarrhal otitis media, non-suppurative otitis media, serotympanum, mucotympanum or glue ear are also used, EOM and secretory otitis media are the most commonly used names [11]. If EOM is not treated, it causes various changes in the TM, which can cause chronic otitis media as well as serious problems, including in the educational and psychological development of the individual [12]. EOM is commonly seen in childhood and is the most common cause of hearing loss in childhood [13]. CSOM, another type of OM, is defined as purulent and involves membrane perforation for at least two weeks. If left untreated, it can also lead to life-threatening infectious complications such as hearing impairment, brain abscesses and sinus thrombosis, and CSOM can last for years [14]. Many of the complications associated with CSOM include cholesteatoma (abnormally cancerous growth of tissue behind the eardrum) [15]. Generally, CSOM and recurrent otitis risk factors are as follows: being male, being white, being young, having the first otitis episode within six months after birth, living in a smoking environment, a family history of middle ear disease, a family or personal allergy history, bottle feeding, bilateral OM, OM history and nasal congestion. Constitutional, genetic and environmental factors can be identified in CSOM [16]. Clinical research on otitis media requires that a researcher be capable of making otoscopic diagnosis with a high degree of accuracy [17]. OM types are examined using devices that give different details of the images in order to achieve diagnosis with otoscopy examination [18].

Computer-aided studies to facilitate the diagnosis and treatment of OM, such as telemedicine, simulation, web-based and classification, have been performed.

Telemedicine is the evaluation of images and clinical reports obtained using biomedical devices between experts [19]. In a study conducted using the telemedicine method, recorded middle ear images of 66 children aged between 9 months and 16 years old were evaluated and diagnosed by experts in different locations, and expert suggestions were brought together [20]. In the medical field, simulations are widely used to develop diagnostic skills. In a study evaluating the accuracy of a traditional otoscope in children with AOM diagnosis compared to a CellScope Oto attached to a smart phone, CellScope Oto was determined to be more functional than the traditional otoscope [21]. In a different study involving medical school students, the effectiveness of a web-based platform was evaluated. The students were randomly divided into two groups, the simulator (21 people) and control group (20 people). After the students were subjected to a pre-test, they were allowed to attend otolaryngology classes and the students were allowed unlimited access to simulators for a week. Then, the test was completed and rated with a different otoscopy video. While there was no significant difference between the groups in the pre-test, in the post-test it was observed that the control group achieved an improvement of 31% after the lessons, while the simulation group achieved a 71% improvement [22]. In a randomized controlled study with medical school students to determine the effectiveness of diagnostic skills in otoscope simulation, web-based learning and classroom learning methods on learning skills were compared for their effect levels. After the students were randomly divided into groups, the diagnostic accuracy and otoscopy skills were determined using a basic test in all groups. The test was repeated immediately after each intervention and three months later. Immediately after the intervention and at the three month follow-ups, an improvement in the diagnostic accuracy level was observed in all groups compared to their baseline status. It was determined that the largest increase was in the otoscope simulation and web-based groups [23]. Various otitis media images obtained with an otoscope device are used for the classification of TM images. An active contour algorithm was used for the segmentation of the membrane region using 135 otitis images of three classes: AOM, EOM and non-effusion OM. It was obtained using a *k*-means clustering algorithm in the translucent region of gray color density in the images. The mean color value and canny edge detection algorithms were used to detect the presence of bubbles in the image. Finally, the hierarchical tree diagram method was used to classify the images. As a result of the classification, an 84% accuracy score was obtained [24]. In another study for the classification of TM images, with three datasets consisting of AOM, EOM and normal TM, images were taken with an otoscope compatible with the phone and then sent to a remote device via an file transfer protocol (FTP). After pre-processing with image processing techniques to identify the membrane area on the image, the active contour segmentation and edge detection techniques were applied and the membrane region was separated, and an attribute vector of this tissue was obtained. Classification was performed with a 70% success rate with a deep first search algorithm [25]. In another similar study, the classification process was carried out for categories of normal, AOM, EOM and COM in a dataset containing 865 TM images. First, to obtain the feature vectors after segmenting the membrane region using active contour segmentation, the grid color moment (GCM) uses color features, a histogram of oriented gradient and a local building pattern algorithm, which are textural feature extraction algorithms. After adding the Gabor feature for each image to this vector, the classification was made using the Adabost method with a success rate of 88.06% [26]. Tympanoplasty is used in the treatment of EOM otitis media. Another study aimed to detect the presence of the tympanoplasty tube in a total of 275 otoscopic TM images, including 235 TM images obtained from the clinic and 40 TM images obtained from the internet. To detect the presence of the tympanoplasty tube, the HOG, SIFT, LBP, GLCM, wavelet of color-based features, edge removal features and a number of advanced feature extraction algorithms were used. Also, the *k*-means algorithm was used to determine the center of the image. After the vector with 22 features was obtained, the classification process was carried out using a support vector machine (SVM). This proposed model was tested on high quality and low quality images and a 90% accuracy rate was obtained in the detection of the tympanostomy tube [27]. In a classification study performed on the visual features of the TM in images in a dataset, the experimental treatment was performed in a dataset with five classes: foreign matter–wax, normal, AOM, EOM and COM. With the help of image processing techniques, 489 images of 562 images in the dataset were handled. After obtaining features, such as perforation, malleus bone, shape, color, wax, liquid and light reflection in the middle ear in the picture, an 81.58% accuracy rate was obtained through classification with a decision tree (DT) method [28]. By using color images, different color channels and color spaces, a total of 186 images, including 111 normal TM and 75 color otitis TM images, the features of the images were represented using HSV color space and HSV color coherence techniques. Then, a performance comparison was made using machine learning algorithms. Classification was carried out using *k*-nearest neighbors (*k*NN), DTs, linear discriminant analysis (LDA), naïve Bayes (NB), multilayer artificial neural networks and SVMs, and 73.11% success was achieved with artificial neural networks and 72.4% with SVMs [29]. The faster-R CNN method was used with 282 TM images to detect and classify the membrane region in middle ear images, and the membrane region was determined with an accuracy rate of 75.85%. The membrane region parts identified for classification of images were given to the input of transfer learning models and 90.25% accuracy rate was obtained in the study [30]. For the classification of normal and AOM TM images in the same dataset, the experimental study was conducted with a total of 223 middle ear images in these classes and the tissue properties were obtained separately from the R, G and B channels and then combined. In addition to the tissue properties obtained by the gray level co-occurance matrix (GLCM), the mean values of each channel of the otoscope image were taken into account to determine whether the otoscope image belonged to the acute or normal class. Finally, this feature set was applied as the input to an artificial neural network (ANN) and 76.14% accuracy was achieved [31]. It was determined that the number of images was increased by using image augmentation techniques for the classification of AOM, Buşon, CSOM and Otit Externa TM images, which contain normal TM images and an abnormal class, and an 82.16% accuracy rate was obtained for the classification made with transfer learning models [32]. Using the same dataset, other images with an abnormal class were also used and a feature set obtained using trained deep transfer learning models was used. Using this feature set, a 99.47% accuracy rate was obtained as a result of the classification made with ANN, *k*NN, DT and SVM classifiers [33]. For classifying normal TM and COM TM images in this dataset, 598 TM images were classified using AlexNet, a deep transfer learning model. A 98.77% accuracy rate was obtained using a ten-fold cross validation method in the study [34]. Normal and COM images were classified using the same dataset and, to test this model's accuracy, five-fold cross validation was divided into 70% training and 30% testing and 50% training—50% testing of the dataset. Classification was carried out using AlexNet, VGG16, VGG19, GoogLeNet and ResNet50 with transfer learning models, and the VGG19 model was classified with a 97.2067% accuracy rate [35]. In a model developed with demographic features and clinical findings from patients as well as TM images, using the age, gender, preoperative audiometric results, ear pathologies and details of the surgical procedure as input data for 150 patients, they developed an interpretive artificial neural network and *k*NN model after the middle ear surgery of individuals with COM disease. The hearing threshold after surgery was accepted as the predicted output and 84% success was achieved using the proposed model and 75.8% success was achieved using *k*NN [36].

In the current study, abnormal tympanic membrane and normal tympanic membrane images are classified in a dataset created with middle ear images collected from volunteer patients. First, images are classified with the AlexNet, VGG16, GoogLeNet and ResNet50 models, which are deep transfer learning models that have proven successful in classifying biomedical images. Then, feature vectors consisting of 1000 features are extracted with fully connected layers of these deep transfer learning models and a feature vector consisting of 4000 features is obtained. In order to determine the most distinctive features among these features, the feature vector is created by selecting 400 features with each using the ReliefF and mRMR feature selection methods. The new feature vector consisting of 800 features is classified using various machine learning algorithms, such as ANN, SVM, *k*NN, NB and DT. A flow diagram of the proposed models can be seen in figure 14.1.

The method and the dataset of TM images used in this study are described in section 14.2. In section 14.3, the transfer learning models and experimental results of the proposed method are given in detail. The discussion and conclusion are provided in the remaining sections.

## 14.2 Materials and methods

### 14.2.1 Database

When looking at the literature studies on the diagnosis of OM, no suitable available dataset was found. First, after obtaining the necessary legal permissions, images were collected via an otoscope device from volunteer patients who came to the otolaryngology service of Özel Van Akdamar Hospital. Between October 2018 and June 2019, images were collected from volunteer patients, and a dataset consisting of 956 images was created and presented to the researchers with open access. The

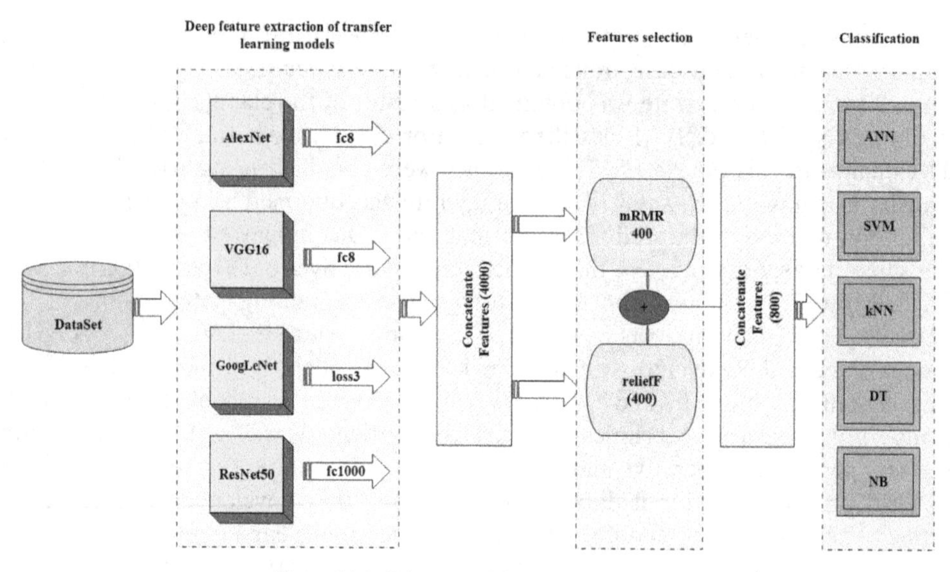


**Figure 14.1.** Diagram of the proposed model.

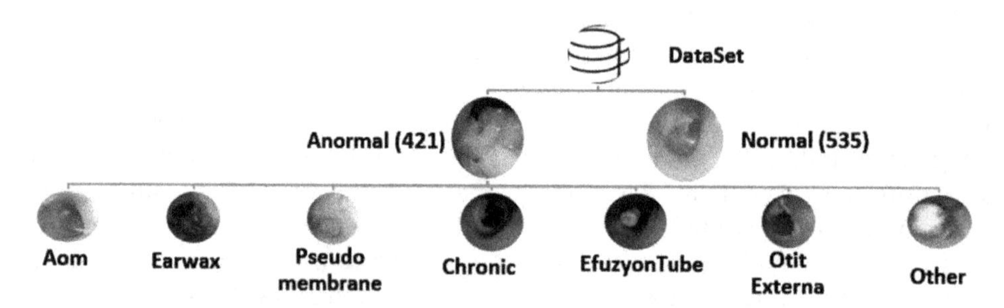


**Figure 14.2.** Tympanic membrane dataset.

labeling of the images was done by three experts and added to separate folders. The images in the dataset consist of normal and abnormal images. Information about the dataset is given in figure 14.2.

### 14.2.2 Convolutional neural network

Deep learning uses deep neural networks that automatically obtain features from raw data without the need to manually extract them [37, 38]. The CNN is a deep neural network that has the ability to extract many features of the image and is widely used in image analysis [39]. It is a special model of multilayer neural networks. As a result of the development of computer systems in recent years, CNN has become more popular and is used very frequently [40, 41]. It has been used to develop the LeNet 5 network for reading numbers on bank checks by Lecun *et al*, which was the first successful CNN model [42]. Successful results have since been obtained in many fields such as the classification of biomedical images, signal processing, detection of lesion regions, botany, astronomy and voice recognition

[43–48]. The CNN consists of sequences of convolution, pooling and fully connected layers, which follow each other and each have a separate task [49].

*Convolution layer.* Deep features of the images are obtained through this layer, known as a kernel. This occurs when filters of various sizes move from top to bottom and from left to right on the image. This layer is connected to the input image with a small number of neurons, called filters or masks, which can have various sizes [50]. The first convolution layer extracts simple features such as edge information. In the subsequent convolution layers that follow, high level properties are detected [51]. This layer does not exist in the traditional ANN. The convolution layer combines features with many convolution kernels to obtain different feature maps [52]. The equation for the convolution operation is

$$(f*h)[m, n] = \sum_j \sum_k h[j, k], f(m - j, n - k). \tag{14.1}$$

Here, the row and column indices of the output matrix, including the input image and the $h$ kernel, are $m$ and $n$, respectively, and $j$ and $k$ are the rows and columns of the new matrix that will occur after the new convolution.

*Pooling layer.* The feature map increases in the convolutional layer and since this will result in high computational costs, the pooling layer is usually located after the convolution layer [52]. The pooling layer reduces the size of the image by combining adjacent pixels in a particular area of the image with a single representative value. The neighboring pixels here are selected from the square matrix and the number of pixels varies depending on the problem. The pooling layer can obtain the maximum value of the input properties as well as the average values. In this way, a summary of the feature map is obtained. The greatest advantage of this layer is that the size of the visual features is reduced and the operation is performed independently of the input image [53].

*Fully connected layer.* In this layer, the properties obtained in the previous convolution and pooling layer are converted into a vector. This layer, which is a typical ANN model, is usually located between the pooling layer and the exit layer. All of the neurons in the previous layer depend on the neurons in the current layer [54]. After the convolution and pooling layers in the previous layers, the data are converted into a single vector [53].

CNN models proven in terms of their generalizability are used for the classification of a large scale dataset. Transfer learning models are such that the trained model can be reused in small datasets for the purpose of training and testing for the intended target [55]. In this study, in order to obtain the feature set of the images included in the OM dataset, a pre-trained model approach of transfer learning is applied. The feature set has been obtained using the AlexNet, VGG16, GoogLeNet and ResNet50 pre-trained transfer learning models.

*AlexNet.* This model was developed by Krizhevsky *et al* to perform the training and classification of the ImageNet database [56]. The AlexNet model, which has achieved very successful results in image classification, has created great excitement [57]. In this context, AlexNet was run on two graphics processors and showed great success by reducing the error rate to 15.3%. Consisting of 25 layers, the AlexNet architecture consists of five convolution and three fully connected layers. In this model, activation maps are created for the next layer using 11×11, 5×5 and 3×3 filters. AlexNet consists of a network with 60 million parameters and 650 000 neurons [57]. There are Fc feature layers in the last layers of this model. The AlexNet architecture has emerged with many innovations such as new dropout techniques and the overcoming the problem of overfitting with the development of ReLU activation function [58].

*VGG16.* This model was designed based on the idea of deeper convolutional neural networks [43]. Developed by Simonyan and Zisserman, this model, which takes the input image size as 224×224 pixels to train the dataset, uses 3×3 filters, a step size (stripe) of 1 and fully connected (FC) layers for feature extraction in its last layers [59, 60]. Similar to the AlexNet architecture, it has a 4096-dimensional activation vector of the first and second FC. The last FC layer is designed to classify 1000 different classes, such as keyboard, animals, etc. The VGG16 architecture was designed by applying smaller filters to reduce the number of trainable parameters [61]. This model, consisting of 41 layers in total, has 13 convolution layers and three fully connected layers [35]. The architecture of the 16 layers of the VGG16 model is given in table 14.1.

*GoogLeNet.* Some changes can be made to the model structure to improve the performance of CNN models. These techniques are to increase the depth and length of the model. To increase the depth of the network, it is necessary to increase the number of filters. The GoogLeNet model does so by making connections called inceptions in the structure of the system, reducing the calculation cost and reducing the number of parameters. While there are approximately 60 million parameters in the AlexNet model, the number of parameters in this model is around 5 million. It

**Table 14.1.** VGG16 model architecture.

| | |
|---|---|
| 1 | Convolution using 64 filters |
| 2 | Convolution using 64 filters + max pooling |
| 3 | Convolution using 128 filters |
| 4 | Convolution using 128 filters + max pooling |
| 5 | Convolution using 256 filters |
| 6 | Convolution using 256 filters |
| 7 | Convolution using 256 filters + max pooling |
| 8 | Convolution using 512 filters |
| 9 | Convolution using 512 filters |
| 10 | Convolution using 512 filters + max pooling |
| 11 | Convolution using 512 filters |
| 12 | Convolution using 512 filters |
| 13 | Convolution using 512 filters + max pooling |
| 14 | Fully connected with 4096 nodes |
| 15 | Fully connected with 4096 nodes |
| 16 | Output layer with softmax activation with 1000 nodes |

has reduced the number of parameters by approximately 12 times compared to the AlexNet model [43]. In the inception architecture, 1×1, 3x3 and 5x5 filters are used to obtain the feature map. The input takes 224×224 pixels as the image size. After these filters are used, the filters are combined in the other layer. The feature map is reduced by using 1×1 filters before 3x3 and 5x5 filters [62].

*ResNet50.* This transfer learning model comes to the fore with its residual blocks and the depth of its structure. The most important element of the ResNet50 model is the residual block. It forms a structure of leftover blocks that feeds the layers that follow each other or feeds the layer a few steps later. The ResNet50 architecture has 49 convolution layers and a fully connected layer. This architecture, consisting of five blocks, has filters of 1×1 and 3x3 size. It has a fully connected layer that ends with 1000 nodes in the last layer of this architecture. This model also takes 224×224 pixels as the input image size, such as VGG16 and GoogLeNet.

### 14.2.3 Feature selection

Feature selection is an approach that is widely used in machine learning applications to reduce the dimensionality before applying any data mining techniques, such as classification, association rules, clustering and regression [63]. The main purpose of the feature selection process is to choose the most appropriate features from the original features without the loss of useful information [64]. In this way, irrelevant and unnecessary features are removed, thus providing the reduction of data dimensionality. The task of feature selection is therefore to find a minimal and optimum feature set that preserves high accuracy while determining the minimal and robust features [65]. This also improves the learning performance and thus reduces the computational cost and training time [66]. Although choosing features has advantages, it has one drawback—it causes data loss [67].

#### *14.2.3.1 Relief*

In this section, we present conceptual definitions of the Relief-based feature selection algorithm in order to comprehend all members of its family. The Relief algorithm, inspired by the specimen-based learning approach [68], was first propounded by Kira and Rendell [69, 70]. This algorithm computes a proxy statistic for each feature that can be used to estimate the endpoint value, which is referred to as the target concept according to the 'quality' or 'relevance' criterion of the features [71]. While it gives a higher value to the characteristics that can be distinguished from the properties in different classes with its neighbors, it clusters the samples in the same class [72]. It sorts the feature vector according to the discrimination weights and a new feature vector is obtained by selecting the features above the threshold value determined by the user [73]. The feature classes selected here predict whether or not it distinguishes the target. This is not a desired feature if it separates samples from the same class [74]. These feature statistics are weighted with a score in the range of −1 (worst) to +1 (best). The pseudo-code of the Relief algorithm is given in algorithm 14.1.

**Algorithm 14.1.** The pseudo-code for the Relief algorithm from [68].

---

**Necessity:** the features and class labels for each training sample should be a vector
  $t \leftarrow$ number of training samples
  $f \leftarrow$ number of feature length
  $m \leftarrow$ number of random training samples out of $t$ used to update $W$
  initialize all feature weights as $W[A]$: = 0.0
**for** $i$: = 1 **to** $m$ **do**
  select a 'target' sample $R_i$, randomly
  find a nearest hit '$H$' and nearest miss '$M$'. (these are samples)
  **for** $A$: = 1 **to** $f$ **do**
    $W[A] := W[A] - \frac{\text{diff}(A, R_i, H)}{m} + \frac{\text{diff}(A, R_i, M)}{m}$
  end for
**end for**
**return** the vector $W$ of feature scores that estimate the quality of features

---

As shown in algorithm 14.1, the Relief algorithm looks for a fixed number of features $m$ determined by the user in the training samples $R_i$. The score vector $W$ assigned to each feature is updated according to differences or similarities in the observed feature value according to the target and neighboring examples. As a consequence, in every loop the distance between the target sample and all other samples is calculated. Relief finds the two closest neighbor samples of the target, one of those is the closest hit $H$ of the same class and the other is the closest miss $M$ of the opposite class. In the last step of the cycle, the weight of the $A$ feature in $W$ is updated, if the feature value differs between $R_i$ and either the nearest $H$ or the nearest $M$. The features that have a different value between $R_i$ and $M$ support the hypothesis that they are informative about the result, so the quality prediction $W[A]$ approaches +1. In contrast, in features with differences between $R_i$ and $H$, the opposite is proved, so the quality prediction $W[A]$ approaches −1. The diff function in algorithm 14.1 calculates the difference of the feature value $A$ between two samples, such as $I_1$ and $I_2$, in the updating process of $W$ [75]. Here, $I_1 = R_i$ and $I_2$ is also either $H$ or $M$. The diff functions for discrete and continuous features are expressed in the following equations, respectively:

$$\text{diff}(A, I_1, I_2) = f(x) = \begin{cases} 0, \text{ if value}(A, I_1) = \text{value}(A, I_2) \\ 1, \quad \text{otherwise} \end{cases} \tag{14.2}$$

$$\text{diff}(A, I_1, I_2) = \frac{\text{value}(A, I_1) - \text{value}(A, I_2)}{\max(A) - \min(A)}. \tag{14.3}$$

The maximum and minimum values of $A$ are searched throughout the entire sample set. Therefore the weights are normalized between 0 and 1 for both continuous and discrete features. In addition, with the division of the diff output to $m$ in equation $W[A]$ in each iteration, all the final weights will be normalized in the range [−1, 1].

*14.2.3.2 Minimum redundancy maximum relevance (mRMR)*

When too many features are used in classification tasks, it can result in low classification accuracy and high computational cost [76, 77]. The mRMR approach has been used by Peng *et al* [78] in the feature selection task in order to overcome this problem. The main purpose of this approach is to maximize the interclass correlation and minimize intraclass correlation. The mRMR method attempts to minimize the redundancy between these attributes, while selecting the most relevant from the labeled class information and feature vector attributes [44, 76]. The features chosen here are expected to have the least common knowledge of each other and to have as much discrimination value as possible. For an otitis media recognition problem, mutual information is exploited to evaluate the similarity or dissimilarity between variables.

For the given two random $X$ and $Y$ variables, their mutual information is expressed by the following equation:

$$I(X;\ Y) = \sum_{x \epsilon X}\sum_{y \epsilon Y} p(x,\ y) \times \log \frac{p(x,\ y)}{p(x) \times p(y)}, \tag{14.4}$$

where $p(x, y)$ represents the joint probabilistic density function of $X$ and $Y$, and $p(x)$ and $p(y)$ denote the probability of coincidence variables of $X$ and $Y$, respectively. When $x_i$ and $c$ represent individual features and classes, respectively, the maximum relevance equation is expressed as follows:

$$\max D(S,\ c),\ D = \frac{1}{|S|}\sum_{x_i \epsilon S} I(x_i;\ c). \tag{14.5}$$

Here $|S|$ is the dimension of the selected feature vector and $I(x_i;c)$ also denotes the mutual information of $x_i$ and $c$.

Although the selected subset feature $S$ with the highest correlation can be found using the maximum relevance criterion, maximum relevance based selected features could have some redundancy. Hence, the minimum redundancy criterion is also considered in order to choose mutually distinguishing features and this is shown in the following equation:

$$\min R(S),\ R = \frac{1}{|S|^2}\sum_{x_i, x_j \epsilon S} I(x_i;\ x_j), \tag{14.6}$$

where $I(x_i; x_j)$ indicates the mutual information of $x_i$ and $x_j$. The algorithm combining the maximum relevance and minimum redundancy criteria is called as mRMR, and is expressed by the following equations in order to optimize $D$ and $R$ synchronically:

$$\max \Phi_1(D,\ R),\ \Phi_1 = D - R \tag{14.7}$$

$$\max \Phi_2(D,\ R),\ \Phi_2 = D/R. \tag{14.8}$$

An incremental search approach is often used to determine optimum features. Suppose that the features $m - 1$ are selected to create the feature set $S_{m-1}$, the following criteria must be met to select the $m$th feature from the remaining feature set $\{X - S_{m-1}\}$:

$$\max_{x_j \epsilon (X - S_{m-1})} \left[ I(x_i; c) - \frac{1}{m-1} \sum_{x_i \epsilon S_{m-1}} I(x_i; x_j) \right] \tag{14.9}$$

$$\max_{x_j \epsilon (X - S_{m-1})} \left[ I(x_i; c) / \frac{1}{m-1} \sum_{x_i \epsilon S_{m-1}} I(x_i; x_j) \right]. \tag{14.10}$$

Although the incremental search approach somewhat increases the computing cost and complexity, it is highly effective in classification with high accuracy. See [78] to obtain more information about the mRMR method.

### 14.2.4 Artificial neural networks

Learning occurs in the human brain through the combination of neurons' synaptic connections. An ANN is a machine learning algorithm inspired the human brain to handle information and produce solutions to nonlinear problems. An ANN consists of three layers: the input layer, the neurons in the intermediate layer (hidden layer) and the output layer [31, 79]. In this learning algorithm, neurons are connected by connections and these connections have a weight value. The network is trained by optimizing the weight values of the connections by updating them. The neuron model of ANN can be seen in figure 14.3.

The inputs ($x_1$, $x_2$, ...$x_n$) found in an ANN neuron are examples that play a role in the learning of the network. These entries are multiplied by the weight values ($w_1$, $w_2$, ...$w_n$). All inputs are multiplied by their weight values to combine in the transfer function. After adding the bias value $b$, it is passed through the activation function $\sigma$. It is calculated mathematically as

$$f(\text{net}) = \sigma\left( \sum_{i=1}^{N} x_i w_i + b \right). \tag{14.11}$$

Activation functions are mathematical functions applied to the linear combination of $x_i$ neurons and $w_i$ neuron weight values and can establish nonlinear relationships between variables [80]. The most used activation functions in ANNs are hyperbolic tangent and sigmoid activation functions [81].

### 14.2.5 Support vector machines

The SVM, which is used in classification and regression problems, was first proposed by Vapnik [82]. This algorithm, which can be trained easily and is also successful in large sized problems, can produce better solutions against overfitting and over-generalization problems compared to neural networks [83]. While there are an

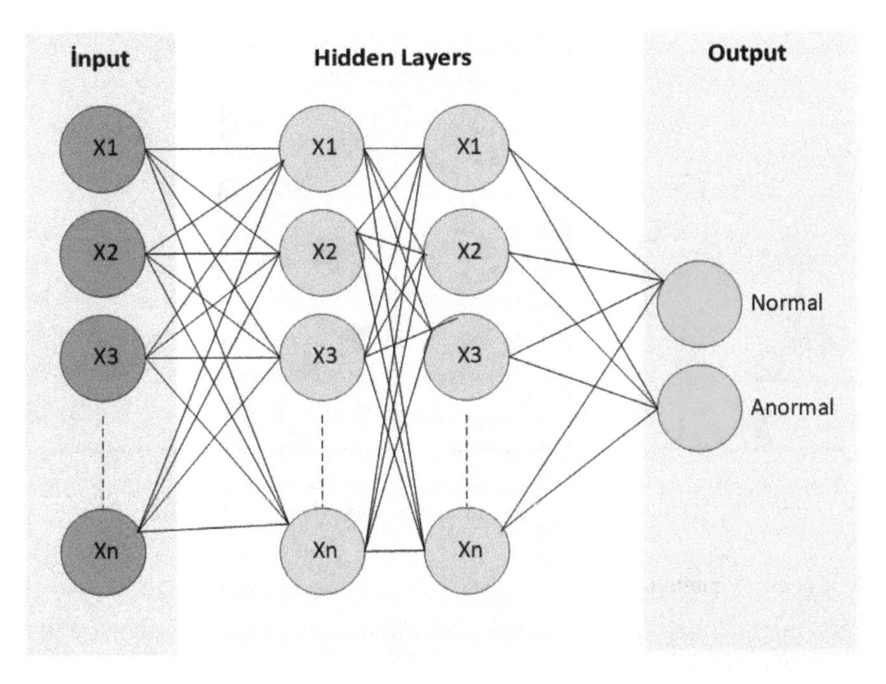


**Figure 14.3.** ANN neuron model.

infinite number of planes separating data from two classes, an SVM tries to find a hyperplane that best separates those data. The SVM chooses the furthest of both classes among these infinite numbers of hyperplanes. The points closest to the hyperplane are called support vectors. With this algorithm, linear problems as well as nonlinear problems can be separated using the kernel function [84]. In the classification process, linear planes may not always perform as desired. Therefore, it may be necessary to use nonlinear parabolic hyperplanes. It is very useful to use parabolic hyperplanes in problems where the properties of the different classes are intertwined in the search space.

The SVM aims to maximize the maximum range to separate the dataset with a hyperplane. Figure 14.4 shows the highest range separation with the SVM. Here, the SVM tries to find the most suitable separator between classes. Assuming that $x_i$ and $y_i$ are the training and testing labels, respectively, in a dataset, $i = 1, 2, \ldots \ell$, $x_i \in R^n$ and $y_i \in \{-1, 1\}$, the SVM needs to find a solution is based on the optimization problem

$$\min \frac{1}{2} \|w\|^2 + C\sum_{i=1}^{\ell} \xi_i \tag{14.12}$$
$$y_i(w_T x_i + b) \geqslant 1 - \xi i,\ i = 1, \ldots .n,\ \xi i \geqslant 0.$$

The regularization parameter $C$ in equation (14.12) balances the complexity of the model with the error of training.

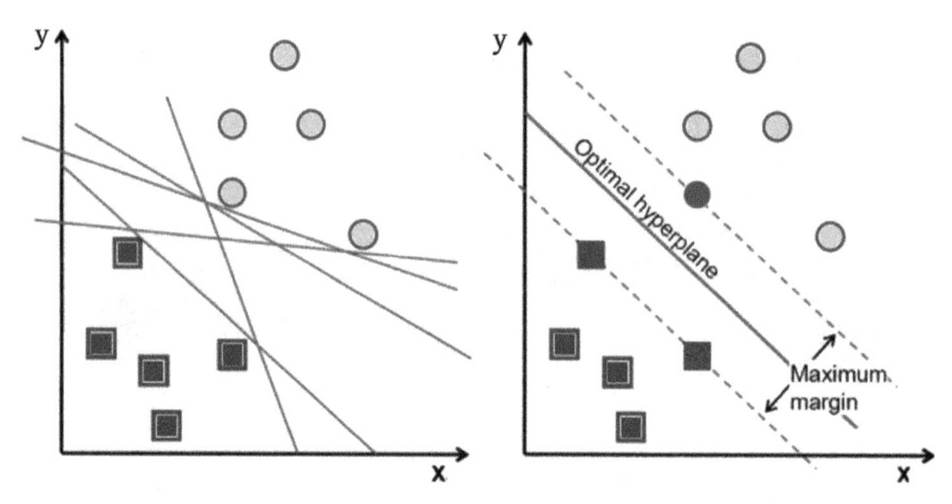


**Figure 14.4.** The SVM.

### 14.2.6 *k*-nearest neighbors

The *k*NN method used the closest neighbors in the training set to estimate the class to which a data point belongs. Here the number of neighbors refers to the number of the closest data points trained to make the *k* classification [85]. There are two major disadvantages to using the *k*NN classifier. One of them is that it works quite slowly when used for a large dataset, because it compares a new example to all the data contained in the dataset. Another disadvantage is that irrelevant samples can be misleading [86]. There are many distance measurement links, such as Euclidean, Manhattan, Minkowski and Chebyshev, to classify the test dataset to the closest training dataset [87]. The mathematical equation for measuring these distances is given in equations (14.13)–(14.16) below.

The Euclidean distance is equal to the square root of the sum of the squares of the differences. The Pythagorean theorem is applied:

$$d_{xy} = \sqrt{\sum_{k=1}^{n} (x_{ij} - y_{ij})^2}. \tag{14.13}$$

In the Manhattan distance, the sum of the absolute differences of the distance between the two points is equal:

$$d_{xy} = \sum_{i=1}^{n} |x_i - y_i|. \tag{14.14}$$

In the Minkoswski distance, when $m = 1$ equals the Manhattan distance, $m = 2$ equals the Euclidean distance:

$$d_{xy} = \left( \sum_{i=1}^{n} |x_i - y_i|^m \right)^{1/m}. \tag{14.15}$$

The Chebyshev distance is the greatest value of the absolute difference between the two points:

$$d_{xy} = \max_{i=1}^{n} |x_i - y_i|. \tag{14.16}$$

### 14.2.7 Decision tree

The DT, another machine learning method used for classification and regression problems, uses the branching method to determine the possible outcomes for a decision [88]. In a DT there are root, branch, leaf and decision nodes [33]. In decision trees, each branch can complete a classification. If no classification takes place at the end of a branch, a decision node is formed at the end of that branch. If a classification occurs at the end of the branch it becomes the leaf node. Each branch shows the result of the test according to the probability distribution on the class [88]. In this way the process starts from the root node and continues to the leaf.

DTs are easy to understand and interpret because they are visualizable, and they can be applied to numerical data as well as categorical data [89].

### 14.2.8 Naïve Bayes

This learning algorithm uses Bayes' theorem. NB shows the relationship between the target variables and independent variables. This algorithm, which specifies a class for each data point, is based on the assumption that all variables are independent. Assuming that $F = \{f_1, f_2, \ldots f_n\}$ the property set is $y_1, y_2, \ldots y_n$ class set. Whenever a class wants to be determined for each sample, Bayes' theorem is applied as follows:

$$P\left(\frac{y}{F}\right) = \frac{P\left(\frac{F}{c}\right)P(c)}{P(F)}. \tag{14.17}$$

Equation (14.17) expresses the probability of $y$ when $P(\frac{y}{F})$ is $F$, while $P(\frac{F}{c})$ expresses the probability that $F$ is corresponding to state $c$. $P(c)$ and $P(F)$ are probability values for $c$ and $F$.

## 14.3 Experimental results

In this study the confusion matrix and receiver operating curve (ROC) were used to evaluate the performance of the model. There are four values in the confusion matrix: true positive (TP), true negative (TN), false positive (FP) and false negative (FN). While the TP and TN classes are correctly predicted values, TN and FN represent wrongly predicted values [90]. Accuracy, sensitivity and specificity values were obtained with the confusion matrix. Accuracy refers to the success of the proposed model on all samples. Sensitivity refers to the ratio of abnormal TM images correctly detected among abnormal TM images, while specificity refers to the accurate classification rate of normal TM among all normal TM images. Accuracy, sensitivity and specificity are calculated, respectively, as

$$\text{Accuracy (Acc)} = \frac{\text{TP} + \text{TN}}{\text{TP} + \text{FP} + \text{TN} + \text{FN}} \tag{14.18}$$

$$\text{Sensitivity (Sn)} = \frac{\text{TP}}{\text{TP} + \text{FN}} \tag{14.19}$$

$$\text{Specifity (Sp)} = \frac{\text{TN}}{\text{TN} + \text{FP}}. \tag{14.20}$$

The ROC curve is one of the graphical methods used to evaluate diagnostic accuracy. There are specificity values on the $x$-axis and sensitivity on the $y$-axis. The area under this curve (AUC) is expected to be as close to one as possible. In the first experimental study for the classification of OM images, transfer learning models were fed with raw OM images. To train our network, the learning rate is $1 \times 10^{-4}$ and the dataset size (mini-batch size) used in each training iteration is 32. The epoch value for which the whole training set was trained once was determined to be 64. As an optimization, the static gradient descent algorithm was determined. The parameters used for the training of the network were determined by trial-spread. The dataset was randomly divided into 70% training and 30% testing to train the network. While 670 of 956 images are reserved for education, 286 images are reserved for testing.

At the end of the training, it was determined that the best learning was realized with a value of 82.86% using the VGG16 model (see table 14.2 for a summary of the results). For the AlexNet, GoogLeNet and ResNet50 models, 81.11%, 82.51% and 81.81% accuracy rates were obtained, respectively. Sensitivity, which is the distinguishing criterion of abnormal TM images, was highest for the VGG16 model at 82.95%. Considering the specificity ratios, which measure the success of distinguishing normal TM images, the best discrimination success was obtained with the VGG16 model. Good classification results were obtained using all the transfer learning models. A ten-fold cross validation method was also used to test the model performance. In this method, the dataset is divided into ten equal parts, and while each dataset is used as a test, the remaining data are used as a training set. Ten-fold cross validation results are given in table 14.3.

When looking at the ten-fold cross validation results, the best results were obtained with the VGG16 model. When looking at other transfer learning models, close results were obtained. While the accuracy rate was 82.16% with the VGG16

**Table 14.2.** Pre-trained CNN models (70% training 30% testing).

| Model | Acc (%) | Sn (%) | Sp (%) |
|---|---|---|---|
| AlexNet | 81.11 | 81.11 | 80.57 |
| VGG16 | 82.86 | 82.95 | 83.19 |
| GoogLeNet | 82.51 | 82.57 | 82.73 |
| ResNet50 | 81.81 | 81.82 | 81.83 |

**Table 14.3.** Ten-fold cross validation pre-trained CNN model results.

| Model | Acc (%) | Sn (%) | Sp (%) |
|---|---|---|---|
| AlexNet | 81.46 | 81.58 | 81.86 |
| VGG16 | 82.16 | 82.30 | 82.62 |
| GoogLeNet | 81.03 | 81.11 | 81.31 |
| ResNet50 | 81.37 | 81.49 | 81.78 |

**Table 14.4.** $k$ value accuracy rate.

| $k$ value | 1 | 2 | 5 | 10 | 50 | 100 |
|---|---|---|---|---|---|---|
| Accuracy (%) | 78.3 | 73.1 | 73.1 | 71.0 | 68.9 | 66.1 |

model, the order accuracy rates of the AlexNet, GoogLeNet and ResNet50 models were 81.46%, 81.03% and 81.37%. The sensitivity criterion, which is the distinguishing criterion of normal TM images, was again determined to be 82.30% with the best result for the VGG16 model. In the AlexNet, GoogLeNet and ResNet50 models, the results are 81.58%, 81.11% and 81.49%, respectively. Similarly, the sensitivity values that distinguish abnormal TM images were obtained with the best result for the VGG16 model of 82.62%. In the AlexNet, GoogLeNet and ResNet50 models, values of 81.86%, 81.31% and 81.78% were obtained, respectively.

In the second experimental study, layers with 1000 features of pre-trained transfer learning models were used. The FC8, FC8, loss3-classifier and FC-1000 layers of the models AlexNet, VGG16, GoogLeNet and ResNet50 were used. The ReliefF and mRMR dimension reduction methods were used on the 4000 feature dataset obtained by combining these features. With these two methods, 400 features were extracted from each and these features were combined. The ANN, SVM, $k$NN, NB and DT methods were used for the classification of the abnormal TM and normal TM images with the new feature sets obtained.

To train using the ANN method, the number of hidden layers is ten, the epoch value is 1000 and the Levenberg–Marquardt algorithm is used as the training algorithm. The dataset is divided into 70% education, 15% testing and 15% validation. From the experimental study, the accuracy rate was obtained as 51.37%. The sensitivity and specificity values, which reflect the success of distinguishing normal TM and abnormal TM images, were 52.14% and 50.43%, respectively.

To classify TM images with the $k$NN method, the number of neighborhood values $k = \{1, 2, 5, 10, 50, 100\}$ and the distance function was set as Euclidean. The dataset is divided into 70% training and 30% testing. When looking at table 14.4 the most effective results were obtained with a $k = 1$ neighborhood value. The accuracy, sensitivity and specificity values were 78.3%, 78.52% and 78.88%, respectively. The accuracy rates of $k$ values are given in table 14.4.

Hyperparameters were optimized and classification automatically adjusted using the decision trees method. The division hyperparameter was set as Gini and the maximum division number was set to 13. The best results for accuracy, sensitivity and specificity values were 75.87%, 76.19% and 76.62%, respectively.

Bayesian optimization and a Gaussian were used as the core in the NB machine learning method. To test this learning algorithm, the dataset was divided into 70% training and 30% testing datasets. The accuracy rate was determined as 72.72%, while the sensitivity value that distinguished the abnormal class was determined as 72.64%, while the value which was the criterion to distinguish normal TM images was measured as 72.46%.

With the dimension reduction feature set, the kernel selected for hyperparameter optimization of the SVM, another machine learning method used for classification of normal TM and abnormal TM images, was determined quadratically. The box constraint level value is determined as 1. The dataset was split as 70% training and 30% testing. While the accuracy value of 81.46% was obtained using the SVM method, the sensitivity and specificity values were determined as 81.59% and 81.86%, respectively. The highest results of the machine learning methods were measured using the SVM method. The number of TM images for which normal TM and abnormal TM were estimated is given in the confusion matrix of table 14.5 and the ROC curve for this method is given in figure 14.5.

In table 14.5, it is seen that the number of images detected correctly from abnormal TM images is lower than the number of normal TM images and the number of incorrectly detected abnormal images is higher than the number of normal TM images. It is known that for machine learning methods biased results are given by unbalanced datasets [91]. The specificity value, which is the discrimination rate of normal TM images, has a higher value than the sensitivity value, which is the rate of discrimination of abnormal TM images.

To evaluate the model's performance, all machine learning methods were examined with the ten-fold cross validation method using the same hyperparameters. Experimental results are given in table 14.6.

When looking at table 14.6, the highest performance results of the proposed model were obtained with the SVM method. While the accuracy value of the model was 83.56% with the SVM method, it separated the abnormal TM images by 83.61%. Normal TM images were separated by 83.77%. With the *k*NN method, the

**Table 14.5.** The confusion matrix of the best analysis accuracy obtained from the proposed method (70% training 30% testing).

| | | Predicted class | |
|---|---|---|---|
| True class | Abnormal | 92 | 34 |
| | Normal | 19 | 141 |
| | | Anormal | Normal |

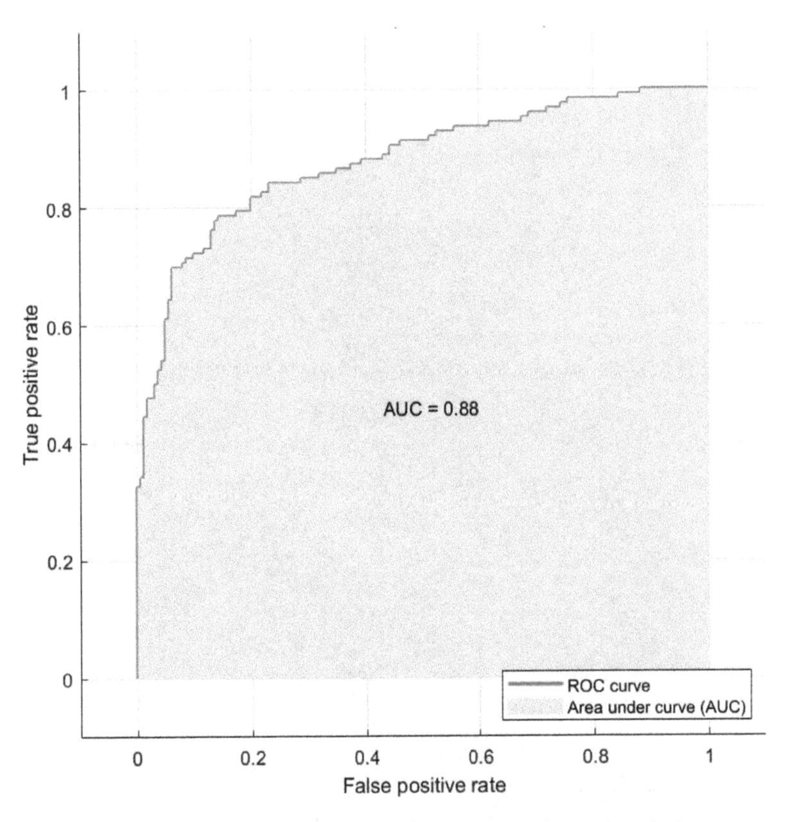


**Figure 14.5.** ROC graph for the SVM (70% training 30% testing).

**Table 14.6.** Machine learning method classification results with ten-fold cross validation.

| ML method | Acc (%) | Sn (%) | Sp (%) |
|---|---|---|---|
| *k*NN (*k* = 1) | 75.46 | 75.52 | 75.57 |
| *k*NN (*k* = 2) | 72.04 | 72.82 | 71.37 |
| *k*NN (*k* = 5) | 75.70 | 76.46 | 77.25 |
| *k*NN (*k* = 10) | 70.78 | 72.36 | 73.45 |
| *k*NN (*k* = 50) | 68.58 | 74.62 | 77.26 |
| *k*NN (*k* = 100) | 67.25 | 74.81 | 77.84 |
| ANN | 51.37 | 52.14 | 50.43 |
| SVM | 83.56 | 83.61 | 83.77 |
| DT | 71.41 | 71.42 | 70.59 |
| NB | 72.25 | 72.12 | 71.69 |

highest accuracy rate was obtained when a $k = 5$ distance to a neighbor was taken. The ANN, DT and NB methods were used to obtain accuracy rates of 51.37%, 71.41% and 72.25%, respectively. Using the proposed model, the SVM method with fewer features obtained better results than the transfer learning models. The ROC

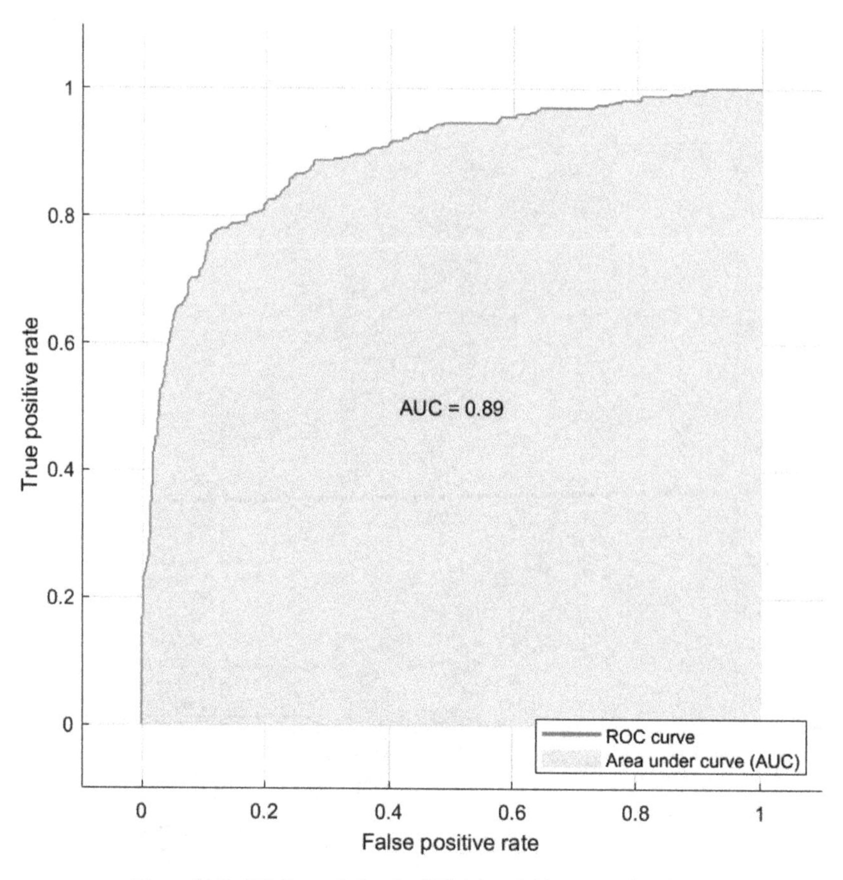


**Figure 14.6.** ROC graph for the SVM (ten-fold cross validation).

**Table 14.7.** The confusion matrix of the best analysis accuracy obtained from the proposed method (ten-fold cross validation).

| | | Predicted class | |
|---|---|---|---|
| True class | Abnormal | 324 | 97 |
| | Normal | 60 | 474 |
| | | Abnormal | Normal |

curve of the SVM method, for which the best results are obtained, is given in figure 14.6 and the confusion matrix in table 14.7.

## 14.4 Discussion

Otitis media is a serious condition that can occur at almost any age and affects the quality of life. This common disease of childhood is the biggest cause of antibiotic prescription [92]. This study proposes a unique model to distinguish between

**Table 14.8.** Comparison of the OM diagnostic models using the same TM image dataset.

| Authors | Methods | Class | Acc (%) |
|---|---|---|---|
| Başaran *et al* [30] | Transfer learning, faster-R CNN | 2 | 90.48 |
| Başaran *et al* [31] | ANN, GLCM, average of color channels | 2 | 76.14 |
| Cömert [32] | Transfer learning | 5 | 82.16 |
| Cömert [33] | Transfer learning, deep features | 5 | 93.05 |
| Başaran *et al* [34] | Transfer learning | 2 | 98.77 |
| Başaran *et al* [35] | Transfer learning | 2 | 97.20 |

diseased TM and normal TM. Transfer learning models are widely used in the diagnosis and classification of biomedical images and successful results have been obtained [93–95]. In this study, the features of deep pre-trained transfer learning models commonly used in the literature were combined and the number of features was reduced by 80% with the dimension reduction method, resulting in better results than for the transfer learning models. While classifying with direct transfer learning models, an 82.86% accuracy rate was obtained with the VGG16 model, while an 83.86% accuracy rate with many fewer features distinguished between normal and diseased TM images using our proposed model. A comparison to studies using the same dataset are given in table 14.8.

Başaran *et al* performed classification using the 282 TM dataset with the diagnosis of the membrane region and the deep membrane learning methods using to detect the membrane parts. The use of many parameters in deep learning can prolong the training time and cause hardware costs [96, 97]. The current study, conducted using the TM dataset, confirms the literature that the education period due to many parameters in deep learning is quite high. Again, when looking at other studies, the images are classified using direct raw data and are given directly to the input of the transfer learning models. Cömert performed classification using transfer learning models with machine learning methods and deep features. However, in this study classification was performed using 1000 featured layers of deep transfer learning and good results were obtained. In this study, good results were obtained by reducing the calculation cost with may fewer features. It presents a high abstraction feature by learning many more detailed features of the images with many layers existing in the deep learning [98]. It is seen that the classification results produced using the color features of the images and the GLCM method, which is the traditional textural feature extraction algorithm for images used by Başaran *et al*, have a lower accuracy rate compared to the studies with deep learning.

## 14.5 Conclusion

In this study, by using the depth features of transfer learning models to diagnose OM, the number of features has been reduced by using dimension reduction algorithms and classification has been carried out using machine learning algorithms and the results have been compared to other studies. The proposed model was also compared to the results obtained using direct transfer learning models and the effectiveness of the models was examined. As a result, the feature set of the proposed model was reduced by 20% compared to the transfer learning models and has been reduced by 80% compared to the feature set obtained with all models. The highest results were obtained with the SVM method at an 83.56% accuracy rate. In addition, the sensitivity and specificity rates were 83.61% and 83.77%, respectively. Normal TM and abnormal TM images were classified with a consistent model recommendation. The original dataset used in this study was presented to other researchers with open access. Multi-class classification studies are planned using the TM dataset in future studies. In addition, examining performance results using other feature reduction algorithms is being considered.

## References


[1] Graydon K, Waterworth C, Miller H and Gunasekera H 2019 Global burden of hearing impairment and ear disease *J. Laryngol. Otol.* **133** 18–25

[2] Chen Y *et al* 2019 Human primary middle ear epithelial cell culture: a novel *in vitro* model to study otitis media *Laryngoscope Investig. Otolaryngol.* **4** 663–72

[3] Libwea J N *et al* 2018 The prevalence of otitis media in 2–3 year old Cameroonian children estimated by tympanometry *Int. J. Pediatr. Otorhinolaryngol.* **115** 181–7

[4] Shah M U, Sohal M, Valdez T A and Grindle C R 2018 iPhone otoscopes: currently available, but reliable for tele-otoscopy in the hands of parents? *Int. J. Pediatr. Otorhinolaryngol.* **106** 59–63

[5] Schilder A G M *et al* 2016 Otitis media *Nat. Rev. Dis. Prim.* **2** 16063

[6] Kim T-H, Jeon J-M, Choi J, Im G J, Song J-J and Chae S-W 2020 The change of prevalence and recurrence of acute otitis media in Korea *Int. J. Pediatr. Otorhinolaryngol.* **134** 110002

[7] Cantarutti A *et al* 2020 Preventing recurrent acute otitis media with *Streptococcus salivarius* 24SMB and *Streptococcus oralis* 89a five months intermittent treatment: an observational prospective cohort study *Int. J. Pediatr. Otorhinolaryngol.* **132** 109921

[8] Marom T, Bobrow M, Eviatar E, Oron Y and Ovnat Tamir S 2017 Adherence to acute otitis media diagnosis and treatment guidelines among Israeli otolaryngologists *Int. J. Pediatr. Otorhinolaryngol.* **95** 63–8

[9] Marchisio P *et al* 2019 Updated guidelines for the management of acute otitis media in children by the Italian society of pediatrics: treatment *Pediatr. Infect. Dis. J.* **38** S10–21

[10] Durgut O and Dikici O S 2019 The effect of adenoid hypertrophy on hearing thresholds in children with otitis media with effusion *Int. J. Pediatr. Otorhinolaryngol.* **124** 116–9

[11] Sürmelioğlu Ö and Dağkıran M O 2013 Otitis media with effusion: diagnosis and treatment *Arch. Med. Rev. J.* **22** 194–208

[12] Kono M *et al* 2020 Features predicting treatment failure in pediatric acute otitis media *J. Infect. Chemother.* **27** 19–25

[13] Sezgin Z 2016 Otitis media with effusion: overview of diagnosis and treatment approaches *Pediat. Pract. Res.* **4** 1–11

[14] Elemraid M A *et al* 2010 Characteristics of hearing impairment in Yemeni children with chronic suppurative otitis media: a case-control study *Int. J. Pediatr. Otorhinolaryngol.* **74** 283–6

[15] Arbağ H *et al* 2002 Diagnostic and therapeutic considerations in chronic otitis media complications *Electron. J. Otolaryngol. – Head Neck Surg.* **1** 3–7

[16] Woo Park H, Ahn J, woong Kang M and Cho Y S 2020 Postoperative change in wideband absorbance after tympanoplasty in chronic suppurative otitis media *Auris Nasus Larynx* **47** 215–9

[17] Stool S E and Flaherty M R N 1983 Validation of diagnosis of otitis media with effusion *Ann. Otol. Rhinol. Laryngol.* **92** 5–6

[18] Coutinho T C, Kim S and Hwang J Y 2020 Discrimination between acute otitis media and otitis media with effusion using a multimode smartphone-based otoscope *Proc. SPIE* **11243** 112431J

[19] Kalender N and Özdemir L 2014 Yaşlılarda sağlık hizmetlerinin sunumunda tele-tıp kullanımı *Anadolu Hemşirelik ve Sağlık Bilimleri Dergisi* 17 (Erzurum: Ataturk University) 50–8

[20] Eikelboom R H, Mbao M N, Coates H L, Atlas M D and Gallop M A 2005 Validation of tele-otology to diagnose ear disease in children *Int. J. Pediatr. Otorhinolaryngol.* **69** 739–44

[21] Mousseau S, Lapointe A and Gravel J 2018 Diagnosing acute otitis media using a smartphone otoscope; a randomized controlled trial *Am. J. Emerg. Med.* **36** 1796–801

[22] Stepniak C, Wickens B, Husein M, Paradis J A, Ladak H M, Kevin M T and Fung S K A 2017 Blinded randomized controlled study of a web-based otoscopy simulator in undergraduate medical education *Laryngoscope 2017* **127** 1306–11

[23] Wu V and Beyea J A 2017 Evaluation of a web-based module and an otoscopy simulator in teaching ear disease *Otolaryngol.—Head Neck Surg.* **156** 272–7

[24] Kuruvilla A, Li J, Yeomans P H, Quelhas P, Hoberman N S A and Kovačević J 2012 Otitis media vocabulary and grammar *19th IEEE Int. Conf. Image Processing (Orlando, FL)* pp 2845–8

[25] Huang Y K and Huang C P 2018 A depth-first search algorithm based otoscope application for real-time otitis media image interpretation *Parallel Distrib. Comput. Appl. Technol. PDCAT Proc.* **2017** 170–5

[26] Shie C K, Chang H T, Fan F C, Chen C J, Fang T Y and Wang P C 2014 A hybrid feature-based segmentation and classification system for the computer aided self-diagnosis of otitis media *2014 36th Annu. Int. Conf. IEEE Eng. Med. Biol. Soc. EMBC 2014* pp 4655–8

[27] Wang X, Valdez T A and Bi J 2015 Detecting tympanostomy tubes from otoscopic images via offline and online training *Comput. Biol. Med.* **61** 107–18

[28] Myburgh H C, van Zijl W H, Swanepoel D, Hellström S and Laurent C 2016 Otitis media diagnosis for developing countries using tympanic membrane image-analysis *EBioMedicine* **5** 156–60

[29] Mironica I, Vertan C and Gheorghe D C 2011 Automatic pediatric otitis detection by classification of global image features *2011 E-Health and Bioengineering Conf. (EHB)* pp 1–4

[30] Başaran E, Cömert Z and Çelik Y 2020 Convolutional neural network approach for automatic tympanic membrane detection and classification *Biomed. Signal Process. Control* **56** 101734

[31] Basaran E, Sengur A, Comert Z, Budak U, Celik Y and Velappan S 2019 Normal and acute tympanic membrane diagnosis based on gray level co-occurrence matrix and artificial neural networks *2019 Int. Artificial Intelligence and Data Processing Symp. (IDAP)* pp 1–6

[32] Cömert Z 2019 Otitis media için evrişimsel sinir ağlarına dayalı entegre bir tanı sistemi *Bitlis Eren Üniversitesi Fen Bilim. Derg.* **8** 1498–511

[33] Cömert Z 2020 Fusing fine-tuned deep features for recognizing different tympanic membranes *Biocybern. Biomed. Eng.* **40** 40–51

[34] Basaran E, Comert Z, Sengur A, Budak U, Celik Y and Togacar M 2019 Chronic tympanic membrane diagnosis based on deep convolutional neural network *UBMK 2019—Proc., 4th Int. Conf. on Computer Science and Engineering*

[35] Başaran E, Cömert Z, Şengür A, Budak Ü, Çelik Y and Toğaçar M 2020 Normal ve kronik hastalıklı orta kulak imgelerinin eEvrişimsel sinir ağları yöntemiyle tespit edilmesi *Türkiye Bilişim Vakfı Bilgi. Bilim. ve Mühendisliği Derg.* **13** 1–10 https://dergipark.org.tr/tr/pub/tbbmd/issue/53711/657649

[36] Szaleniec J *et al* 2013 Artificial neural network modelling of the results of tympanoplasty in chronic suppurative otitis media patients *Comput. Biol. Med.* **43** 16–22

[37] Ullo S L, Khare S K, Bajaj V and Sinha G R 2020 Hybrid computerized method for environmental sound classification *IEEE Access* **8** 124055–65

[38] Demir F, Şengür A, Bajaj V and Polat K 2019 Towards the classification of heart sounds based on convolutional deep neural network *Heal. Inf. Sci. Syst.* **7** 16

[39] Ding C 2020 Convolutional neural networks for particle shape classification using light-scattering patterns *J. Quant. Spectrosc. Radiat. Transf.* **245** 106901

[40] Kujawa S, Mazurkiewicz J and Czekała W 2020 Using convolutional neural networks to classify the maturity of compost based on sewage sludge and rapeseed straw *J. Clean. Prod.* **258** 120814

[41] Bajaj V, Taran S, Tanyildizi E and Sengur A 2019 Robust approach based on convolutional neural networks for identification of focal EEG signals *IEEE Sensors Lett.* **3** 1–4

[42] Lecun Y, Bottou L, Bengio Y and Haffner P 1998 Gradient-based learning applied to document recognition *Proc. IEEE* **86** 2278–324

[43] Altuntaş Y, Cömert Z and Kocamaz A F 2019 Identification of haploid and diploid maize seeds using convolutional neural networks and a transfer learning approach *Comput. Electron. Agric.* **163** 104874

[44] Toğaçar M, Ergen B and Cömert Z 2020 Detection of lung cancer on chest CT images using minimum redundancy maximum relevance feature selection method with convolutional neural networks *Biocybern. Biomed. Eng.* **40** 23–39

[45] Guo Y, Budak Ü and Şengür A 2018 A novel retinal vessel detection approach based on multiple deep convolution neural networks *Comput. Methods Programs Biomed.* **167** 43–8

[46] Toğaçar M, Cömert Z and Ergen B 2020 Classification of brain MRI using hyper column technique with convolutional neural network and feature selection method *Expert Syst. Appl.* **149** 113274

[47] Anders F, Hlawitschka M and Fuchs M 2020 Automatic classification of infant vocalization sequences with convolutional neural networks *Speech Commun.* **119** 36–45

[48] Cecil D and Campbell-Brown M J 2020 The application of convolutional neural networks to the automation of a meteor detection pipeline *Planet. Space Sci.* **186** 104920

[49] Chaudhary S, Taran S, Bajaj V and Sengur A 2019 Convolutional neural network based approach towards motor imagery tasks EEG signals classification *IEEE Sens. J.* **19** 4494–500

[50] Shanthi T, Sabeenian R S and Anand R 2020 Automatic diagnosis of skin diseases using convolution neural network *Microprocess. Microsyst.* **76** 103074

[51] Liu Y H 2018 Feature extraction and image recognition with convolutional neural networks *J. Phys. Conf. Ser.* **1087** 62032

[52] Zhao G, Liu G, Fang L, Tu B and Ghamisi P 2019 Multiple convolutional layers fusion framework for hyperspectral image classification *Neurocomputing* **339** 149–60

[53] Sarıgül M, Ozyildirim B M and Avci M 2019 Differential convolutional neural network *Neural Netw.* **116** 279–87

[54] Wang F *et al* 2019 Generative adversarial networks and convolutional neural networks based weather classification model for day ahead short-term photovoltaic power forecasting *Energy Convers. Manag.* **181** 443–62

[55] Kumari P and Seeja K R J 2020 Periocular biometrics for non-ideal images: with off-the-shelf deep CNN and transfer learning approach *Procedia Comput. Sci.* **167** 344–52

[56] Krizhevsky A, Sutskever I and Hinton G E 2012 ImageNet classification with deep convolutional neural networks *Proc. of the 25th Int. Conf. on Neural Information Processing Systems* **1** 1097–105

[57] Lu S, Lu Z and Zhang Y-D 2019 Pathological brain detection based on AlexNet and transfer learning *J. Comput. Sci.* **30** 41–7

[58] Liu Z 2020 Soft-shell shrimp recognition based on an improved AlexNet for quality evaluations *J. Food Eng.* **266** 109698

[59] Toğaçar M, Özkurt K B, Ergen B and Cömert Z 2019 BreastNet: a novel convolutional neural network model through histopathological images for the diagnosis of breast cancer *Physica* A **545** 123592

[60] Rauf H T, Lali M I U, Zahoor S, Shah S Z H, Rehman A U and Bukhari S A C 2019 Visual features based automated identification of fish species using deep convolutional neural networks *Comput. Electron. Agric.* **167** 105075

[61] Abdalla A *et al* 2019 Fine-tuning convolutional neural network with transfer learning for semantic segmentation of ground-level oilseed rape images in a field with high weed pressure *Comput. Electron. Agric.* **167** 105091

[62] Wang R, Li W and Zhang L 2019 Blur image identification with ensemble convolution neural networks *Signal Process.* **155** 73–82

[63] Vithya M and Sangaiah S 2020 Recommendation system based on optimal feature selection algorithm for predictive analysis *Emerging Research in Data Engineering Systems and Computer Communications* (Berlin: Springer) pp 105–19

[64] Khare S K and Bajaj V 2020 Time-frequency representation and convolutional neural network-based emotion recognition *IEEE Trans. Neural Networks Learn. Syst.* 1–9

[65] Ladha L and Deepa T 2011 Feature selection methods and algorithms *Int. J. Comput. Sci. Eng.* **3** 1787–97

[66] Sutha K and Tamilselvi J J 2015 A review of feature selection algorithms for data mining techniques *Int. J. Comput. Sci. Eng.* **7** 63

[67] Kilicarslan S, Adem K and Celik M 2020 Diagnosis and classification of cancer using hybrid model based on ReliefF and convolutional neural network *Med. Hypotheses* **137** 109577

[68] Aha D W, Kibler D and Albert M K 1991 Instance-based learning algorithms *Mach. Learn.* **6** 37–66

[69] Kira K and Rendell L A 1992 The feature selection problem: traditional methods and a new algorithm *Assoc. Advance. Artific. Intell.* **2** 129–34

[70] Kira K and Rendell L A 1992 A practical approach to feature selection *Machine Learning Proceedings 1992* (Amsterdam: Elsevier) pp 249–56

[71] Urbanowicz R J, Meeker M, La Cava W, Olson R S and Moore J H 2018 Relief-based feature selection: introduction and review *J. Biomed. Inform.* **85** 189–203

[72] Zhang X, Zhang Q, Chen M, Sun Y, Qin X and Li H 2018 A two-stage feature selection and intelligent fault diagnosis method for rotating machinery using hybrid filter and wrapper method *Neurocomputing* **275** 2426–39

[73] Huang Y, McCullagh P J and Black N D 2009 An optimization of ReliefF for classification in large datasets *Data Knowl. Eng.* **68** 1348–56

[74] Reyes O, Morell C and Ventura S 2015 Scalable extensions of the ReliefF algorithm for weighting and selecting features on the multi-label learning context *Neurocomputing* **161** 168–82

[75] Robnik-Šikonja M and Kononenko I 2001 Comprehensible interpretation of relief's estimates *Machine Learning: Proc. of the Eighteenth Int. Conf. on Machine Learning (ICML2001)* (Williamstown, MA: Morgan Kaufmann) pp 433–40

[76] Yan X and Jan J M 2019 Intelligent fault diagnosis of rotating machinery using improved multiscale dispersion entropy and mRMR feature selection *Knowledge-Based Syst.* **163** 450–71

[77] Cıbuk M, Budak U, Guo Y, Ince M C and Sengur A 2019 Efficient deep features selections and classification for flower species recognition *Measurement* **137** 7–13

[78] Peng H, Long F and Ding C 2005 Feature selection based on mutual information criteria of max-dependency, max-relevance, and min-redundancy *IEEE Trans. Pattern Anal. Mach. Intell.* **27** 1226–38

[79] Cömert Z and Kocamaz A D 2017 A study of artificial neural network training algorithms for classification of cardiotocography signals *Bitlis Eren Univ. J. Sci. Technol.* **7** 93–103

[80] Paneiro G and Rafael M 2020 Artificial neural network with a cross-validation approach to blast-induced ground vibration propagation modeling *Undergr. Space* https://doi.org/10.1016/j.undsp.2020.03.002

[81] Yadav A M, Chaurasia R C, Suresh N and Gajbhiye P 2018 Application of artificial neural networks and response surface methodology approaches for the prediction of oil agglomeration process *Fuel* **220** 826–36

[82] Vapnik V 1998 The support vector method of function estimation *Nonlinear Modeling* (Berlin: Springer) pp 55–85

[83] Sahana M, Rehman S, Sajjad H and Hong H 2020 Exploring effectiveness of frequency ratio and support vector machine models in storm surge flood susceptibility assessment: a study of Sundarban Biosphere Reserve, India *CATENA* **189** 104450

[84] Papandrea P J, Frigieri E P, Maia P R, Oliveira L G and Paiva A P 2020 Surface roughness diagnosis in hard turning using acoustic signals and support vector machine: a PCA-based approach *Appl. Acoust.* **159** 107102

[85] Cömert Z and Kocamaz A F 2017 Using wavelet transform for cardiotocography signals classification *25th Signal Processing and Communications Applications Conf. (SIU)* pp 1–4

[86] Müller P *et al* 2019 Scent classification by *k* nearest neighbors using ion-mobility spectrometry measurements *Expert Syst. Appl.* **115** 593–606

[87] Arian R, Hariri A, Mehridehnavi A, Fassihi A and Ghasemi F 2020 Protein kinase inhibitors' classification using *k*-nearest neighbor algorithm *Comput. Biol. Chem.* **86** 107269

[88] Gohari M and Eydi A M F 2020 Modelling of shaft unbalance: modelling a multi discs rotor using *k*-nearest neighbor and decision tree algorithms *Measurement* **151** 107253

[89] Ochiai Y, Masuma Y and Tomii N 2019 Improvement of timetable robustness by analysis of drivers' operation based on decision trees *J. Rail Transp. Plan. Manag.* **9** 57–65

[90] Toğaçar M, Ergen B and Cömert Z 2020 Waste classification using AutoEncoder network with integrated feature selection method in convolutional neural network models *Measurement* **153** 107459
[91] Devarriya D, Gulati C, Mansharamani V, Sakalle A and Bhardwaj A 2020 Unbalanced breast cancer data classification using novel fitness functions in genetic programming *Expert Syst. Appl.* **140** 112866
[92] Demant M N, Jensen R G, Bhutta M F, Laier G H, Lous J and Homøe P 2019 Smartphone otoscopy by non-specialist health workers in rural Greenland: a cross-sectional study *Int. J. Pediatr. Otorhinolaryngol.* **126** 109628
[93] Toğaçar M, Ergen B and Cömert Z 2020 BrainMRNet: brain tumor detection using magnetic resonance images with a novel convolutional neural network model *Med. Hypotheses* **134** 109531
[94] Pathak Y, Shukla P K, Tiwari A, Stalin S, Singh S and Shukla P K 2020 Deep transfer learning based classification model for COVID-19 disease *IRBM*
[95] Swati Z N K *et al* 2019 Brain tumor classification for MR images using transfer learning and fine-tuning *Comput. Med. Imaging Graph.* **75** 34–46
[96] Abu Mallouh A, Qawaqneh Z and Barkana B D 2019 Utilizing CNNs and transfer learning of pre-trained models for age range classification from unconstrained face images *Image Vis. Comput.* **88** 41–51
[97] Qin J, Pan W, Xiang X, Tan Y and Hou G 2020 A biological image classification method based on improved CNN *Ecol. Inform.* **58** 101093
[98] Currie G, Hawk K E, Rohren E, Vial A and Klein R 2019 Machine learning and deep learning in medical imaging: intelligent imaging *J. Med. Imaging Radiat. Sci.* **50** 477–87

**IOP** Publishing

Modelling and Analysis of Active Biopotential Signals in Healthcare, Volume 2

**Varun Bajaj and G R Sinha**

# Chapter 15

# Modelling and analysis for active infrared thermography for breast cancer screening

**Ravibabu Mulaveesala[1], Geetika Dua[2], Vanita Arora[3] and Anshul Sharma[1]**

[1]*InfraRed Imaging Laboratory (IRIL), Department of Electrical Engineering, Indian Institute of Technology Ropar, Bara Phool, Birla Seed Farms, Rupnagar, Punjab, INDIA 140001*

[2]*Electronics and Communication Engineering, Thapar Institute of Engineering and Technology, P.O. Box 32, Bhadson Road, Patiala, Punjab, Pin -147004, India*

[3]*School of Electronics, Indian Institute of Information Technology Una, Himachal Pradesh, India - 177005*

Breast cancer is a major disease of females in almost all countries. Early diagnosis of the disease at its preliminary stage is very important. Early diagnosis is of considerable interest to the medical community as it increases the chances of survival of the patient. The approaches, such as mammography, biopsy, magnetic resonance imaging, ultrasound scanning, etc, which are commonly used in the diagnosis of the disease have limited applicability due to the involvement of harmful radiation, discomfort to the patient, complexity of data interpretation, bulky test equipment, high cost, etc. Thus there is great demand to develop a safe, economical and portable screening approach for the timely detection of breast tumours. Active infrared thermography is one such screening technique, which is remote, fast, safe, comfortable, economical and non-invasive and can be complementary to other methods such as mammography, ultrasound scanning, etc, for the early identification breast cancer, irrespective of the patient's age, size and type of breast (fat or dense). This chapter deals with the numerical analysis for the active infrared thermographic technique, to spot the presence of hidden tumours inside a simulated breast sample. Furthermore, frequency domain based phase and time domain based phase and amplitude thermal images are produced using post-processing analysis of the data obtained from the numerical simulation of the proposed models. The studies reported in this chapter show the potential of the proposed approach in the early detection of hidden tumours in fatty as well as dense breasts.

doi:10.1088/978-0-7503-3411-2ch15  

## 15.1 Introduction

Active infrared thermography (AIRT) is a promising non-invasive screening approach for the timely detection of cancer in the breast [1, 2]. Currently, breast cancer is a serious health issue throughout the world, in particular for women. However, this disease can be curable if detected in its early stages, i.e. the rate of survival is related to the stage at which it is diagnosed [3–6]. The expected survival rate is around 98% for about 10 years if the disease is detected at stage 0 whereas it is around 65% if diagnosed at stage III [7]. Thus, in order to enhance the survival rate timely screening is essential.

In recent years, researchers have reported several screening approaches for breast cancer detection which include biopsy, mammography, computerized tomography, magnetic resonance tomography and ultrasound scanning. The main requirements for a good screening technique are accuracy, easy deployment, high sensitivity and comfort of the patient. However, the aforementioned techniques are, in general, costly, time-consuming, non-portable and uncomfortable which limit the screening and diagnosis capabilities. Also, some of these techniques are limited in their practical use due to various reasons such as the harmful radiation involved, the discomfort caused to the patient during screening and the complexity involved in the interpretation of the results, etc. In order to overcome these limitations, there is an essential requirement for deploying a highly sensitive, remote, safe and whole-field screening modality for the timely identification of breast tumours. AIRT is emerging as a popular diagnostic method and possesses almost all the required characteristics [8].

AIRT was developed to be a non-contact, non-invasive, portable, wide area monitoring, safe, cost-effective, easy to analyse, radiation free and comfortable diagnostic method for the detection of tumours at an early stage with enhanced sensitivity and resolution. It involves mapping of the thermal profiles over the breast by detecting the infrared radiation emanating over it. It is a safe and sensitive imaging modality to track the thermal gradients over the breast under examination [9]. In general, cancerous regions exhibit relatively higher thermal gradients over the breast regions in comparison to healthy regions because of their higher metabolic rate. Thus these can be located as hotspots in the thermal images. Also, cancerous cells produce a chemical substance which facilitates the inception of blood vessels, i.e. angiogenesis, and favours the dilation of normal blood vessels providing extra blood to increase the tumour growth which ultimately leads to the generation of more heat. A several works have even proved that the growth rate of a tumour is almost proportional to its temperature. AIRT detects the asymmetrical thermal gradients to locate hot and cold regions over the breast to explore the presence of tumours inside the breast [10, 11].

Active infrared thermographic techniques can be classified as optically excited, radio frequency excited and ultrasound excited thermographic techniques. Optically excited infrared thermography (OEIRT) is the most popular among these due to its fast, remote, non-contact and broad area monitoring capabilities. AIRT has proved to be a promising emerging screening modality to provide information about hidden tumours at a very early stage. With optical excitation, an external heat stimulus is directed upon the surface of the breast under study to create a relevant thermal difference between the features of interest (tumour) and the background. As a result, thermal waves diffuse into the breast sample and alter the temperature over it depending on the tumour present inside it. These time dependent thermal gradients are utilized and further processed to obtain quantitative information regarding the tumours [12].

Depending on the type of heat flux employed, OEIRT can be classified as a pulse or modulated (periodic or aperiodic) infrared thermographic technique. Among the pulse based thermographic techniques, pulse and pulse phase thermography (PT and PPT) are more prominent methods. In these, the breast is exposed to a short duration pulse shaped optical excitation and the corresponding resultant temporal thermal distribution is recorded and analysed [13]. This technique can be implemented in almost no time due to the use of short duration safe exposure. However, the recorded thermal distribution in PT is significantly affected by non-uniform illumination and the surface conditions of the breast. These perturbations may cause difficulty in locating the tumours. PPT differs from PT as it involves post-processing (frequency domain analysis) of the recorded thermal distribution data to obtain magnitude and phase thermal images. Phase thermal images are of particular interest to many researchers as they not only provide qualitative representation of the tumour but they also help in providing quantitative information. However, PPT provides phase thermal images which are no longer influenced by the surface features and emissivity changes of the breast surface. Further, the short duration excitation used in these techniques demands high peak power heat sources as a means to keep the average energy supplied the same. This may cause a feeling of discomfort to the patient who is undergoing screening.

As a means to overcome the problems associated with high heat inputs over the breast, continuous modulated OEIRT techniques are preferred. Modulated OEIRT techniques can be periodic and aperiodic. If the incident flux is modulated at a predefined single frequency, then such a thermographic technique is called lock-in thermography (LT) [14, 15]. The resultant temporal thermal profile is recorded during the heating and is further post-processed to obtain the phase thermal images as in PPT. Even though LT detects anomalies that are located relatively deep even with use of reduced peak power heat inputs, single frequency thermal excitation still does not improve its test resolution. The tumour location and its lateral dimensions inside the breast are unknown, thus this technique demands reiteration every time with a different frequency in order to resolve the hidden tumour. This makes LT a very time-consuming screening procedure. Thus, to overcome the limitations associated with pulse based techniques (high peak power requirement) and the periodic modulated approach (repetitiveness), this chapter presents the applicability of another of the available aperiodic OEIRT techniques, i.e. linear frequency modulated thermal wave imaging (LFMTWI) [16]. Aperiodic thermal excitation ensures improved detection resolution and sensitivity by using comparably low peak power heat sources in a relatively shorter experimentation time.

This chapter highlights the application of LFMTWI for identifying tumours at early stage in all types of breast tissue samples (fat or dense). Dense breasts have less fat compared to breasts that are not dense. Breasts that have a high proportion of fat compared to the gland region are considered under the category of fatty breasts. Dense breasts have a larger gland tissue region relative to the amount of fat. LFMTWI utilizes a frequency modulated thermal stimulus which is imposed over the breast surface (both dense and fatty) with a pre-specified frequency band with almost equal energy at each frequency. The resultant thermal waves propagate inside the breast producing a similar temporal thermal variation over it. The tumours present within the breast vary the flow of heat resulting in temporal thermal changes over the breast surface, which is further recorded and processed using frequency and time domain analysis to construct magnitude and phase images [17–20]. The applied post-processing methods emphasize the potential of the proposed screening approach in detecting tumours present in breast samples. Also, it focuses on providing quantitative results by taking the signal-to-noise ratio (SNR) as a quality factor. The following sections describe the numerical modelling of both dense and fatty breasts as well as the adopted post-processing analysis schemes.

## 15.2 Numerical modelling and simulation

In this section, the numerical modelling of a human breast is performed using a finite element method (FEM). Human (mainly female) breast three-dimensional models (one a dense breast and other a fatty breast) are developed representing a semi-ellipsoid multi-layered structure with seven tissue layers with different thermophysical properties. The considered tissue layers and the corresponding thermophysical properties (thermal conductivity ($k$), specific heat ($c$), density ($\rho$)) of these layers and of the tumour are as given in table 15.1. Here, $\omega_b$ and $Q_m$ denote the blood perfusion and the metabolic heat generation rate, respectively [21–24].

### 15.2.1 Modelling of the dense breast

The schematic diagram of the simulated dense breast with every tissue layer dimension is as shown in figure 15.1. Four simulations are performed by varying the depth to diameter ratio for a tumour which is artificially induced inside the

**Table 15.1.** Thermophysical properties [21–24].

| Various tissue layers | $\rho$ (kg/m$^3$) | $k$ (W/m·K) | $c$ (J/kg·K) | $\omega_b$ (m$^3$/s/m$^3$) | $Q_m$ (W/m$^3$) |
|---|---|---|---|---|---|
| Epidermis | 1200 | 0.235 | 3589 | 0 | 0 |
| Reticular dermis | 1200 | 0.445 | 3300 | 0.0013 | 368.1 |
| Papillary dermis | 1200 | 0.445 | 3300 | 0.0002 | 368.1 |
| Fat | 930 | 0.21 | 2770 | 0.0002 | 400 |
| Gland | 1050 | 0.48 | 3770 | 0.0006 | 700 |
| Chest wall | 1100 | 0.48 | 3800 | 0.0009 | 700 |
| Muscle | 1100 | 0.48 | 3800 | 0.0009 | 700 |
| Tumour | 1050 | 0.48 | 3852 | 0.12 | 10000 |

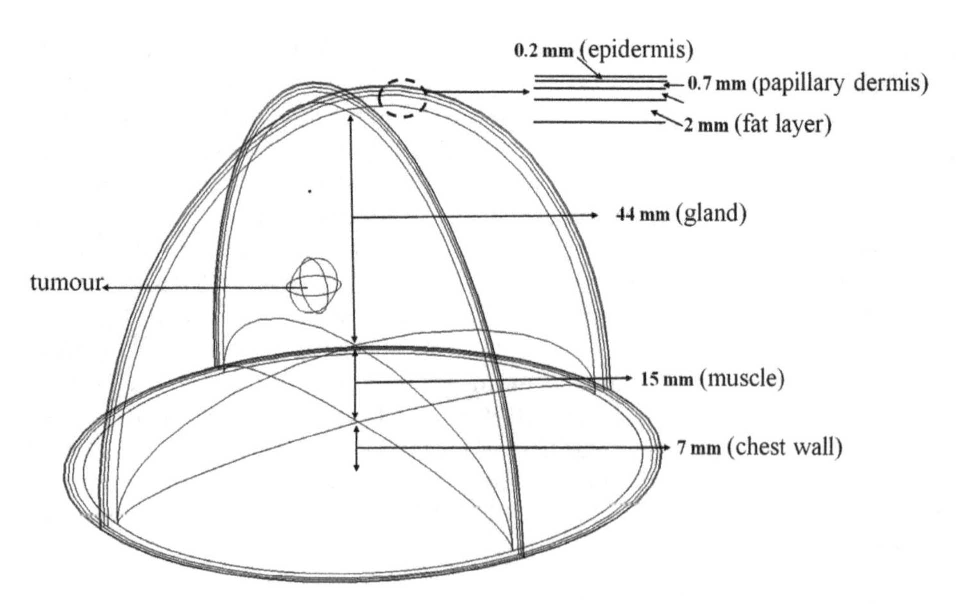


**Figure 15.1.** Schematic diagram of the modelled dense breast.

**Table 15.2.** Tumour variations corresponding to dense breast.

| Simulation No. | Diameter (mm) | Depth from the front surface (mm) | Depth to diameter ratio |
|---|---|---|---|
| 1 | 5 | 15 | 3 |
| 2 | 7 | 14 | 2 |
| 3 | 9 | 13 | 1.45 |
| 4 | 10 | 12.5 | 1.25 |

modelled breast sample. The variations of the tumour for different simulations are as given in table 15.2. The schematic representation corresponding to each simulation model for the dense breast is as illustrated in figure 15.2.

### 15.2.2 Fatty breast

The dense and fatty breast models are differentiated on the basis of the fat and gland layer thicknesses. A schematic diagram of the simulated fatty breast with every tissue layer and its dimension is as shown in figure 15.3.

For this model four simulations are also performed by varying the depth to diameter ratio for a tumour which is artificially induced inside the modelled fatty breast. The variations of the tumour parameters for different simulations are as given in table 15.3. The schematic representation corresponding to each simulation model for the fatty breast is as illustrated in figure 15.4.

All the models are generated with a normal mesh (a maximum element size of 11.3 mm and minimum element size of 2.03 mm) using tetrahedral elements for the thermal image analysis to predict the presence of tumours.

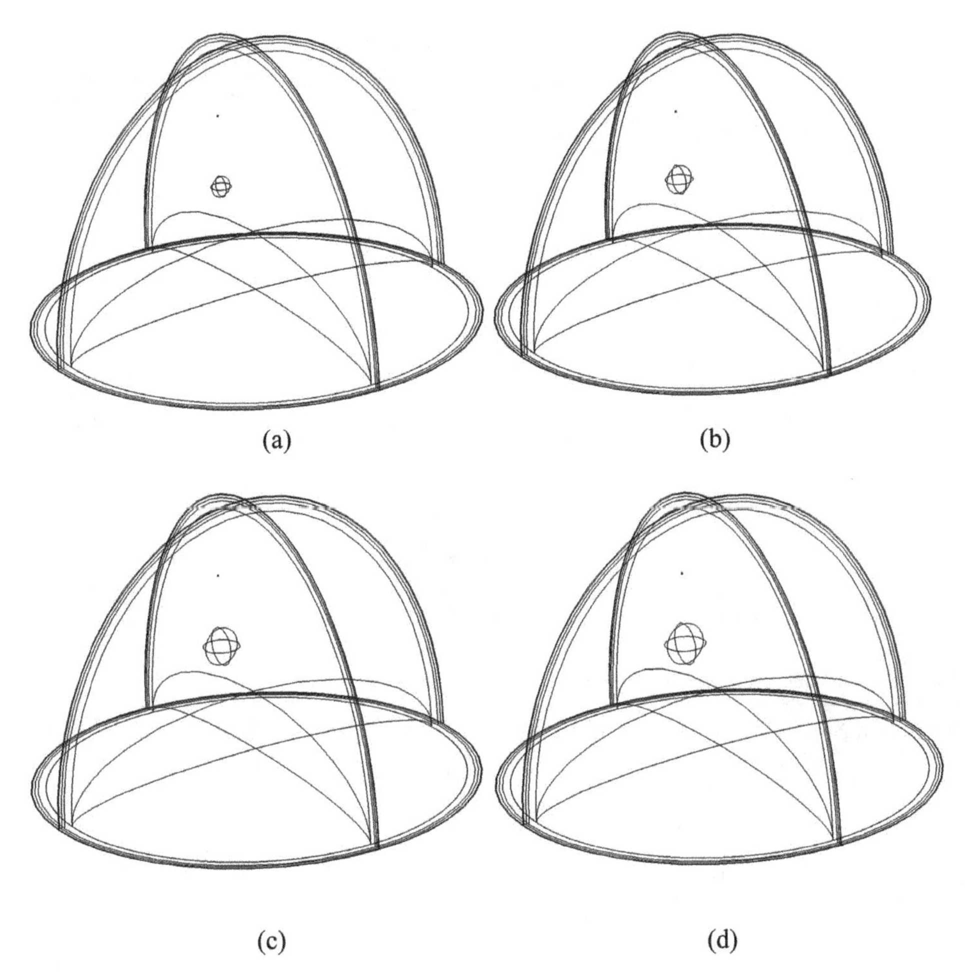

**Figure 15.2.** Simulation model for the dense breast: (a) simulation no. 1, (b) simulation no. 2, (c) simulation no. 3 and (d) simulation no. 4.

### 15.2.3 Simulation

The 3D finite element analysis (FEA) is accomplished by exposing the front surface of the modelled breast tissue specimens to a linear frequency modulated (LFM) heat flux of 50 W/m$^2$. The imposed LFM thermal signal has frequency varying in the range of 0.002 Hz–0.2 Hz for the duration of 500 s. The corresponding LFM signal is as shown in figure 15.5.

The thermal distribution profile within the soft tissue is described by Pennes bio-heat equation [25–28] as

$$\rho \cdot c\frac{\partial T}{\partial t} = \nabla(k \cdot \nabla T) + \omega_b \cdot c_b \cdot \rho_b(T_a - T) + Q_m, \tag{15.1}$$

where $\omega_b$ and $Q_m$ represents the blood perfusion and the metabolic heat generation rate, $c_b$ and $\rho_b$ represent the specific heat and density of the blood, respectively, $T_a$ represents

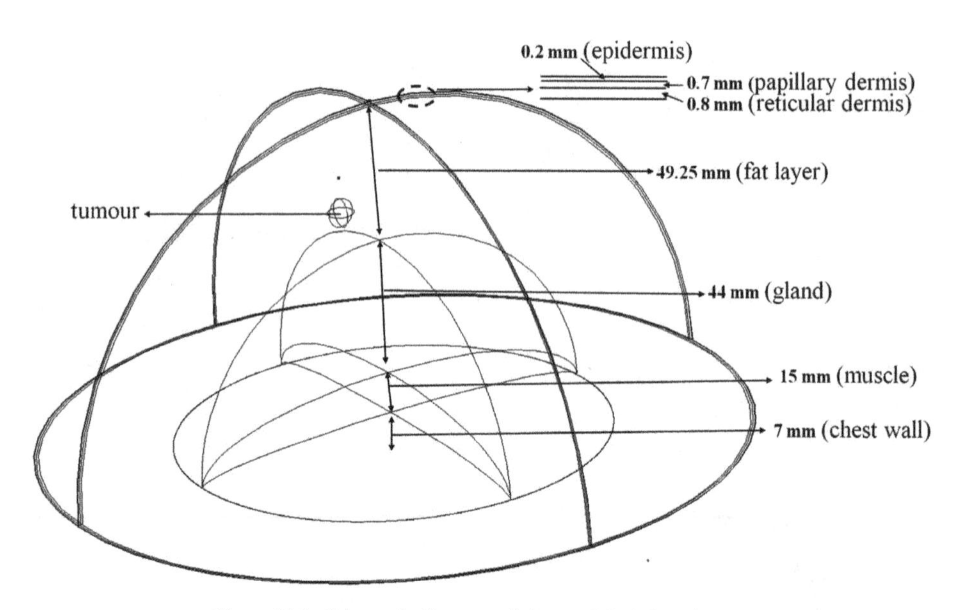


**Figure 15.3.** Schematic diagram of the modelled fatty breast.

**Table 15.3.** Tumour variations corresponding to the fatty breast.

| Simulation No. | Diameter (mm) | Depth from the front surface (mm) | Depth to diameter ratio |
|---|---|---|---|
| 1 | 5 | 15 | 3 |
| 2 | 7 | 14 | 2 |
| 3 | 9 | 13 | 1.45 |
| 4 | 10 | 12.5 | 1.25 |

the arterial blood temperature and $T$ is the local temperature of the breast tissue. The values considered for the density and specific heat of blood are 1056 kg m$^{-3}$ and 3660 J kg$^{-1}$ K, respectively. The value of $T_a$ is taken as the core human body temperature, i.e. 310.15 K. All the simulations are performed under adiabatic boundary conditions. During active heating, the produced temporal thermal distribution profile at the breast skin is captured and recorded at a frame rate of four frames per second. The typical representation of the schematic diagram showing the experimental set-up for this numerical study is as shown in figure 15.6.

The recorded thermal data corresponding to each simulation is pre-processed to attain the mean zero thermal response by employing an appropriate polynomial fit. The flowchart illustrating the computation of zero mean data is as shown in figure 15.7.

Here, ($T(x, y, t)$) denotes the recorded temporal thermal distribution profile ($x$, $y$ refer to the spatial co-ordinates and $t$ represents the index of the image sequence) and $T_{\text{zeromean}}(x, y, t)$ denotes the zero mean thermal distribution profile. The resultant zero mean data are further post-processed using frequency and time domain post-analysis schemes as described in the next section.

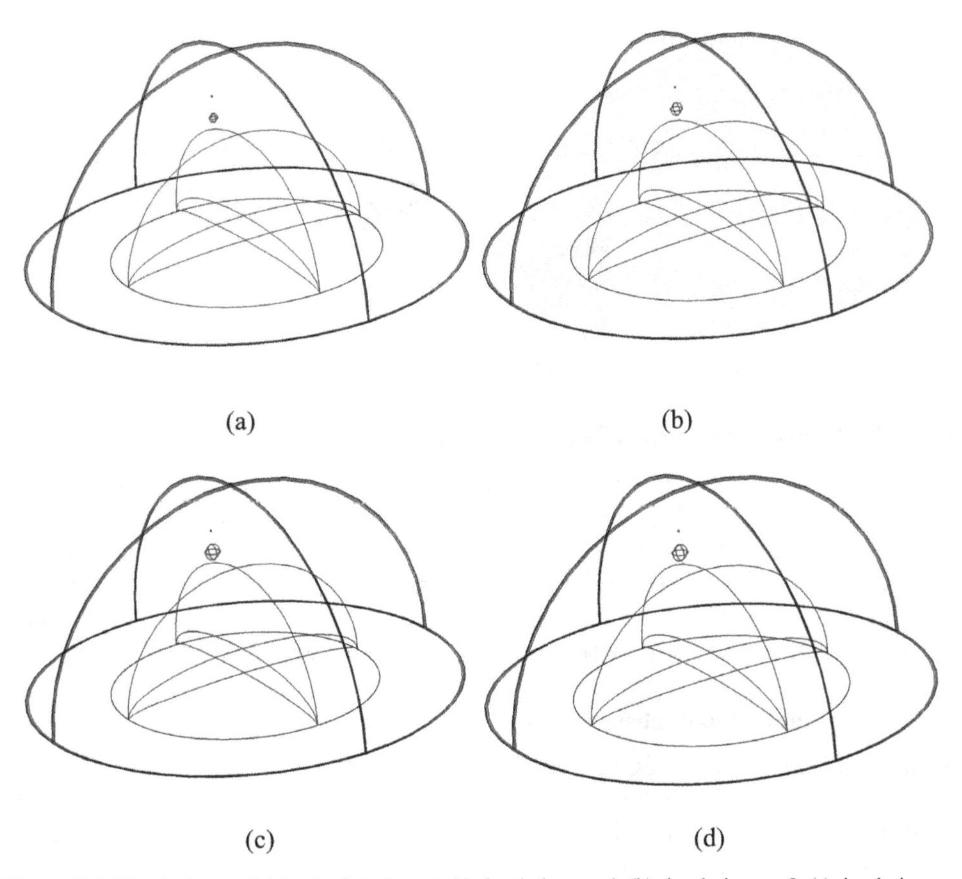


**Figure 15.4.** Simulation model for the fatty breast: (a) simulation no. 1, (b) simulation no. 2, (c) simulation no. 3 and (d) simulation no. 4.

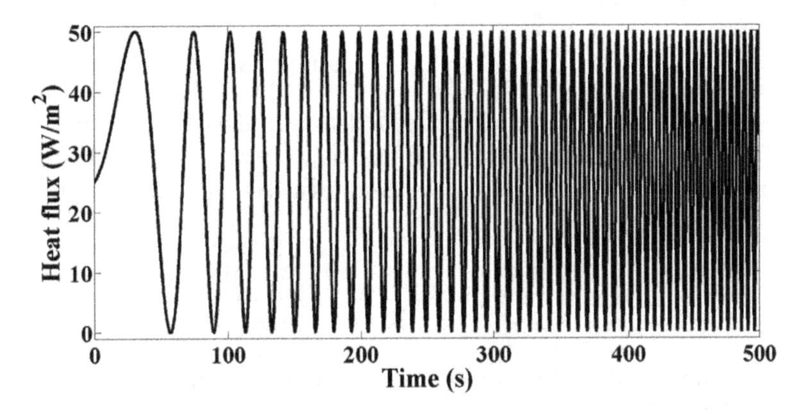


**Figure 15.5.** Imposed LFM heat flux.

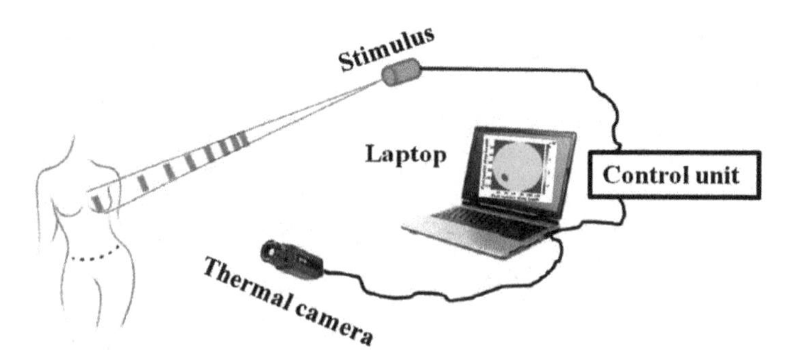


**Figure 15.6.** Schematic representation of the experimental set-up.

## 15.3 Post-processing analysis schemes

Appropriate post-processing of the resultant zero mean thermal distribution data is necessary, in particular if quantification of the defects is required. A wide variety of schemes has been introduced, keeping in mind the objective of improving the test resolution, sensitivity and simplifying data analysis [29–34]. In this chapter, the phase and correlation coefficient (amplitude) thermal images are reconstructed using time- and frequency domain based analysis schemes as described in the following.

### 15.3.1 Frequency domain phase approach

In frequency domain analysis, the computed zero mean temporal thermal distribution data $T_{\text{zeromean}}(x, y, t)$ are analysed by computing the discrete Fourier transform (DFT) to construct the phase and magnitude sequence. One-dimensional DFT ($T_f(x, y, k)$) is computed for every single pixel in the field of observation using the formula

$$\begin{aligned} T_f(x, y, k) &= \sum_{n=0}^{N-1} T_{\text{zeromean}}(x, y, t) e^{-\frac{j2\pi kn}{N}} \\ &= \text{Re}(T_f(x, y, k)) + j\text{Img}(T_f(x, y, k)), \end{aligned} \tag{15.2}$$

where $N$ refers to the total number of frames, $k$ denotes the bin number, $\text{Re}(T_f(x, y, k))$ and $\text{Img}(T_f(x, y, k))$ represents the real and imaginary parts of $T_f(x, y, k)$, respectively. Further, the phase thermal image sequence is constructed from the computed real and imaginary parts using the formula

$$\angle T_f(x, y, k) = \tan^{-1}\left(\frac{\text{Img}(T_f(x, y, k))}{\text{Re}(T_f(x, y, k))}\right). \tag{15.3}$$

The flowchart representing the computation of the frequency domain phase is as illustrated in figure 15.8.

### 15.3.2 Time domain approach

The time domain amplitude (TDA) and time domain phase (TDP) are computed as described in the flowchart in figure 15.9. To compute the TDP, the first requirement

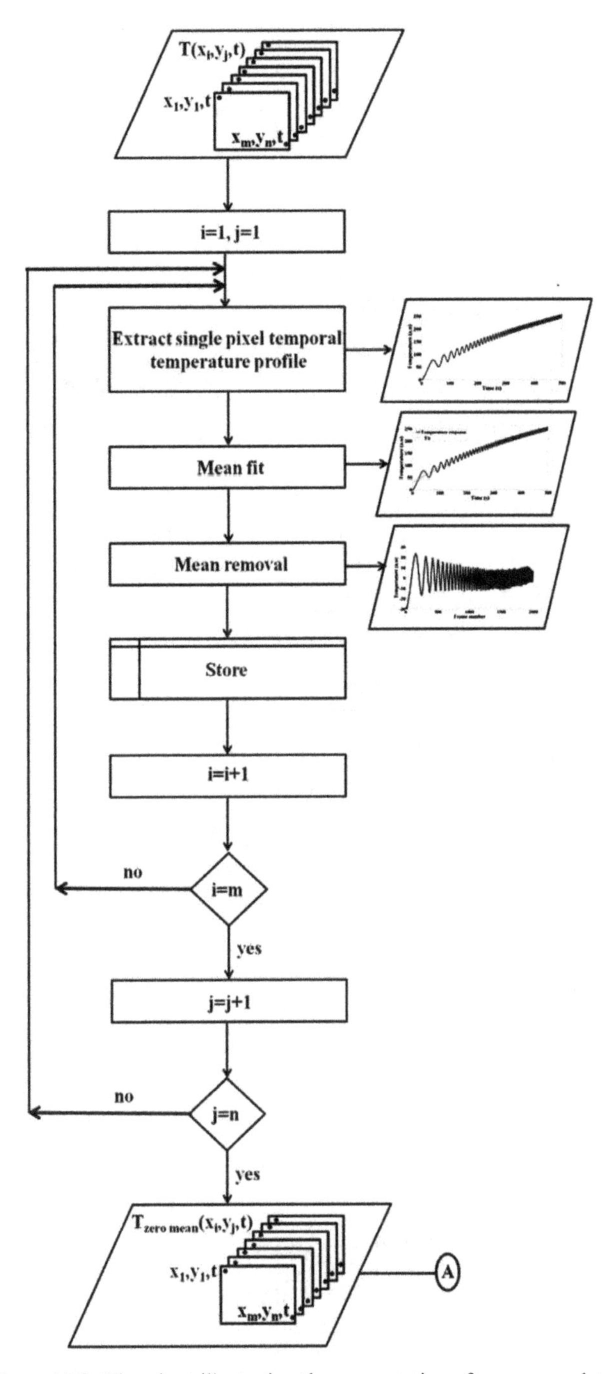


**Figure 15.7.** Flowchart illustrating the computation of zero mean data.

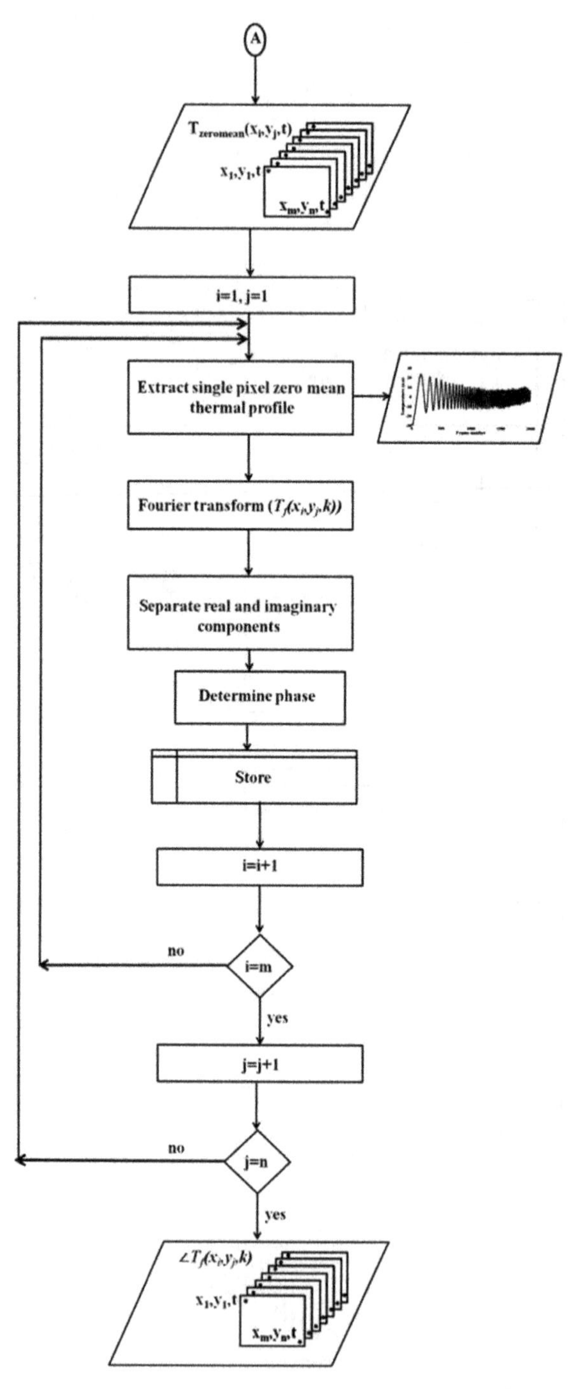


**Figure 15.8.** Flowchart illustrating the computation of the frequency domain phase.

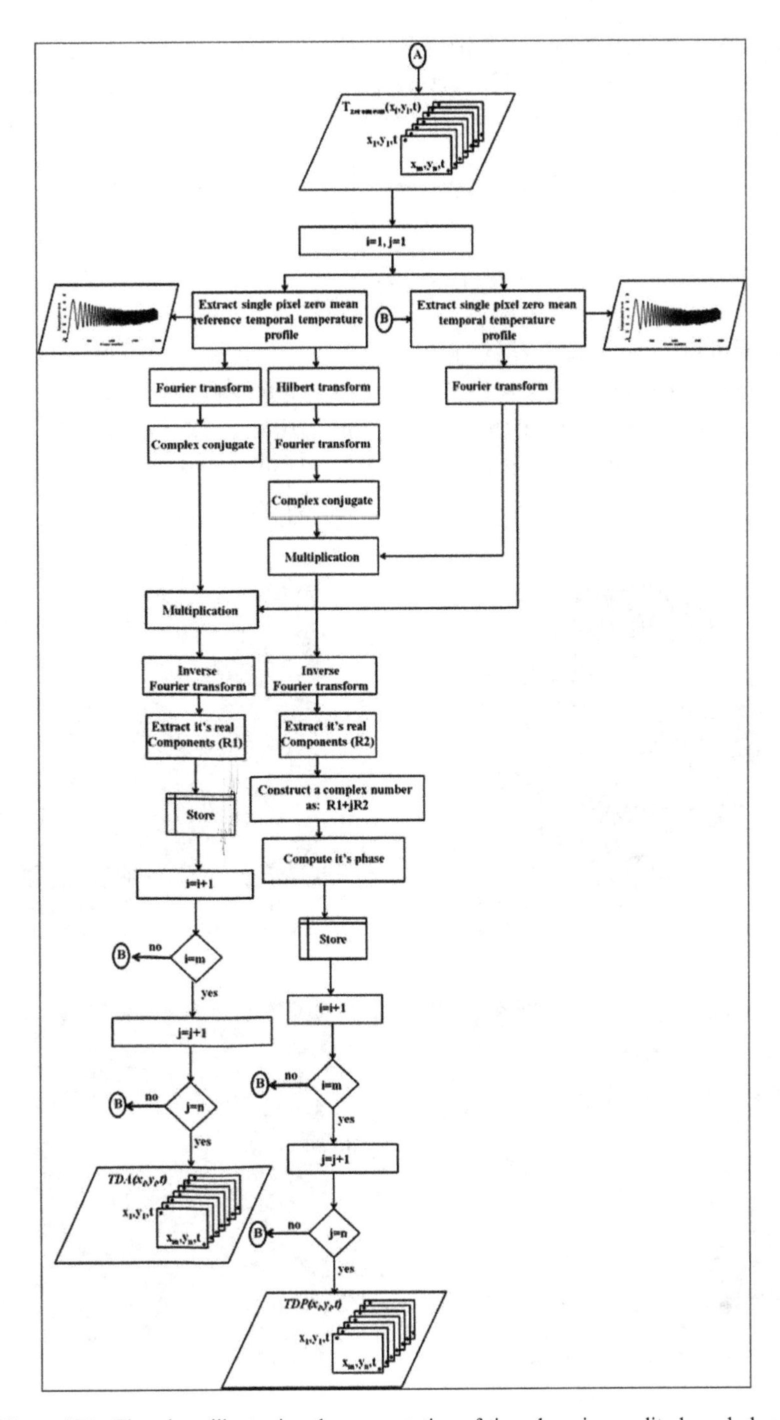


**Figure 15.9.** Flowchart illustrating the computation of time domain amplitude and phase.

is to obtain an analytic signal. Further, to obtain the analytic signal, the Hilbert transform is implemented. The Hilbert transform ($T_H(x, y, t)$) applied to a reference zero mean thermal profile ($T_{Ref}(x, y, t)$) is defined as

$$T_H(x, y, t) = H[T_{Ref}(x, y, t)] = \frac{1}{\pi}\int_{-\infty}^{\infty} \frac{T_{Ref}(x, y, \tau)}{t - \tau} d\tau \quad (15.4)$$

$$= T_{Ref}(x, y, t) \times \frac{1}{\pi t}. \quad (15.5)$$

Thus the Hilbert transform $T_H(x, y, t)$ of $T_{Ref}(x, y, t)$ can be referred to as the convolution (*) of $T_{Ref}(x, y, t)$ with the signal $\frac{1}{\pi t}$. Here, the convolution is computed in the frequency domain and then again converting back the signal in the time domain. Further, the analytic signal ($T_a(x, y, t)$) is obtained as

$$T_a(x, y, t) = T_{zeromean}(x, y, t) + jT_H(x, y, t). \quad (15.6)$$

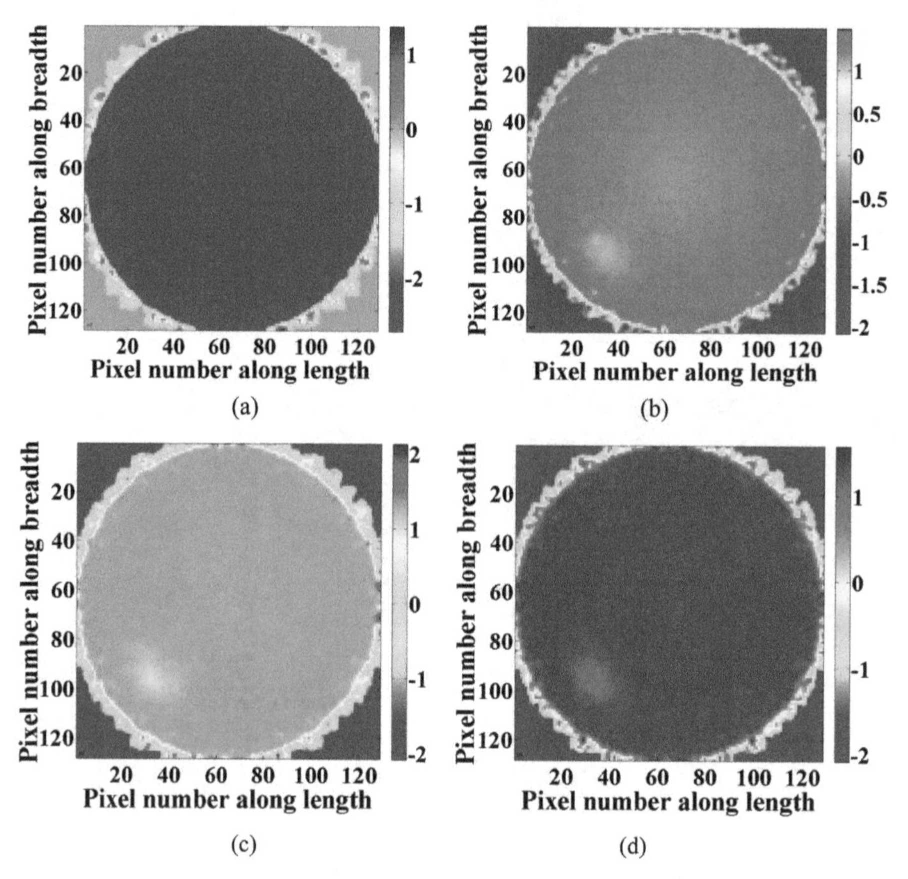


**Figure 15.10.** Frequency domain phase thermal images for different depth to diameter ratios of (a) 3, (b) 2, (c) 1.45 and (d) 1.25 obtained at a time instant of 0.5 s.

The phase thermal image is then reconstructed as

$$\mathrm{TDP} = \tan^{-1}\frac{T_{\mathrm{H}}(x,\, y,\, t)}{T_{\mathrm{zeromean}}(x,\, y,\, t)}. \tag{15.7}$$

TDA is computed to obtain pulse compressed pulses as

$$\mathrm{TDA} = T_{\mathrm{zeromean}}(x,\, y,\, t) * T_{\mathrm{Ref}}(x,\, y,\, t), \tag{15.8}$$

where * denotes the convolution operator. It is again computed using complex multiplication in the frequency domain.

## 15.4 Results and discussion

This section provides the results obtained after applying the proposed post-processing analysis schemes corresponding to both dense and fatty breasts for tumour variations (depth to diameter ratio).

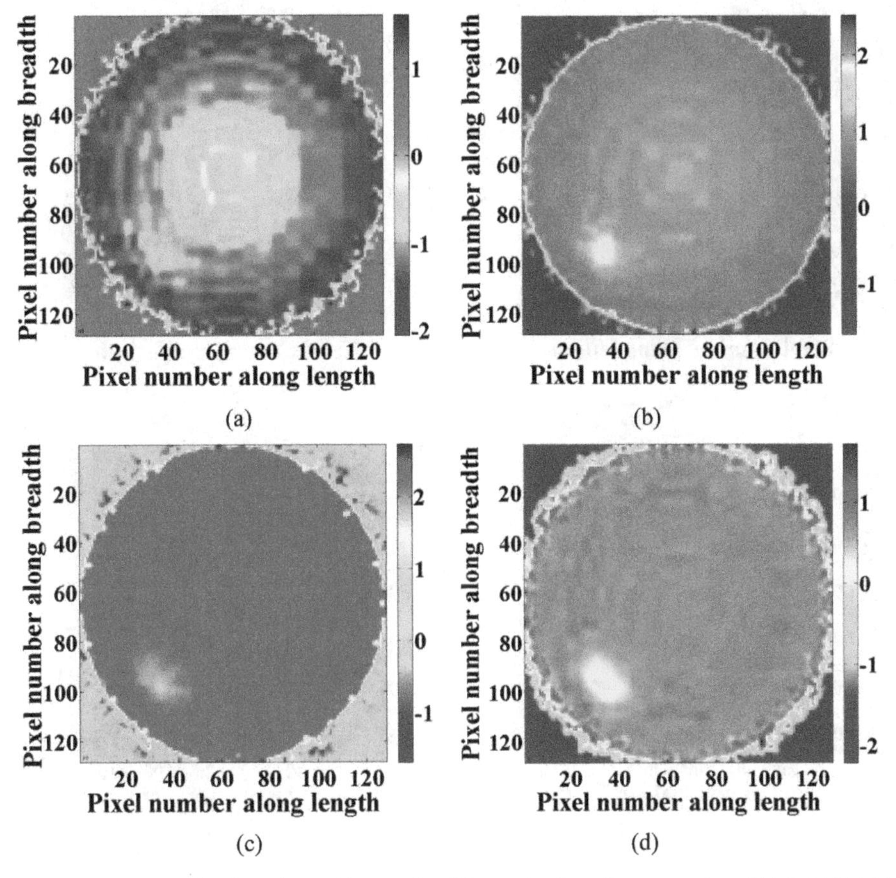


**Figure 15.11.** Time domain phase thermal images for different depth to diameter ratios of (a) 3, (b) 2, (c) 1.45 and (d) 1.25 obtained at time instants of 215.5 s, 293 s, 273.5 s and 354.5 s, respectively.

### 15.4.1 Results obtained for the dense breast

*15.4.1.1 Frequency domain phase (FT phase) for different depth to diameter ratios*

Figure 15.10 shows the frequency domain phase images corresponding to dense breast for different depth to diameter ratios of 3, 2, 1.45 and 1.25 obtained at a time instant of 0.5 s.

*15.4.1.2 Time domain phase for different depth to diameter ratios*

Figure 15.11 shows the time domain phase images corresponding to dense breast for different depth to diameter ratios of 3, 2, 1.45 and 1.25. The image relative to ratio 3 is obtained at the time instant of 215.5 s, for ratio 2 it is obtained at 293 s, for ratio 1.45 it is obtained at 273.5 s and for ratio 1.25 it is acquired at an instant of 354.5 s.

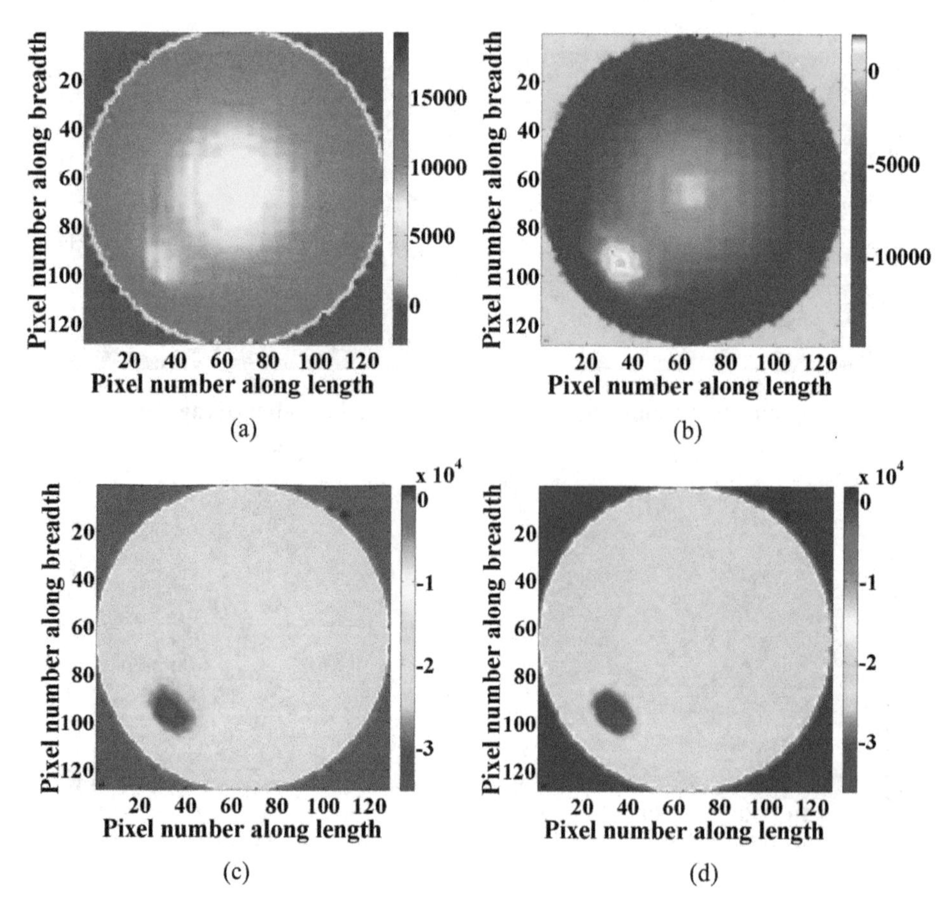


**Figure 15.12.** Time domain amplitude thermal images for different depth to diameter ratios of (a) 3, (b) 2, (c) 1.45 and (d) 1.25 obtained at a time instant of 261.5 s, 237 s, 266.5 s and 266.5 s, respectively.

*15.4.1.3 Time domain amplitude for different depth to diameter ratios*
Figure 15.12 shows the time domain amplitude images corresponding to dense breast for different depth to diameter ratios of 3, 2, 1.45 and 1.25 acquired at time instants of 261.5 s, 237 s, 266.5 s and 266.5 s, respectively.

### 15.4.2 Results obtained for the fatty breast

*15.4.2.1 Frequency domain phase for different depth to diameter ratios*
Figure 15.13 shows the frequency domain phase images corresponding to fatty breast for different depth to diameter ratios of 3, 2, 1.45 and 1.25 acquired at a time instant of 0.5 s.

*15.4.2.2 Time domain phase for different depth to diameter ratios*
Figure 15.14 shows the time domain phase images corresponding to fatty breast for different depth to diameter ratios of 3, 2, 1.45 and 1.25 obtained at a time instant of 129 s, 129.5 s, 321.5 s and 146 s, respectively.

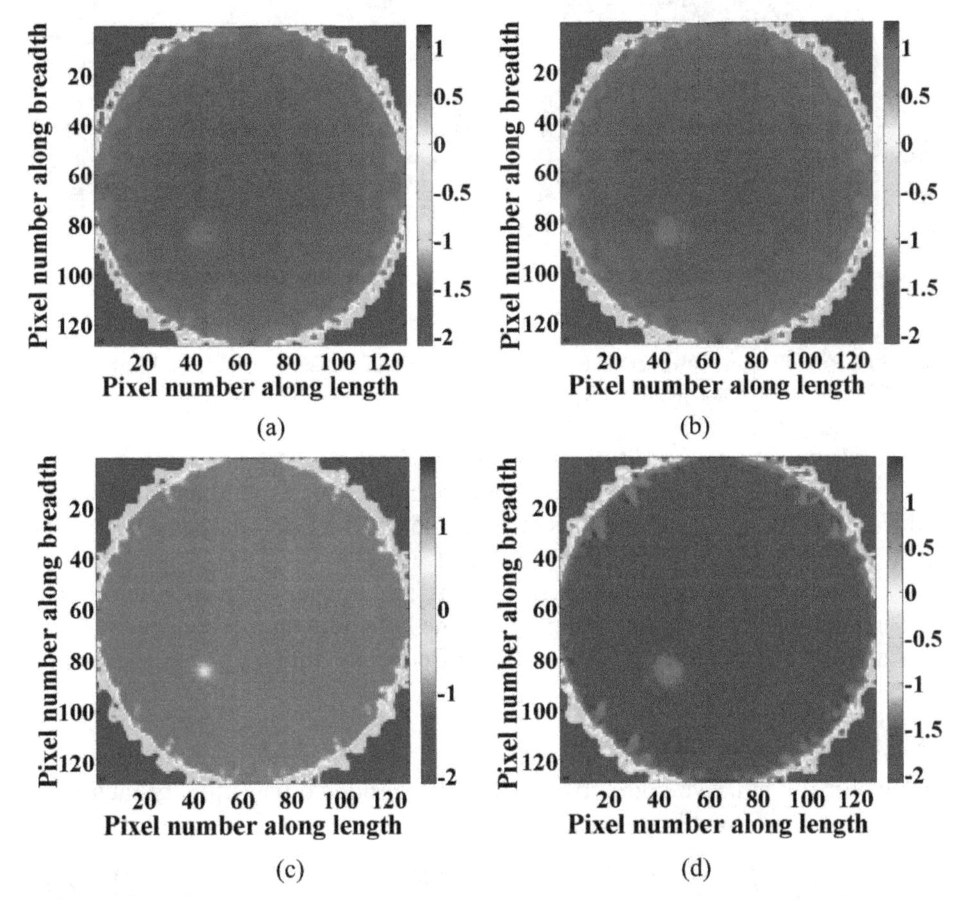


**Figure 15.13.** Frequency domain phase thermal images for different depth to diameter ratio of (a) 3, (b) 2, (c) 1.45 and (d) 1.25 obtained at a time instant of 0.5 s.

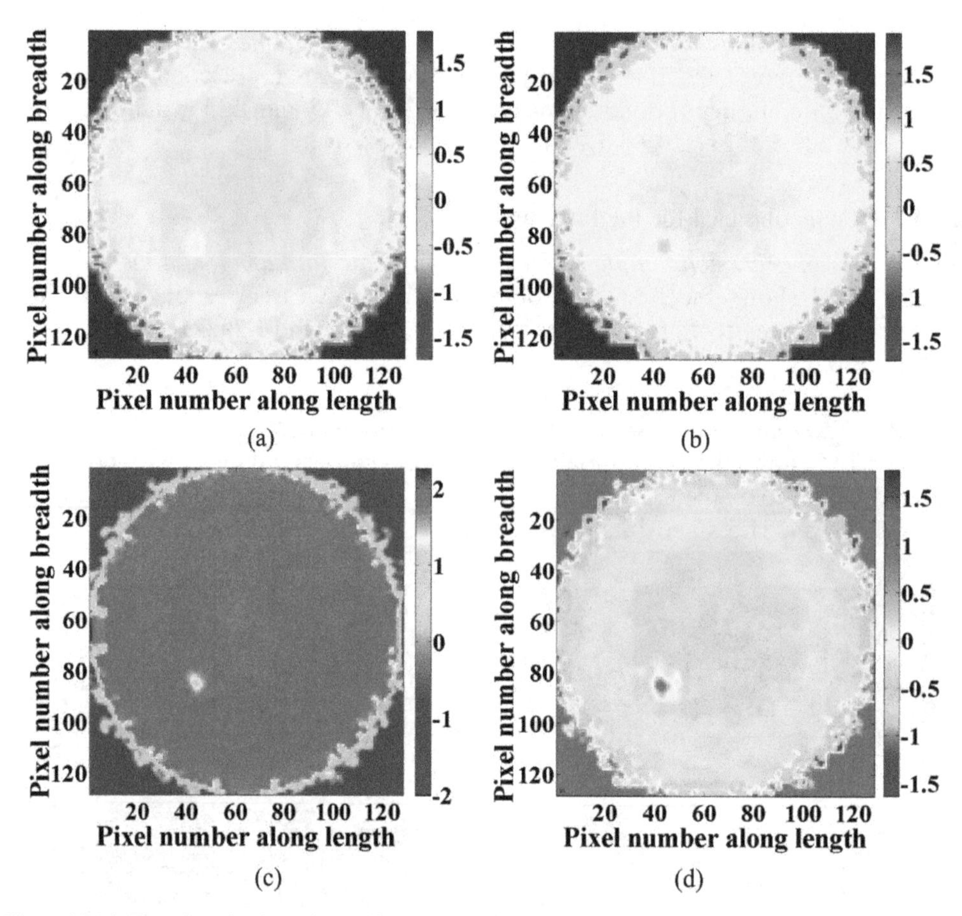


**Figure 15.14.** Time domain phase thermal images for different depth to diameter ratios of (a) 3, (b) 2, (c) 1.45 and (d) 1.25 obtained at a time instant of 129 s, 129.5 s, 321.5 s and 146 s, respectively.

*15.4.2.3 Time domain amplitude for different depth to diameter ratios*

Figure 15.15 shows the time domain amplitude images corresponding to fatty breast for different depth to diameter ratios of 3, 2, 1.45 and 1.25 obtained at a time instant of 311.5 s, 168.5 s, 255.5 s and 326 s, respectively.

It can be observed from all the resultant thermal images obtained after applying the aforementioned processing schemes that the time domain methods prove to be a better approach relative to the conventional frequency domain phase approach.

### 15.4.3 Signal-to-noise ratio (SNR)

Further, for quantitative comparison, SNR is considered as a quality factor. The values of SNR are calculated using the equation

$$\mathrm{SNR} = 20\mathrm{Log}\left(\frac{\text{mean of defective area} - \text{mean of non-defective area}}{\text{standard deviation of non-defective area}}\right). \quad (15.9)$$

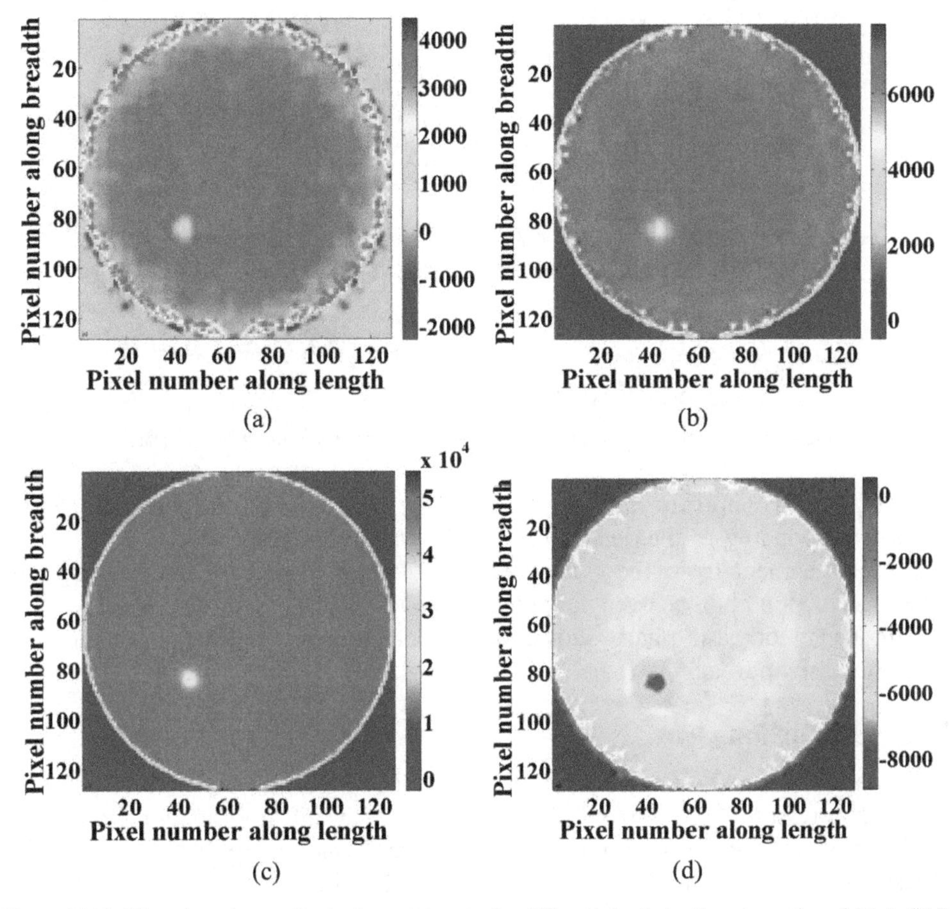


**Figure 15.15.** Time domain amplitude thermal images for different depth to diameter ratios of (a) 3, (b) 2, (c) 1.45 and (d) 1.25 obtained at a time instant of 311.5 s, 168.5 s, 255.5 s and 326 s, respectively.

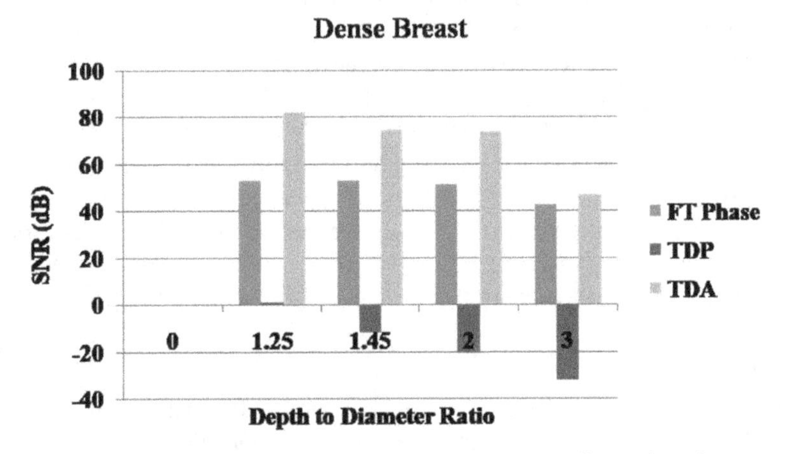


**Figure 15.16.** SNR versus depth to diameter ratio corresponding to dense breast.

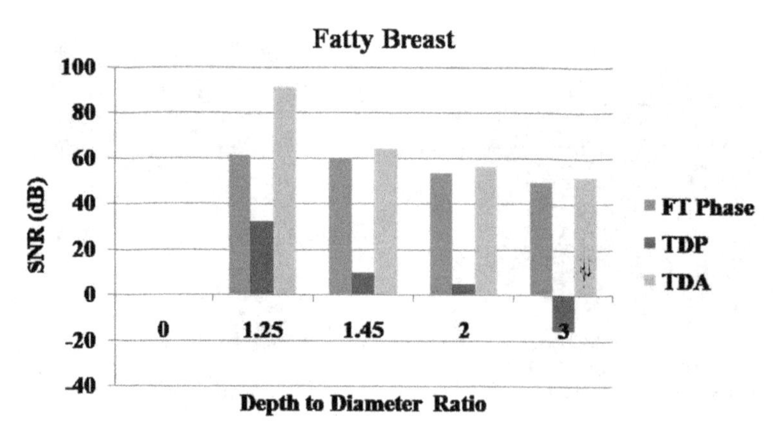


**Figure 15.17.** SNR versus depth to diameter ratio corresponding to fatty breast.

The SNR versus depth to diameter ratio is plotted as shown in figures 15.16 and 15.17 corresponding to the dense breast and fatty breast, respectively.

It seems evident from the computed SNR values that the TDA has superior tumour detection abilities over the phase approach. Further it is also observed that the frequency domain phase shows better detection capabilities over the time domain phase approach.

## 15.5 Conclusion

This chapter proposes a full-field, remote, non-invasive, safe and patient-friendly pulse compression favourable frequency modulated infrared imaging screening modality for the timely detection of breast cancer, which is applicable irrespective of the age of the person and the type of breast. Further, it is emphasized that the results of the correlation based pulse compressed amplitude images show better detection capabilities than those of the phase images. By comparing various processing and post-processing schemes by taking the SNR into account as the figure of merit, it is concluded that the results obtained from TDA show higher tumour detection sensitivity than those of the time and frequency domain phase images. Even though the US Food and Drug Administration (FDA) issued a warning on 25 February 2019 that infrared thermography should be used in conjunction with and not as a replacement for mammography, due to its merits as a full-field, remote, safe and patient-friendly screening modality along with the portability of the screening system makes active infrared thermography a promising approach for the early detection of breast cancer in both dense as well as fatty breasts in all ages of women, even if they are pregnant or nursing.

## Acknowledgements

This work was supported financially by the Global Innovation and Technology Alliance (GITA) from the project entitled ‘The Development of a Portable THERMOgraphy-based Health DeTECTion System (THERMOTECT) in breast cancer screening’ with reference number 2016UK0202022 IN - UK RFP 2016.

## References

[1] Ng E Y K and Kee E C 2007 Advanced integrated technique in breast cancer thermography *J. Med. Eng. Technol.* **32** 103–14

[2] Keyserlingk J R, Ahlgren P D, Yu E, Belliveau N and Yassa M 2000 Functional infrared imaging of the breast *IEEE Eng. Med. Bio.* **19** 30–41

[3] Williams K L and Handley R S 1961 Infrared thermometry in the diagnosis of breast disease *Lancet* **2** 1378–81

[4] Usuki H, Onoda Y, Kawasaki S, Misumi T, Murakami M, Komatsubara S and Teramoto S 1990 Relationship between thermographic observations of breast tumors and the DNA indices obtained by flow cytometry *Biomed Thermol.* **10** 282–5

[5] Sinha G R and Bhagwati C P 2014 *Medical Image Processing: Concepts and Applications* (Delhi: Prentice Hall) pp 1–280

[6] Koay J, Herry C H and Frize M 2004 Analysis of breast thermography with artificial neural network *Proc. 26th IEEE EMBS Conf. (San Francisco, CA, USA, Sept. 2000)* pp 1159–62

[7] Nover A B, Jagtap S, Waqas A, Yegingil H, Shih W Y, Shih W and Brooks A D 2009 Modern breast cancer detection: a technological review *Int. J. Biomed. Imaging* **2009** 1–14

[8] Head J F and Elliott R L 2002 Infrared imaging: making progress in fulfilling its medical promise *IEEE Eng. Med. Biol. Mag.* **21** 80–5

[9] Bhagwati C P and Sinha G R 2014 Abnormality detection and classification in computer-aided diagnosis (CAD) of breast cancer images *J. Med. Imag. Health Informat.* **4** 881–5

[10] Mulaveesala R and Dua G 2016 Non-invasive and non-ionizing depth resolved infra-red imaging for detection and evaluation of breast cancer: a numerical study *Biomed. Phys. Eng. Exp.* **2** 1–8

[11] Ng E Y K and Sudharsan N M 2001 An improved three dimensional direct numerical modelling and thermal analysis, of a female breast with tumour *Proc. Inst. Mech. Eng.* H **21** 25–38

[12] Maldague X P V 2001 *Theory and Practice of Infrared Technology for Non Destructive Testing* (New York: Wiley)

[13] Almond D P and Lau S K 1994 Defect sizing by transient thermography I: an analytical treatment *J. Phys. D: Appl. Phys.* **27** 1063–9

[14] Busse G and Eyerer P 1983 Thermal wave remote and nondestructive inspection of polymers *Appl. Phys. Lett.* **43** 355–7

[15] Busse G, Wu D and Karpen W 1992 Thermal wave imaging with phase sensitive modulated thermography *J. Appl. Phys.* **71** 3962–5

[16] Mulaveesala R and Tuli S 2005 Implementation of frequency modulated thermal wave imaging for non-destructive sub-surface defect detection *Insight: Non-Destr. Test. Condition Monitoring* **47** 206–8

[17] Vavilov V, Maldague X, Dufort B, Fobitaille J and Picard J 1993 Thermal nondestructive testing of carbon epoxy composites: detailed analysis and data processing *NDT & E Int.* **26** 85–95

[18] Ghali V S, Mulaveesala R and Takei M 2011 Frequency-modulated thermal wave imaging for non-destructive testing of carbon fiber-reinforced plastic materials *Meas. Sci. Technol.* **22** 1–4

[19] Busse G 1979 Optoacoustic phase angle measurement for probing a metal *Appl. Phys. Lett.* **35** 759–60

[20] Mulaveesala R and Tuli S 2006 Theory of frequency modulated thermal wave imaging for nondestructive subsurface defect detection *Appl. Phys. Lett.* **89** 191913

[21] Mulaveesala R and Dua G 2016 Non-invasive and non-ionizing depth resolved infra-red imaging for detection and evaluation of breast cancer: a numerical study *Biomed. Phys. Eng. Expr.* **2** 1–8

[22] Bhagwati C P and Sinha G R 2015 Gray level clustering and contrast enhancement (GLC–CE) of mammographic breast cancer images *CSI Trans. Springer* **2** 279–86

[23] Wang J, Chang K J, Chen C Y, Chien K L, Tsai Y S, Wu Y M, Teng Y C and Shih T T 2010 Evaluation of the diagnostic performance of infrared imaging of the breast: a preliminary study *Biomed. Eng. Online* **9** 1–10

[24] Dua G and Mulaveesala R 2018 Applicability of active infrared thermography for screening of the human breast: a numerical study *J. Biomed. Opt.* **23** 1–12

[25] Pennes H H 1948 Analysis of tissue and arterial blood temperatures in the resting human forearm *J. Appl. Physiol.* **1** 93–122

[26] Sudarshan N M, Ng E Y K and Teh S I 1995 Surface temperature distribution of a breast with and without tumour *Comput. Methods Biomech. Biomed. Eng.* **2** 187–99

[27] Ring E F J and Ammer K 2015 The technique of infrared imaging in medicine *Infrared Imaging: A Casebook in Clinical Medicine* (Bristol: IOP Publishing) pp 1-1–1-10

[28] Tseng Y P, Bouzy P, Stone N, Pedersen C and Tidemand-Lichtenberg P 2018 Long wavelength identification of microcalcifications in breast cancer tissue using a quantum cascade laser and upconversion detection *Proc. SPIE 10490, Biomedical Vibrational Spectroscopy 2018: Advances in Research and Industry (USA, February 2018)* pp 1–9

[29] Figueiredo A A A, Fernandes H C and Guimaraes G 2018 Experimental approach for breast cancer center estimation using infrared thermography *Infrared Phys. Technol.* **95** 100–12

[30] Negied N K 2019 Infrared thermography-based breast cancer detection-comprehensive Investigation *Int. J. Pattern Recognit. Artif. Intell.* **33** 1957002

[31] Mance M, Bulic K, Antabak A and Milosevic M 2019 The influence of size, depth and histologic characteristics of invasive ductal breast carcinoma on thermographic properties of the breast *EXCLI J.* **18** 549–57

[32] Tabatabaei N, Mandelis A and Amaechi B T 2011 Thermophotonic radar imaging: an emissivity-normalized modality with advantages over phase lock-in thermography *Appl. Phys. Lett.* **98** 163706

[33] Dua G and Mulaveesala R 2013 Applications of barker coded infrared imaging method for characterisation of glass fibre reinforced plastic materials *Electron. Lett.* **49** 1071–3

[34] Bhagwati C P and Sinha G R 2010 An adaptive *k*-means clustering algorithm for breast image segmentation *Int. J. Comput. Appl.* **10** 35–8

**IOP** Publishing

Modelling and Analysis of Active Biopotential Signals in Healthcare, Volume 2

**Varun Bajaj and G R Sinha**

# Chapter 16

# Photoacoustic microscopy: fundamentals, instrumentation and applications

**Mayanglambam Suheshkumar Singh, Anjali Thomas and Souradip Paul**

*Biomedical Instrumentation and Imaging Laboratory (BIIL), School of Physics (SoP), Indian Institute of Science Education and Research Thiruvananthapuram (IISER-TVM), Thiruvananthapuram, India - 695551*

This chapter discusses a detailed account of photoacoustic microscopy (PAM) (both from technological and application aspects) and its associated challenges. PAM imaging has been evolving and advancing significantly in the past two decades. Recently, studies have demonstrated that the achievable spatial (lateral) resolution reaches the sub-microscopic scale (~100 nm) and this regime of photoacoustic imaging is sometimes known as photoacoustic nanoscopy (PAN). The advent of the sub-diffraction limited PAN imaging modality permits extensions of the applications of PAM not only to cellular biology and its applications but also to neuronal brain imaging (for pre-clinical studies). There are continuous efforts to explore further advances in this technology and its associated applications.

## 16.1 Introduction

Photoacoustic microscopy (PAM) is one of the two sub-categories of the photoacoustic imaging (PAI) modality, which has proven a promising imaging technology. At a similar pace as photoacoustic tomography (PAT), which is the other PAI modality, PAM has been evolving and advancing significantly [1] (both from the technological and application aspects) for the past two decades. It results in different imaging configurations or modalities. Each of the various modalities has its own advantageous and disadvantageous features. This means to say that to draw maximal benefit from a particular imaging modality (more specifically, PAM), one is mandated to configure the imaging system according to its specific requirements and/or applications. In addition to the technological aspects, studies have

doi:10.1088/978-0-7503-3411-2ch16

been focusing not only on exploiting its pre-clinical and clinical applications but also on instrumentation (including data acquisition) and post-processing of the collected data (including improving the processing speed and derivation of patho-physiological information of clinical interest). The aim of this chapter is to discuss the details of PAM, from its fundamental to technological and application aspects, its associated challenges and its scope for future development.

It is accepted by everyone in the research domain of imaging that PAI has an unprecedented and unique capability of providing various types of patho-physiological information (of clinical interest, say Hb, $HbO_2$, SO and total Hb) using a single imaging unit. In addition to its high spatial resolution (~micrometres) achievable at an imaging depth beyond the optical transport mean free path for physiological information (~ 1 mm), it also offers a platform for the development of multi-modal imaging combining two or more imaging modalities in a single unit. The dual-modality of ultrasound imaging and PAI, which provides structural (from ultrasound imaging) and functional information (from PAI) of biological tissues, is already available for commercial use, while many studies on multi-modal imaging (including MRI and PAI, and light sheet microscopy and PAI, etc) are reported in the literature. These multi-modal imaging technologies facilitate combining the advantageous features of individual imaging modalities. Unfortunately, despite the unique features of PAI, the accessibility of PAI for research studies and clinical applications has been restricted. This is solely because of the high cost of the device and technology. Thus only a handful of laboratories (either research laboratories or clinics around the globe) have the privilege of house a PAI imaging system. For this reason, almost a decade ago, the light-emitting diode (LED) based PAI modality was developed and this device for laboratory and pre-clinical research studies has been successfully commercialized around the globe by CYBERDYNE, Inc. In LED-based PAI, the expensive pulsed (~nanoseconds) laser source that is conventionally employed in PAI systems to induce PA signals is replaced by an array of high-power LEDs, which is highly cost-effective. This LED-based technology (more specifically, LED-based PAI) greatly reduces the cost. To date, research on LED-based PAM has not been reported, which means to say that existing LED-based PAI is, technologically, based on PAT. From a thorough understanding of the PAM working principles and technology (as is discussed in sections 16.2 and 16.3), LED-based AR-PAM can be easily translated from the existing technology of laser-based AR-PAM by replacing the source of the pulsed optical beam from a pulsed (~nanoseconds) laser by an array of high-power LEDs, as is done in LED-based PAT. The translation of LED-based OR-PAM from conventional pulsed laser-based OR-PAM may be questionable due to the challenge of obtaining a tightly focusing optical spot (with a spot diameter ~100 nm [2] to ~1 $\mu$m) from a highly expanded and incoherent optical beam (or a collection of LED light sources). In addition, an LED-based PAI system is limited in achieving a high PA signal strength from sub-surface tissue, which is one of the major drawbacks of the technology. This is because of the limited optical energy (say, a fraction of a millijoule) obtainable from commercially available high-power LEDs, or even from a collection of high-power LEDs. This challenge of low signal strength can be addressed, partially, by adopting our recently reported opto-thermal based technique [3], where a continuous wave (CW) optical beam, in addition to an optical pulsed beam (pulse width ~nanoseconds) for transient photo-excitation, illuminates the tissue region of interest in a non-invasive and non-destructive manner.

## 16.2 The theoretical basis of PAM

Photoacoustic microscopy (PAM) imaging is the generation or reproduction of images (2D, 3D, 4D, or higher dimension) which are representatives of the strength of an initial photoacoustic pressure ($P_0$) or its derivatives (patho-physiological information) distributed over a sample material of interest. In contrast to PAT, in which the photoacoustic signal arriving at the sample boundary is collected by a single or an array of unfocused transducers, a tightly focusing ultrasound transducer is employed to pick up the initial pressure increase (represented by $P_0$, not $P$) selectively from the narrow focal region of the focusing ultrasound transducer.

Similarly to the case of PAT, the entire process of PAM imaging can be divided into four distinct stages: (i) irradiation of a given tissue sample with electromagnetic waves, (ii) generation of initial pressure waves ($P_0$), (iii) detection and acquisition of $P_0$ remotely from the tissue sample boundary and (iv) recovery of physical or patho-physiological information derived from the measured $P_0$ and its subsequent generation of representative images (2D, 3D or 4D). The first stage—that constitutes irradiation of the tissue material with electromagnetic waves, i.e. propagation of EM waves in the tissue medium—is governed by Maxwell's theory of electromagnetic waves. This implies that this stage is characterized by the optical properties of the propagating medium (the optical absorption coefficient ($\mu_a$), the optical scattering coefficient ($\mu_s$) and the refractive index ($n$)). In PAM, irradiation by these electromagnetic waves is carried out generally in two ways. (i) An expanded optical (pulse) beam (beam width ~centimeters) is delivered to the tissue surface so as to cover the entire region of interest for imaging. This occurs in a similar way to that of PAT. Primarily, this expanded beam illumination is adopted in the acoustic resolution photoacoustic microscopy (AR-PAM) imaging modality. (ii) A tightly focusing optical beam with a narrow focal zone or spot (spot diameter ~100 nm–1 $\mu$m [2]) is delivered to remotely irradiate a pre-specified (narrow) spot deep inside the tissue. The optical resolution photoacoustic microscopy (OR-PAM) imaging modality adopts this method of photo-excitation to induce initial pressure waves ($P_0$). Because of this tightly focusing nature of optical beam delivery to a tissue, for a given optical power or energy the effective optical energy received at a pre-specified location on the tissue sample is much higher in the case of OR-PAM than in AR-PAM, i.e. at the cost of complexity in the instrumentation of the PAI system, OR-PAM demands a lower power pulse laser source in comparison to that of AR-PAM which results not only in reducing the cost of the PAI system but also in photo-damage of the intervening tissue medium (the intervening tissue between the surface and the deep-seated targets of interest for imaging). However, in OR-PAM the achievable imaging depth, which is characterized by the penetration depth up to which the optical beam can be tightly focused, is limited to the optical transport mean path (given by $\ell_s = \ell\,(1/1 - g)$ [4, 5], where $\ell$ is the optical mean free path, $g$ (= $\langle\cos\theta\rangle$) is the optical anisotropy factor and $\theta$ is the optical scattering angle). The optical mean free path gives the optical path length between two consecutive scattering events of the propagating photons in the (tissue) medium and it is given by $\ell = nl$ (where $l$ is the physical distance). These two length scales (namely $\ell$ and $\ell_s$) are related to the scattering coefficient (given by $\mu_s$ (= $1/\ell$)) and the reduced scattering coefficient (given by $\mu_s'$ (= $1/\ell_s$)). In addition, tissue is an optically turbid medium, i.e. the propagation of light in tissue is highly diffusive at the length scale of $\ell_s$. This means to say that at the length scale of the mean free path ($\ell$), the propagation of light is characterized by rectilinear (or sometimes ballistic) propagation, while at a length scale of $\ell_s$ the propagation of light is highly diffusive in nature and propagating photons completely lose the history of their optical paths. In other words, at this length scale ($\ell_s$) the probability of scattering of the incident photons is uniformly or equally distributed in all possible directions of an entire solid angle. In this way it is not possible to bring a beam (even a collimated beam) of propagating photons to a point at a length scale beyond the transport mean free path ($\ell_s$). Conclusively, from the theoretical perspective there exists a maximal limit at which the incident optical beam can be tightly focused to a point. For soft tissue this is measured to be 1 mm (for $\ell = 0.1$ mm and $g = 0.89$ for soft tissues [4, 5]). The second stage that involves the generation of high-frequency ultrasound signals (known as photoacoustic signals) with a frequency of the order of megahertz to gigahertz, results due to the transient deposition of optical energy followed by rapid heating. More specifically, due to the intrinsic (optical) absorption coefficient and its spatial distribution $\mu_a(\vec{r})$ in the medium, the sample material absorbs light energy and the absorbed optical energy is converted into heat through vibrational and oscillational relaxation of the constituting particles/molecules, i.e. there exists a localized increase of temperature over the laser-irradiated tissue region, and this temperature increase ($\Delta T(\vec{r})$) is characterized by a spatial distribution of the optical absorption coefficient ($\mu_a(\vec{r})$). In this way, due to rapid heating associated with short pulsed light illumination (with a duration less than the time scales of the thermal and stress confinements), the light irradiated tissue undergoes thermoelastic expansion which is subsequently followed by a transient increase in pressure (we call the initial pressure increase or initial PA signals $P_0$ [6–8]). Assuming that the absorbed optical energy is converted completely into heat energy, whereby non-thermal effects such as fluorescence is neglected, the initial pressure is written as [6–8]

$$P_0(\vec{r}, t, T) = \frac{\beta(\vec{r}, T)}{\kappa_T(\vec{r}, T)} \frac{1}{\rho(\vec{r}, T)c_V(\vec{r}, T)} \mu_a(\vec{r})\varphi(\vec{r}, t), \tag{16.1}$$

$$= \frac{\beta(\vec{r}, T)v_{ac}^2(\vec{r}, T)}{c_P(\vec{r}, T)} \mu_a(\vec{r})\varphi(\vec{r}, t), \tag{16.2}$$

$$= \Gamma(\vec{r}, T)\mu_a(\vec{r})\varphi(\vec{r}, t), \tag{16.3}$$

$$= \Gamma(\vec{r}, T)A_e(\vec{r}, t), \tag{16.4}$$

where $\Gamma = \frac{\beta v_{ac}^2}{c_P} = \frac{\beta}{\kappa_T} \frac{1}{\rho c_V}$ is the Gruneisen parameter, $c_P$ is the specific heat capacity at constant pressure, $c_V$ is the specific heat capacity at a constant volume, $\beta$ is the thermal coefficient of expansion, $v_{ac}$ is the speed of sound in the medium, $\varphi$ is optical fluence which is energy per unit area per unit time, $A_e(\vec{r}, t)(= \mu_a(\vec{r})\varphi(\vec{r}, t))$ is a specific or volumetric optical absorption that measures the optical energy being absorbed by the tissue sample per unit time per unit volume and $T$ is the absolute temperature measured in kelvins. For a given optical energy from a source, due to the focusing nature, the optical fluence ($\varphi$) and hence the strength of the initial PA-pressure ($P_0$) is measured as being higher for OR-PAM than that of AR-PAM. Equations (16.1)–(16.4) present various ways of expressing the initial PA signals ($P_0$) in terms of the physical parameters that we generally come across in the literature or books on PAI. The third stage comprises the detection and acquisition of the initial pressure increase ($P_0$) deep inside the tissue by positioning an acoustic sensor or an array of acoustic sensors in the tissue sample boundary, i.e. the detection of $P_0$ deep inside the tissue from the sample boundary. In PAM (either AR-PAM or OR-PAM), an ultrasound transducer is employed to pick up $P_0$ from its narrow focal zone, i.e. a time-resolved $P_0$ signal appearing at the focal zone is detected and subsequently acquired by an array of transducer elements distributed over the entire surface of the transducer unit (for the case of a single element transducer, one may consider that the entire space of the transducer surface is composed of an infinite number of transducer elements). From the engineering aspects, the measurement of an output signal from a transducer unit is an average of measurements obtained from all of these elements in the transducer surface. This implies that the output signal from a transducer unit gives a measure of $P_0$ averaged over a cross section (with the thickness being defined by the sampling period of the data acquisition and the width being defined by the focal spot diameter) in the focal zone of the tightly focusing transducer. In short, in PAM the spatial resolution of the output signal ($P_0$) measured from a focusing transducer is characterized by the focal spot size of the transducer unit. In OR-PAM, instead of an expanded beam, a tightly focusing pulsed (~nanoseconds) optical beam is introduced into the sample material to induce photo-excitation (more specifically $P_0$) in a very narrow region dictated by the focal spot size of the focusing optical beam. This implies that in this (OR-PAM) imaging modality there exist two focal zones, one corresponds to the focusing optical beam while the other corresponds to the focusing ultrasound beam. These two focal zones are co-aligned in such a way that not only are the axes of the beams co-aligned, but the focal zones are also located at the same point. This kind of arrangement is called a confocal configuration. In such a confocal configuration, the achievable spatial resolution is determined by the corresponding beam with a narrower focal spot. For OR-PAM the achievable spot size of the optical beam reaches an order of 100–1000 nm which is far smaller than that of the ultrasound beam (~25 $\mu$m or more), i.e. the obtainable spatial resolution in OR-PAM is directed by the optical diffraction limit (discussed in section 2.2) as is suggested by the name of the imaging modality. For the case of AR-PAM, the focal spot size of the optical beam is chosen to be wider than that of the ultrasound beam and thus the spatial resolution obtainable from the imaging modality is given by the focal spot size of the focusing ultrasound transducer. Generally, an expanded (instead of focusing) optical beam is employed in the AR-PAM imaging modality. Conclusively, in the OR-PAM imaging modality the spatial resolution as well as penetration depth (sometimes called imaging depth) are dictated by the optical beam and its characteristics (relative to the optical properties of the tissue medium). In contrast, in AR-PAM both the spatial resolution and imaging depth are determined by the acoustic beam and its characteristic properties relative to the acoustic properties of the tissue.

Equations (16.1)–(16.4) clearly demonstrate that the photoacoustic (initial) pressure increase ($P_0$) is dependent not only on the incident pulsed laser beam (more specifically, the optical energy per unit area per unit time that is called optical fluence ($\phi$)) but also on various physical parameters of the tissue sample (thermal or thermodynamic, mechanical or visco-elastic, acoustic and optical properties). In addition, the physiological activities and thus the patho-physiological states of a biological specimen are closely related to and thus characterized by the physical and/or chemical properties of tissues. This implies that from the (experimental) measurement of $P_0$ one can derive various patho-physiological parameters that are of clinical interest for the diagnosis and therapeutic treatment of various diseases. It has been a focus of research for the last two decades to develop various non-invasive and/or non-destructive techniques to derive physical quantities individually, which appear on the right-hand side of equations (16.1)–(16.4) from the experimental measurement of $P_0$, which is a measure of the collective effect of all the quantities on the right-hand side. With a given optical fluence ($\phi$), recovery of the optical absorption coefficient ($\mu_a$) was reported and has been practiced for the past two decades. Further, employing a spectral un-mixing technique, recovery of the physiological information (concentration of Hb, $HbO_2$, total Hb and SO) from experimentally measured $\mu_a$ has been reported. Recently, several techniques were reported in the literature for the estimation of tissue elastic properties, temperature ($T$) and acoustic velocity, etc. From experimentally measured photoacoustic signals or its derivatives, one can generate a matrix (an array of data in 2D, 3D and 4D) and subsequently corresponding image representations (2D, 3D and 4D) of the photoacoustic signal and its derivatives can be obtained.

### 16.2.1 The resolution of PAM

The resolution of any imaging modality can be considered from two aspects (namely, spatial resolution and temporal resolution, which are collectively considered as spatio-temporal resolution). Spatial resolution defines the smallest possible object or target an imaging device can discern. It is also defined as the closest possible separation between two targets that can be observed by the device as separate ones. In this way, the spatial resolution can further be separated into lateral resolution and axial resolution that give the respective measures of the closest distance between objects being kept in a perpendicular direction or along the axis of the imaging device. From wave theory (more specifically, the theory of diffraction) these parameters (lateral or axial resolution) are characterized by the diffraction limit which is dependent on the parameter of the interrogating signal [9] (say the wavelength ($\lambda$) or frequency ($\gamma$ (= $c/\lambda$), where $c$ is the speed of propagation of the interrogating signal in the medium), the property of the medium (refractive index ($n$) for EM waves and acoustic impedance for acoustic waves) and the physical characteristics of the device (the size of the numerical aperture relative to the distance of the aperture to the screen or detector). The lateral spatial resolution is characterized by the wavelength of the interrogating signals while the axial resolution is dependent on the bandwidth of the interrogating signals or waves.

Temporal resolution is the minimum time interval that is required for the generation of two sequential images (2D or 3D or higher dimension) which are, in the case of PAI, representative of the photoacoustic signals or their physical derivatives. This is an inverse of the frame rate, which is the number of sequential frames generated per unit time. The temporal resolution is collectively determined by the pulse repetition period of the pulsed optical beam, the scanning speed and dimension of scanning, the data acquisition and its transfer to the system and the speed of post-processing. It is described in section 16.3, where a detailed account of the instrumentation is given.

### 16.2.2 The spatial resolution of OR-PAM

The maximum lateral (spatial) resolution achievable for OR-PAM is given by [9]

$$\alpha_{\text{lat}} = 0.51\frac{\lambda_{\text{opt}}}{\text{NA}_{\text{opt}}} \tag{16.5}$$

$$= 0.51\frac{c}{\text{NA}_{\text{opt}}\gamma_{\text{opt}}}, \tag{16.6}$$

where $\lambda_{\text{opt}}$ is the wavelength of the interrogating (pulse) laser beam ($\gamma_{\text{opt}}(= c/\lambda_{\text{opt}})$ is the corresponding frequency), $c$ is the speed of propagation of the optical beam and $\text{NA}_{\text{opt}}$ is the numerical aperture of the imaging device (more specifically, the optical system or the part adopted for photo-excitation to induce initial the PA signal ($P_0$), i.e. an optical system for delivering a tightly focusing pulsed laser beam in the case of the OR-PAM imaging modality).

Again, the axial (spatial) resolution obtainable in OR-PAM is given by [9]

$$\alpha_{\text{axial}} = 0.88\frac{v_{\text{ac}}}{\text{NA}_{\text{ac}}\Delta\gamma_{\text{ac}}}, \tag{16.7}$$

where $\gamma_{\text{ac}}$ is the bandwidth of the acoustic beam which is employed as a detection signal in PAM (either OR-PAM or AR-PAM), $V_{\text{ac}}$ is the speed of the propagation of acoustic waves and $\text{NA}_{\text{ac}}$ is the numerical aperture of the imaging system (more specifically, the tightly focusing ultrasound transducer unit in either the OR-PAM or AR-PAM imaging modality).

### 16.2.3 The spatial resolution of AR-PAM

The maximum lateral (spatial) resolution that can be achieved in AR-PAM is given by [9]

$$\alpha_{\text{lat}} = 0.71\frac{\lambda_{\text{ac}}}{\text{NA}_{\text{ac}}} \tag{16.8}$$

$$= 0.71\frac{v_{\text{ac}}}{\text{NA}_{\text{ac}}\gamma_{\text{ac}}}, \tag{16.9}$$

where $\lambda_{ac}$ is the optical wavelength of the interrogating (pulse) laser beam ($\gamma_{ac}(= v_{ac}/\lambda_{ac})$ is the corresponding frequency), $v_{ac}$ is the speed of propagation of the acoustic beam and $NA_{ac}$ is the numerical aperture of the imaging device (more specifically, the acoustic system or the part adopted for the detection of the light induced an initial PA signal ($P_0$)). The axial resolution of AR-PAM is similar to that of OR-PAM (as shown in equation (16.7)) [9].

## 16.3 The instrumentation of PAM

As is explained in section 16.1, for both the OR-PAM and AR-PAM imaging modalities there exists a very close similarity in physical fundamentals and these processes are broadly sub-grouped into four separate stages: irradiation of a given tissue sample with electromagnetic waves, generation of initial pressure increase ($P_0$), detection and acquisition of $P_0$ from the (tissue) sample boundary, and recovery of physical or patho-physiological information from the measured $P_0$ and the subsequent generation of representative images. From the perspective of instrumentation (i.e. design and development) there exists a sharp difference between OR-PAM and AR-PAM. The primary difference arises from the design of the optical system to deliver an optical pulsed laser beam for the requirement of photo-excitation to induce the initial PA signals ($P_0$). Some of the sub-classes or variants of PAM imaging modalities (AR-PAM and OR-PAM) are partially diversified from one another in terms of the collection of time-resolved data while generating its counterpart 2D and 3D data. Several technological developments have been advanced and this remains one of the focuses of research in this field.

### 16.3.1 AR-PAM

AR-PAM can be further sub-grouped into two main classes, as is shown in figures 16.1 and 16.2. One is based on the expanded beam illumination of a pulsed laser (which we call expanded beam AR-PAM (Xb-AR-PAM)) while the other employs a focusing beam for illumination (abbreviated as Fb-AR-PAM). These two modalities have their own advantages and disadvantages. Because of the focusing nature of the illuminating beam, Fb-AR-PAM demands a pulsed laser source of lower optical energy output so as to obtain the required optical fluence being delivered to a deep-seated tissue target. Thus the cost of the device can be reduced. However, this is at the cost of complexity in the instrumentation of the device.

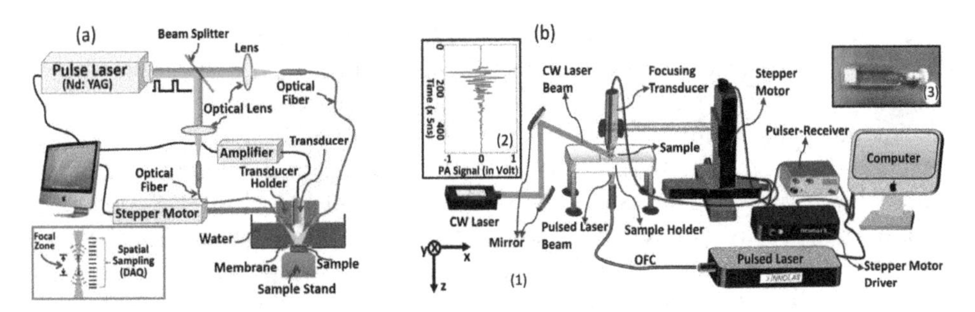


**Figure 16.1.** Schematic diagram of expanded beam AR-PAM (Xb-AR-PAM) in the (a) reflection mode and (b) transmission mode.

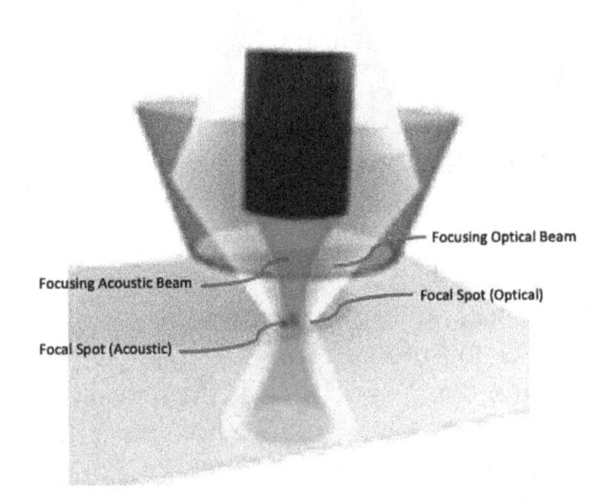


**Figure 16.2.** Schematic diagram of focusing beam AR-PAM (Fb-AR-PAM). Adapted with permission from [9]. Copyright 2019 Elsevier.

### *16.3.1.1 Expanded beam AR-PAM*

Expanded beam AR-PAM (Xb-AR-PAM) has two commonly used configurations. (i) The reflection mode in which (pulse) optical beam illumination and acoustic wave detection are situated on the same side of the imaging sample (generally both of them are housed in a single holder). A typical schematic diagram is shown in figure 16.1(a). (ii) The transmission mode in which a beam of optical pulses (pulse width ~nanoseconds) for transient photo-excitation is located opposite to the ultrasound transducer unit that is employed for the detection and collection of photoacoustic waves ($P_0$), whereby the imaging sample is positioned in between (a schematic diagram is shown in figure 16.1(b)). For both configurations, a train of short duration pulses of an optical beam from a laser source is coupled by one or more optical fibres, i.e. the output optical beam from a pulsed laser source is coupled (using optical fibres) to irradiate the tissue sample of interest over a pre-specified region. For biomedical applications, generally an optical window (with a wavelength in the range ~600–1000 nm) is employed where the imaging devices achieve higher penetration depth due to the relatively low (optical) absorption coefficient of the skin in this optical window. A tightly focusing ultrasound transducer is kept in position—facing the optical illumination on the other side of the imaging sample (for the transmission mode)—to pick up the transient light induced PA signals ($P_0$) selectively from its narrow focal zone. Since the detected photoacoustic signal strength is very weak (measured voltage ~microvolts [7, 10]), PA signals are amplified and then acquired by using a data acquisition system attached to a computer system. From a scanning position, where the optical illumination and ultrasound detection are kept fixed in a position, one can obtain an A-scan (or 1D time-resolved) data, which are a measure of time-resolved $P_0$ along a line (in the narrow focal zone). From this time-resolved A-scan, depth-resolved data are obtained by employing a distance–time relationship ($s = vt$, where $s$ is the distance travelled in a time interval ($t$) with a given (constant) speed ($v$)). Then, one can generate 2D or 3D data representative of PA signals by raster scanning in 1D or 2D, respectively. The counterpart 4D is obtained through the collection of 3D data at different time points, i.e. 4D is generated by extending 3D along with time points as another dimension. Generally, a high precision translation stage is employed for raster scanning in which a focused transducer is driven to translational movement with a step size of ~micrometres while the pulsed optical beam and imaging sample are kept fixed in position, i.e. the sample of interest over a pulsed laser-irradiated region is scanned by the translational movement of a tightly focusing US-transducer. This mode of detection is for the case of transmission mode Xb-AR-PAM. In contrast, for the reflection mode Xb-AR-PAM, the focusing ultrasound transducer and optical fibre are housed together in a single holding system, i.e. the pulsed laser beam and ultrasound transducer are operating in reflection mode. In this reflection mode the holding system is driven by translational movement for raster scanning while the imaging sample is kept in a fixed position. For both cases of transmission and reflection, individual components—light illumination, ultrasound detection and acquisition—are controlled in a synchronous fashion. From experimentally measured data (more specifically, a matrix with each element representing $P_0$ or its derivatives), one can generate photoacoustic representative images (2D, 3D and 4D) employing graphic-based tools or software. During experiments, for the coupling of acoustic beam transmission from the tissue surface to the ultrasound transducer, an acoustic coupling medium is introduced between the imaging sample and ultrasound transducer.

#### *16.3.1.2 Focusing beam AR-PAM (Fb-AR-PAM)*

The construction of focusing beam AR-PAM (Fb-AR-PAM) is similar to that of OR-PAM (presented in section 16.3.2) with the exception that the focal spot size of the acoustic beam is narrower than that of the optical focal spot. Again, in contrast to Xb-AR-PAM where the optical beam is expanded (not focusing), the optical beam is focused in the case of Fb-AR-PAM. Except for this difference in pulsed laser irradiation, the acquisition of $P_0$ through raster scanning and the generation of images is identical to Xb-AR-PAM.

### 16.3.2 OR-PAM

Figure 16.3 depicts a representative schematic diagram of OR-PAM (adapted from [11]). OR-PAM primarily differs from AR-PAM in the illumination of a tightly focusing optical beam and its required optical arrangement. A highly collimated optical beam is generated by using an optical system (consisting of converging lenses, pin-holes and an iris) and subsequently a tightly focusing optical beam (a focal spot size of the order of 100–1000 nm) is achieved by employing a microscopic objective (with a high numerical aperture). Further, a combination of two prisms sandwiched by a thin layer of silicone oil for acoustic–optical coaxial alignment is employed as a waveguide for both optical and acoustic signals. This optical component serves as the reflector for acoustic waves but as an optically neutral medium. Acoustic waves emerging from the target are diverted with an angular deflection of 90° while optical waves (directed towards the imaging sample surface) propagate without deviation.

Generally, either a combination of two right angle prisms (shown in the inset of figure 16.3) or combination of one right angle prism and one rhomboid prism is employed for such acoustic–optical coaxial alignment. An acoustic lens is fixed at the bottom of the lower prism (rhomboid prism as it is shown in the schematic diagram), which enables to obtain tightly focusing acoustic beam (with a focal spot size of the order of 25–100 $\mu$m). The optical and acoustic foci are aligned con-focally as well as coaxially to maximize the sensitivity in the detection of photoacoustic signals ($P_0$). In OR-PAM, 1D data are obtained by acquiring 1D time-resolved $P_0$ signals (as is done in AR-PAM) while 2D or 3D data are obtained by raster scanning in which the acousto-optic foci are kept fixed in a position while the imaging target is driven to move in a line or a horizontal plane. Another method is also adopted in which the optical unit consists of prisms rotated on a particular axis that facilitates guiding the optical illumination beam as well as the acoustic beam. In this way, the

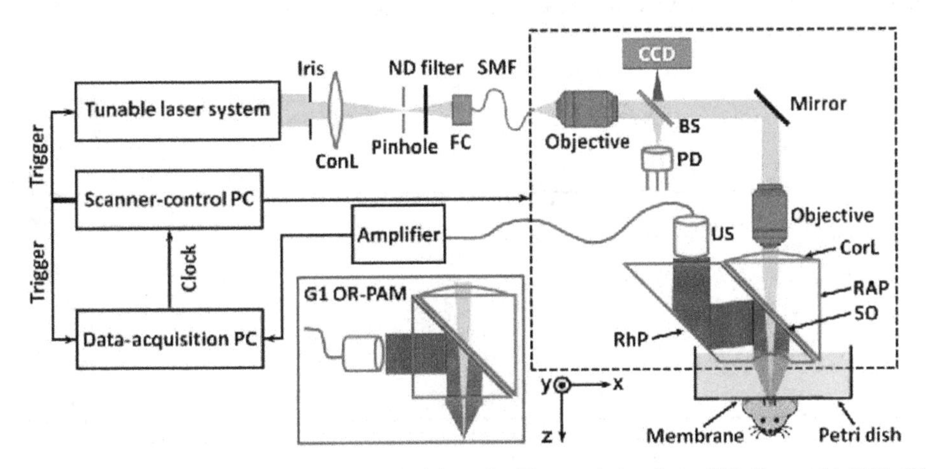


**Figure 16.3.** Schematic diagram of OR-PAM. Adapted with permission from [12]. Copyright 2011 OSA Publishing.

field of view (FOV) obtainable from OR-PAM is very limited (in addition to the limited penetration depth).

Several imaging modalities for PAM have been studied. The comparison of various existing PAM imaging modalities is provided in figure 16.4. The figure demonstrates that each imaging modality has its own advantages and disadvantages.

As discussed above, 2D or 3D PA data and consequent images are generated using raster scanning for both AR-PAM and OR-PAM. A raster scanning point corresponds to the delivery of one pulse and thus the total number of matrix elements for generating a PA representative image corresponds to the total number of pulses from a pulsed laser source. This implies that the time taken for acquisition of data is governed by the pulse repetition frequency which is an inverse of the pulse repetition period. In short, the imaging speed or frame rate of PAM is characterized by the pulse repetition frequency of the laser source, and hence it can be significantly improved by increasing the pulse repetition rate of the laser (say ∼100 kHz) [12].

## 16.4 Applications

Recent studies have focused on the exploration of potential applications of PAM imaging modalities. Several promising applications have been demonstrated. The applications of PAM may be considered from two different aspects, namely biological and pre-clinical studies: (i) PAM imaging modalities and their technologies can be adapted for microscopic biological specimens as well as a specimen of interest for pre-clinical studies (small animals including the rat and mouse) and (ii) the PAM imaging modalities give a signal contrast to the specimen of interest for biological and pre-clinical applications. For clinical studies and their applications, PAM imaging modalities are not feasible, which is mainly because of its limited penetration depth (achievable only to an order of ∼1–4 mm, which is very low for clinical applications). In other words PAM is restricted in its adoption as a non-destructive scanning device in the clinic for human subjects.

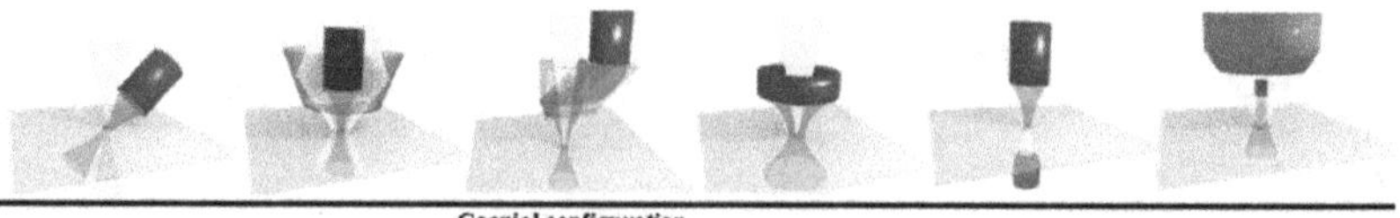

| Key feature | Off-axis alignment | Dark-field Illumination | Coaxial configuration based on an opto-ultrasound beam combiner | Coaxial configuration based on a ring transducer | Transmission mode | Reflection mode with a large optical NA |
|---|---|---|---|---|---|---|
| *In vivo* imaging | Δ | O | O | O | X | O |
| Coaxiality | X | O | O | O | O | O |
| SNR | X | O | O | O | O | O |
| AR/OR Modes | OR only | AR only | Both | OR | OR | OR only |

X: Bad, Δ: Intermediate, O: Good

**Figure 16.4.** Comparison of various existing PAM imaging modalities. Adapted from with permission from [9]. Copyright 2019 Elsevier.

Studies [13, 14] have demonstrated that a sub-micrometre lateral resolution (of the order of 100 nm) could be achieved by employing high-NA objective lenses in OR-PAM imaging system. Experiments were conducted in healthy mouse red blood cells (RBCs) to demonstrate that OR-PAM technology achieved the milestone of imaging samples at the cellular level. A 3D PA image of a single RBC is presented in figure 16.5(a). The 3D image gives a volumetric rendering of a single RBC and it shows that the RBCs take a bi-convex or doughnut shape. From images obtained from OR-PAM imaging technology, the thickness of a single RBC is estimated to be 3.84 $\mu$m. The study demonstrates a unique achievement of scientific importance (more specifically in PAI), extending the limit of achievable spatial resolution in 3D that enables the characterization of the morphology of a single cell (more specifically an RBC) and its patho-physiological states (size, shape and intrinsic optical absorption properties). PAM is non-invasive in nature and there is no requirement for introducing contrast dye in the biological imaging specimen, which is uniquely different to other optical-based imaging modalities which require staining of the biological specimen. Even though the optical-based imaging modalities can provide a higher resolution image, a 3D image of a single cell cannot be achieved [13]. Figure 16.5(b) depicts a PA image of melanoma cells. Melanin pigments, which are molecules found in the melanosome organelle (within melanocytes), provides signal contrast for PAI [14]. This result clearly shows that diffraction limited OR-PAM facilitates sufficient spatial resolution to discern not only the shape of melanoma

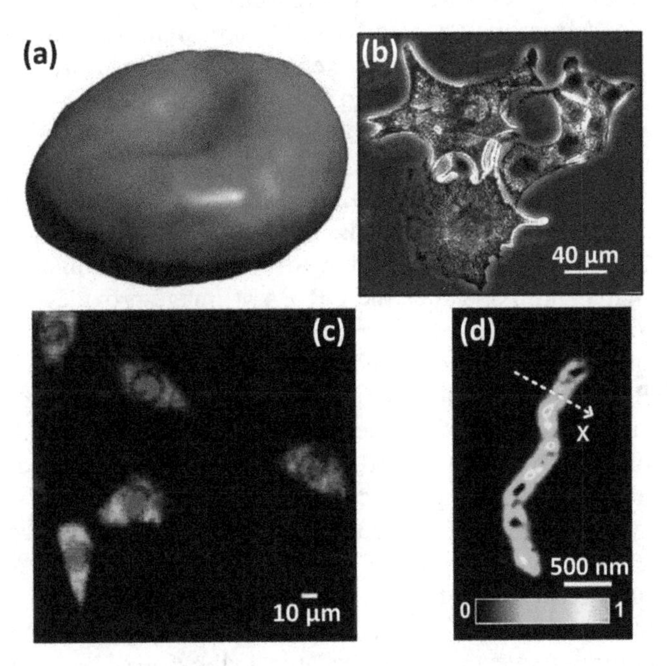


**Figure 16.5.** (a) 3D reconstructed image of an RBC obtained from OR-PAM. (b) OR-PAM image of melanoma cells. (c) OR-PAM image of fibroblasts. (d) OR-PAM image of mitochondria. (a) Adapted with permission from [13]. Copyright 2015 OSA Publishing. (b) and (d) Adapted with permission from [15]. Copyright 2014 SPIE. (c) Adapted with permission from [12]. Copyright 2013 SPIE.

cells but also the details within the cell. Figure 16.5(c) provides a label-free (i.e. unstained) image of a fixed fibroblast, which was obtained using diffraction limited OR-PAM. The image displays fibroblast cytoplasms and fibroblast nuclei. From figure 16.5(c) one can observe that individual mitochondria are not resolved. This is due to the low contrast between the mitochondria and other parts of the cytoplasm, in addition to the insufficient axial resolution of OR-PAM, i.e. with an enhanced spatial resolution it may be possible to observe individual mitochondria. Figure 16.5(d) shows a tubular-shaped mitochondrion in a fibroblast (NIH 3T3) in melanoma cells, which was obtained using OR-PAM imaging technology. Cytochromes, which are mitochondrial hemoproteins, have a strong optical absorption coefficient and thus constitute a primary absorber in melanotic melanoma cells. In this way, cytochromes serve as endogenous pigments which are highly specific to the inner mitochondrial membrane and this feature facilitates giving high contrast to PA signals. Figure 16.6 presents the experimental results of pre-clinical studies where experiments were conducted in small animal models. Figure 16.6(a) displays an image of the vascular anatomy of the whole ear of a mouse obtained by diffraction limited OR-PAM. Experiments were conducted on a living nude mouse (Hsd:Athymic Nude-FoxnlNU, Harlan) which was 10 weeks old. A magnified image (shown in the inset of figure 16.6(a)) depicts that densely packed capillary beds, along with discrete RBCs travelling along capillaries, can be resolved with OR-PAM. Figure 16.6(b) shows a transcranial image of the brain of a living (adult) mouse obtained using OR-PAM. In the experiments, the skull vessels in a three-month-old Swiss Webster mouse (Hsd:ND4, Harlan) were perfused with methylene blue dye. Methylene blue-perfused skull vessels and RBC-perfused cortical vessels provide strong photoacoustic signals and they are shown in different colours in the depth-encoded image (figure 16.6(a)). The figure shows that the skull vessels and the cortical vessels in living mice can be clearly differentiated. Figure 16.6(c) depicts a label-free ophthalmic angiography of the iris microvasculature of a living adult Swiss Webster mouse. The inset in the figure gives a magnified image of the square boxed area in figure 16.6(c). Haemoglobin is the predominant light absorber in an intrinsic micro-circulation and thus the OR-PAM image provides a map of blood vessels (more specifically RBCs as carriers of haemoglobin). This implies that the OR-PAM images provide patho-physiological information on tissues through the recovery of RBC-perfused capillaries responsible for supplying oxygen to tissues.

Research continues to explore further potential applications of PAM (including the recovery of elastic properties and blood flow velocity, both amplitude and direction) with experiments being limited to tissue-mimicking samples. Experiment-based studies were reported [7] for the investigation of the signal contrast of PA signals with the elastic coefficient of the targets. The study was extended [8] to show that the frequency response of the ultrasound-detecting transducer shifts with the change in contrast of either the optical or mechanical properties of the targets, i.e. a change in the natural frequency of the mechanical vibration of the tissue targets with a change in the physical properties of imaging targets (tissue contrast). Experimental studies were performed in an agar-based tissue-mimicking phantom, shown in figure 16.7(a), where various targets (T) of different contrast in terms of optical and

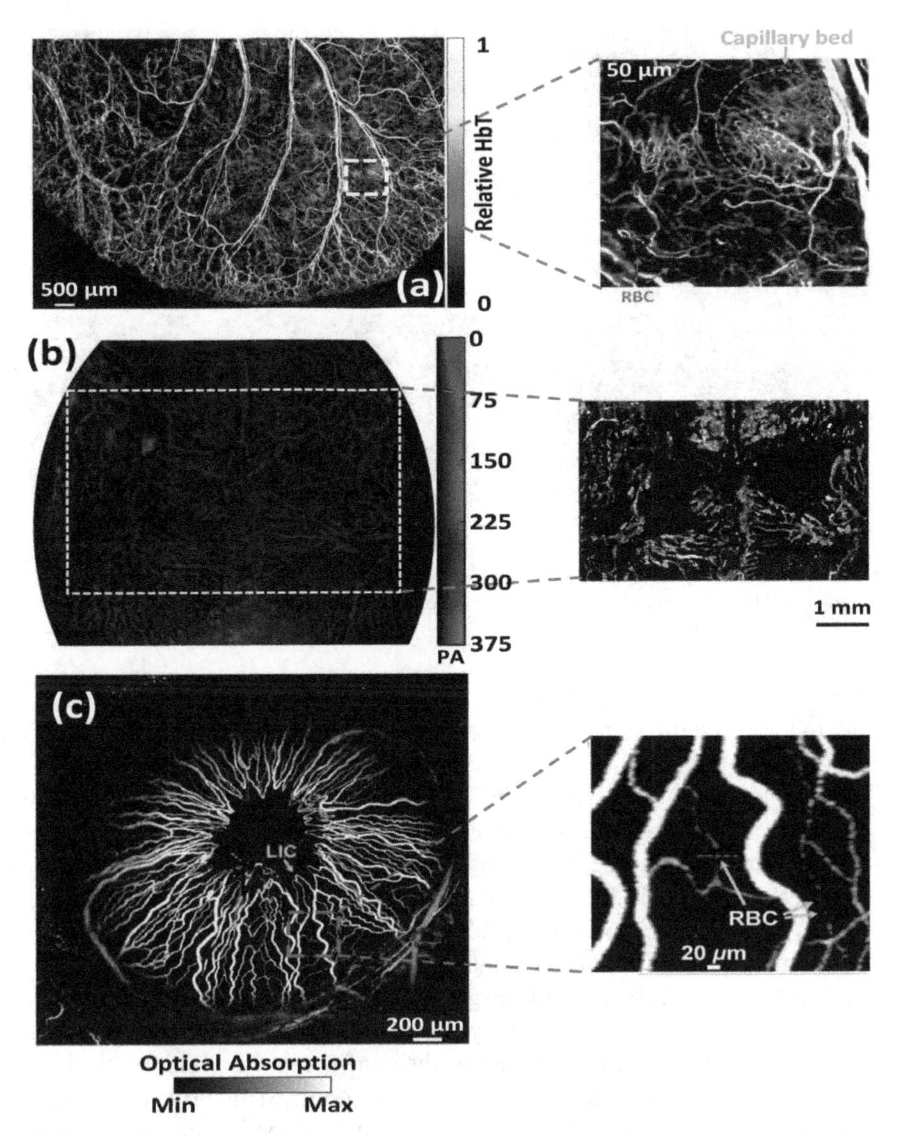


**Figure 16.6.** (a) OR-PAM image representative of relative HbT in a living mouse ear. A densely packed capillary bed (corresponding to square box) is displayed on the right. (b) OR-PAM image of a living adult mouse brain (transcranial) depicting methylene blue-perfused skull vasculature and blood-perfused cortical vasculature (at 570 nm) while the methylene blue-perfused skull vasculature (at 650 nm) is shown on the right. (c) OR-PAM image of angiography of the iris microvasculature of a mouse. The experiment was performed in an live adult Swiss Webster mouse. A magnified image corresponding to the square shown in (c) is depicted on the right. (a) and (b) Adapted with permission from [11]. Copyright 2011 OSA Publishing. (c) Adapted with permission from [16]. Copyright 2010 OSA Publishing.

mechanical properties were embedded in the background sample. In figure 16.7(a), for all targets ($T_1$–$T_6$) the optical absorption coefficient ($\mu_a$) and optical scattering coefficient ($\mu_s$) were tailored similarly to the background (0.01 and 1.20 mm$^{-1}$, respectively) with the exception of $T_2$. For $T_2$, $\mu_a = 0.02$ mm$^{-1}$. The elastic

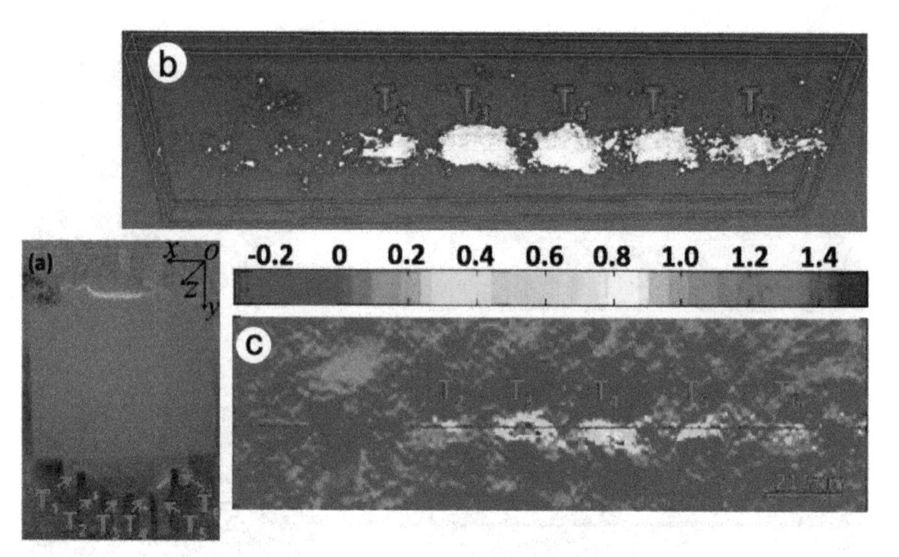


**Figure 16.7.** (a) An agar-based tissue-mimicking phantom where six different targets are embedded in the background. (b) 3D reconstructed image representative of PA signal strengths (measured by AR-PAM [7, 8, 10]) from (c) a sequence of 2D PA representative images.

coefficients (more specifically, Young's modulus ($E$)) were tailored to be 345, 269, 202 and 143 kPa for different targets ($T_3$ to $T_6$, respectively), while $E = 94$ kPa for the background. A detailed description of the agar sample preparation is provided in [8]. These tailored elastic coefficients are within the range of elastic properties corresponding to cancerous soft tissues. Figures 16.7(b)–(c) depict experimental results that demonstrate the feasibility of recovery of the elastic properties of biological tissue. Figure 16.7(b) presents a 3D representative image of laser-induced PA signals, while figure 16.7(c) provides a 2D reconstructed image. The image demonstrates that the detected PA signals give a signal contrast for targets ($T_2$–$T_6$) with respect to the background (but target ($T_1$) is not detectable, i.e. $T_1$ gives no PA signal contrast with respect to the background). The observed contrast in PA signal strength corresponding to $T_2$ is because of the contrast in $\mu_a$, while for $T_1$ the signal contrast is not observable because of there is no contrast in both the optical and elastic coefficients with respect to the background. Experiments were conducted using AR-PAM in transmission mode [7, 8, 10].

## 16.5 Summary

Photoacoustic microscopy provides a higher spatial resolution but lower penetration depth. Several PAM imaging modalities have been reported. Each modality has its own advantages and disadvantages. PAM has so far been exploited for biological applications of imaging cells and organelles (components of a cell) as well as for some pre-clinical research studies. There is potential to replace the expensive pulsed laser sources with affordable high-power LED sources, for the requirements of transient photo-excitation to induce photoacoustic signals. In turn, one may adopt photo-thermal based techniques for the enhancement of the weak photoacoustic signal strength that results from the replacement of a high-power laser source by LED sources.

## References

[1] Wang L V and Hu S 2012 Photoacoustic tomography: *in vivo* imaging from organelles to organs *Science* **335** 1458–62

[2] Strohm E M, Moore M J and Kolios M C 2016 Single cell photoacoustic microscopy: a review *J. Sel. Top. Quant.* **22** 6801215

[3] Thomas A, Naidu A, Varghese R and Singh M S 2019 Enhancement of photoacoustic signal from contrast agent with pre-illumination *Proc. SPIE Photonic West Photons Plus Ultrasound: Imaging and Sensing (San Francisco, CA, 2019)* 1087833

[4] Singh M S, Rajan K and Vasu R M 2011 Estimation of elasticity map of soft biological tissue mimicking phantom using laser speckle contrast analysis *J. Appl. Phys.* **109** 104704

[5] Singh M S 2019 Optical polarization technique for enhancement of image quality in speckle contrast-based perfusion imaging: a characterization study *AIP Advs.* **9** 075003

[6] Singh M S and Thomas A 2019 Photoacoustic elastography imaging: a review *J. Biomed. Opts.* **24** 040902

[7] Singh M S and Jiang H 2014 Elastic property attributes to photoacoustic signals: an experimental phantom study *Opt. Lett.* **39** 3970–3

[8] Singh M S and Jiang H 2016 Ultrasound (US) transducer of higher operating frequency detects photoacoustic (PA) signals due to the contrast in elastic property *AIP Adv.* **6** 025210

[9] Jeon S, Kim J, Lee G, Baik J W and Kim C 2019 Review on practical photoacoustic microscopy *Photoacoustics* **15** 100141

[10] Singh M S and Jiang H 2014 Estimating both direction and magnitude of flow velocity using photoacoustic microscopy *Appl. Phys. Lett.* **104** 253701

[11] Hu S, Maslov K and Wang L V 2011 Second-generation optical-resolution photoacoustic microscopy with improved sensitivity and speed *Opt. Lett.* **36** 1134–6

[12] Zhang C, Zhang Y S, Yao D K, Xia Y and Wang L V 2013 Label-free photoacoustic microscopy of cytochromes *J. Biomed. Opt.* **18** 020504

[13] Dong B, Li H, Zhang Z, Zhang K, Chen S, Sun C and Zhang H F 2015 Isometric multimodal photoacoustic microscopy based on optically transparent micro-ring ultrasonic detection *Optica* **2** 169–76

[14] Strohm E M, Berndl E S L and Kolios M C 2013 High frequency label-free photoacoustic microscopy of single cells *Photoacoustics* **1** 49–53

[15] Danielli A, Maslov K, Uribe A G, Winkler A M, Li C and Wang L V 2014 Label-free photoacoustic nanoscopy *J. Biomed. Opt.* **19** 086006

[16] Hu S, Rao B, Maslov K and Wang L V 2010 Label-free photoacoustic ophthalmic angiography *Opts. Lett.* **35** 1–3

**IOP** Publishing

Modelling and Analysis of Active Biopotential Signals in Healthcare, Volume 2

**Varun Bajaj and G R Sinha**

# Chapter 17

# Rigorous performance assessment of computer-aided medical diagnosis and prognosis systems: a biostatistical perspective on data mining

**Marjan Mansourian[1,2], Hamid Reza Marateb[3], Mahsa Mansourian[4], Mohammad Reza Mohebbian[5], Harald Binder[6,7] and Miguel Ángel Mañanas[1,8]**

[1]*Biomedical Engineering Research Centre (CREB), Automatic Control Department (ESAII) Universitat Politècnica de Catalunya-Barcelona Tech (UPC), Barcelona, Spain*

[2]*Epidemiology and Biostatistics Department, Health School, Isfahan University of Medical Sciences, Isfahan, Iran*

[3]*Biomedical Engineering Department, Engineering Faculty, University of Isfahan, Isfahan, Iran*

[4]*Medical Physics Department, Isfahan University of Medical Sciences, Isfahan, Iran*

[5]*Department of Electrical and Computer Engineering, University of Saskatchewan, SK, Canada*

[6]*Institute of Medical Biometry and Statistics, Faculty of Medicine, Medical Center University of Freiburg, Freiburg, Germany*

[7]*Freiburg Center for Data Analysis and Modeling, University of Freiburg, Freiburg, Germany*

[8]*Biomedical Research Networking Center in Bioengineering, Biomaterials, and Nanomedicine (CIBER-BBN), Spain*

A lot of computer-aided medical diagnosis and prognosis systems have been suggested in the literature in areas such as oncology, cardiology, psychology and diabetes. Such systems have also been integrated with signal processing in the analysis of biopotentials in healthcare. They can provide second opinions to physicians and are assumed to improve accuracy. However, these systems often have not been used widely by clinicians due to their complexity and incomplete evidence for their reliability. Questions such as the reliability of a diagnosis system for medical doctors, its proper comparison with the state-of-the-art, avoiding incorrect interpretation and reporting proper performance indices are not only linked to pattern recognition but also to the characteristics typically considered in biostatistics. Based on our experience in designing medical diagnosis systems, teaching biostatistics and research methods to biomedical engineers for many years,

doi:10.1088/978-0-7503-3411-2ch17  

and our reviews of such systems in international journals, in this chapter we provide an overview of such issues.

## 17.1 Introduction

A large amount of data is analysed (i.e. 'mined') by data mining methods to identify valuable knowledge [1]. This is part of the field of knowledge discovery in databases (KDDs) in the literature [2]. The KDD process goes from data to knowledge, which includes selection, preprocessing, transformation, data mining and interpretation or validation (figure 17.1). Data mining is a multidisciplinary field, including information science, machine learning, statistics, database systems and visualizations. Data mining methods are usually classified as predictive and descriptive methods [3]. Predictive models use supervised learning for the classification of a new object into previously defined classes, while, for example, clustering methods (i.e. unsupervised learning) are used in a descriptive manner to identify the number of classes in the data [4]. The following exemplary, partly overlapping tasks are performed in data mining: classification, regression, clustering, summarization, dependency modeling and trend analysis [5]. Data mining methods could also be categorized based on specific applications. One of the most important applications of data mining methods is in healthcare [3, 6, 7]. Due to the tremendous amount of data in medicine, and the research questions related to prediction and ontology, data mining methods could be efficiently used to answer such questions. Some of the applications of data mining methods in medicine are (but are not limited to) cancer prognosis [8, 9], cardiology [10–13], diabetes [14, 15], pediatrics [16, 17], gastrology [18], nutrition [19], COVID-19 diagnosis and prognosis [20], and x-ray image classification [21].

Whenever a discrete or continuous quantity of interest (i.e. an outcome) is required to be characterized based on input variables (i.e. classification and regression), the problem is suited for data mining. In such tasks, the computer output can be considered as a 'second opinion' in the final decisions made by clinicians [22]. Computer-aided diagnosis [23], i.e. the identification of a disease, and prognosis [24], i.e. prediction of the course of the disease, are two main categories in which medical data mining is used. Although the performance of such systems has been improved in the literature, many of these systems have not received much attention by clinicians due to their complexity and incomplete evidence for their reliability.

In this chapter, we will provide insights on the experimental design, validity measures and required clinical aspects for data miners who would like to work on medical data. We also provide worked examples using MATLAB® version 9.4 (MATLAB 2018a, The MathWorks, Inc., Natick, MA, USA) and IBM SPSS® version 22 (Statistics for Windows, Version 22.0. Armonk, NY, USA; IBM Corp.). However, alternative packages could also be used, such as the open-source statistical environment R (R Core Team, R: A language and environment for statistical computing, R Foundation for Statistical Computing, Vienna, Austria; https://www.R-project.org/) to perform such tasks. The structure of the chapter is briefly shown in figure 17.2.

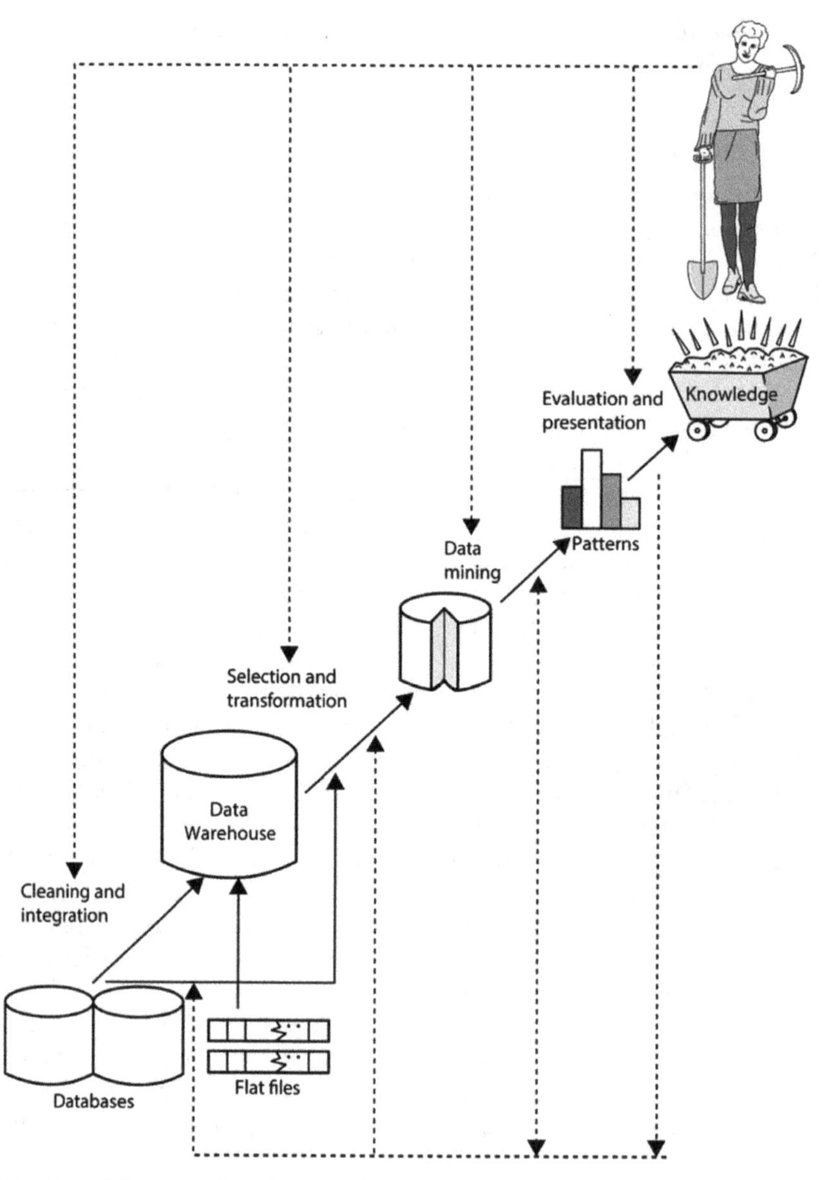


**Figure 17.1.** Data mining as one the main steps of knowledge discovery. Reproduced with permission from [5]. Copyright 2011 Morgan and Kaufmann.

## 17.2 Methods

### 17.2.1 Experimental design

Perhaps the best comprehensive and compact checklist required for prediction models implemented for medical research questions is TRIPOD [25], which stands for 'Transparent Reporting of a multivariable prediction model for Individual Prognosis Or Diagnosis'. The checklist is shown in figure 17.3.

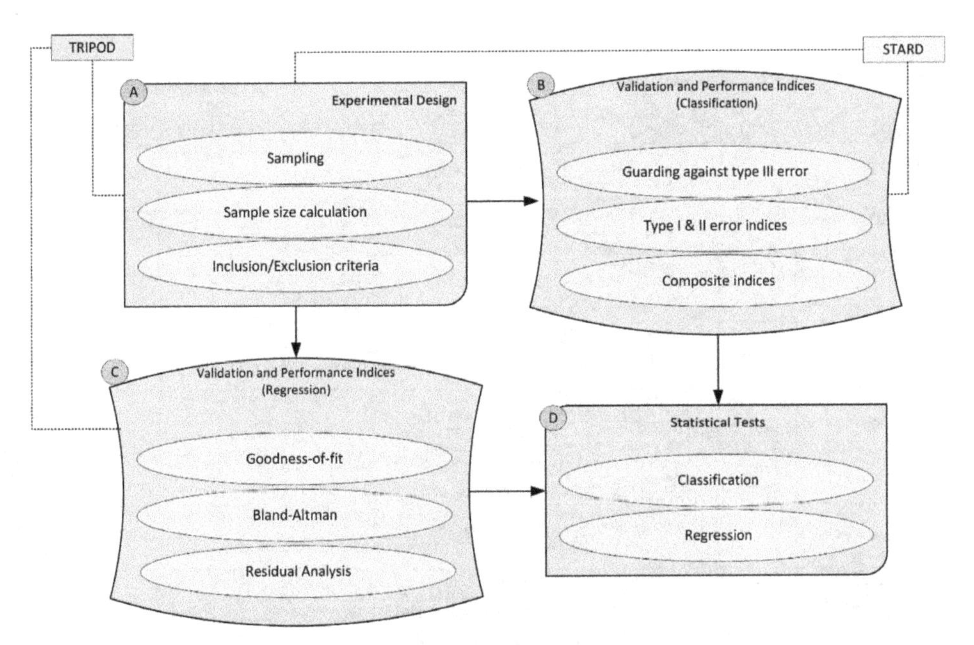


**Figure 17.2.** The structure of the chapter at a glance. (A) The experimental design concepts required for medical data mining algorithms conforming to the STARD and TRIPOD guidelines (https://www.equator-network.org/). The validation framework and performance indices required for (B) classification and (C) regression problems, and (D) statistical tests useful for investigating whether a proposed algorithm significantly outperforms the state-of-the-art.

In addition to the items discussed in the title, abstract, introduction, discussion and other sections, that are useful in any clinical scientific report, the sections on methods (items 4–12) and results (items 13–17) have valuable information not only to improve the quality of the work but also to improve the transparency of the prediction model study. Figure 17.2(A) contains some of the critical points of the TRIPOD checklist, i.e. 5.a (participants) 'sampling method', 5.b 'inclusion/exclusion criteria' and 8 'sample size calculation'. A medical data mining study not only must focus on the data mining algorithms but also provides enough clinical information to reduce the risk of bias.

Randomization not only reduces the subjective selection bias, it also provides a basis for statistical inference [26]. Non-representative sampling, i.e. convenience sampling, is not acceptable in many epidemiological studies [27]. When the dataset is recorded based on this sampling method, its data mining results cannot be generalized to the entire population. The authors must thus mention the sampling mechanism in the study report. Sampling methods not only have applications in the experimental design, they are also used in database sampling when big data are analysed. Proper sampling methods must be used in such cases [28]. It is also necessary to identify the properties of the population of interest in medical data mining studies. It includes the inclusion criteria, i.e. the eligibility criteria for participants, and the exclusion criteria, which are conditions for which the subjects who entered the study are excluded during the study [29].

TRIPOD Checklist: Prediction Model Development and Validation

| Section/Topic | Item | | Checklist Item | Page |
|---|---|---|---|---|
| **Title and abstract** | | | | |
| Title | 1 | D;V | Identify the study as developing and/or validating a multivariable prediction model, the target population, and the outcome to be predicted. | |
| Abstract | 2 | D;V | Provide a summary of objectives, study design, setting, participants, sample size, predictors, outcome, statistical analysis, results, and conclusions. | |
| **Introduction** | | | | |
| Background and objectives | 3a | D;V | Explain the medical context (including whether diagnostic or prognostic) and rationale for developing or validating the multivariable prediction model, including references to existing models. | |
| | 3b | D;V | Specify the objectives, including whether the study describes the development or validation of the model or both. | |
| **Methods** | | | | |
| Source of data | 4a | D;V | Describe the study design or source of data (e.g., randomized trial, cohort, or registry data), separately for the development and validation data sets, if applicable. | |
| | 4b | D;V | Specify the key study dates, including start of accrual; end of accrual; and, if applicable, end of follow-up. | |
| Participants | 5a | D;V | Specify key elements of the study setting (e.g., primary care, secondary care, general population) including number and location of centres. | |
| | 5b | D;V | Describe eligibility criteria for participants. | |
| | 5c | D;V | Give details of treatments received, if relevant. | |
| Outcome | 6a | D;V | Clearly define the outcome that is predicted by the prediction model, including how and when assessed. | |
| | 6b | D;V | Report any actions to blind assessment of the outcome to be predicted. | |
| Predictors | 7a | D;V | Clearly define all predictors used in developing or validating the multivariable prediction model, including how and when they were measured. | |
| | 7b | D;V | Report any actions to blind assessment of predictors for the outcome and other predictors. | |
| Sample size | 8 | D;V | Explain how the study size was arrived at. | |
| Missing data | 9 | D;V | Describe how missing data were handled (e.g., complete-case analysis, single imputation, multiple imputation) with details of any imputation method. | |
| Statistical analysis methods | 10a | D | Describe how predictors were handled in the analyses. | |
| | 10b | D | Specify type of model, all model-building procedures (including any predictor selection), and method for internal validation. | |
| | 10c | V | For validation, describe how the predictions were calculated. | |
| | 10d | D;V | Specify all measures used to assess model performance and, if relevant, to compare multiple models. | |
| | 10e | V | Describe any model updating (e.g., recalibration) arising from the validation, if done. | |
| Risk groups | 11 | D;V | Provide details on how risk groups were created, if done. | |
| Development vs. validation | 12 | V | For validation, identify any differences from the development data in setting, eligibility criteria, outcome, and predictors. | |
| **Results** | | | | |
| Participants | 13a | D;V | Describe the flow of participants through the study, including the number of participants with and without the outcome and, if applicable, a summary of the follow-up time. A diagram may be helpful. | |
| | 13b | D;V | Describe the characteristics of the participants (basic demographics, clinical features, available predictors), including the number of participants with missing data for predictors and outcome. | |
| | 13c | V | For validation, show a comparison with the development data of the distribution of important variables (demographics, predictors and outcome). | |
| Model development | 14a | D | Specify the number of participants and outcome events in each analysis. | |
| | 14b | D | If done, report the unadjusted association between each candidate predictor and outcome. | |
| Model specification | 15a | D | Present the full prediction model to allow predictions for individuals (i.e., all regression coefficients, and model intercept or baseline survival at a given time point). | |
| | 15b | D | Explain how to the use the prediction model. | |
| Model performance | 16 | D;V | Report performance measures (with CIs) for the prediction model. | |
| Model-updating | 17 | V | If done, report the results from any model updating (i.e., model specification, model performance). | |
| **Discussion** | | | | |
| Limitations | 18 | D;V | Discuss any limitations of the study (such as nonrepresentative sample, few events per predictor, missing data). | |
| Interpretation | 19a | V | For validation, discuss the results with reference to performance in the development data, and any other validation data. | |
| | 19b | D;V | Give an overall interpretation of the results, considering objectives, limitations, results from similar studies, and other relevant evidence. | |
| Implications | 20 | D;V | Discuss the potential clinical use of the model and implications for future research. | |
| **Other information** | | | | |
| Supplementary information | 21 | D;V | Provide information about the availability of supplementary resources, such as study protocol, Web calculator, and data sets. | |
| Funding | 22 | D;V | Give the source of funding and the role of the funders for the present study. | |

*Items relevant only to the development of a prediction model are denoted by D, items relating solely to a validation of a prediction model are denoted by V, and items relating to both are denoted D;V. We recommend using the TRIPOD Checklist in conjunction with the TRIPOD Explanation and Elaboration document.

**Figure 17.3.** The TRIPOD checklist for medical prediction algorithms. Reproduced with permission from [25]. Copyright 2015 BMC Medicine.

Missing value analysis is also essential. Missing values are unavoidable in clinical studies. Not all of the variables for the entire subjects are usually available. It must be mentioned whether only subjects with complete information were analysed or whether missing data imputation was used. Such missing data could be missing not at random (MNAR), missing at random (MAR) and missing completely at random (MCAR), and resulting biases can be addressed by using sensitivity analysis [30]. In the first case, the missing values depend on the hypothetical values, and removing observations with missing values can result in bias in the model. Many methods have been proposed in the literature for missing data analysis in pattern classification (figure 17.4). It is necessary to know their underlying assumptions to reduce the risk of biases [31–35]. Multiple imputation methods (figure 17.4) were proposed in the literature, and there are guidelines (e.g. [35]) whether to use multiple imputations or not based on the MNAR and MCAR conditions. In clinical studies, the imputation method (if used; figure 17.3 item 9) and the sensitivity analysis (STROBE statement, https://www.equator-network.org/reporting-guidelines/strobe/) are two critical issues to address.

Sample size calculation is a critical issue that is usually addressed in clinical studies. However, in (biomedical) engineering projects, it is not common to describe why, for example, 15 or 30 subjects or cases were analysed [36]. In clinical studies, however, there are rigorous sample size estimation formulas based on the statistical

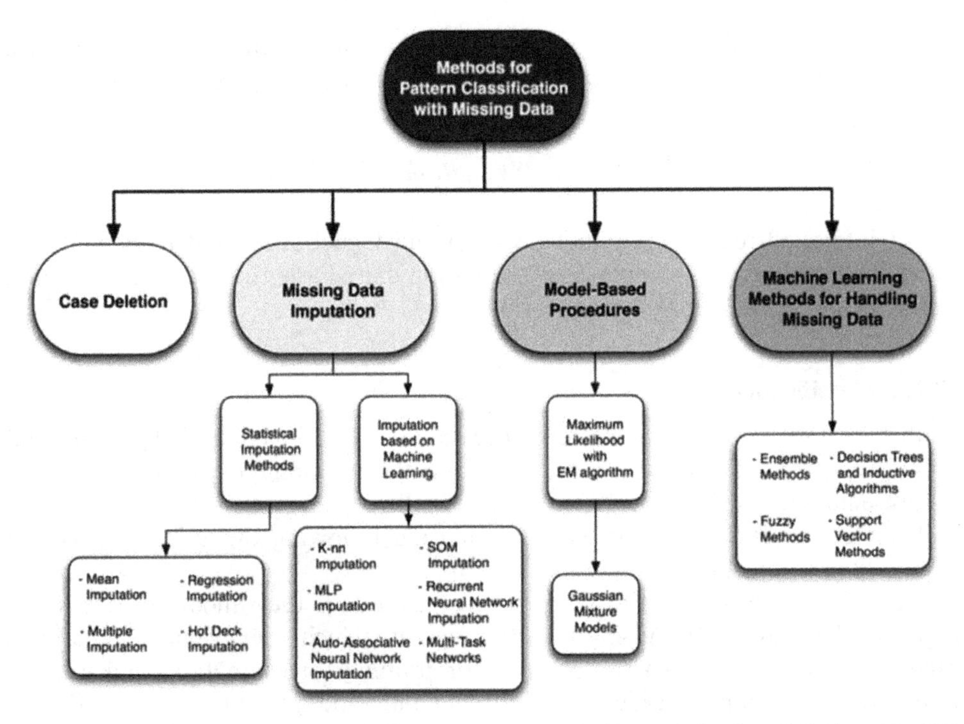


**Figure 17.4.** Pattern classification with missing data. Reproduced with permission from [31]. Copyright 2010 Springer.

hypothesis and the subsequent statistical tests (see for example [37], or implementations such as G*Power, https://www.psychologie.hhu.de/arbeitsgruppen/allgemeine-psychologie-und-arbeitspsychologie/gpower.html [38, 39]). It is also necessary to report the sample size calculation details based on the TRIPOD guidelines (figure 17.3 item 8 and figure 17.2(A)). Although sample size calculation has been extensively studied in medical research [37, 40–44], few studies have focused on its procedures for data mining [45–47]. We provide example 17.1 from the data mining study performed by Marateb *et al* [11] in which the sample size estimation formula based on the desired type I and II error levels proposed by Buderer [45] and Hajian-Tilaki [46] was used.

**Example 17.1.** In a case-control setting, a data mining algorithm is used to predict if a subject has dyslipidemia (i.e. lipoprotein metabolism disorder) or not using anthropometric indicators, the family history of disease and gene mutations [11]. The total sample size ($N$) is calculated using the following equation based on the expected sensitivity ($\mathrm{Se_e}$), specificity ($\mathrm{Sp_p}$) and the prevalence of dyslipidemia in the population (Prev), that were set to 70%, 77% and 42%, respectively, based on the literature [48–50]:

$$N = \max\left(\frac{z_{\alpha/2}^2 \times \mathrm{Se_e} \times (1 - \mathrm{Se_e})}{d^2 \times \mathrm{prev}}, \frac{z_{\alpha/2}^2 \times \mathrm{Sp_e} \times (1 - \mathrm{Sp_e})}{d^2 \times (1 - \mathrm{prev})}\right), \tag{17.1}$$

where $\alpha$ is the significance level and $d$ is the maximum marginal error, which were both set to 0.05 [37]. The number of subjects in the case ($n_{\mathrm{case}}$) and control ($n_{\mathrm{control}}$) groups were estimated as

$$n_{\mathrm{control}} = N \times (1 - \mathrm{prev}); \quad n_{\mathrm{case}} = N - n_{\mathrm{control}}. \tag{17.2}$$

For interested readers, Hajian-Tilaki [46] and Obuchowski and Hillis [47] also provided formulas for sample size estimation based on the receiver operating characteristic (ROC) and other scenarios.

### 17.2.2 Classification

In this section, a variety of performance indices used in medical diagnosis and prognosis systems is discussed. We also provide the additional information required for presentation based on the clinical standards. Binary and multi-class classification performance indices and their clinical extension following the STARD [51] and TRIPOD [25] guidelines are discussed.

One of the critical issues of data mining methods is the validation framework (figures 17.2 and 17.3). A simple validation, in which the available dataset is randomly divided into two subsets, one for training (or estimation) and one for testing (or validation), is called 'hold-out'. The method not only makes inefficient use of the data, but it also gives a pessimistically biased error estimate [52, 53]. Additionally, it does not guard against testing hypotheses suggested by the data (type III errors [54]) as it is possible to permute the data until we have acceptable accuracy on the training and test sets in a 'hold-out' setting. Thus, other validation frameworks, such as repeated hold-out, cross-validation, .632+ bootstrap and leave-one-out validation, could be considered based on the properties of the data, such as the number of samples [55]. These issues are also discussed in the TRIPOD guidelines from a clinical perspective [25].

In a cross-validation setting, not only is the cross-validated confusion matrix reported for all the samples used as test sets, but the standard deviation (SD) values of the performance indices must also be reported in the analysed folds in addition to the MEAN values, showing the consistency of the performance of the method. Also, the Confidence Interval (CI) 95% of the performance indices must be reported based on the STARD and TRIPOD guidelines (figure 17.3 item 16). This issue is addressed in detail in section 17.2.2.1.

*17.2.2.1 Binary classification*

After the generation of a confusion matrix (i.e. the contingency table or error matrix [56]), the signal detection theory parameters [57] are assessed (table 17.1). Let $E$ be the event that the test result is positive and $D$ be the event that the tested person has the disease as confirmed by the gold standard. It is thus possible to provide the relationship between signal detection theory parameters and related probabilities [8, 58]:

$$\begin{cases} \mathrm{TP}/N = P(ED) = P(E\backslash D) \times P(D) \\ \mathrm{TN}/N = P(E^cD^c) = P(E^c\backslash D^c) \times P(D^c) \\ \mathrm{FP}/N = P(ED^c) = P(E\backslash D^c) \times P(D^c) \\ \mathrm{FN}/N = P(E^cD) = P(E^c\backslash D) \times P(D) \end{cases} . \tag{17.3}$$

Knowing that $P(D)$ is the prevalence of the disease (Prev), it is possible to provide the following statements based on the conditional probabilities:

$$\begin{cases} \mathrm{Se} = P(E\backslash D) \\ \mathrm{Sp} = P(E^c\backslash D^c) \\ \alpha = 1 - \mathrm{Sp} = P(E\backslash D^c) \\ \beta = 1 - \mathrm{Se} = P(E^c\backslash D) \end{cases} \tag{17.4}$$

**Table 17.1.** The confusion matrix of a two-class classification system.

| | | Condition (as determined by the 'gold standard') | |
|---|---|---|---|
| | Total population | Condition positive event ($D$) | Condition negative event ($D^c$) |
| **Test outcome** | Test outcome positive event ($E$) | True positive (TP) | False positive (type I error) (FP) |
| | Test outcome negative event ($E^c$) | False negative (type II error) (FN) | True negative (TN) |

For an event $A$, its complement is shown as $A^c$.

where $\alpha$ (= false alarm (FA)) and $\beta$, are type I and II errors, and Se and Sp are sensitivity and specificity, respectively. It must be mentioned that not only $P(E\backslash D)$, i.e. Se, is essential but also the posterior probability $P(D\backslash E)$, i.e. precision (= positive predictive value (PPV)) is critical in a medical diagnosis or prognosis system. It is, in fact, the disease probability given that the patient test result is positive. It is possible to calculate the posterior probability using the Bayes' formula [58] as follows:

$$P(D\backslash E) = \frac{\text{Se} \times \text{prev}}{\text{Se} \times \text{prev} + (1 - \text{Sp}) \times (1 - \text{prev})}. \tag{17.5}$$

Similarly, it is possible to estimate the posterior probability $P(D^c\backslash E^c)$, i.e. negative predictive value (NPV) as the following:

$$P(D^c\backslash E^c) = \frac{\text{Sp} \times (1 - \text{prev})}{\text{Sp} \times (1 - \text{prev}) + (1 - \text{Se}) \times \text{prev}}. \tag{17.6}$$

This is the probability that a subject is healthy, given that the test result is negative. It is also possible to define traditional signal detection theory parameters based on the confusion matrix (table 17.1) as follows:

$$\text{Se} = \text{Recall} = \text{TP}/(\text{TP} + \text{FN}) \tag{17.7}$$

$$\text{Sp} = \text{TN}/(\text{TN} + \text{FP}) \tag{17.8}$$

$$\text{Precision} = \text{PPV} = \text{TP}/(\text{TP} + \text{FP}) \tag{17.9}$$

$$\text{Accuracy} = (\text{TP} + \text{TN})/(\text{TP} + \text{TN} + \text{FP} + \text{FN}). \tag{17.10}$$

Moreover, the following performance indices can also be calculated:

$$\text{LR}^{+} = \text{Se}/(1 - \text{Sp}) \tag{17.11}$$

$$\text{LR}^{-} = (1 - \text{Se})/\text{Sp} \tag{17.12}$$

$$\text{DOR} = \text{LR}^{+}/\text{LR}^{-} = \text{TP} \times \text{TN}/(\text{FP} \times \text{FN}) \tag{17.13}$$

$$\text{AUC} = (\text{Se} + \text{Sp})/2 \tag{17.14}$$

$$\begin{aligned} \text{F1S} &= 2 \times \text{precision} \times \text{recall}/(\text{precision} + \text{recall}) \\ &= 2 \times \text{TP}/(2 \times \text{TP} + \text{FN} + \text{FP}) \end{aligned} \tag{17.15}$$

$$\begin{aligned} \text{MCC} = {} & (\text{TP} \times \text{TN} - \text{FP} \times \text{FN}) \\ & /\sqrt{((\text{TP} + \text{FP}) \times (\text{TP} + \text{FN}) \times (\text{TN} + \text{FP}) \times (\text{TN} + \text{FN}))} \end{aligned} \tag{17.16}$$

$$\text{DP} = \left(\sqrt{\frac{3}{\pi}}\right) \times \log(\text{DOR}) \tag{17.17}$$

$$K(C) = 2 \times (TP \times TN - FP \times FN)/((TP + FP) \times (FP + TN) + (TP + FN) \times (FN + TN)), \quad (17.18)$$

where DOR is the diagnostic odds ratio, AUC is the area under the ROC curve (i.e. the balanced diagnostic accuracy) [59], F1S is the F1-score as the harmonic mean of the precision and recall, MCC is the Matthews correlation coefficient [60, 61] (i.e. phi coefficient [62]), DP is the discriminant power [63, 64] and K(C) is the Cohen's kappa coefficient as the agreement rate between the predicted class labels and the gold standard [65, 66].

Among the proposed indices, Sp and Se are related to type I and II errors, while accuracy, F1S, AUC and MCC are composite indices (figure 17.2 (B)). When the dataset is imbalanced (i.e. Prev is not close to 0.5), the accuracy is biased toward the majority class and is not a proper performance index. Moreover, when Prev is higher than 0.5, F1S is also biased and must thus be avoided. It has also been shown in the literature that MCC is a more informative and truthful score compared to accuracy and F1S in binary classification problems [67]. Among the composite indices, the AUC and Cohen's kappa could also provide the related *p*-value measures supporting the statistical interpretation. Moreover, when using the following composite indices, it is possible to provide some sentences indicating the performance of the prediction algorithm (e.g. in the abstract of the paper): AUC [59], Cohen's kappa [65], MCC [68, 69] and DP [63, 64]. Such interpretations were provided by Marateb *et al* [11], which are also available at https://ars.els-cdn.com/content/image/1-s2.0-S2001037017300880-mmc2.docx and are provided in example 17.2.

As indicated in equations (17.5) and (17.6), the posterior probabilities PPV and NPV are dependent not only on the Se and Sp but also on Prev [70]. As an example, if the prediction system has Se and Sp indices of 80% and 95%, respectively, the PPV is 94% and 64%, with Prev of 50% and 10%. Thus, the prevalence parameter in the test dataset must resemble the realistic condition of the population. When the prevalence of the disease is 10%, and the prediction system is tested on a balanced dataset (Prev = 50%), its precision is reduced in the population down to 64%. Thus, the single indices list must include the parameters Se, Sp and PPV. The proposed composite indices list contains the parameters DOR, AUC, MCC, DP and Cohen's kappa. Moreover, a diagnosis/prognosis system could, for example, be considered clinically reliable if the minimum Se and Sp are 80% and 95%, respectively [71], AND the minimum DOR is 100 [72], AND the maximum false discovery rate (FDR = 1 − PPV) is 5% [11, 73]. As a complementary condition, a minimum NPV of 95% could be mentioned.

Following the STARD and TRIPOD guidelines, it is necessary to provide the CI 95% of the performance indices indicating how the prediction from the analysed samples is generalized to the entire population. These concepts have also been used in the regulation of medical devices. For example, in automated external defibrillator (AED) devices, there is a diagnosis algorithm to identify whether an ECG rhythm is shockable or not. When the rhythm is shockable, the AED device

**Table 17.2.** The performance indices of the cancer diagnosis system in example 17.2.

| Index | Se (%) | Sp (%) | PPV (%) | NPV (%) | AUC | MCC | $LR^+$ | $LR^-$ | DOR | DP | Kappa |
|---|---|---|---|---|---|---|---|---|---|---|---|
| **Value** | 78 | 95 | 91 | 88 | 0.87 | 0.76 | 17 | 0.23 | 74 | 1.83 | 0.76 |
| **95% CI** | 74–83 | 94–97 | 88–94 | 86–91 | 0.84–0.89 | 0.73–0.79 | 12–25 | 0.19–0.28 | 46–120 | 1.62–2.03 | 0.71–0.80 |

automatically performs the cardiac defibrillation [74]. The AED device must meet the IEC-60601–2–4 standard in which the 90% one-sided lower confidence limit of the Se and Sp parameters must be higher than some thresholds in the shockable or non-shockable categories (table 17.2, [75]).

Here, we provide the CI formula for Se (equation (17.19)) and Sp (equation (17.20)). Similarly, it is possible to calculate the CI of the other indices based on the CI of proportions [76]. We provided a MATLAB function ('calc_performance_indices.m') to calculate the performance indices as well as their CI 95% for the interested reader, which is available at https://doi.org/10.6084/m9.figshare.12763589.v2.

$$SE_{Se} = \sqrt{\frac{Se \times (1 - Se)}{TP + FN}};\ Se \pm z_{1-\alpha/2} \times SE_{Se} \tag{17.19}$$

$$SE_{Sp} = \sqrt{\frac{Sp \times (1 - Sp)}{TN + FP}};\ Sp \pm z_{1-\alpha/2} \times SE_{Sp}. \tag{17.20}$$

**Example 17.2.** In a cross-validated confusion matrix, after comparing the result of the cancer diagnosis system with the gold standard, we obtain TP = 250, TN = 520, FP = 25 and FN = 70. Provide the performance assessment indices and related 95% CI. Is the designed prediction system clinically reliable? Also, provide the statements related to the performance of the proposed method.

The results are listed in table 17.2 when using the ('calc_performance_indices.m') MATLAB function.

The analysed diagnosis system is not clinically reliable as it violates many conditions, including the statistical power (Se) less than 80%. The statements based on AUC, MCC, DP and kappa are as follows:

'The proposed prediction system has *very good* balanced diagnosis accuracy, with a *high* agreement between predicted and observed class labels, *limited* discriminant power and *excellent* class labeling agreement rate.'

**Table 17.3.** The multi-class classification performance indices. Reproduced with permission from [77].

| Measure | Formula | Evaluation focus |
|---|---|---|
| **Average accuracy** | $\frac{\sum_{i=1}^{l}\frac{TP_i+TN_i}{TP_i+FN_i+FP_i+TN_i}}{l}$ | The average per-class effectiveness of a classifier. |
| **Error rate** | $\frac{\sum_{i=1}^{l}\frac{FP_i+FN_i}{TP_i+FN_i+FP_i+TN_i}}{l}$ | The average per-class classification error. |
| **Precision$_\mu$** | $\frac{\sum_{i=1}^{l}TP_i}{\sum_{i=1}^{l}(TP_i+FP_i)}$ | Agreement of the data class labels with those of a classifier if calculated from sums of per-text decisions. |
| **Recall$_\mu$** | $\frac{\sum_{i=1}^{l}TP_i}{\sum_{i=1}^{l}(TP_i+FN_i)}$ | Effectiveness of a classifier to identify class labels if calculated from sums of per-text decisions. |
| **F-score$_\mu$** | $\frac{(\beta^2+1)\text{Precision}_\mu\text{Recall}_\mu}{\beta^2\text{Precision}_\mu+\text{Recall}_\mu}$ | Relations between the data's positive labels and those given by a classifier based on sums of per-text decisions. |
| **Precision$_M$** | $\frac{\sum_{i=1}^{l}\frac{TP_i}{TP_i+FP_i}}{l}$ | An average per-class agreement of the data class labels with those of a classifier. |
| **Recall$_M$** | $\frac{\sum_{i=1}^{l}\frac{TP_i}{TP_i+FN_i}}{l}$ | An average per-class effectiveness of a classifier to identify class labels. |
| **F-score$_M$** | $\frac{(\beta^2+1)\text{Precision}_M\text{Recall}_M}{\beta^2\text{Precision}_M+\text{Recall}_M}$ | Relations between the data's positive labels and those given by a classifier based on a per-class average. |

The number of classes is $l$ which is equal to $K$ in the sigma.

### *17.2.2.2 Multi-class classification*

In a multi-class classification problem with $K$ classes, a $K \times K$ confusion matrix $C = [c_{ij}]$ is obtained, where $c_{ij}$ is the number of cases assigned to class ($i$) by the algorithm and belongs to the gold standard class ($j$). In addition to the *overall accuracy*, which is the summation of the diagonal elements divided by the number of samples, the other performance indices required [77] are listed in table 17.3 ($l = K$). For each class, $TP_i = c_{ii}$, $FP_i = \sum_{i\neq j} c_{ij}$, $FN_i = \sum_{t\neq i} c_{ti}$ and $TN_i = \sum_{(t,j)\neq i} c_{tj}$ are calculated. The notations $\mu$ and M denote $\mu$- and macro-averaging, and the parameter $\beta$ is 1 in the F-score formulas.

It must be noted that the $\mu$-average is not sensitive to the predictive performance for individual classes, and can be misleading when the class distribution is imbalanced. Also, since $TN_i$ ($I = 1, \ldots, K$) is much higher than other multi-class signal detection parameters, the 'average accuracy' and 'error rate' are usually avoided. Thus, $\text{Precision}_M$, $\text{Recall}_M$ and $\text{F-score}_M$ are reported in addition to the overall accuracy parameter.

As an example, for the confusion matrix $C = [140, 13, 31; 23, 170, 12; 10, 31, 240]$, the multi-class performance indices were calculated using the MATLAB function 'calc_multi_class_indices.m' which is available at https://doi.org/10.6084/m9.

figshare.12763589.v2. The overall accuracy, macro-averaged precision, recall, and F-score are 82%, 81%, 82% and 82%, respectively. Moreover, Cohen's kappa could also be calculated in multi-class problems. Using the MATLAB function 'Kappa.m' from https://doi.org/10.6084/m9.figshare.12763589.v2, kappa is 0.73 [CI 95%: 0.68–0.77], showing an excellent class labeling agreement rate.

Gorodkin proposed a generalization of the MCC for multi-class problems [78, 79]. The concept of AUC was also generalized to the multi-class problems in the literature [80].

### 17.2.3 Regression analysis

In regression analysis, the relationship between the outcome (i.e. dependent variable) and a set of predictors (i.e. independent variables) is investigated. Although the outcome could be of any measurement scale, we only focus on cases where the outcome is a continuous variable, i.e. categorical variables are not considered. In this way, the conditional distribution of a continuous variable is estimated based on some inputs. In engineering, this topic is covered in system identification, while regression analysis is used in statistical modeling. In addition to the goodness-of-fit of a model, regression diagnostics is required in which the residual signal is analysed. In clinical applications, visual examination of the residual graphs is critical. Thus, in any regression analysis, for example, the three following outputs should be provided (figure 17.2(C)): the goodness-of-fit (e.g. $R$-squared [81]), the residual graph (e.g. the Bland–Altman plot [82]) and the regression diagnostics (e.g. normality, homoscedasticity and multicollinearity [83]). These outputs are complementary to each other. In other words, if one of these outputs is not provided, it is difficult to judge whether the regression model is suitable or not. Additionally, the statistical tests in section (17.2.4) (or figure 17.2(D)), may be used to investigate whether the regression model is suitable or not.

$R$-squared, i.e. the coefficient of determination, is one of the many generic goodness-of-fit measures used in the literature. It identifies the proportion of the sample variance in the dependent variable that is explained by the independent variables:

$$R^2 = 1 - \frac{\sum_i (y_i - \tilde{y}_i)^2}{\sum_i (y_i - \bar{y})^2}, \tag{17.21}$$

where $y_i$ is the samples of the observed data $y$ (gold standard), $\tilde{y}_i$ is the samples of the predicted outcome and $\bar{y}$ is the mean of the observed data.

Since the $R$-squared is biased upward for population variance proportion estimation, i.e. having enough variables it tends to 1, it could be corrected, and the following measure is provided:

$$\text{adj} \cdot R^2 = 1 - (1 - R^2) \times \frac{N - 1}{N - k - 1}, \tag{17.22}$$

where $k$ is the number of predictor variables (and $N$ is the sample size) in the analysis.

In fact, for a small sample size, when the number of predictors increases, the adjusted $R$-squared decreases. It must be mentioned that in model selection when we have models with different tunable parameters, indices such as the Akaike information criterion (AIC) or the Bayesian information criterion (BIC) are used instead of $R$-squared. Also, in some applications, other goodness-of-fit indices are required. For example, when the risk charts of survival models are generated, the Nam-D'Agostino $\chi^2$ index is usually reported as the goodness-of-fit in survival settings [12, 84]. Moreover, for any goodness-of-fit index, the CI 95% must be provided in clinical applications following the TRIPOD guidelines (figure 17.3 item 16). The following example is provided to show whether $R$-squared could indicate underestimation or overestimation of the model.

**Example 17.3** Suppose that we have a sinusoid signal defined as $y = \sin(2 \times \text{pi} \times 50 \times [0{:}.001{:}3 \times 20e^{-3}])$ and the reconstructed signal is $f_1 = y + 2 + 0.01 \times \text{randn}(\text{size}(y))$. The $R$-squared value is −7.14 showing that the model fit is not satisfactory. Note that negative $R$-squared occurs when we do not have a constant term in the regression or in nonlinear regressions or a very bad fit [85]. Now, if we have $f_2 = y + 0.2 + 0.01 \times \text{randn}(\text{size}(y))$. The $R$-squared value is 0.92, which indicates much better performance. Yet, the $R$-squared values do not indicate whether we have 'significant' overestimation or underestimation. That is why statistical tests, such as paired $t$-tests, are used in a complementary manner (section 17.2.4). Note that the correlation coefficients are not sensitive to the offset values and would report an excellent correction in $f_1$ prediction.

The Bland–Altman plot, i.e. the Tukey mean difference plot, is frequently used in medical statistics to identify the agreement between two different quantitative variables [86, 87]. In regression analysis, it is used to show the agreement between the outcome and its prediction as a scatter plot between their average ($x$-axis) versus their difference ($y$-axis). The Bland–Altman plot [87] has been so frequently used in the literature that it became number 29 in a list of the top 100 most-cited papers of all time with over 23 000 citations by 2014 [88] (cited 46508 times by July 2020). It evaluates a difference between the mean differences and provides an agreement interval [82]. An example is provided in figure 17.5.

Figure 17.5 shows that the mean difference is not zero (−469). If we want to identify whether the difference is significant or not, we need to use the one-sample $t$-test with zero references on the residual signal. The difference between the two measurements is not homogeneous (the bias for lower $x$-axis values is higher than that of higher $x$-axis values), and if regression is performed on the Bland–Altman plot, a regression line, not in parallel with the $x$-axis, could be identified. Moreover, the difference surpassed the higher and lower limits. Thus, the agreement between the measured and predicted variables is not acceptable.

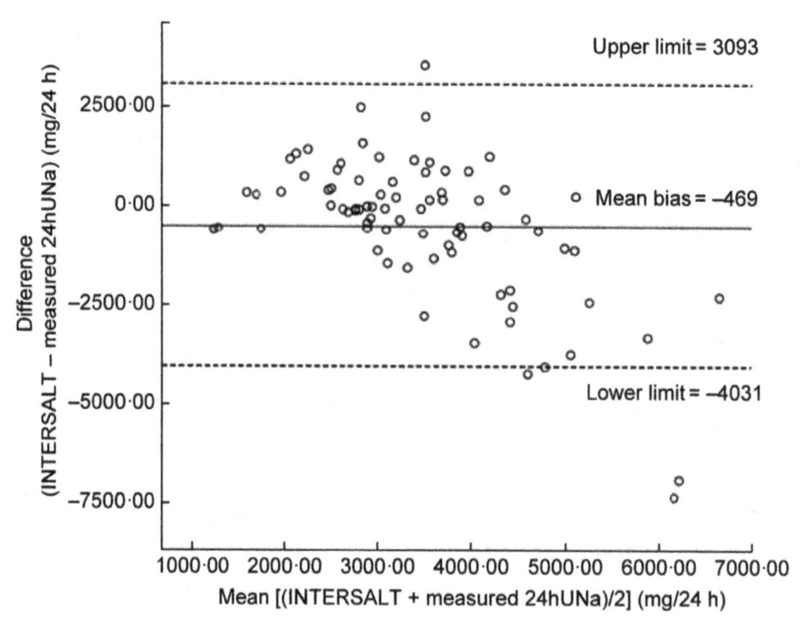


**Figure 17.5.** The Bland–Altman plot of the mean bias between predicted 24 h urinary sodium excretion (24hUNa) using the INTERSALT formula and measured 24hUNa from evening spot urine samples. Reproduced with permission from [19]. Copyright 2020 Cambridge University Press.

In addition to the Bland–Altman plot that could be regarded as one of the visualization methods for the residual signal analysis, regression diagnostics have been well studied in statistics and system identification. Multicollinearity, e.g. when some of the predictors are highly correlated, must be identified as it affects the regression procedure. When it happens, the data redundancy could increase the risk of overfitting and some ill-conditioned matrices could be seen during regression analysis. Some regression methods (such as multiple linear regressions) provide a variance inflation factor (VIF) as the amount of multicollinearity [89]. In such cases, highly correlated predictors must be combined or selected to reduce multicollinearity.

Moreover, the homoscedasticity (i.e. finite equal variances) of the residual signal must be assessed (e.g. using the Breusch–Pagan test [90]), and the absence of serial correlation (i.e. autocorrelation at lag 1) in the residual signal should be analysed (e.g. using Durbin–Watson tests [91]). It should be mentioned that outliers usually affect the regression method unless the method is robust [92]. Thus, outlier detection methods should be used to identify such samples [93, 94]. In system identification, however, we must verify if the residuals are uncorrelated with the inputs (i.e. no unmodeled dynamics are left between the input and the residuals) and the residual signals are uncorrelated (whiteness test; the autocorrelation of the residual signal plotted together with the 99% uncertainty interval around zero) [95]. Although there are overlaps between the residual signal analysis in statistics and system identification, the entire residual diagnostics methods must be performed to make sure that the prediction is reliable.

### 17.2.4 Statistical tests

Inferential statistics is essential in many areas of biomedical sciences, e.g. regarding the proper interpretation of the results in a scientific paper or rigorous comparison of the diagnosis/prognosis method with the state-of-the-art. Specific statistical tests provide such inferential procedures. Although there is some biostatistical material for biomedical engineers in the literature [96, 97], there are some specific challenges that we would like to cover in this section (figure 17.2 (D)). They are related to the following critical question [98, 99]:

How can we assess whether a classification/regression method significantly outperformed the other method?

In addition to the randomization test in which the performance of a classifier is assessed when the gold standard labels are randomly permuted and the validation framework, which are two prerequisites of classification validation, there are critical concerns when an author claims whether a method outperformed the state-of-the-art. It is not possible to claim such statements when we only consider the MEAN (and SD) of the performance indices of two analysed methods on the same dataset, in particular if the boxplots of the performance indices of those two methods (over the analysed test folds; assuming that the folds are the same in both methods) overlap. How can we prove that the superiority is not by chance? How can we assume that the superiority we found on the samples could be generalized to the population?

In the first scenario, we have a binary classification problem. We want to identify which method *significantly* outperformed the other method on a group of $N$ cross-validated samples. Here, we only have two classifiers.

The analysed dataset (https://archive.ics.uci.edu/ml/datasets/heart+Disease) [100] in the SPSS format and the classifier labels are available at https://doi.org/10.6084/m9.figshare.12763589.v2. Subjects with complete features (age, gender, chest pain type, serum cholesterol, diabetes, number of major vessels (0–3) colored by fluoroscopy and thallium-201 stress scintigraphy) were used as predictors, and binary diagnosis of heart disease (angiographic disease status) was used as the gold standard. The classification was performed using some classifiers, and their labels for 278 cross-validated subjects were recorded. We would like to check whether the classifier 1 *significantly* outperformed classifier 2.

McNemar's test (i.e. the Gillick test) [8, 11, 53, 99] could be used in this scenario. Briefly, the $z$-statistic is calculated as follows:

$$z = \frac{|x - y| - 1}{\sqrt{x + y}}, \tag{17.23}$$

where $x$ ($y$) is the number of cases that were misclassified by the first (second) classifier but not the second (first) classifier, respectively. The null hypothesis of having the same amount of error in both classifiers is rejected if $| z | > 1.96$ ($p < 0.05$) [53, 101]. When the null hypothesis is rejected, the classifier with higher MCC value can be considered to significantly outperform the other classifier. Using the MATLAB function 'run_mcnemar_test.m' (available at https://doi.org/10.6084/

m9.figshare.12763589.v2), it is seen that none of the classifiers significantly outperformed the other ($z = 0.79$, MCC1 = 0.51, MCC2 = 0.56). Now, if we wanted to compared the performance of classifiers 1 and 3 (https://archive.ics.uci.edu/ml/datasets/heart+Disease), in the first place, the results were ($z = 2.56$, MCC1 = 0.51 and MCC3 = 0.64): 'Classifier 3 significantly outperformed classifier 1!'. In the analysed dataset, we have three classifiers. We need to perform three pairwise comparisons and, finally, a classifier that significantly outperforms the others, or with the minimum number of losses, would rank the first. The problem is that when we repeat a statistical test, it increases the type I error [102]. The probability of not obtaining a significant result is 1–0.05 = 0.95 when we run a test once. If we run the test three times, the probability decreases to $1-(1-0.05)^3 = 0.14$. Thus, the chances of a type I error are about 1 in 7 instead of 1 in 20. One of the compensation methods is using the adjusted $p$-values or test-wise significance level (e.g. the Bonferroni correction [103]), where 0.05 is divided by the number of comparisons and is used as a significance criterion. A complementary strategy is using Cochran's Q-test to check whether any significant difference could be seen in comparing three (or more) classifiers, and if the null hypothesis is rejected, McNemar's pairwise comparison with Bonferroni correction is used.

Similar questions must be answered in regression analysis, such as whether a regression method significantly overestimated or underestimated, or whether a proposed regression method significantly outperformed the state-of-the-art. To answer the first question, we calculate the residual signal. If it is normally distributed, a paired $t$-test (or one-sample $t$-test of the residual signal with zero references) could be used. When the null hypothesis is rejected, the sign of the mean difference identifies whether we have significant overestimation or underestimation. When the residual signal is not normally distributed, the Wilcoxon signed-rank test is used.

As an example, the simulated dataset in example 17.3 (the '$y$' and '$f_2$' variables) is analysed (available at https://doi.org/10.6084/m9.figshare.12763589.v2). The residual signal was first calculated and its normality was assessed using the Shapiro–Wilk test (analyse → descriptive statistics → explore). Since the residual signal was normally distributed (the Shapiro–Wilk test; $p = 0.375$), the paired $t$-test was used (analyse → compare means → paired-samples $t$-test), it shows significant differences based on the mean difference; it is shown that the estimated method significantly overestimated ($p < 0.001$) which was not shown using goodness-of-fit measures. When we have multiple comparisons (i.e. the second question), we first use repeated-measures ANOVA (Friedman's nonparametric test) [104]. If the result is significant, then we use a paired $t$-test (Wilcoxon signed-rank nonparametric test) with Bonferroni correction for multiple comparisons. Note that when using repeated-measures ANOVA, its assumptions must be checked [105].

## 17.3 Conclusion and future scope

In this chapter, we focused on the biostatistical concepts useful for assessing computer-aided diagnosis/prognosis systems, including experimental design, performance indices, goodness-of-fit and related statistical tests (figure 17.2). Moreover, guidelines such as STARD and TRIPOD were reviewed (figure 17.3). Some worked examples and MATLAB functions were also provided for the interested reader. Implementation of these concepts can strengthen the validity of medical data mining algorithms.

However, we did not review specific performance indices used in medical image diagnosis (e.g. overlap, relative area difference, Dice's coefficient) [106]. Moreover, not only can the rigorous validation of the data mining methods improve their acceptability to clinicians, the data mining methodology is also critical. Generally, (clinically interpretable) rule-based systems are preferred compared to black-box methods since the former can provide explicit rules and can make use of the expert's knowledge [14, 107]. Engineers must pay attention to the fact that such methods, even if properly evaluated (as described in this chapter), do not replace clinicians and they are just second opinions to help the clinicians in better diagnosis and prognosis.

## Acknowledgements

The research leading to these results has received funding from the European Union's Horizon 2020 research and innovation programme under the Marie Skłodowska-Curie grant agreement No 712949 (TECNIOspring PLUS) and from the Agency for Business Competitiveness of the Government of Catalonia, University of Isfahan and Isfahan University of Medical Sciences. This work was also supported by the Ministry of Economy and Competitiveness (MINECO), Spain, under contract DPI2017–83989-R.

## References

[1] Yang J, Li Y, Liu Q, Li L, Feng A, Wang T, Zheng S, Xu A and Lyu J 2020 Brief introduction of medical database and data mining technology in big data era *J. Evid. Based Med.* **13** 57–69

[2] Fayyad U, Piatetsky-Shapiro G and Smyth P 1996 From data mining to knowledge discovery in databases *AI Mag.* **17** 37–7

[3] Jothi N, Rashid N A and Husain W 2015 Data mining in healthcare—a review *Procedia Comput. Sci.* **72** 306–13

[4] Kantardzic M 2019 *Data Mining: Concepts, Models, Methods, and Algorithms* (New York: Wiley)

[5] Han J, Kamber M and Pei J 2011 Data mining concepts and techniques 3rd edn *The Morgan Kaufmann Series in Data Management Systems* (Waltham, MA: Morgan Kaufmann) pp 83–124

[6] Bou Rjeily C, Badr G, Hajjarm El Hassani A and Andres E 2019 Medical data mining for heart diseases and the future of sequential mining in medical field *Machine Learning Paradigms: Advances in Data Analytics* ed G A Tsihrintzis, D N Sotiropoulos and L C Jain (Cham: Springer International Publishing) pp 71–99

[7] Lavrač N and Zupan B 2005 Data mining *Medicine Data Mining and Knowledge Discovery Handbook* ed O Maimon and L Rokach (Boston, MA: Springer US) pp 1107–37

[8] Mohebian M R, Marateb H R, Mansourian M, Mañanas M A and Mokarian F 2017 A hybrid computer-aided-diagnosis system for prediction of breast cancer recurrence (HPBCR) using optimized ensemble learning *Comput. Struct. Biotechnol. J.* **15** 75–85
[9] Kourou K, Exarchos T P, Exarchos K P, Karamouzis M V and Fotiadis D I 2015 Machine learning applications in cancer prognosis and prediction *Comput. Struct. Biotechnol. J.* **13** 8–17
[10] Marateb H R and Goudarzi S 2015 A noninvasive method for coronary artery diseases diagnosis using a clinically-interpretable fuzzy rule-based system *J. Res. Med. Sci.* **20** 214–23
[11] Marateb H R *et al* 2018 Prediction of dyslipidemia using gene mutations, family history of diseases and anthropometric indicators in children and adolescents The CASPIAN-III study *Comput. Struct. Biotechnol. J.* **16** 121–30
[12] Sarrafzadegan N *et al* 2017 PARS risk charts: a 10-year study of risk assessment for cardiovascular diseases in Eastern Mediterranean region *PLoS One* **12** e0189389
[13] Pavithra V and Jayalakshmi V 2020 A review on predicting cardiovascular diseases using data mining techniques *Proceeding of the Int. Conf. on Computer Networks, Big Data and IoT (ICCBI–2019)* (Cham: Springer International Publishing) pp 374–80
[14] Marateb H R, Mansourian M, Faghihimani E, Amini M and Farina D 2014 A hybrid intelligent system for diagnosing microalbuminuria in type 2 diabetes patients without having to measure urinary albumin *Comput. Biol. Med.* **45** 34–42
[15] Kavakiotis I, Tsave O, Salifoglou A, Maglaveras N, Vlahavas I and Chouvarda I 2017 Machine learning and data mining methods in diabetes research *Comput. Struct. Biotechnol. J.* **15** 104–16
[16] Mansourian M, Marateb H R, Kelishadi R, Motlagh M E, Aminaee T, Taslimi M, Majdzadeh R, Heshmat R, Ardalan G and Poursafa P 2012 First growth curves based on the World Health Organization reference in a nationally-representative sample of pediatric population in the Middle East and North Africa (MENA): the CASPIAN-III study *BMC Pediatr.* **12** 149
[17] Kelishadi R, Marateb H R, Mansourian M, Ardalan G, Heshmat R and Adeli K 2016 Pediatric-specific reference intervals in a nationally representative sample of Iranian children and adolescents: the CASPIAN-III study *World J. Pediatr.* **12** 335–42
[18] Mansourian M, Marateb H R, Keshteli A H, Zadeh H D, Mananas M A and Adibi P 2020 Symptom-based ordinal scale fuzzy clustering of functional gastrointestinal disorders, medRxiv: 2020.05.11.20098376
[19] Mohammadifard N, Marateb H, Mansourian M, Khosravi A, Abdollahi Z, Campbell N R, Webster J, Petersen K and Sarrafzadegan N 2020 Can methods based on spot urine samples be used to estimate average population 24 h sodium excretion? Results from the Isfahan Salt Study *Public Health Nutr.* **23** 202–13
[20] Wynants L *et al* 2020 Prediction models for diagnosis and prognosis of COVID-19 infection: systematic review and critical appraisal *Brit. Med. J.* **369** m1328
[21] Ozturk T, Talo M, Yildirim E A, Baloglu U B, Yildirim O and Rajendra Acharya U 2020 Automated detection of COVID-19 cases using deep neural networks with x-ray images *Comput. Biol. Med.* **121** 103792
[22] Doi K 2007 Computer-aided diagnosis in medical imaging: historical review, current status and future potential *Comput. Med. Imaging Graph.* **31** 198–211

[23] Jalalian A, Mashohor S, Mahmud R, Karasfi B, Saripan M I B and Ramli A R B 2017 Foundation and methodologies in computer-aided diagnosis systems for breast cancer detection *EXCLI J.* **16** 113–37

[24] Madabhushi A, Agner S, Basavanhally A, Doyle S and Lee G 2011 Computer-aided prognosis: predicting patient and disease outcome via quantitative fusion of multi-scale, multi-modal data *Comput. Med. Imaging Graph.* **35** 506–14

[25] Collins G S, Reitsma J B, Altman D G and Moons K G M 2015 Transparent reporting of a multivariable prediction model for individual prognosis or diagnosis (TRIPOD): the TRIPOD Statement *BMC Med.* **13** 1

[26] Smith T M F 1983 On the validity of inferences from non-random samples *J. R. Stat. Soc.* A **146** 394–403

[27] Tyrer S and Heyman B 2016 Sampling in epidemiological research: issues, hazards and pitfalls *BJPsych. Bull.* **40** 57–60

[28] Tan P-N, Steinbach M, Karpatne A and Kumar V 2019 *Introduction to Data Mining* 2nd edn (Upper Saddle River, NJ: Pearson)

[29] Patino C M and Ferreira J C 2018 Inclusion and exclusion criteria in research studies: definitions and why they matter *J. Bras. Pneumol.* **44** 84

[30] Sterne J A C, White I R, Carlin J B, Spratt M, Royston P, Kenward M G, Wood A M and Carpenter J R 2009 Multiple imputation for missing data in epidemiological and clinical research: potential and pitfalls *Brit. Med. J.* **338** b2393

[31] García-Laencina P J, Sancho-Gómez J-L and Figueiras-Vidal A R 2010 Pattern classification with missing data: a review *Neural Comput. Appl.* **19** 263–82

[32] Huque M H, Carlin J B, Simpson J A and Lee K J 2018 A comparison of multiple imputation methods for missing data in longitudinal studies *BMC Med. Res. Methodol.* **18** 168

[33] Hayati Rezvan P, Lee K J and Simpson J A 2015 The rise of multiple imputation: a review of the reporting and implementation of the method in medical research *BMC Med. Res. Methodol.* **15** 30

[34] Harel O, Mitchell E M, Perkins N J, Cole S R, Tchetgen Tchetgen E J, Sun B and Schisterman E F 2018 Multiple imputation for incomplete data in epidemiologic studies *Am. J. Epidemiol.* **187** 576–84

[35] Jakobsen J C, Gluud C, Wetterslev J and Winkel P 2017 When and how should multiple imputation be used for handling missing data in randomised clinical trials—a practical guide with flowcharts *BMC Med. Res. Methodol.* **17** 162

[36] Balki I *et al* 2019 Sample-size determination methodologies for machine learning in medical imaging research: a systematic review *Can. Assoc. Radiol. J.* **70** 344–53

[37] Machin D, Campbell M J, Tan S B and Tan S H 2009 *Sample Size Tables for Clinical Studies* (New York: Wiley)

[38] Faul F, Erdfelder E, Lang A-G and Buchner A 2007 G*Power 3: a flexible statistical power analysis program for the social, behavioral, and biomedical sciences *Behav. Res. Methods* **39** 175–91

[39] Faul F, Erdfelder E, Buchner A and Lang A-G 2009 Statistical power analyses using G*Power 3.1: tests for correlation and regression analyses *Behav. Res. Methods* **41** 1149–60

[40] Jones S R, Carley S and Harrison M 2003 An introduction to power and sample size estimation *Emerg. Med. J.* **20** 453–58

[41] Schmidt S A J, Lo S and Hollestein L M 2018 Research techniques made simple: sample size estimation and power calculation *J. Invest. Dermatol.* **138** 1678–82

[42] Bouman A C, ten Cate-Hoek A J, Ramaekers B L T and Joore M A 2015 Sample size estimation for non-inferiority trials: frequentist approach versus decision theory approach *PLoS One* **10** e0130531

[43] Desu M M 2012 *Sample Size Methodology* (Amsterdam: Elsevier Science) pp 1–135

[44] Hajian-Tilaki K 2011 Sample size estimation in epidemiologic studies *Caspian J. Intern. Med.* **2** 289–98

[45] Buderer N M 1996 Statistical methodology: I. Incorporating the prevalence of disease into the sample size calculation for sensitivity and specificity *Acad. Emerg. Med.* **3** 895–900

[46] Hajian-Tilaki K 2014 Sample size estimation in diagnostic test studies of biomedical informatics *J. Biomed. Inform.* **48** 193–204

[47] Obuchowski N A and Hillis S L 2011 Sample size tables for computer-aided detection studies *AJR Am. J. Roentgenol.* **197** W821–8

[48] Costanza M C and Paccaud F 2004 Binary classification of dyslipidemia from the waist-to-hip ratio and body mass index: a comparison of linear, logistic, and CART models *BMC Med. Res. Methodol.* **4** 7

[49] Hovsepian S, Kelishadi R, Djalalinia S, Farzadfar F, Naderimagham S and Qorbani M 2015 Prevalence of dyslipidemia in Iranian children and adolescents: a systematic review *J. Res. Med. Sci.* **20** 503–21

[50] Tabatabaei-Malazy O, Qorbani M, Samavat T, Sharifi F, Larijani B and Fakhrzadeh H 2014 Prevalence of dyslipidemia in Iran: a systematic review and meta-analysis study *Int. J. Prev. Med.* **5** 373–93

[51] Bossuyt P M *et al* 2015 STARD 2015: an updated list of essential items for reporting diagnostic accuracy studies *Brit. Med. J.* **351** h5527

[52] Devijver P A and Kittler J 1982 *Pattern Recognition: A Statistical Approach* (Upper Saddle River, NJ: Prentice Hall) pp 1–448

[53] Webb A R 2011 *Statistical Pattern Recognition* (Hoboken, NJ: Wiley) pp 1–666

[54] Mosteller F 2006 A *k*-sample slippage test for an extreme population *Selected Papers of Frederick Mosteller* ed S E Fienberg and D C Hoaglin (New York: Springer) pp 101–9

[55] Koutroumbas K and Theodoridis S 2008 *Pattern Recognition* (New York: Academic) pp 1–984

[56] Stehman S V 1997 Selecting and interpreting measures of thematic classification accuracy *Remote Sens. Environ.* **62** 77–89

[57] McNicol D 2005 *A Primer of Signal Detection Theory* (New York: Taylor and Francis) pp 1–242

[58] Ross S M 2014 *Introduction to Probability and Statistics for Engineers and Scientists* (San Diego, CA: Elsevier Science) pp 1–686

[59] Šimundić A-M 2009 Measures of diagnostic accuracy: basic definitions *EJIFCC* **19** 203–11

[60] Boughorbel S, Jarray F and El-Anbari M 2017 Optimal classifier for imbalanced data using Matthews correlation coefficient metric *PLoS One* **12** e0177678

[61] Matthews B W 1975 Comparison of the predicted and observed secondary structure of T4 phage lysozyme *Biochim. Biophys. Acta* **405** 442–51

[62] Cramér H 1999 *Mathematical Methods of Statistics* (Princeton, NJ: Princeton University Press) pp 1–575

[63] Sokolova M, Japkowicz N and Szpakowicz S 2006 Beyond accuracy, F-score and ROC: a family of discriminant measures for performance evaluation *AI 2006: Advances in Artificial Intelligence* (Berlin: Springer) pp 1015–21

[64] Mert A, Kılıç N, Bilgili E and Akan A 2015 Breast cancer detection with reduced feature set *Comput. Math. Methods Med.* **2015** 265138

[65] Fleiss J L, Levin B and Paik M C 2003 *Statistical Methods for Rates and Proportions* (New York: Wiley) pp 1–768

[66] Delgado R and Tibau X-A 2019 Why Cohen's kappa should be avoided as performance measure in classification *PLoS One* **14** e0222916

[67] Chicco D and Jurman G 2020 The advantages of the Matthews correlation coefficient (MCC) over F1 score and accuracy in binary classification evaluation *BMC Genomics* **21** 6

[68] Mukaka M M 2012 Statistics corner: a guide to appropriate use of correlation coefficient in medical research *Malawi Med. J.* **24** 69–71

[69] Baldi P, Brunak S, Chauvin Y, Andersen C A and Nielsen H 2000 Assessing the accuracy of prediction algorithms for classification: an overview *Bioinformatics* **16** 412–24

[70] Mercaldo N D, Lau K F and Zhou X H 2007 Confidence intervals for predictive values with an emphasis to case-control studies *Stat. Med.* **26** 2170–83

[71] Ellis P D 2010 *The Essential Guide to Effect Sizes: Statistical Power, Meta-Analysis, and the Interpretation of Research Results* (Cambridge: Cambridge University Press) pp 1–173

[72] Ghosh A K 2008 *Mayo Clinic Internal Medicine Review* 8th edn (Boca Raton, FL: Taylor and Francis) pp 1–1121

[73] Colquhoun D 2014 An investigation of the false discovery rate and the misinterpretation of *p*-values *R. Soc. Open Sci.* **1** 140216

[74] Figuera C, Irusta U, Morgado E, Aramendi E, Ayala U, Wik L, Kramer-Johansen J, Eftestøl T and Alonso-Atienza F 2016 Machine learning techniques for the detection of shockable rhythms in automated external defibrillators *PLoS One* **11** e0159654

[75] Kerber R E *et al* 1997 Automatic external defibrillators for public access defibrillation: recommendations for specifying and reporting arrhythmia analysis algorithm performance, incorporating new waveforms, and enhancing safety. A statement for health professionals from the American Heart Association Task Force on Automatic External Defibrillation, Subcommittee on AED safety and efficacy *Circulation* **95** 1677–82

[76] Newcombe R G 1998 Two-sided confidence intervals for the single proportion: comparison of seven methods *Stat. Med.* **17** 857–72

[77] Sokolova M and Lapalme G 2009 A systematic analysis of performance measures for classification tasks *Inf. Process. Manag.* **45** 427–37

[78] Gorodkin J 2004 Comparing two *K*-category assignments by a *K*-category correlation coefficient *Comput. Biol. Chem.* **28** 367–74

[79] Kautz T, Eskofier B M and Pasluosta C F 2017 Generic performance measure for multiclass-classifiers *Pattern Recognit.* **68** 111–25

[80] Hand D J and Till R J 2001 A simple generalisation of the area under the ROC curve for multiple class classification problems *Mach. Learn.* **45** 171–86

[81] Balakrishnan N, Colton T, Everitt B, Piegorsch W, Ruggeri F and Teugels J L 2014 *R* squared, adjusted *R* squared *Wiley StatsRef: Statistics Reference Online* (Chichester: Wiley)

[82] Giavarina D 2015 Understanding Bland Altman analysis *Biochem. Med.* **25** 141–51

[83] Montgomery D C, Peck E A and Geoffrey Vining G 2012 *Introduction to Linear Regression Analysis* (New York: Wiley) pp 1–642

[84] Demler O V, Paynter N P and Cook N R 2015 Tests of calibration and goodness-of-fit in the survival setting *Stat. Med.* **34** 1659–80
[85] Alexander D L J, Tropsha A and Winkler D A 2015 Beware of $R^2$: simple, unambiguous assessment of the prediction accuracy of QSAR and QSPR models *J. Chem. Inf. Model.* **55** 1316–22
[86] Altman D G and Bland J M 1983 Measurement in medicine: the analysis of method comparison studies *J. R. Statist. Soc. Ser.* D **32** 307–17
[87] Martin Bland J and Altman D 1986 Statistical methods for assessing agreement between two methods of clinical measurement *Lancet* **327** 307–10
[88] Van Noorden R, Maher B and Nuzzo R 2014 The top 100 papers *Nature* **53** 550–3
[89] Allison P D 1999 *Multiple Regression: A Primer* (Newbury Park, CA: Pine Forge) pp 1–202
[90] Breusch T S and Pagan A R 1979 A simple test for heteroscedasticity and random coefficient variation *Econometrica* **47** 1287–94
[91] Durbin J and Watson G S 1971 Testing for serial correlation in least squares regression. III *Biometrika* **58** 1–19
[92] Beck A 2014 *Introduction to Nonlinear Optimization: Theory, Algorithms, and Applications with MATLAB* (Philadelphia, PA: SIAM) pp 1–276
[93] Aggarwal C C 2016 *Outlier Analysis* (Basel: Springer International) pp 1–466
[94] Cook R D 2011 Cook's distance *International Encyclopedia of Statistical Science* ed M Lovric (Berlin: Springer) pp 301–2
[95] Schoukens J, Pintelon R and Rolain Y 2012 *Mastering System Identification in 100 Exercises* (New York: Wiley) pp 1–282
[96] King A and Eckersley R 2019 *Statistics for Biomedical Engineers and Scientists: How to Visualize and Analyze Data* (Cambridge, MA: Elsevier Science) pp 1–274
[97] Marateb H R, Mansourian M, Adibi P and Farina D 2014 Manipulating measurement scales in medical statistical analysis and data mining: a review of methodologies *J. Res. Med. Sci.* **19** 47–56
[98] de Leeuw J, Jia H, Yang L, Liu X, Schmidt K and Skidmore A K 2006 Comparing accuracy assessments to infer superiority of image classification methods *Int. J. Remote Sens.* **27** 223–32
[99] Dietterich T G 1998 Approximate statistical tests for comparing supervised classification learning algorithms *Neural Comput.* **10** 1895–923
[100] Detrano R, Janosi A, Steinbrunn W, Pfisterer M, Schmid J J, Sandhu S, Guppy K H, Lee S and Froelicher V 1989 International application of a new probability algorithm for the diagnosis of coronary artery disease *Am. J. Cardiol.* **64** 304–10
[101] Demšar J 2006 Statistical comparisons of classifiers over multiple data sets *J. Mach. Learn. Res.* **7** 1–30
[102] Mittelhammer R C, Judge G G and Miller D J 2000 *Econometric Foundations Pack with CD-ROM* (Cambridge: Cambridge University Press) pp 1–784
[103] Miller R G Jr 2012 *Simultaneous Statistical Inference* (New York: Springer) pp 1–299
[104] Hajiaghababa F, Marateb H R and Kermani S 2018 The design and validation of a hybrid digital-signal-processing plug-in for traditional cochlear implant speech processors *Comput. Methods Programs Biomed.* **159** 103–9
[105] Zhao J, Wang C, Totton S C, Cullen J N and O'Connor A M 2019 Reporting and analysis of repeated measurements in preclinical animals experiments *PLoS One* **14** e0220879

[106] Gonçalves V M, Delamaro M E and Nunes F de L dos S 2014 A systematic review on the evaluation and characteristics of computer-aided diagnosis systems *Rev. Bras. Eng. Agric. Ambient./Braz. J. Agric. Environ. Eng.* **30** 355–83

[107] Leondes C T 2002 *Expert Systems: The Technology of Knowledge Management and Decision Making for the 21st Century* (New York: Academic) pp 1–1947

CPSIA information can be obtained
at www.ICGtesting.com
Printed in the USA
BVHW051857050221
599342BV00003B/6

9 780750 334099